Living
with the
Earth *Third*
Edition

Concepts in
Environmental
Health Science

Living
with the
Earth
Third Edition

Concepts in Environmental Health Science

Gary S. Moore

CRC Press
Taylor & Francis Group
Boca Raton London New York

CRC Press is an imprint of the
Taylor & Francis Group, an informa business

CRC Press
Taylor & Francis Group
6000 Broken Sound Parkway NW, Suite 300
Boca Raton, FL 33487-2742

© 2007 by Taylor & Francis Group, LLC
CRC Press is an imprint of Taylor & Francis Group, an Informa business

International Standard Book Number-10: 0-8493-7998-9 (Hardcover)
International Standard Book Number-13: 978-0-8493-7998-7 (Hardcover)

Library of Congress Cataloging-in-Publication Data

Moore, Gary S.
    Living with the earth : concepts in environmental health science / Gary S. Moore. -- 3rd ed.
        p. cm.
        Includes bibliographical references (p.)
    ISBN 978-0-8493-7998-7 (alk. paper)
    1. Environmental health--Computer-assisted instruction. 2. Medicine, Industrial--Computer-assisted instruction. 3. Industrial hygiene--Computer-assisted instruction. 4. Industrial safety--Computer-assisted instruction. I. Title.

RA565.M665 2007
362.1--dc22   2007060502

Visit the Taylor & Francis Web site at
http://www.taylorandfrancis.com

and the CRC Press Web site at
http://www.crcpress.com

Preface

INTENT OF THE BOOK

This book is a college-level textbook for introductory courses in environmental health and environmental sciences. It is suitable for both undergraduate and graduate level courses for students of environmental health, environmental sciences, community health education, epidemiologists, civil engineers, community medicine, general medicine, and public health. It is also highly recommended for members of boards of health, health officers and health inspectors, and citizens who are members of proactive environmental groups such as the Sierra Club, public interest research groups, Clean Water Action, Environmental Defense Fund, and many others.

SYNOPSIS OF THE BOOK

This book contains 14 carefully illustrated and narratively exciting chapters on the subject of environmental health sciences. The contents are derived from a core course Dr. Moore instructs at the University of Massachusetts titled "Environmental Health Practices," and this book incorporates both the traditional concepts associated with environmental health, along with new, emerging, and controversial issues associated with environmental threats to human health. Emphasis is placed on biological, chemical, and physical sources of pollution, and the methods of controlling or limiting those exposures. Careful attention is paid to presenting a balanced view representing opposing scientific views on major issues ranging from the greenhouse effect to reproductive problems associated with endocrine disruptors.

KEY FEATURES

- A wide offering of instructor-support materials including PowerPoint presentation slides in digital format, a Web-enhanced course site, and an examination (test) bank
- Numerous author-created, detailed technical illustrations and photographs throughout the text and the Web-enhanced course site
- New chapters on emerging diseases and emergency preparedness, including the most significant reasons for their emergence, and the major etiological agents associated with emerging diseases and bioterrorism
- Measures of population dynamics including the contrasts of various cultural philosophies regarding "overpopulation," and the new problems of "under-population" in the developed nations

- New developments in areas of:

  The emerging problems of asthma and associated air pollutants

  The genetic basis of cancer and the roles of behavior and pollution in cancer risk

  The biology of genetics, mutation and birth defects, and much more

USING THE WEBSITE

The Website has been constructed to provide maximum usefulness for the student and the instructor. The site has been actively tested by student use and modified to meet student and instructor expectations. This is an ongoing and dynamic effort, and readers are encouraged to suggest improvements in the Website that would make it more useful and interesting. The site has been designed to work with most popular browsers and is fast-loading, with most images linked to text and selected for viewing by the user. This design is intended to speed up the use of the site and let users get to where they want quickly without being overwhelmed by graphics that automatically load and slow down the process.

STUDENTS

The Website address is http://www-unix.oit.umass.edu/-envhl565/. There will be areas on the Website that you will need a username and password. Students will use "ulearn" for the username and "mar99" for the password. Once you connect with the site and enter the correct ID and password, the possible use of the site will be immediately obvious. The most useful section to you will likely be the site location featuring "A Book Study Guide." This section features chapter-by-chapter detailed notes and associated Web links with each chapter. Just scroll down to the chapter of interest, and clicking on the highlighted chapter number (view text with images, **Chapter 1.html**) will bring the detailed notes section for that chapter to your screen. You may print these out directly. The associated figures for the chapter are in full color and may be displayed by clicking on the highlighted figure number (Figure 1.1). These may also be printed for your use. They are copyrighted figures and are not to be used for electronic or printed distribution in any form or manner without the publisher's permission.

Associated with each chapter are links to Websites that will take you to some of the most useful sites maintained on that subject matter by the U.S. government, international and national organizations, academic institutions, and some private groups. You may take a trip through the Amazon rain forest with both still and video images by entering the National Aeronautics and Space Administration Site (NASA) at "Live from the Rainforest" under the Chapter 1 Web links. If you want to know what industries close to your home are producing and/or emitting hazardous chemicals, and the amount and type of chemicals, enter the site called "Scorecard Home." These and other sites offer a wealth of information, which the student is encouraged to explore. There are also a number of searchable databases where you will be able to enter and find most any environmental topic by keyword.

INSTRUCTOR MATERIALS

The instructor for this course will have all of the features available to students and will also have access to chapter-by-chapter PowerPoint files which they may download directly from the Website. There are nearly 1100 slides with more than 250 images in full color taken directly from the book. The slides are based on the detailed course notes condensed from each chapter. The PowerPoint files may be shown using an LCD projection system in combination with a computer, printed on transparencies, or sent to a service bureau for digital imaging to 35 mm slides. The digital files may be modified by the instructor in PowerPoint to meet his/her needs in the classroom. I am also including in the instructor materials section a test bank of exam questions on a chapter-by-chapter basis. The questions include multiple choice, true and false, and short essay questions. The answers are supplied along with the page numbers in the text where the answers may be found. The test bank materials are in rich text format (RTF) that may be opened with most any common popular word processor. The instructor will also benefit from the searchable databases provided as Web links to obtain the latest information on environmental health topics. Additionally, I will also be providing a "hot-topics" section where new materials on important environmental issues will be placed, making the book and its associated Website current. The instructors are encouraged to contact me with information they believe should be included in the "hot-topics" section, and recommendations for the Website they believe would increase its usefulness. This Website is password protected. Instructors may acquire the user name and password by submitting a request to GMoore43@CHARTER.NET. Please include your academic affiliation, phone number, and academic email address.

Acknowledgments

My wife Lucille has provided unwavering support during my long periods of absence from her side, accepting my dedication to this project and keeping me fed and clothed in the interim. I extend sincere thanks to all the students mentioned above who wrote some of the text for selected chapters and provided much needed proofreading for most of the book. These students all benefited enormously from the experience, and grew in confidence and capabilities. I provided instruction in the areas of writing and organization, page layout, and graphic design. Each student took something of importance with them. John Edman, an entomology professor at the University of Massachusetts, provided slides for most of the insects discussed in Chapter 6. The pictures featuring African children and scenes were provided by Marie Krause Cote, who spent several months in Africa assisting as an educator to tribeswomen in producing cloth goods for market. Many of the pictures on wildlife, wetlands, and biomes came from U.S. government departments including the U.S. Fish and Wildlife Services (USFWS), the National Aeronautics and Space Administration (NASA), and the U.S. Environmental Protection Agency (USEPA).

Author

Gary Moore, M.S., Dr.P.H., is a faculty member at the Yale School of Public Health (clinical associate professor) where his primary responsibilities include the development and presentation of Web-based courses on emergency preparedness for the health professions. He is a visiting scientist at Harvard School of Public Health, Public Health Practice, Center for Public Health Preparedness, where he is responsible for the development and presentation of Web-based courses on emergency preparedness for the public health and health professions. He is a faculty member at the University of Massachusetts Medical School, Department of Family Practice. His primary responsibilities include the development and presentation of Web-based courses and creation of a distance learning program for the delivery of physician-created courseware. Dr. Moore is an emeritus faculty member in the Department of Environmental Health Sciences in the School of Public Health and Health Sciences at the University of Massachusetts where he has taught several graduate level environmental courses including courses on air quality assessment, environmental regulations and compliance, and environmental health practices. He has many publications (over 75) in the area of health effects from environmental insults, with many of his recent books having Web-enhanced components including:

- *You CAN Teach Online — Building Creative Learning Environments: A Web-Enhanced Resource* (with K. Winograd and D. Lange).
- *Environmental Compliance — A Web-Enhanced Resource*. This has a dedicated interactive Website.
- *Living with the Earth, Concepts in Environmental Health Science, A Web-Enhanced Book*. This book, which has a dedicated interactive Website, is one of *Choice* magazine's "Outstanding Academic Books of 1999," judged on distinguished scholarship. *Choice* is a publication of the Association of College and Research Libraries.

Dr. Moore has received several awards for his efforts in the area of providing training and certification programs to boards of health in Massachusetts during 1998 and 1999, including recognition from the Massachusetts Senate, the Massachusetts House of Representatives, the Massachusetts Association of Health Boards (MAHB), and the Robert C. Huestis/Eric W. Mood from the New England Public Health Association (NEPHA). The certification and training programs for boards of health have been very successful. Since 1995 Dr. Moore has assisted MAHB in establishing the groundwork for this first-in-the-nation program, providing practical educational opportunities to people serving every community in the state.

Dr. Moore is committed to providing the maximum learning experience and highest possible profile for his graduate students. Some of those graduate students have contributed significantly to the preparation of the first edition and are listed below.

KATHLEEN BELL, M.P.H.

Kathleen Bell received her bachelor of arts degree in biology from Cornell University. Her work experience includes environmental claim management. She received her master of public health degree in environmental health science from the University of Massachusetts School of Public Health. Currently, Kathleen lives in New England, where she enjoys hiking, running, and skiing.

CHRISTOPHER LANDRY, M.P.H.

Christopher Landry received his B.S. in biology from the University of Alabama in 1993. As an undergraduate he was invited to serve as a project supervisor for a Mexican mission. He then went to El Salvador where he served as a water sanitation technician for the Peace Corps. He received his master's degree in public health from the University of Massachusetts at Amherst. Currently, he is pursuing a career as a public health advisor to Latin American countries.

AMY TOMASELLO, M.P.H.

Amy Tomasello graduated from the University of Rhode Island, Kingston, where she received her bachelor of science in pharmacy. She worked as a consultant pharmacist for two years before returning to school. In 1998 she received her master of public health in environmental health sciences from the University of Massachusetts at Amherst. She resides in Massachusetts where she enjoys outdoor recreational activities.

SECOND EDITION

The following students are credited with contributing to updating chapters in the second edition:

Book Chapter	Contributor(s)	Book Chapter	Contributor(s)
Chapter 1: Ecosystems and Energy Flow	Sean Jobin, Glenna Vitello	Chapter 8: Foodborne Illness	Heather Burelle
Chapter 2: Human Populations	Danielle Bachand, April McNally	Chapter 9: Water and Wastewater	Nick Constantino, Ian Cambridge

Book Chapter	Contributor(s)	Book Chapter	Contributor(s)
Chapter 3: Environmental Degradation and Food Security	Xiaohan Zou	Chapter 10: Air Quality and Radiation	Keith Rego, Tim Lockhart
Chapter 4: Environmental Disease	Kenneth Huber	Chapter 11: Solid and Hazardous Waste	Mary Rothermich
Chapter 5: Toxicity and Toxins	Pradeep Sukumar	Chapter 12: Risk Assessment and Management	Mary Wicks, Jason Jenkins
Chapter 6: The Trouble with Pests	Fred Hooven	Chapter 13: Environmental Regulations	Amy Crosby, Pam Hogan
Chapter 7: Emerging Diseases	Vipion Aggarwal, Erin Grenier		

Contents

1 Ecosystems and Energy Flow

CHAPTER OBJECTIVES

A student, after reading this chapter, should be able to:

1. Discuss and define the concepts of *biosphere* and *climate*.
2. List and explain the factors influencing climate.
3. Define the term *biome*. List the major global biomes and discuss their primary features.
4. Describe the flow of energy through ecosystems.
5. Describe and explain the various trophic levels.
6. List and explain the various nutrient cycles including the carbon, nitrogen, and phosphorous cycles.
7. Define the term *succession*, explain the mechanisms of succession, and discuss the types of human intervention that interfere with succession.

INTRODUCTION

We are immersed in life. We breath it in, we walk on it, we touch it. Each footstep on a fertile lawn or forest mat will send tremors to trillions of bacteria, millions of algae, fungi, and protozoa, and hundreds of insects and worms. The skin on our bodies, when viewed microscopically, is a teeming matrix of tiny caverns filled with bacteria, viruses, and mites. So dense are the unseen life-forms on our bodies that they form an almost complete shell about each of us. Every breath draws in untold numbers of fungal spores, bacteria, viruses, and other microbes. Life abounds most any place inhabited by humans. Life thrives on the nutrients in soil and water, the oxygen and carbon dioxide in the air, and on the sunlight that ultimately powers most life. In those areas of the earth or its environs where nutrients are depleted, oxygen is rare, sunlight is extinguished, or moisture is diminished, life becomes reduced or is absent. Conditions for most life are found in a layer about the globe that extends from approximately 5 mi in the atmosphere, where some microbial spores and insects may be found, to 5 mi below the ocean surface, where some unusual life-forms adapted to darkness and high pressure, survive. This theoretical "layer of life" is called a *biosphere* because life is thought not to exist outside this area. Most life occurs in a much narrower layer extending from about a 600-ft depth in the ocean, where sunlight is able to penetrate, to the summer snow line of high mountain peaks, where a thin layer of soil supports plant life such as lichens and mosses. Within this biosphere, the forms and quantities of life vary dramatically. Surface- or land-based life may be categorized into major regions known as *biomes*. Biomes are based on the dominant types of vegetation that are strongly correlated with regional climate patterns. We will explore the various types of biomes in short order, but first let us look at the driving force of these regional differences — the climate.

CLIMATE

WHAT IS IT?

Climate varies in latitudinal bands around the globe. Relatively constant warm and humid weather characterizes most equatorial countries, and as you travel north or south, warm months decrease in number and freezing months increase. Climate can be viewed as average weather within a geographical area viewed over years, or even centuries. Climate, like weather, includes temperature, precipitation, humidity, wind velocity and direction, cloud cover, and associated solar radiation. Climate does change, but over centuries or millennia. Some of these climatic changes are dramatic. There have been many ice ages with the formation of glaciers, resulting in major climatic changes globally. These climatic changes may be due to different phenomena, which include (1) changes in ocean temperatures, (2) changes in the earth's orbital geometry, (3) volcanic activity with increased atmospheric dust and reduced sunlight penetration, (4) variations in solar radiation, or (5) increases in atmospheric gases that absorb heat energy.[1]

Volcanic eruptions have been thought to be a major factor in climate change, based on a range of studies that examine changes in the environment. Volcanic eruptions can cause sudden and severe environmental changes. Some eruptions emit large volumes of sulfur and particulate matter that can change climate and depress temperatures around the world. The impact of nine volcanic eruptions between 1883 and 1982 was analyzed, and the study discovered that volcanic eruptions are associated with surface temperature anomalies that usually fall within the normal seasonal statistical variation and that they may be related to temperature and circulation anomalies. However, questions are still unanswered as to the scale and patterns of the cooling.[2] Many of these gases are produced by human activities such as the burning of fossil fuels, which release large amounts of heat-absorbing carbon dioxide.

Scientists from the U.S. and Australia are reporting that during the last 100 million years there may have been a sudden release of greenhouse gases from the ocean floor. It is believed that gradual warming caused a change in ocean currents, which resulted in warm water from the surface plunging deeper into the ocean. The solid methane trapped in the ocean sediments became gaseous. The methane then exploded from the sediments, triggering mudslides. These mudslides allowed the methane to escape into water and then into the atmosphere. The methane reacted with the oxygen and formed CO_2, causing heat to be trapped in the atmosphere. The end result was an increase in ocean temperature by 5 to 9°F. This warming was so severe that it wiped out many species of marine life. Even this happened without humans being around, it seems that today we are creating the same conditions that caused this climate change. A huge amount of heat-trapping greenhouse gases are being released into the atmosphere at a relatively fast rate. Based on work by the National Academy of Sciences, they conclude that the Earth's surface temperature has risen by about 1°F in the past century, with accelerated warming during the past two decades. They also report that there is new and stronger evidence that most of the warming over the last 50 years is attributable to human activities.[3] The U.S. Environmental Protection Agency (EPA) reports that human activities have altered the chemical composition of the atmosphere through the buildup of greenhouse gases, primarily carbon dioxide, methane, and nitrous oxide.[3] The main greenhouse gas that caused this was CO_2. Since the Industrial Revolution, CO_2 has been released, causing its atmospheric content to rise by 30% since the 18th century.[4] These climate-changing gases will be discussed in more detail in Chapter 10. Do you want to learn more about climate change and policies? Point your browser to http://www-unix.oit.umass.edu/~envhl565/weblinks.html and click on any of the Chapter 1 Web links associated with climate change including:

- The Climate Diagnostics Center
- Climate Prediction Center
- Northeast Regional Climate Center
- United Nations Framework Convention on Climate Change, UNFCCC

HOW IS CLIMATE AFFECTED?

Climate is most affected by temperature. By the end of the next century, scientists predict that atmospheric CO_2 levels will be double that of the preindustrial era. They also suggest that it could have been much greater if not for a continuous evolution of energy sources away from carbon producers such as wood and coal to more efficient sources like natural gas and hydrogen. Regardless of the decreasing CO_2 to energy ratio, population explosions and economy increases are causing a continuous rise in CO_2 production. As a result we can expect extreme changes in the world's climate, weather, and lifestyle.[5]

The power that modifies, influences, and controls climate originates in the sun. The amount of sunlight striking a particular area of the globe determines the level of warmth and, ultimately, the movement of air and the amount of precipitation. Because the earth is a globe that rotates about an axis that is tilted, the amount of sunlight striking the earth varies by region and time. The intensity or energy from the sun diminishes with the distance from the sun. More specifically, the intensity declines inversely with the square of the distance from the sun. Consequently, our distance of 93 million miles from the sun means we receive a very small fraction of the sun's intensity (about two billionth of the sun's energy output is pointed in this direction).[6]

The earth's distance from the sun has little to do with the seasons. The seasons are caused by the tilt of the earth on its axis as it revolves around the sun. The earth is tilted at a 23.5° angle from a vertical axis drawn perpendicular to the plane of the earth's orbit around the sun. This tilt causes some parts of the earth to get more slanting rays of sunlight some of the year and more vertical rays of sunlight at other times. When a hemisphere of the earth is tilted toward the sun, it is summer in that hemisphere. Moreover, because the earth is tilted, seasonal differences will occur as the earth rotates about the sun. As seen in Figure 1.1, the equator receives vertical rays of the sun so that the largest amount of solar energy is absorbed in a band circumventing the globe at the equator. Further, the angle at the equator does not vary much as the earth rotates about the sun so that there are no significant seasonal changes, and temperatures remain fairly warm and constant throughout. As an example, New England is tilted more toward the sun in the summer months and receives vertical rays, but receives slanted rays in the winter because of the earth's tilted axis (Figure 1.1). Conversely, Australia, which is in the Southern Hemisphere, experiences just the opposite seasons. The sun impacts the earth in bands of decreasing energy extending north and south from the equator. Warmer air can hold more moisture, and hence may deposit that moisture as precipitation. Therefore, as you

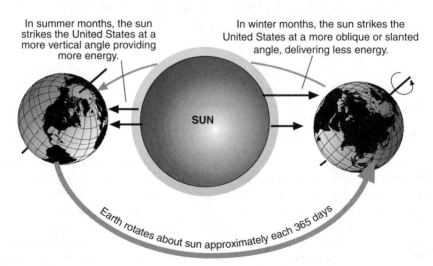

FIGURE 1.1 Variations in the angle of the sunlight striking the earth influences the seasons.

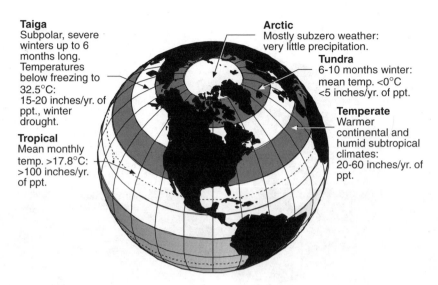

Taiga
Subpolar, severe winters up to 6 months long. Temperatures below freezing to 32.5°C: 15-20 inches/yr. of ppt., winter drought.

Tropical
Mean monthly temp. >17.8°C: >100 inches/yr. of ppt.

Arctic
Mostly subzero weather: very little precipitation.

Tundra
6-10 months winter: mean temp. <0°C <5 inches/yr. of ppt.

Temperate
Warmer continental and humid subtropical climates: 20-60 inches/yr. of ppt.

FIGURE 1.2 Variations in global regional temperatures and precipitation.

move farther from the equator, the seasons develop, and months of colder and drier air increase, producing major climatic regions following approximate latitudinal bands around the earth (Figure 1.2). Sir George Hadley provided the first written explanation of global air circulation in 1735.[1] Although modified in recent years, the basic concept still holds. The sun warms air at the equator. As the molecules of air absorb heat energy, they spread apart, and the gases expand and become less dense. This expanded air rises, traveling away from the equator. The warm air loses its heat by radioactive and convective cooling and falls back to the earth, where it flows toward the equator and completes the Hadley cell (Figure 1.3A). More recent models show that there are multiple Hadley cells known as the *three-zone model*.[1] In this model, Hadley-type cells are formed on both sides of the equator (Figure 1.3B). Thermal energy is lost and subsidence of cool dry air occurs at about 30°

Polar easterlies

Westerlies

Northeast tradewinds
Equatorial doldrums
Southeast tradewinds

Hadley Model (A)   **Three-Zone Model (B)**

FIGURE 1.3 Hadley (A) and three-zone (B) models of global circulation.

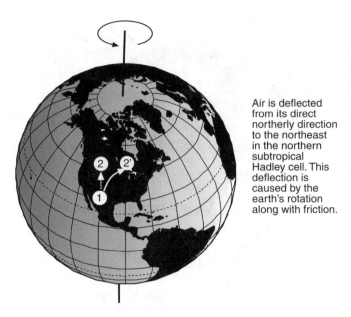

Air is deflected from its direct northerly direction to the northeast in the northern subtropical Hadley cell. This deflection is caused by the earth's rotation along with friction.

FIGURE 1.4 The Coriolis effect produces an easterly or westerly deflection in the expected trajectory of the wind. (Adapted from Godish, T., *Air Quality*, 3rd ed., Lewis Publishers, New York, 1997, chap. 4.)

north and south latitudes. The falling dry air compresses and warms. The warmer dry air absorbs moisture, creating some of the driest land areas on earth at these latitudes. The Sahara desert in North Africa and the Chilean coastal deserts are representative of these areas. There are other reasons for desert formation, and these are discussed later in this chapter under the section Desert Biomes. Moving north or south of these latitudes brings you into the temperate latitudes, including the bulk of the U.S., Central Asia, and Europe. The atmospheric flows in these middle latitudes are highly variable because there is a convergence of polar and tropical fronts with many migrating high- and low-pressure systems that cause frequent weather changes. The collision of warm and cold fronts often produces storm systems, resulting in precipitation and, ultimately, highly productive agricultural land. As can be seen in Figure 1.3B, the large-scale wind flow patterns in the temperate climates are not directly north or south, but tend to be deflected to the east in the Northern Hemisphere and to the west in the Southern Hemisphere. The deflection of air masses to the east or west is a result of the earth's rotation, causing the deflection of air from its northerly or southerly path, and this is known as the *Coriolis effect* (Figure 1.4).[1] The effect seen in the temperate North American zone is prevailing winds that flow from the west to the east, and these are known as *westerlies*. Because of this effect, there is a tendency for storm systems to travel from the west coast to the east coast of the U.S. We saw hurricanes like Floyd throughout the 1940s, 1950s, and 1960s. Scientists are now investigating why they seem to be returning. Many meteorologists agree that the climate of the earth shifts naturally and that this trend is normal. Others, however, believe that there is another factor involved, a human-made factor called *global warming*. As we release more and more greenhouse gases into the atmosphere, both the air and ocean temperatures will rise significantly. Hurricanes gain their strength from thermal energy found in air and water; the effect of these increasing temperatures create the perfect conditions for what scientists refer to as *hypercanes*. Storms of this size and power would dwarf hurricane Floyd. Such storms have been linked to the possible destruction of the dinosaurs 65 million years ago. Strong winds that approach 150 to 160 mph, they stir up cold waters from deep below the ocean's surface. This causes a cooling effect that eventually destroys the storm. For this reason we almost never see a class five hurricane. However, warming of the ocean temperatures is suspected in the series of hurricanes that pounded the southeastern American coastline in 2005, including the category 5 hurricane Katrina, which became a category 3 storm just as it overwhelmed New Orleans and surrounding areas, producing one of the deadliest natural disasters in U.S. history. By August 30, 2005, one day after the category 3 storm made landfall, three quarters of the city of New Orleans became flooded with as much as 20 ft (6 m) of water in some locations.[7] Several levees breached, causing the flood, which was called "the largest civil engineering disaster in the history of the United States."[8] This event continues to have huge repercussions for much of the population, including damage to the economy and impact on politics in the entire U.S. We hope not to see frequent hypercanes, but it is true that temperatures are rising, ice caps are melting, and most of the U.S. population resides in coastal communities.[8] Ultimately, climates are determined by the position of the earth with respect to the sun. This influences how much solar energy is absorbed, what wind patterns are established, and how much moisture is deposited. Temperature and moisture combine to become very important factors in establishing climates, which determine the major regional grouping of plants and animals known as biomes.

ECOSYSTEMS AND BIOMES

ECOSYSTEMS

Ecosystems are often a component of a biome. The relationship between biosphere, biomes, ecosystems, and populations is shown in Figure 1.5. A system may be viewed as a collection of components that are interdependent, or working together. For example, a collection of gears and springs as you might find in a wall clock or nondigital wrist watch may be viewed as a system that functions in a very precise way to monitor time. Changes or alterations to any component gear may alter or halt this function. The concept of system can be made very inclusive, such as a society with a system of government, or an individual with a system of organs or tissues such as the immune system. The level at which you wish to focus depends on what you wish to study. Ecosystems refer to identifiable areas within nature in which the organisms interact among themselves and their physical environment and exchange nutrients largely within that system. An ant farm inside a glass aquarium is an ecosystem. Maggots and microbes causing decay of garbage in a collection container constitute an ecosystem. A pond, or tidal pool, a river valley, or the planet itself can be considered an ecosystem. All these ecosystems consist

FIGURE 1.5 Relationship of biosphere, biome, ecosystem, and populations.

of biotic and abiotic components. The biotic components include living organisms and their products, including waste products and decaying matter such as urine, feces, decaying leaves and twigs, bones, and flesh. The physical or abiotic components of the ecosystem include such things as water, air, sunlight, minerals, and their interaction to produce climate, salinity, turbulence, and other conditions that influence the physical conditions under which organisms survive. The interaction of abiotic and biotic components of the environment is so closely intertwined that they cannot be easily separated for study. A pound of soil, for instance, consists of fragments of stone and pulverized rock, water, decaying plants and animals, and numerous living organisms such as bacteria, fungi, protozoa, insects, and worms. Without the biotic components of soil, it could not support the growth of plants and would be lifeless dust.

BIOMES

Biomes may be seen as groupings of plants and animals on a regional scale whose distribution patterns depend heavily on patterns of climate. The biome is identified by the climax vegetation or community. A climax community is identified as one that forms in an undisturbed environment and continues to grow and perpetuate itself in the absence of further disturbance. For instance, a mature climax forest of hardwoods (oak, beech, elm, maple, etc.) may take a century or more to develop from an abandoned northeastern farm field as a series of different plants succeed each other to this final stage. The biome is a community of interacting vegetation, soil types, and animal populations that are adapted to the physical environment of the region. The major biomes of the earth include tundra, taiga, temperate forest and grassland, deserts, and tropical biomes (Table 1.1).[6,9,10] These major biomes are described briefly in Chapter 3 under the topic of soils, with greater attention paid to conditions of soil depth and fertility. In this section we will discuss the biomes in more detail with regard to climate conditions, typical climax vegetation, and predominant mammal populations.

TABLE 1.1
Major Earth Biomes

BIOME and LOCATION	RAINFALL	TEMPERATURE	CLIMAX VEGETATION	ANIMALS
Tundra High latitudes of N. hemisphere in a belt around Arctic Ocean.	Less than 5 inches per year, strong drying winds.	6 to 10 weeks growing season. Winters are dark and long (6 to 10 months). Mean monthly temp. below freezing.	Small shrubs, mosses, sedges, lichens, some grasses.	Muskox, arctic hare, arctic fox, migratory waterfowl, and caribou.
Taiga Boreal forest in N. hemisphere forming a continuous belt of coniferous trees across N. America through Eurasia. Overlying formerly glaciated area	15 to 20 inches annually. Low evaporation rates make area humid in summer. Winter drought.	Subpolar. Severe winter up to 6 months. Extreme temperatures of below freezing to +32°C.	Needleaf, coniferous trees (spruce, fir, pine).	Fur-bearing predators such as lynx, wolverine, fisher, and sable. Mammalian herbivores such as snowshoe hares, lemmings, and voles. Large herbivores such as elk, moose, and beaver.
Temperate Forest Western and central Europe, eastern Asia, eastern North America.	20 to 60 inches per year.	Warmer, continental and humid subtropical climates. Drought during cold winters.	Temperate broadleaf deciduous forest. Oak, maple, beech, hickory, elm. Autumn leaf fall provides for abundant, rich humus.	Herbivores such as white-tail deer, gray squirrels, and chipmunks. Omnivores such as raccoons, skunks, and black bears. Carnivores largely eliminated such as timber wolves, mountain lions, and bobcats.
Temperate Grasslands N. America in central lowlands and high plains. Steppes of Ukraine, eastward through Russia and Mongolia. Pampas of S. America. Veldt of Republic of S. Africa.	10 to 20 inches per year. Much falls as snow.	Semiarid, warm to hot summer.	Rich mix of perennial grass underlain by fertile soils. Members of sunflower and pea families. Scattered woodlands in wetter areas.	Low in diversity. Bison and pronghorn in N. America. Rodent herbivores including ground squirrels and gophers. Carnivores/omni-vores include coyotes, badgers, and black-footed ferrets.
Tropical Savannas E. Africa, Venezuela, Columbia, Brazil, Central America.	Dry and wet seasons. 30 to 50 inches with dry season less than 4 inches a month.	Mean monthly temperature above 17.8°C.	Perennial grasses, dry zones, with sporadic woods in wetter areas.	Elephants, zebras, and wildebeests. Hoofed mammals such as antelopes, gazelles, and giraffes. Carnivores such as lions, leopards, jackals, and hyenas.

Continued

TABLE 1.1 *(Continued)*

BIOME and LOCATION	RAINFALL	TEMPERATURE	CLIMAX VEGETATION	ANIMALS
Tropical Forest Broadleaf forest or rainforest includes areas in Central and S. America, Central Africa, South and Southeast Asia.	In excess of 100 inches a year.	Mean monthly temp. above 17.8°C.	Widely spaced emergent trees of 100 to 120 feet. A closed canopy of 60 to 80 foot trees. Air plants, woody vines, climbers.	Most complex biome with numerous birds, insects, and primates.
Deserts West coast of continents between 20° and 30° latitudes including Baja, Ca., Atacama in S. America, and Namib in Southwest Africa. Interior positions of continents such as Gobe of Mongolia, Australian desert, and the Great Basin desert in the U.S. Rainshadow areas of high mountain ranges such as Death Valley, CA and Peruvian desert of S. America.	Less than 10 inches precipitation per year, arid climate in which evaporation exceeds precipitation annually.	May exceed 37.8°C in summer afternoons but drop to below freezing at night. Winters may be cool to cold.	Evergreen, or small-leaved deciduous plants, typically with thorns or spines and shallow, extensive root systems. Some plants with long tap roots (20 to 30 feet) to reach groundwater.	Animals with small body sizes and long appendages. Bright plumage to reflect sunlight. Reptiles have waterproof skin, produce uric acid instead of urine, and hard-shelled eggs.

TUNDRA

The tundra is limited to the upper latitudes of the Northern Hemisphere and forms a belt around the Arctic Ocean (Figure 1.6). The continents do not extend southward sufficiently to create a similar biome in the southern latitudes. The tundra appears as a barren and treeless land consisting mostly of mosses, lichens, sedges, dwarf shrubs, and perennial forbs. The scattered shrubs and woody bushes lie close to the ground, where they can preserve some warmth. The growing season is very short, lasting 6 to 10 weeks. The winters are long, dark, and cold with mean monthly temperatures below freezing (<0°C), and precipitation less than 5 in. per year. Dryness is further encouraged by strong, drying winds. Beneath the thin layer of topsoil lies a permanent layer of ice or frozen soil known as *permafrost*. The permafrost, along with the cold weather, prevents the growth of larger plants and trees, prevents drainage, and leads to the formation of bogs and marshes.

The soils are shallow and poor in fertility because of the limited growing season combined with the slow decomposition of organic matter in extreme cold temperatures. Consequently, the soils and the associated plant life tend to be very fragile, and damage done to the tundra is usually long-lasting.

FIGURE 1.6 Typical tundra biome and locations.

In 1998, a team of researchers at Vostok, Antarctica, drilled 3263 m into the ice and brought up ancient ice samples. From these samples it was possible to gather information such as age, temperature at the time the ice formed, and composition of the atmosphere. The composition of ancient atmospheres is shown through the analysis of tiny air bubbles trapped in the ice. Glaciologist Claude Lorius found a strong positive relation between CO_2 and methane. Both these chemicals are known as greenhouse gases. They have shown the rapid temperature changes of the last ice age. It also relates to the current increase in global temperature. This could be due to the increased levels of greenhouse gases in the atmosphere. Computers have been very useful when making paleoclimate models and finding climate mechanisms.[9]

Scientists were able to date the ice core. This was done by using an ice flow model using electrical conductivity measurements, ice accumulation changes, and correlation with other paleoclimate records. The data on greenhouse gases led Lorius to conclude that the greenhouse gases and the ice sheets "played a significant part in the glacial/interglacial climate changes." There is also a doubt as to which came first, the greenhouse gases or the climate changes. At this time it is thought that climate change is a result of the increased greenhouse gas concentration.[9,11]

The Arctic supports great amounts of wildlife; numbers are large but species are very few. These species have adapted to the extremely harsh conditions by morphological, physiological, and behavioral adaptations. Many animals have developed large, heavy bodies with a thick layer of fur (muskox) or feathers. Some species accumulate fat during a shortened growing season (bear), whereas others adapt by experiencing periodic population crashes and explosions, as in the lemming population. Some species such as caribou and waterfowl migrate southward from the tundra during the harsh winter conditions. Some of the better known Arctic inhabitants include muskox, caribou, grizzly bears, polar bears, wolves, foxes, snowshoe hares, weasels, and mink. A few species of birds (about six or seven) such as the snowy owl and ptarmigan also survive here. A number of tundra species can also be found at higher elevations in the mountainous regions of the Northern Hemisphere, where thin soils and cold temperatures prevent the growth of most middle-latitude trees and allow tundra species to grow above the tree line. Here the alpine tundra and alpine grasses flourish with sedges, perennial grasses, and dwarf shrubs.

TAIGA

The taiga, or boreal forest, is an almost continuous band of coniferous (cone-bearing) trees extending in a giant arc from Alaska and Canada, through Europe and Siberia.[9] It exists over areas that were once glaciated (Figure 1.7). Such areas tend to have long, severe winters lasting up to 6 months with temperatures below freezing and growing seasons between 50 and 100 d. The rainfall tends to be 15 to 20 in. annually, but poor evaporation and poor drainage result in the formation of bogs, which are often overgrown with sphagnum moss. The cone-bearing trees include fir, spruce, and pine. These needle leaf trees are adapted to the taiga through their conical shape, which helps shed snow, and the waxy, thin needles, which reduce water loss and protect against drying winds. The dark green needles absorb heat and are not shed, so that photosynthesis can start immediately with the beginning of the shortened growing season.[10]

Fur-bearing predators including wolverines, sable, fisher, mink, and lynx seek mammalian herbivores such as snowshoe hares, red squirrels, lemmings, and voles. Many of these species develop white coats during the winter to blend with the snow. A number of larger herbivores may also be found in or near the taiga, including moose, elk, and red deer. Year-round birds of the taiga include finches, sparrows,

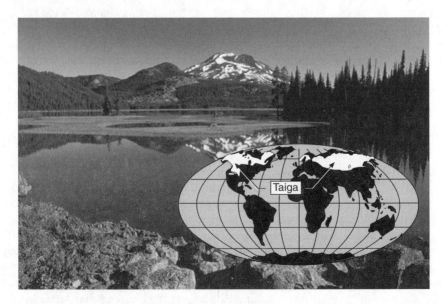

FIGURE 1.7 Typical taiga biome and locations.

and ravens. Some birds such as the grosbeak and crossbill become migratory during years when pinecones are reduced in numbers.

TEMPERATE AREAS

Temperate Broadleaf Deciduous Forest

The temperate broadleaf deciduous forests (Figure 1.8) are located in western and central Europe, and eastern Asia and eastern North America. Climax deciduous forests include oak, maple, hickory, chestnut, elm, linden, and walnut trees. The temperate forests receive 20 to 60 in. of precipitation distributed evenly throughout the year. Snow and ice conditions during winter months make water unavailable and create droughts during the cold season. The broadleaf trees lose their leaves (deciduous) during the winter to conserve moisture and resist the cold. As the cold fall weather settles in, the leaves begin the deciduous process (leaf fall) and the leaves turn to brilliant reds, oranges, and yellows as chlorophyll formation is halted and the leaves finally die and fall to the ground, forming a litter that decays further to a rich organic mat called *humus*.

Mammals in this biome include acorn and nut feeders such as squirrels and chipmunks; omnivores such as raccoons, skunks, black bear, and opossum; and rare carnivores such as mountain lions, bobcats, and wolves. Except for the recent emigration of coyotes from the grasslands, the carnivores have been mostly eliminated by habitat destruction and hunting. Birds include insect eaters such as hummingbirds, thrushes, warblers, and wrens. Omnivores such as ravens and cavity nesters such as woodpeckers and chickadees also abound.[9,10] Most of the bird species

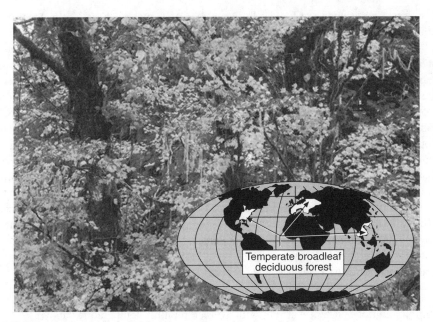

FIGURE 1.8 Typical broadleaf temperate forest biome and locations.

are migratory and fly south for the winter, and a number of mammalian species enter into hibernation.

Temperate Evergreen Forests

In temperate areas where soil is poor and droughts and fires are frequent, the predominant species tend to be evergreens, including coniferous, needle-leafed trees with some broad-leafed evergreens. In such areas, spruce, fir, and ponderosa pine proliferate. These are known as *temperate evergreen forests*. In cool coastal climates where there is considerable rainfall or frequent heavy fogs, one is likely to see forests of giant trees such as redwoods, which characteristically reach heights of 198 ft (60 m) to 297 ft (90 m). These are temperate rain forests and such areas exist along the Pacific coast of North America stretching from northern California to the Sitka spruce in Alaska. Chile and New Zealand also have these temperate forests.

Chaparrals

Areas of moderately dry climate and limited summer precipitation are further characterized by small (0.9 to 4.5 m) shrubs with leathery leaves that contain aromatic and flammable substances. Consequently, these chaparrals often catch fire, allowing shrubs to regrow from parts of the plant that survive near the ground. Unfortunately, when such growth is allowed to proliferate near houses that are rapidly expanding into these areas, they likely become fuel for wildfires.

Temperate Grasslands

Temperate grasslands are known by a number of names such as North American prairie, Asian steppes, South African veldt, and the South American pampas (Figure 1.9).[9,10] These semiarid regions tend to be located inland along the middle latitudes and experience only 10 to 20 in. of precipitation a year, much of which falls as snow. The predominant plant forms are perennial grasses, forbs, and members of the sunflower and pea families.[10] The perennial grasses grow from buds just below the ground surface and are adapted to fire, drought, and cold, and the well-developed root systems are able to trap moisture and nutrients. The more humid grasslands are populated by sod-forming grasses that have rhizomes or stems under the ground, from which new plants are formed. Areas that are drier tend to favor bunch grasses that reproduce by seed. Biodiversity in temperate grasslands is limited. Most commonly seen in temperate grasslands are the herbivore rodents, including ground squirrels, prairie dogs, and pocket gophers. Large animals such as bison, pronghorn, grizzly bears, and wolves have largely been confined to isolated preserves or the mountains, as humans destroy more habitats.[9]

THE TROPICAL RAIN FOREST

The tropical rain forest is complex in terms of species diversity and structure. The combination of constant warmth, with average monthly temperatures above 17.8°C and precipitation greater than 100 in. per year encourages rapid plant growth through-

FIGURE 1.9 Typical temperate grassland biome and locations.

out the year. The high humidity and continued warmth promote rapid growth and great diversity. There are relatively few numbers of any particular species, but a myriad of species exist in the tropical rain forest. These species account for nearly three quarters of the total number of species on earth. Further, more than 40% of the world's plants and animals grow in the tropical rain forest, which account for less than 7% of the earth's surface.[9,10] There are no seasons in the rain forest, so there is no annual rhythm, and each species flowers or fruits at different times of the year on their own cycle. The majority of plants are tall trees with thin trunks that produce branches and leaves near the top. Many of these leaves are elongated at the tip to encourage dripping or shedding of water ("drip-tip" leaves), which reduces mold and insect attack.

The life of the forest occurs in the canopy. The canopy is an almost unbroken sea of green produced by leaves at 70 to 100 ft above the forest floor (Figure 1.10). Most of the sunlight is absorbed in the canopy, and it is where the bulk of photo-synthesis occurs, resulting in more than 10 tons of new growth per acre per year. Much of the animal life also lives in the canopy. The leathery green leaves sprouting from branches at the top of trees help shield the forest floor from the direct impact of intense showers. Occasionally, a taller tree penetrates the lower canopy to form an umbrella-like canopy over the trees below. These trees reach heights of 130 ft and are called *emergent trees*. Temperatures in the canopy often reach 35.5°C (96°F) with a humidity of 50 to 60%, whereas the temperature on the floor of the forest are normally lower, at 27.7°C (82°F) with a humidity of 90%.

Researchers in Costa Rica have discovered evidence that increasing temperatures have slowed down the growth of tropical forests. It has been suggested that the tropical rain forests have actually added billions of tons of greenhouse gases into the atmosphere during the 1990s. Although trees take in CO_2 and then release O_2 during photosynthesis, they also release some CO_2 as a by-product of respiration.

FIGURE 1.10 Typical tropical rain forest biome and locations.

When growing rapidly, plants take up more than they can produce. But, if growth is slow, then the balance will shift. In 1984, researchers Deborah Clark and David Clark began researching the growth rates of adult tropical rain forest trees. The samples included six different tree species with both fast- and slow-growing tree types. They found that the growth of the trees changed each year. In years with high levels of CO_2, the trees had grown slowly, and in years when the level was small, the trees grew rapidly.

Less than 10% of the sunlight reaches the forest floor, and there is intense competition for available energy at this level. There is sparse vegetation in the inner sections of the undisturbed rain forest, and these sections have to compete intensely in an area with little sunlight and poor soil. This part of the rain forest is known as the understory (Figure 1.10). The soils are often soggy and quite shallow because the tremendous biological activity quickly decomposes organic matter and converts it into new life. The soils may be only a few inches deep, covering red clay. This has caused many trees to adapt by forming shallow root systems and bases that spread at the bottom of the tree much like a buttress. This growth provides support in thin soil and may enhance the ability of the tree to secure nutrients over a broader area.

The reduced light beneath the canopy has encouraged a number of plant strategies to reach the top of the rain forest. Air plants, or epiphytes, grow on branches high in the trees. Being supported by the tree limbs, epiphytes extract moisture from the air and trap leaf fall and wind-blown dust for nutrients. Bromeliads (pineapple family), orchids, cactuses, and ferns are all represented in the canopy as epiphytes. Many plants such as members of the fig family begin life as epiphytes in the canopy and send their root systems downward to the forest floor. Such plants are then called *stranglers*. Many woody vines grow quickly up the tree trunks when a temporary gap opens in the canopy. Here they flower and fruit in the treetops. These woody vines are called *lianas* and are often deciduous. Few animals exist on the forest floor. Beetles, butterflies, and other insects abound in the canopy. A rich array of birds, monkeys, and frogs also inhabit the canopy.

Tropical rain forests have been linked to possible cures for diseases and viruses. This is also true for the ocean. Animals from the ocean such as horseshoe crabs, starfish, bacteria, and ocean molds have provided a few potential cures for cancers, Alzheimer's disease, and other diseases. The ocean covers almost 70% of the earth's surface. The organisms living in the ocean have primitive life systems and are built for survival. They are examples of the early stages of life. They have simple mechanisms to fight diseases. Biologists William Fenical and Bob Jacobs from the University of California at Santa Barbara have been looking to the ocean as a source of possible cures of diseases. They have been going to reefs in the Caribbean. They are studying animals that are "loaded with biologically active molecules" which they think have natural defense mechanisms. Fenical has already discovered a substance from a sea fan that reduces inflammation from sunburn or chemical burns. This substance is much more effective than current hydrocortisone creams found in drug stores. These pseudopterosins may offer other cures such as arthritis and psoriasis and may even limit damage from the sun. Fenical has also tested a substance found in the sea off of Australia, eleuth-erobin, which could possibly cure a cancer by disrupting malignant cell division (http://cancer.ucsd.edu/Research/summaries/wfenical.asp).

Other marine biologists have found an anticancer chemical, halichondrin B. It comes from a "yellow-slimy" sponge found off the New Zealand coast. Another, Bugula, is an organism that is found on the bottom of boats; it is being checked as a possible treatment for leukemia and kidney cancer.[12,13]

Testing and getting FDA approval for these possible cures takes years and lots of research. So, the marketing of these possible drugs takes time, which can be discouraging. Fenical is optimistic and continues to research medical cures from the ocean.[12,13]

Typical rain forests circumvent the globe in a band from 20° S to 30° N and are found in Central and South America, Eastern and Central Africa, Malaysia, Indonesia, the Philippines, Papua New Guinea, and the Solomon Islands (Figure 1.10).

Deserts

Deserts are defined by their arid climates, which average less than 10 in. of precipitation a year and where evaporation tends to exceed this precipitation, causing a negative annual water budget. When rain falls, it tends to be very localized and unpredictable as to when it will occur. Further, much more rainfall may occur in some years than in others. Deserts are not always hot. They can easily reach temperatures higher than 37.8°C (100°F) on summer days, plummeting to 6.7°C (20°F)

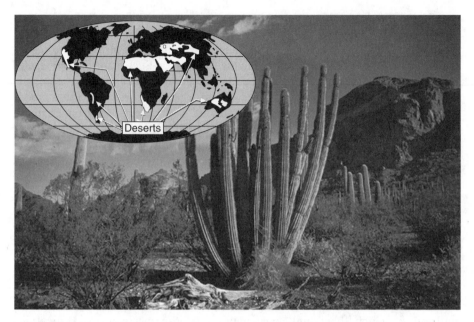

FIGURE 1.11 Typical desert biome and locations.

at night. Some deserts may not experience frost, whereas others may have extended periods of subzero temperatures and even snow.[9,10]

Deserts are created by specific geographic and climate conditions:

1. Within the latitudes of 20° S to 30° N and along the western coasts of continents, the winds tend to be easterly (coming from the east; see Figure 1.3B) and consequently keep moist air rising off the oceans from reaching the coast. This causes some of the world's driest deserts such as the Sahara in Africa, Baja California in North America, and the Atacama in South America (Figure 1.11; Figure 1.11).

2. Near the 30° latitude, subtropical air descends in association with the Hadley cell (Figure 1.3B), and as it descends it compresses, causing the formation of heat and dry, warm air. This air then absorbs moisture, resulting in little precipitation and produces such areas as the Sahara and Australian deserts.

3. Temperate deserts are generally located in areas known as *rainshadows*. Rainshadows occur on the lee side of high coastal mountain ranges (Figure 1.12). As moisture-laden air approaches the windward side of the mountains, the air rises, cools, and creates a low-pressure condition causing precipitation. The windward side receives large amounts of rain and snow. The air falls on the leeward side, compresses and warms, and so absorbs moisture, producing an arid environment and creating such deserts as Death Valley in California or the Peruvian desert.

FIGURE 1.12 Rainshadow deserts are created on the lee side of coastal ranges. (Adapted from Turk, J., and Turk, A., *Environmental Science*, Saunders College Publishing, Philadelphia, PA, 1988.)

4. The Gobi desert of Mongolia and the Great Basin desert of the U.S. are located in the interiors of continents, where they are distant from sources of moist air and where the rainshadow effect may also be important.[9,10]

The deserts are not completely lacking in water, and many life-forms have adapted to the arid conditions. The major plant life in deserts are evergreen shrubs that have small drought-resistant leaves, often have spines or thorns, and sweet-smelling oils. The root systems usually extend outward by great distances to capture whatever moisture falls. Other plants have taproots that extend to 30 ft below the ground to reach underground water. Still other plants store water in their tissues to be used during dry spells. These plants are known as *succulents*. Depending on the species, water may be stored in different parts of the plant including stem, leaf, fruit, or roots. The familiar American cactus is a stem succulent. The aloe plant is an African leaf succulent.

Desert fauna have also adapted to aridity. Body sizes are small, with long appendages that maximize heat radiation. Animals may have skin without sweat glands. They may concentrate urine into uric acid, have deposits of fat in tails or humps, or glands that secrete salt without moisture loss. They may be nocturnal or may even be dormant during summer months.

ENERGY FLOW

ENERGY SOURCE

All life requires energy, and ultimately the energy comes from the sun. Sunlight travels across the galaxy at 186,000 mi/sec to reach us, and in a period of less than 10 min it loses all but the smallest fraction of its radiation intensity. Close to 99.9% of the sun's energy reaching the earth is reflected into space, absorbed as heat, or evaporates as water. Plants absorb a portion (about 0.1%) of this incoming light energy, particularly in the blue and red wavelengths of the visible

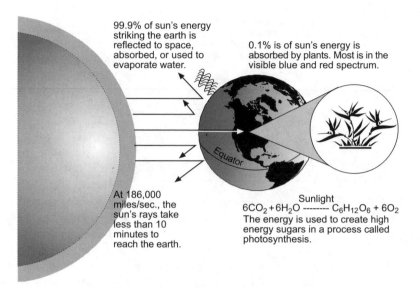

FIGURE 1.13 Distribution of sun's energy striking the earth.

spectrum. This energy is used by the plants for photosynthesis to create simple sugars from carbon dioxide and water with the release of oxygen (Figure 1.13). The simple sugars are used to store chemical energy, and the amount of energy stored by the plant is normally expressed as *kilocalories of chemical energy stored per square meter of plant tissue per year* (kcal/m²/yr). Some 20% of this stored energy is used by the plants for respiration and it therefore escapes in the form of heat to the environment. The rest (from 50 to 80%) is converted to growth of the plant, thereby increasing the mass of the plant. The flow of energy and the conservation of matter must obey certain fundamental physical laws (Table 1.2). The accumulation of organic material in an ecosystem is referred to as *biomass*. Some ecosystems, such as tropical rain forests or marshes, are highly efficient in producing new plant growth because of the availability of nutrients, sunlight, warmth, and moisture. Consequently, the biomass in these areas is much greater than in a desert.

Within the biosphere, energy flows through the interconnecting web of life from organism to organism. The plant produced may be available to a variety of insects and animals, including humans, who must consume the plant to obtain the energy-rich molecules as animals cannot photosynthesize. These consumers, or heterotrophs, convert about 10% of the consumed kilocalories into animal flesh or organic matter (Figure 1.14). The 90% of the remaining consumed energy is used by heterotrophs in respiration necessary for the energy of motion, the maintenance of body functions, or is excreted.

As energy is transferred through the food chain, about 90% of that available energy is lost with each transfer, and the consumer uses one tenth. Consequently, an animal such as a wolf, which consumes deer or rabbits that eat grass, would be a secondary consumer and would receive (1/10 × 1/10 = 1/100) of the available

TABLE 1.2

SOME NATURAL PHYSICAL LAWS

Law of Conservation of Matter: While undergoing normal changes in physical or chemical reactions, matter cannot be created or destroyed. Practically, this means the earth has finite amounts of chemicals and nutrients that must be recycled if life is to continue. Oxygen can exist as iron oxides, carbon dioxide, nitrogen oxides, molecular oxygen, water, alcohols, within organic compounds and numerous other forms–but the total amount of oxygen available on the planet remains the same. The Law of Conservation of Matter means that everything is accounted for and that there is a mass balance. As we breath in oxygen and catabolize plant sugars in the process of respiration, we give off carbon dioxide and water. The plants are able to use the carbon dioxide and water in the process of photosynthesis which results in the production of sugars and oxygen.

Photosynthesis: $6CO_2 + 6H_2O$ -------- $C_6H_{12}O_6 + 6O_2$

Respiration: $C_6H_{12}O_6 + 6O_2$ -------- $6CO_2 + 6H_2O$ + energy

Without this exhange or recycling of oxygen, the oxygen would soon be used up and life as we know it would be gone. As another example, if we burn 1000 tons per day of solid waste in an incinerator, we will get 1000 tons per day out of the incinerator. The materials discharged will be ash, some unburnable solids, particulates, and gases–but the total matter will weigh the same as that which entered.

The First Law of Thermodynamics: Energy is defined as the capacity or potential to perform work. There are many forms of energy including kinetic, gravitational, chemical and thermal energy. Energy can be transformed from one form to another but cannot be created or destroyed. In practical terms, the energy of the sun is converted by plants to chemical energy in the form of sugars and plant tissues that can be used by herbivores as a source of energy. Plants use about 0.1% of the sun's energy reaching the earth. The rest is radiated back to space, absorbed as heat, or used to evaporate water. Animals that eat the plants convert about 10% for their use, with about 1% of this converted to animal tissue and the remainder used in respiration to maintain organ systems, provide motion, and produce heat. The energy used in respiration escapes as thermal or heat energy. Thus, energy was converted from light energy to chemical energy to thermal energy–so the total amount of energy remains the same. However, the energy so produced is not useful to plants or animals for growth in its final form. This relates to the Second Law of Thermodynamics.

The Second Law of Thermodynamics: This law is based on the concept of **entropy**, which means that matter or energy will disperse toward maximum disorder in an undisturbed environment. Just as dye particles will evenly distribute in a cup of water, so will energy move toward a greater state of dispersion that is less useful or unavailable. As light energy is converted to chemical energy and then to heat energy, the heat is radiated ultimately to space where the heat is distributed uniformly throughout. Consequently, there must be a constant supply of energy from the sun to drive the life forces on Earth. Although the total energy has not declined, it moves unidirectionally through the world to become transformed and dispersed into unuseful forms.

energy in the plant (Figure 1.14). In an example that demonstrates the efficiency of eating at a lower trophic level, 3000 lb of corn and grain would support the energy requirements of one 1500-lb steer, which in turn would support the energy requirements of one 150-lb human (Figure 1.15). If the grain were eaten directly by humans, it could likely support up to 20 people or more in the same amount of time. Only the wealthy countries can support a human population where meat is one of the main food consumed. A primarily vegetarian society would be able to feed many times the number of humans with the same amount of grain that is used to feed cattle. Countries such as India, China, and most of the lesser-developed countries (LDCs), as in Africa, are predominantly vegetarian for this reason.

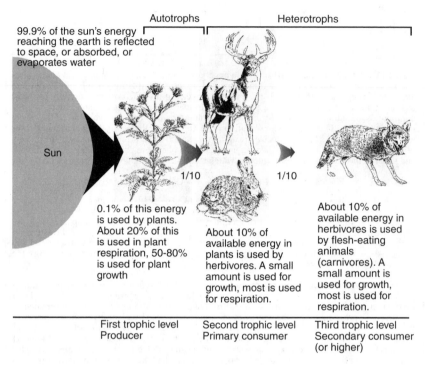

FIGURE 1.14 Relationships of energy flow in an ecosystem.

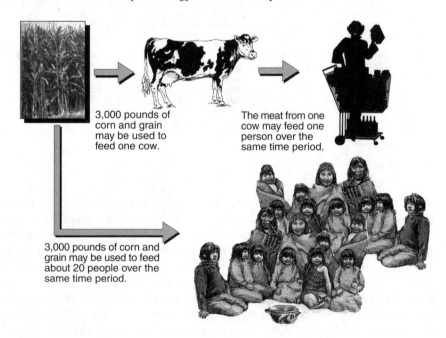

FIGURE 1.15 Comparison in efficiencies of feeding at various trophic levels.

CONSUMPTION TYPES AND TROPHIC LEVELS

Animals that eat only plants are *herbivores*. Other animals that eat primarily animal flesh often consume herbivores. These animals are called *carnivores* and may include lions, tigers, hyenas, frogs, or even ladybugs (they feed on aphids). Animals that eat plants and animals are termed *omnivores* and include rats, bears, humans, hogs, and foxes.

All energy comes from the sun, and it flows in one direction from plants, which are producers and belong to the first trophic level. Primary consumers or herbivores belong to the second trophic level. Caterpillars, grasshoppers, cattle, and elephants are primary consumers because they are herbivores that feed primarily on plants. Secondary consumers (carnivores) belong to the third trophic level (or higher). Each stage through which this energy travels is called a trophic level. Humans, along with many other animal species, are omnivores and may consume plants or animals. This means that many species may belong to several trophic levels. Bears, for instance, may belong to trophic levels two, three, four, and five. Bears will consume berries (herbivore, trophic level two), mice (trophic level three), frogs (trophic level four), or shrews (trophic level five). An example of this complex interrelationship is given in Figure 1.16. These patterns of consumption tend to be complicated and the term *food web* has been used to refer to these complex patterns of consumption among many trophic levels. This is important from a survival point of view, because those animals capable of feeding at many trophic levels will be able to survive if portions

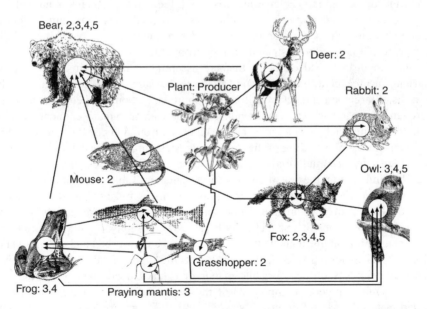

FIGURE 1.16 Simplified food web showing some of the possible trophic levels (represented by numbers) for various species. The arrows show direction of consumption. (Adapted from Turk, J. and Turk, A., *Environmental Science*, Saunders College Publishing, Philadelphia, PA, 1988, chap. 5.)

of their food supply are decreased or eliminated. Because the amount of available energy (kcal) is reduced by a factor of ten at each trophic level, then there is a practical limit to the number of trophic levels that can be sustained. This number seems to be limited to five. As you move from primary consumers to higher trophic levels, the numbers and biomass (dry weight per unit area) of land organisms decreases with each successive level. The plants, or producers, far outnumber the next trophic level, or herbivores, which in turn far outnumber and have greater biomass than carnivores. This is because at each higher trophic level, there is less useful energy available. Consider that there are fewer grizzly bears than caribou, or fewer foxes than rabbits. As previously mentioned, more animals (including humans) can survive as herbivores than as carnivores.

NUTRIENTS

Recycling

Although there is a constant supply of sunlight, the nutrients tied up in these trophic levels need to be recycled. Some animals may eat, and even prefer, dead or decaying animal matter. Many of us have seen crows or ravens settle down on a road to consume the remains of a small dead animal. We know that vultures are also primarily scavengers because they prefer to feed on the dead remains of animals. Some animals that are predominantly omnivores or carnivores may also become opportunistic scavengers. Hyenas and foxes will feed on the remains of animals that have died or have been killed. The author has witnessed hogs consuming dead rats, which also blurs the separation of scavenger from omnivores or carnivores. Unlike energy, which is rapidly dissipated through trophic levels, nutrients consist of 40 of the 92 known chemical elements that are essential to most forms of life. When nutrients are recycled, the process is called *biogeochemical cycling*. Nearly all life consists of large percentages of sulfur, phosphorous, carbon, oxygen, hydrogen, and nitrogen. These are known as *macronutrients*. There are also many elements required in tiny amounts such as zinc, manganese, chlorine, iron, and copper that are termed *trace elements*.[6] The cycling of these nutrients is essential because they would soon be depleted by accumulation in dead matter, precluding the development of new life. Remember from the law of conservation of matter (Table 1.2) that matter cannot be created or destroyed. Another group of organisms essential to the process of biogeochemical cycling are the decomposers. Decomposers are insects, bacteria, fungi, and protozoans that sequentially break down complex organic materials into low-energy mineral nutrients that once again may be reabsorbed and used by plants. This process is normally sequential: those specific types of saprophytic organisms that feed primarily on dead organic matter first attack dead plants or animals. As these organisms process components of the dead tissues, they leave behind undi-gested materials that are in turn processed by the next group of decomposers, until finally there is nothing left but energy-poor mineral nutrients. This sequence may be so predictable that the science of forensic entomology has developed, and prac-titioners can judge the time of death of a human from the types of decomposers evident at the time of discovery of the body. Remember the television show *C.S.I*? Check out http://www.forensicentomology.com/index.htm.

NUTRIENT CYCLES

The recycling of nutrients from the inorganic to organic and back again in a living system was previously defined as biogeochemical cycling. This cycling may take two forms. The nutrients may be sedimentary and originate from soil or rocks, or gaseous and originate from air or water. The gaseous cycle is very efficient compared to the sedimentary cycle and this may be why the major constituents of most life consist of carbon, nitrogen, hydrogen, and oxygen. Representative substances from the gaseous and sedimentary cycles are discussed in the following text.

Carbon Cycle

Humans, along with most animals and plants, are carbon-based life-forms. It is not surprising then that carbon is available in such a wide spectrum of forms including the physical states of gas, liquid, or solid, and the chemical forms, including organic and inorganic (Figure 1.17). Inorganic sources of carbon includes gaseous carbon dioxide, which occurs in the atmosphere at about 0.03% (300 ppm). Carbon dioxide is released to the atmosphere from: (1) the respiratory processes of animals and

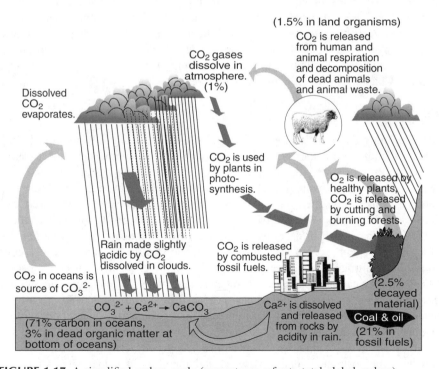

FIGURE 1.17 A simplified carbon cycle (percentages refer to total global carbon).

plants, which consume oxygen and release carbon dioxide and water; (2) the combustion of fossil or organic fuels such as coal, wood, oil, and gas; and (3) the decomposition of organic matter (leaves, carcasses, and animal waste) by insects and microbes. Carbon dioxide is also dissolved in water, where it forms a weakly acidic solution. Small amounts of carbonate and bicarbonate ions are produced, and more than 99% of the carbon dioxide remains dissolved in the water without conversion. Therefore, the water and precipitation are only slightly acidified by carbon dioxide. However, the amount of carbon dioxide in the water is enormous, and it is estimated that the amount of carbon dissolved in the oceans is 70 to 100 times greater than that present in the atmosphere.[6,9] The carbonate ions formed in the water can react with calcium to form calcium carbonate, which mineralizes to form limestone. Limestone can also be formed from shellfish that create an insoluble calcium.

Plants convert inorganic carbon and water to useful energy-rich sugars through photosynthesis. The plants then become food for many forms of animal life, which in turn create new and more complex organic carbon compounds. The carbon is returned to the atmosphere through respiration and the decomposition of plant or animal matter or animal waste products. Although much of the decay of plant and organic matter may occur over weeks or months, it may sometimes be incomplete because the materials are buried and lying in peat bogs or moorlands, the bottom of oceans, or beneath the ground, where geological processes convert this material into the fossil fuels of coal, oil, or natural gas. These latter materials may be recovered and burned to release carbon dioxide back into the atmosphere. Fossil fuels and forests are releasing greater and greater quantities of carbon dioxide into the atmosphere. Many scientists and world leaders think the increase in atmospheric carbon dioxide is contributing to global warming. This topic is covered in more detail in Chapter 10, which is on air pollution.

Nitrogen Cycle

Although nitrogen is the major component of the earth's atmosphere (constituting 78% of the air we breathe), it cannot be used directly by animals and most plants. Useable forms of nitrogen are nitrates (NO_3), nitrites (NO_2), and ammonia ions (NH_3^+). Therefore, the natural mechanisms must exist that convert atmospheric nitrogen to useable forms. There are two known major natural mechanisms (Figure 1.18). One method involves the conversion of atmospheric nitrogen to nitrates (fixation) by: (1) bacteria such as *Rhizobium* sp., which live symbiotically in the root nodules of leguminous plants; or (2) some species of organobacteria that can convert gaseous atmospheric nitrogen to useable nitrate forms. The other natural conversion occurs when lightning causes atmospheric nitrogen to react with oxygen, producing nitrates in a number of steps. Nitrates are also produced in the atmosphere from nitrogen oxides created within the high-temperature confines of boilers or internal combustion engines. Nitrates are also released from the erosion of nitrate-rich rocks into soil and water. Nitrogenous compounds are continually being recycled from animal wastes and the decomposition of organisms, which soil microbes act on to produce ammonia. Ammonia is then converted to nitrates by nitrifying bacteria. Nitrates are a form of nitrogen useable by plants. Nitrates and ammonia are made by the oxidation of

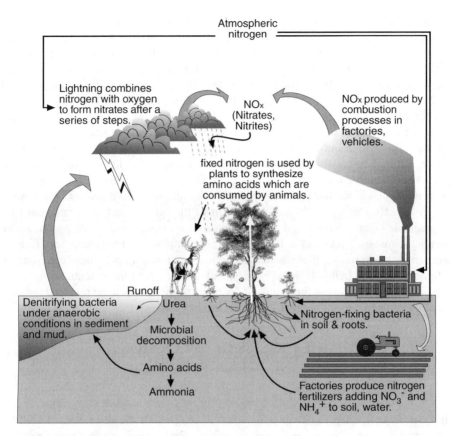

FIGURE 1.18 The basic nitrogen cycle. (Adapted from Turk, J. and Turk, A., *Environmental Science*, Saunders College Publishing, Philadelphia, PA, 1988, chap. 5.)

ammonia from nitrogen and hydrogen to create agricultural fertilizers, which are added to soils to increase fertility. This process is energy consuming and consequently is an expensive process. Once nitrogen is formed into complex organic molecules, it may be recycled back to atmospheric nitrogen by denitrifying bacteria, which live in the mud and sediment of lakes, ponds, streams, and estuaries, where they carry out these activities anaerobically (in the absence of oxygen).

Inland waterways, estuaries, and coastal ecosystems around the world have been subjected to a long-term increase in organic matter that has been washed off the land. A major consequence of this is an oxygen-depleted zone, such as the one off the coast of Louisiana. Many experts say that over fertilization from organic runoff has surpassed pollutants released from industries as a threat to marine ecosystems.[14]

Phosphorous Cycle

Calcium, iron, and phosphorous are examples of nutrients that flow through living systems in a pattern of recycling that is measured in eons of time or on a geological scale. These substances are gradually leached from sedimentary rock by the actions of rain or erosion from water in a process referred to as the *sedimentary cycle*. The phosphorous cycle is described here as being representative of this type of cycle. Phosphorous is the main element in compounds such as adenosine triphosphate (ATP). ATP is important in the transfer of chemical energy in animals. Phosphates are also a major component of deoxyribonucleic acid (DNA) and ribonucleic acids (RNA) and may be found in large amounts in shells, bones, and even teeth. Phosphates do not occur as gas, but are bound in sedimentary rocks, where they are released in small amounts from weathering. Plants take up the released phosphates in the roots, which may then be consumed by animals. Animal waste and decomposing animal carcasses release phosphorous back to the soil for reuse by plants in a fairly efficient manner when not disturbed. However, the processes of mining and agriculture can erode soil and carry phosphorous into streams, estuaries, and oceans, where it settles in a mostly insoluble form to the bottom and becomes generally unavailable. Consequently, adding phosphate fertilizers, which are produced by mining phosphate rock, enriches most agricultural soils. Phosphate rocks are a nonrenewable resource that was created millions of years ago. Therefore, sources of phosphates are being depleted rapidly, and we are facing the potential of infertile soils, because it may take thousands of years for phosphorous to be replaced through geological processes of weathering sedimentary rock.

THE PROCESS OF SUCCESSION AND RETROGRESSION

Succession

In a subsequent chapter (Chapter 3) you will read about the degrading of land by overgrazing or overuse, leading to retrogression. Retrogression is a negative process that reduces the diversity of an ecosystem, decreasing the number of organisms and so weakening the food web and placing greater numbers of species at risk. The normal undisturbed process of nature is one of succession. *Succession* refers to the predictable and gradual progressive changes of biotic communities toward the establishment of a climax community. A *climax community* is one that perpetuates itself with no further succession within an undisturbed ecosystem. Succession usually begins when a disturbance has occurred that kills many or all the life-forms within a region. Complete destruction of areas can occur in association with volcanic eruptions, retreat of glaciers, or impacts from large meteorites. In these events, all life, including that in the topsoil, is destroyed and primary succession must take place by first creating soil on the barren lava or exposed rock surfaces. This is accomplished initially through the weathering action of rain and dissolved carbon dioxide, which create fragments and dissolved minerals. Dust is captured in cracks and crevices along with microscopic organisms and seeds carried by the wind or deposited by small animals and birds. Mixtures of fungi and algae grow together and are known as *lichens*. These are highly resilient life-forms capable of surviving under sparse nutrient conditions. They gain a foothold and excrete organic acids that further deteriorate the rock surface, releasing more minerals. The organic material provided by decaying lichens and the inorganic minerals released from the rocks serve as nutrients for mosses that gradually replace the lichens, creating a denser growth that supports insects and other invertebrates while adding more organic matter to the mass. Eventually, enough organic matter is present to grow the seeds of some rooted plants. After many years, perhaps hundreds, greater size and varieties of plants and animals move into the once-barren area, and eventually a climax community may develop. The reason for the succession of plants is that each stage of plants fails to compete successfully with the next stage. The tall grasses have more extensive root systems that draw moisture and other nutrients away from the wildflowers, whereas the grasses grow taller and reduce the available sunlight to these same flowers. Ultimately the grasses are in turn overshadowed by the new-growth tress, and so on.[6,9]

When soil conditions are disrupted but there remains topsoil and some limited vegetation, succession can take place much more quickly. This process has been termed *secondary succession*. A common example of this is the return of an abandoned farm field to a climax forest. Because the fertile soil is already present, seeds and rhizomes present in the soil quickly grow. Plants rapidly move in, sprouting from the present seeds or rhizomes, or carried to the site by wind, birds, and grazing animals. These early plants are also known as *pioneer plants*, and may include wildflowers, followed by tall grasses and compact woody bushes. These grasses will eventually crowd out the wildflowers, but new-growth trees such as dogwood and sumac eventually replace themselves. After a time, perhaps a century or two, the climax forest trees will move into a typical temperate forest and include such trees as oak, maple, hickory, and elm. These trees will replace the new forest trees.[6,9]

RETROGRESSION

Not all disrupted ecosystems undergo a progressive return to stability with climax vegetation. Human intervention, in particular, has threatened the stability of ecosystems throughout the world. Activities that cause changes in natural cycles or population such as destroying species or adding new ones can be a threat to environmental stability. Stable ecosystems are ones in which materials are constantly recycled within the system through growth, consumption, and decomposition. These processes tend to balance each other so that there is little net loss over long periods of time in a process called *dynamic equilibrium*. Although minor disruptions occur to most ecosystems and a true steady state is seldom achieved, there is a long-term balance within stable ecosystems whereby nutrients revolve within that system with little overall loss. However, when the climate changes, soils are poisoned or eroded, or species are eliminated or introduced, the internal compensating mechanisms of the ecosystem may not be sufficient to return the ecosystem to its previous stable state, and a new combination of plant and animal life may be established.

In September two hurricanes released almost 3 ft of rain on eastern North Carolina. This rain turned the region into polluted floodwaters. This was caused by decomposing chickens and hogs, rotting farm fields, and ruined neighborhoods. This was one of the most disastrous floods in this region's history. Scientists now feel that the flood has also created an ecological situation that could change the biological rich waters of the Outer Banks from the mainland of Carolina. These bays and inlets are the second-largest estuaries in the country. The main problem is that the flood picked up huge amounts of organic matter, which included decomposing vegetation, topsoil, fertilizers, raw sewage, and waste created from hogs and farms. This water went directly into the estuary, which turned the water into a coffee color. This flood may have already killed many marine creatures like fish, worms, crustaceans, and crabs. No matter what happens in years to come, it may take years to return to normal, if the estuary has this ability.[14] As plots of land are overgrazed or intensively farmed, tropical forests burned and then farmed, water resources depleted, or other poor land management techniques employed, complex ecosystems are often replaced. This new ecosystem state may result in fewer overall species by a process called *retrogression*. Further, the species remaining may be less desirable from a human point of view. Simple systems in which fewer species exist or diversity reduction takes place are at greater risk of failure. The lack of cross-links in an ecosystem food chain means fewer alternatives for feeding, and when a species is reduced in number or dies off, its predators also die off in the absence of other alternatives.

The earth's climate has been changing rapidly, and the temperature has been increasing. As the climate changes, owing to human influence, there is an increased danger to human health. There are direct and indirect effects on human health. Direct effects are those like heat stroke, hypothermia, and deaths from hurricanes, floods, and tidal waves. Some indirect effects are the result of changing environments and ecosystems such as crop failure, algae blooms in the world's seas, and mental health disorders. Diseases such as malaria, dengue fever, and cholera are moving into northern countries due to the warming of climates and ecosystems. These threats are mostly due to human actions. Even with immediate action, much of what is happening will continue to occur into the future.[15]

The human population is exerting enormous pressure on ecosystems throughout the world as it continues to multiply at logarithmic rates and develop energy-intensive technologies, resulting in the discharge of dramatic levels of toxic substances into the air, water, and land. Most biotic communities are proving unable to respond to the unrelenting pressures of disruption, causing major losses in species, soil degradation, desertification, contaminated water, possible climate changes, and other changes in global ecosystems that retard human survival and quality of life.

REFERENCES

1. Godish, T., *Air Quality*, 3rd ed., Lewis Publishers, New York, 1997, chap. 4.
2. Sadler, J.P. and Grattan, J.P., Volcanoes as agents of past environmental change, *Glob. Planet. Change*, 21, 181, 1999.
3. U.S. Environmental Protection Agency, Global Warming: Climate, last modified on Friday, January 7, 2000, http: yosemite.epa.gov/oar/globalwarming/content/Climate.html.
4. Stevens, W., Lessons from Ancient Heat Surge, *The New York Times*, November 23, 1999.
5. Stevens, W.K., Carbon Fuel Use Down Globally, *Sunday Republican*, October 31, 1999, p. A-24.

6. Nadakavukaren, A., *Our Global Environment*, 4th ed., Waveland Press, Prospect Heights, IL, 1995, chap. 1.
7. Lush, T., Times staff writer, For forecasting chief, no joy in being right: Max Mayfield strives for accuracy, but worries about complacency, http://www.sptimes.com/2005/08/30/State/For_forecasting_chief.shtml, August 30, 2005.
8. Bradshaw, L. and Slonsky, L.B., A First Person Account of New Orleans After the Hurricane Struck, *Socialist Worker*, September 9, 2005, http://www.socialistworker.org/2005-2/556/556_04_RealHeroes.shtml.
9. Turk, J. and Turk, A., *Environmental Science*, Saunders College Publishing, Philadelphia, PA, 1988, chap. 5.
10. Woodward, S., Introduction to Biomes, Geography, 235, June 1998, http://www.runet.edu/~swoodwar/classes/geog235/biomes.
11. Schoen, D., Learning From Polar Ice Core Research, *Environmental Science and Technology News*, April 1, 1999.
12. Sargent, W., *The Year of the Crab, Marine Animals in Modern Medicine*, Norton and Company, New York, 1998.
13. Grouse, L., Turning Molecules into Medicine; The Role of the National Cancer Institute's Developmental Therapeutics Program in Drug Development, National Cancer Institute, U.S. National Institutes of Health, February 2006, http://www.cancer.gov/newscenter/benchmarks-vol6-issue1/page1.
14. Glanz, J., After the Storm, an Ecological Bomb, *New York Times*, November 30, 1999.
15. Kingsnorth, P., Human Health on the Line, *The Ecologists*, 29, March–April 1998.

2 Human Populations

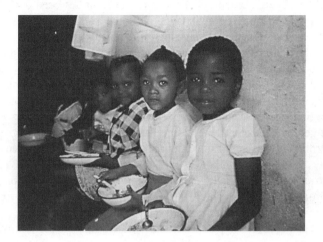

CHAPTER OBJECTIVES

A student, after reading this chapter, should be able to:

1. Define the attributes of populations, including birthrates and death rates, growth rate, density, and mobility (immigration and emigration).
2. Calculate rate of natural increase from birthrates and death rates, and mathematically demonstrate the effects of age–sex composition on a population.
3. Define biotic potential and maximum growth rate, and list the various limits to growth.
4. Identify, list, and explain the population growth forms.
5. Recognize and explain the concept population explosion with respect to complete and incomplete demographic transition.
6. Define population implosion and discuss the conditions that lead to this phenomenon.
7. Explain the role of urbanization in influencing sustainability of populations.
8. Explain global population projections and differentiate between developed and lesser-developed countries (LDCs) with respect to these projections.
9. List and discuss the various options for fertility control methods, contrasting the effectiveness, risks, and benefits of each type.

INTRODUCTION   ·

This is a text on environmental health. Why is the subject of human population treated here? The answer is brief and startlingly cold-blooded. Human numbers have risen so rapidly that the carrying capacity of sections of the earth has been overwhelmed and the sociopolitical structures of civilization in some parts of the world have been threatened. We must understand the environment and realize that environmental destruction is the national security issue of the early 21st century. Billions of impoverished people are surging in an enormous tidal wave into the urban centers around the globe, desperately seeking relief from the crumbling ecosystems of rural life. These hordes are fleeing from deforestation, soil erosion, water depletion, natural disasters, hunger, emerging diseases, and hopelessness, only to find themselves enmeshed in uncontrolled settlements of never-ending shanty towns built with corrugated metal shacks or cardboard. Armies of undernourished and potbellied children run through puddled streets with floating garbage where malarial mosquitoes swarm in great black clouds. Electricity, sewage system, and running water are rare. It is little wonder that in these environs criminal anarchy is the endemic illness. Overpopulation, infectious disease, unprovoked crime, few resources, and the influx of more refugees undermines nation states, leading to the empowerment of private armies, security firms, and international drug cartels. This is a vision of the early 21st century unfolding in many parts of the LDCs, and threatens to expand along with the growth of human populations.[1]

Because you are reading this text, you are probably sharing in the educational resources of the well-fed and technology-rich Western society, where recent studies indicate that overpopulation may no longer be a problem. In much of Western Europe and Japan, declining populations are causing major concerns about the economy, on account of a diminishing workforce, falling real estate prices, and a dwindling economy. So, where will the industrial workers come from? Who will care for the aged? In the U.S., demographic shifts are reshaping the racial landscape. There is an increasing divergence of racial generations, creating a gap that pits the *grays* against the *skateboarders*. In 2025, the number of elderly white Americans will increase by 67% to 47 million, and less than half the children will be white. During the end of the same period, 38% of the working American population will be nonwhite (Asian, Black, Hispanic, etc.). The mostly white elderly population will be expecting retirement and medical benefits to be paid by a racially diverse group that must yield a greater proportion of their limited income to pay for sharply increasing costs in caring for the elderly.

The world is becoming increasingly bifurcated into the *haves* and *have-nots*, and there is a growing realization that surging populations, environmental degradation, and ethnic conflict are strongly intertwined.[1] Therefore, understanding the dynamics of human populations is the first order of business in the study of environmental health.

THE CHARACTERISTICS OF POPULATIONS

It is usually easy to recognize a population. Constituent individuals have similar physical characteristics that identify them as a species and tend to inhabit specific

Birthrate equals the number of live children born in a year per 1000 total population

$$\text{Birthrate in year Y} = \frac{\text{Number of live children born in year Y}}{\text{Midyear population in year Y}}$$

$$\text{Birthrate in year 1998} = \frac{4,345,600 \text{ (children born in 1998)}}{271,600,000 \text{ (population in mid-1998)}} = 16/1000$$

$$\text{Death rate in year Y} = \frac{\text{Number of deaths in year Y}}{\text{Midyear population in year Y}}$$

$$\text{Death rate in year 1998} = \frac{2,172,800 \text{ (deaths in 1998)}}{271,600,000 \text{ (population in mid-1998)}} = 8/1000$$

Rate of natural increase in year 1998 = (Birthrate − Death rate) = 8/1000 or 0.8 %*
* These are approximate numbers for the U.S. used only for example.

FIGURE 2.1 Calculations of birthrate, death rate, and rate of natural increase.

areas. A species is normally considered to be a group of organisms that can breed together with the production of viable and fertile offspring. However, just because members of a species can breed together does not mean that they do. Normally, organisms breed in much smaller collections of like organisms called *populations*. A population is considered to be the breeding group for an organism. Different species not only have differing physical attributes, but also differ in their population characteristics. Most species of rats will give birth to several litters a year and have a relatively short life span of 1 to 2 years. Elephants, on the other hand, have long gestation periods lasting nearly 2 years, may live more than 70 years, and rarely give birth to more than 1 at a time. Each population has characteristics that help to identify it. Some of these characteristics are birthrate, death rate, age distribution, and sex ratio. Birthrate refers to the number of individuals added to a population through reproduction (live births) and is normally expressed as the number of live births per 1000 population (counting the population at the midpoint of the year). The example provided in Figure 2.1 shows a birthrate of 16/1000. Similarly, death rate is also calculated using total deaths divided by the midyear total population. The rate of natural increase is determined by subtracting the death rate from the birthrate. It reflects the growth rate in which migration is not considered. The growth of a population in the absence of migration must depend on the birthrate being higher than the death rate. The calculation of the rate of natural increase depends on other factors beyond this simple calculation. A more accurate estimation of population growth may be made if the sex ratio and age distribution of the population are known. The age–sex composition of the population has a profound effect on the birth and death rates of a country. This is so because the probability of dying or giving birth within any given year depends on the age and sex of the population members. Clearly, an aging population is going to have a higher death rate and lower birthrate than a population with many young women of reproductive age. The sex ratio refers to the

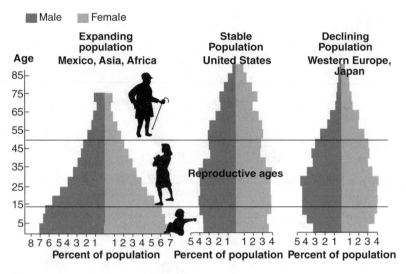

FIGURE 2.2 Age distribution types in human populations.

number of males relative to the number of females in the population, whereas age distribution describes the percentage of the population in various age categories from birth to death. In most societies, it is useful to separate the age groups into prereproductive juveniles, reproductive adults, and postreproductive adults. Some approximate differences in age–sex composition for different countries are shown in Figure 2.2. These graphs are created by plotting the number of males and females for age groups in 5- or 10-year increments within a population. LDCs of the world, such as Mexico, Africa, and much of Asia, more closely approach the broad-based pyramid-shaped graph in Figure 2.2. The distribution for these countries has a very broad base, indicating that birthrates have been very high in recent years. Further, there are many young children in the population who will likely reach reproductive age in the next few years and so rapidly expand the population of these countries in the next generation. It has been estimated that the group of young women entering childbearing years will be the largest ever. Additionally, there will be over 1 billion young people between the ages of 15 and 24 and 3 billion under age 25.[2] Those countries with more stable populations, such as the U.S. or France, have approximately an equal number of people in the age groups from 0 to 35. As these younger groups mature, there will be approximately the same number entering the reproductive years so that the population will remain fairly constant. This is complicated in the U.S. because of the more than 1 million people immigrating into the U.S. each year. It is the only industrialized country in which large population increases are projected.[3] Many Western European countries and Japan are experiencing a decline in growth rates. This is characterized by an age distribution with fewer infants and children than young adults. In the years to follow, there will be fewer people in the reproductive age groups and a population decline will occur in these countries if current trends continue.

Even with this information, predicting what will happen to future populations is a difficult task. Imagine that a large population of youngsters enters the pyramid at the base and travels upward in a group of diminishing size (because of deaths) into reproductive age, and then in even smaller numbers into postreproductive age. If this were a constant behavior of the population, predictions would be easy. However, wars, disease, famine, and reproductive intervention may cause the next generation to be reduced in number. This creates a constrictive formation or pinching at the base as seen in Western Europe (Figure 2.2). Because it is not easy to predict war, disease, or even human behavior with regard to reproduction, estimating future populations is a matter of some difficulty. Many countries are grappling with issues of unexpected population explosion or implosion right now.

Although it is apparent from this discussion that accurate population predictions may be difficult, there are other indicators useful in estimating future population changes. These include emigration, immigration, infant mortality, and total fertility rate.

In nature, when the density of organisms becomes too great, the intense competition for food, water, and other resources reduces the population. Some species have the ability to disperse or migrate out of the area, and in doing so, temporarily relieve the overcrowding. This process is called *emigration*. The pressure for resources may be the consequence of rapid population growth, and the disappearance of essential resources brought on by natural phenomena such as drought. When species emigrate from an area, they must *immigrate* or enter into another area. Driven by natural disasters, war, disease, and disappearing resources, the numbers of refugees worldwide may exceed 15 million, with about 880,000 to 1.4 million immigrants entering the U.S. each year, including more than 200,000 who enter illegally. The number of immigrants is likely to remain high for many years. This is in contrast with an American fertility rate that has been below replacement level for the 20 years. Immigration has, therefore, had a disproportionate impact on population growth in the U.S. The consequences of immigration to the U.S. will be discussed later in this chapter.

As discussed earlier, death rates are calculated by dividing the total number of deaths by the total midyear population. However, deaths of persons in the pre- and reproductive age groups will have a greater effect on future population growth than deaths in the postreproductive age group. Deaths of the young will mean a loss of potential births. Because of this impact, infant mortality is often a measured population characteristic that is used in conjunction with death rates. Infant mortality can be determined by the following relationship:

Infant mortality (%) = (No. of infant deaths/No. of live births) × 100%

The birthrate is useful for determining how many births are occurring in the total population, but does not tell what the current fertility patterns are. A useful predictor of reproductive behavior in a population is total fertility rate (TFR). TFR represents the number of children a woman in a given population is likely to bear during her reproductive lifetime, provided that birthrates remain constant for at least a generation. Recent figures show that the world TFR is 2.7, with the more developed nations having a TFR of 1.6 and the less developed ones, excluding China, having a TFR of 3.5.[4-6] The TFR is useful in determining the replacement level, which is defined as a TFR that corresponds to a population exactly replacing itself. The replacement TFR level for most countries is accepted as being 2.1, although some LDCs have such a high infant mortality because of poor nutrition, disease, and medical services that the TFR may be as high as 2.7. Almost all European countries are experiencing below-replacement fertility rates, with levels as low as 0.8 in Macao SAR (special administrative region), China. Whereas many of the LDCs have reduced the TFRs, most are still well above the replacement level of 2.1, with the highest recorded TFR of 8 for the year 2005 in Niger.[6]

If a person knows how many females of reproductive age are in a population based on the age–sex distribution and also knows the average number of births per female from the TFR, then it is easy to predict the next generation population provided the birthrates do not change. Of course, birthrates often do change as a consequence of changing social behavior, wars, famine, disease, or migration. The inability to predict these phenomena or their effects on the TFR contributes to inaccurate demographic predictions.

POPULATION DYNAMICS

There are periodic upsurges in many populations that lead to overwhelming numbers. Whether these population explosions occur in rabbits, lemmings, soldier ants, or locusts, there are always some natural pressures that bring the population back into balance with the natural surroundings. Bacteria reproducing by binary fission can grow to millions of organisms within a few hours given the right conditions, whereas many fungi grow cubically and could easily grow to fill an entire room overnight if pressures did not exist to limit their growth. Fish and amphibians produce thousands of eggs, plants release millions of seeds, and many animal species produce multiple litters of offspring in a year. Every species has an inherent reproductive capacity to produce offspring that is generally many times greater than needed to replace the parents that die.

Clearly, if these organisms and offspring all survived, we would soon be over-whelmed by the sheer numbers of these species. In fact, even the human population growing at a rate of 1.6% for the next few hundred years would completely cover every square inch of the earth and its oceans. The unrestricted growth of populations resulting in the maximum growth rate for a particular population is called its *biotic potential*. The biotic potential of species differs markedly and is influenced by: (1) the frequency of reproduction, (2) the total number of times the organism repro-duces, (3) the number of offspring from each reproductive cycle, and (4) the age at which reproduction starts. Some bacteria reproduce every 20 min by splitting into two (binary fission), whereas polar bears give birth only once every 3 or 4 years giving them a much lower biotic potential. This biotic potential is never actually reached for any species, because the growth is limited by one or more factors that are collectively referred to as *environmental resistance*. Environmental resistance refers to those pressures that limit population and may include such factors as disease, wars, predatory behavior, toxic waste accumulation, or species competition. The actual rate of increase is then described by the biotic potential minus the environmental resistance as shown in Figure 2.3.[7] The sum of these limiting factors serves to control the numbers of a species that can survive in a defined area over

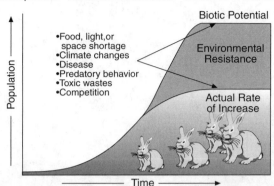

Biotic potential – Environmental resistance = Actual rate of increase

FIGURE 2.3 Biotic potential – Environmental resistance = Actual rate of increase.

time, and this represents the carrying capacity of that area for a particular species. The carrying capacity of a defined system can be demonstrated in the laboratory by cultivating a bacterial organism in a nutrient broth at 37°C and removing the organisms hourly to count them. This is accomplished by making serial dilutions of the original broth culture and placing 1.0 ml of each dilution onto separate nutrient agar plates. Incubation of these plates results in the growth of discrete colonies that can be counted to determine the population at hourly intervals (Figure 2.4). By plotting

Calculation: Number of colonies on plate x reciprocal of dilution of sample = bacteria/ml. In this example, there are 20 colonies on the plate of 1:100,000 dilution = 2 million bacteria/ml. A growth curve can be constructed if the original inoculum is counted by this process hourly for 24 to 48 hours.

FIGURE 2.4 The plate-counting technique is a method that may be used to count bacteria to generate a bacterial growth curve.

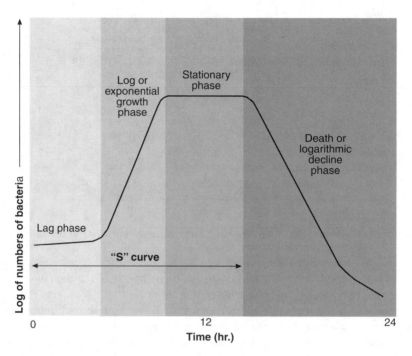

FIGURE 2.5 Bacterial growth curve showing various phases of growth including lag, exponential, stationary, and death phases.

these numbers, a typical sigmoid, or S-shaped, growth curve can be demonstrated (Figure 2.5). The growth curve initially shows a lag phase, in which the organisms show no increase in growth rate but are undergoing changes that ready them for rapid proliferation. This is followed by a rapid increase in growth in which the population doubles every few hours. The population increases at a rate proportional to its size, or exponentially. This is known as the *exponential growth phase*.

To demonstrate logarithmic growth, consider that a single bacteria splits every 30 min. If a single bacteria starting at time zero divides in 30 min to form 2 bacteria, then in another 30 min, each of the 2 divides to form 4 bacteria. If this were to continue unabated for 24 h, there would be 48 generations for a population of over 280 trillion organisms. Can you determine the number that would be produced if growth continued uninterrupted another 4 h?

Eventually, the limiting environment of the culture tube restricts the growth of the bacteria by an accumulation of toxic wastes and lowered pH. The number of organisms dying eventually comes to equal the number of living organisms so that a stationary level is reached. This also represents the carrying capacity of this culture tube environment. If left to grow without removing wastes or adding nutrients, the bacterial population will die off rapidly in what is known as the *death phase*. This S-shaped growth curve was described in 1932 by the Russian ecologist G.F. Gausse when he studied the growth of experimental populations of *Paramecium caudatum*, a freshwater protist.

FIGURE 2.6 J-shaped growth curve showing growth extending beyond the carrying capacity of environment.

If an organism grows too rapidly and the population escalates beyond the carrying capacity of the environment in which it is located, a J-shaped growth curve may develop (Figure 2.6). This curve is somewhat misleading, because it is truncated before the outcome of the event is revealed, and may more accurately be referred to as *geometric growth*. There are many examples in nature of insects, mammals, or other life-forms experiencing an explosive growth in population that far exceeds the carrying capacity of their habitat. The consequence of this is a precipitous decline in the population once food supplies are depleted, disease intervenes, or the habitat is changed in some other way to create hostile conditions. This behavior sometimes oscillates every few years as in the case of lemmings that inhabit the Arctic tundra north of the Canadian forest. Every 3 to 4 years the population explodes, then crashes the following year, followed by a 2-year cycle of slow recovery (Figure 2.7).[8] There are other patterns of growth as well. Predator–prey interactions such as occur with the lynx and rabbit are likely to produce an oscillating pattern for both species, with the predator showing increases and declines following such changes in the prey population. As the rabbit population increases, the lynx will have more prey to feed on, and its population will grow. Eventually, there will be sufficient predators to cause a reduction in the rabbit population, and its decline will cause a decline in the predator lynx population soon after, as seen in Figure 2.8A. Another type of growth curve is seen in Figure 2.8B, for a population heading toward extinction. This type of curve can be drawn for more and more species everyday as human intervention eliminates species and habitats, spreads pesticides, and hunts down

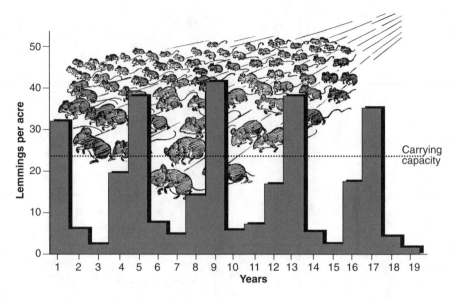

FIGURE 2.7 Typical population crashes seen in lemming populations every 3 to 4 years.

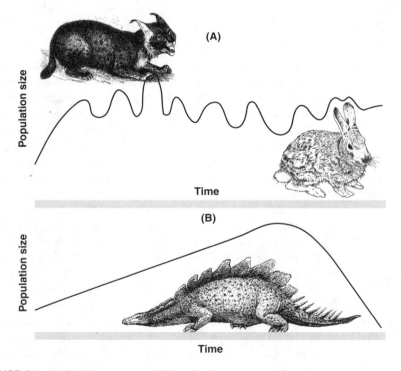

FIGURE 2.8 (A) Predator–prey and (B) extinction-type growth curves.

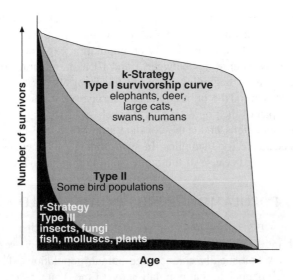

FIGURE 2.9 Characteristic survivor curves showing Type I (prosperous human population), Type II (impoverished human population and certain bird populations), and Type III (fish, plants, fungi, etc.). (Adapted from Turk, J. and Turk, A., *Environmental Science*, Saunders College Publishing, Philadelphia, PA, 1988, chap. 7.)

larger species for human uses. The disappearance of wildlife and species extinction will be discussed in more detail in Chapter 3.

The science of demography has categorized two different reproductive strategies for living organisms. One strategy is known as a *k-strategy*, in which large organisms with relatively long life spans have only a few offspring, but devote their energies to protecting and nurturing the offspring to enhance their individual survival until they can reproduce. Examples falling into this group would be elephants, deer, large cats, swans, and humans. These populations normally stabilize at the carrying capacity of their environments, at which point density-dependent factors limit growth. Density-dependent factors include food supply, which becomes more limiting as the size of the population grows. The pattern for mortality of these species can be demonstrated by plotting a survivorship curve, in which the fate of young individuals is followed to the point of death in order to describe mortality at different ages.[9] This curve can be used to predict the probability that a newborn will reach a specific age, which is useful in determining the life expectancy within a population The Type I curve typical for the k-strategy survival population shows that the majority of infants reach sexual maturity and many survive to maturity, eventually dying of diseases from old age (Figure 2.9).This is in sharp contrast to the *r-strategy* populations, which are typically small, short-lived organisms that produce large numbers of offspring and receive little or no parental care. This is characteristic of species of invertebrates, fish, mollusks, plants, fungi, gypsy moths, locusts, grasshoppers, and some rodents. Large numbers are produced but few survive. There is high mortality among the offspring as depicted by the Type III survivor curve (Figure 2.9). These organisms are limited by density-independent factors such as a drought that dries up a pond, or sudden climatic changes such as El Nino, which alters the temperature of the water, making it uninhabitable for certain species. Such populations often demonstrate wildly fluctuating numbers with population explosions and crashes.

A Type II survival curve lies between I and III and is characterized by a high mortality in the infant or beginning stages of life from disease, genetic defects, or predation. When this initial die-off is passed, the survivors die at a smaller but constant rate over the life span of the population.[9] This is characteristic for some bird species and for human populations under the constant stress of poor sanitation, infectious diseases, and inadequate food supplies. In such populations, there is a tendency to produce more offspring because of the higher mortality in infants or offspring.

POPULATION TRENDS IN THE WORLD

Demographers use the information on population size, fertility rates, migration, birthrates, death rates, growth rates, infant mortality, density, age–sex composition, and other factors to statistically characterize human populations. Their purpose is to predict what will happen to that population over time. You may question why this is so important. What is the value of this information? The statistical knowledge of human populations certainly dates back thousands of years to the beginnings of civilization when support systems for maintaining large groups of interdependent people became necessary. Government structures developed, and there was a need to raise taxes, maintain armies, harvest and store foodstuffs, provide education or training, build roads, and promote commerce. These activities provide the infrastructure on which society depends to this day. Even in the smallest towns, the local officials must have census data to know how to plan possible expansions in schools, sewer and septic systems, road construction, power lines, building lots, and land conservation. Humans require significant infrastructure to maintain a healthy, educated, and physically comfortable lifestyle. This takes significant planning and the knowledge of how many people to account for. There is little that goes on in society today that does not depend on a knowledge of human numbers. Marketing products, predicting enrollments, estimating tax incomes, analyzing insurance risks, determining future costs for elderly care, etc., are just some examples of the uses for population numbers by age, race, sex, interests, disease, etc. Understanding the growth patterns of human populations is vital to a nation's interest and perhaps to the survival of the human race.

HISTORICAL TRENDS

Core samples taken from the Arctic ice and from deep-sea beds reveal a chaotic history of extremes in the earth's surface temperature over several hundred thousand years. Then, as if almost by chance, the earth's temperature warmed and stabilized about 10,000 years ago. Humans began to develop agriculture and domesticate animals. The ability to feed larger groups of people from planted crops and domesticated herds allowed civilizations to develop. The inefficiency of a nomadic existence was no longer necessary, and soon the world's population began to increase as population groups could sustain more children. Although accurate predictions of

FIGURE 2.10 History of human population from early man, *Homo habilis*, to the present, showing exponential growth beginning around the Industrial Revolution. (Adapted from Turk, J. and Turk, A., *Environmental Science*, Saunders College Publishing, Philadelphia, PA, 1988, chap. 7.)

human populations are questionable before circa 1700, estimates of primitive civilizations around 8000 BCE (before the Common Era) place the number of humans at about 5 million (Figure 2.10). The advent of agriculture is thought to have promoted the growth of human populations to about 500 million by 1650. The world's population then doubled in the next 200 years to about 1 billion people. During this time the industrial revolution began, introducing mechanization into agriculture with increased production and delivery of foodstuffs, the production of fabrics and hardware, and the construction of domiciles. The ability to house and feed more people caused an eruption in the birthrates and numbers of people. By 1930, there were nearly 2 billion on the planet, and it took less than 80 years to double the world's population. In the late 1800s, awareness of sanitation led to the chlorination of public water supplies. The introduction of public sewers, pasteurization of milk, and food protection programs soon followed. Certainly by the early 1900s, authorities in most major cities in the U.S. and Europe understood these concepts and employed the principles of sanitation. Consequently, the spread of communicable diseases through milk, water, and food was dramatically reduced. Death rates plummeted, and so growth rates exploded into the 20th century. New discoveries in antibiotic therapy, vaccines, and medical treatment saved many more lives and increased life span. The consequence was alarming in its effect on the growth rate. In 1975, the world's population had increased to 4 billion people. It took less than 45 years for the population to double from 1930. In the fastest population spurt in history, it took fewer than 40 years for the population to double; the world reached 6 billion people on October 12, 1999. Today the world's population is estimated to be greater than 6500 million.[10] Consider that the world's population in 1943 was about 2 billion people, and therefore in my lifetime (so far) it has more than tripled.

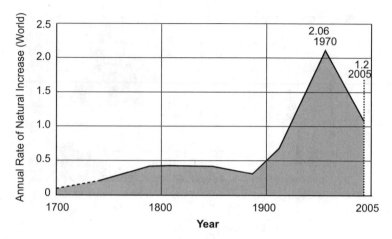

FIGURE 2.11 The world population growth rate from 1700 to 1997.

The rate of growth of the world population from the mid-1700s until the present is shown in Figure 2.11. The growth rate peaked in 1970 at 2.06% and has declined since then to 1.2%[5,6] This decline in growth rate is directly related to declines in births prompted by international efforts, which began in the 1960s, to stem the tide of the growing global population. This concern escalated from private groups, spread to national governments, and then to international agencies. The majority of the population increases were being seen in the LDCs, and it became clear that the population increases were so rapid and so large that they were surpassing the carrying capacity of these countries and threatening their stability. The majority of these nations are racing to reverse this trend and stabilize their populations before major calamities occur. However, the efforts at achieving population stability should have begun decades earlier. The combination of population momentum and a large population base in pre- and reproductive ages makes this effort very difficult. In order to understand how we reached this point, it is helpful to look at the relationships between growth rate, doubling time, demographic transition, and incomplete demographic transition.

Growth Rate

A rate is normally expressed as one quantity in relation to another. Runners like to keep track of the time it takes to run a 10-km course in minutes per mile. This is a rate measurement, and runners can study their progress by decreasing the minutes per mile that it takes to run the course. In a similar way, growth rate can be used to assess additions or deletions to the population. A population is increased by births or immigration and decreased by deaths or emigration. The rate of births is the ratio of births to the population, and death rates represent the ratio of deaths to the population. Growth rate is then determined by the birthrate minus the death rate. If the growth rate is positive, there will be an increase in population. Even if the rate of growth is slowed (declines), the population will accumulate. Similarly, even if the runner's pace is slowed, he or she will still accumulate miles. Consequently, we are still contributing large numbers of people to the population despite a declining growth rate. This addition

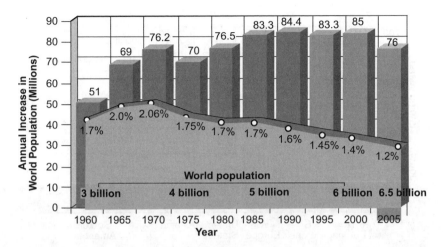

FIGURE 2.12 Annual increase in world population vs. global growth rate.

becomes exaggerated when the base of people is larger. In 1975, the growth rate was close to 1.75% with a global population of 4 billion people, leading to 4 billion × 0.0175 = 70 million new infants each year. On October 12, 1999, the population reached 6 billion people with a growth rate of 1.4%, leading to 6 billion × 0.014 = 84 million new infants per year. Today, world population is growing by 76 million people per year (about 240,000 people per day), and we have surpassed 6.5 billion people. The latest UN midrange estimate suggests that there will be 8.9 billion people on the earth by 2050, with total population shrinking over the next 100 years.[11] Despite the successes in reducing the worldwide growth rates, the population has grown so much that even the smaller growth rates lead to additions of larger numbers of people to the global population (Figure 2.12). This is referred to as *population momentum*. For example, in Botswana, a country experiencing a drop in life expectancy, the population is expected to double by 2050 because of population momentum.[7] Added to the problem of population momentum are the differences in age–sex distribution in LDCs vs. the developed countries. Figure 2.2 compares the age–sex distribution for the lesser- and more-developed countries. The very broad base of LDCs indicates that birthrates have been high and that young children represent the largest percentage of the population. As these children reach reproductive age and bear their own children, the population will rapidly expand. The combined population momentum of huge populations that have a large percentage of young prereproductive people will lead to significant population increases through the next generation.

Doubling Time

Another useful way to demonstrate growth rate is to present it as doubling time, or the number of years for a human population to double its size. The doubling time can be calculated according to the following relationship:

$$\text{doubling time} = 0.70/\text{growth rate} \times 100$$

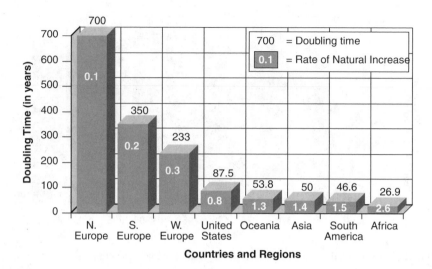

FIGURE 2.13 Population doubling times and rates of natural increase comparing certain developed and lesser-developed regions and countries.

This relationship is derived from the formula for calculating compound interest at the bank and is an approximation. It is useful to remember that a growth rate in any population compounds itself as the individuals entering the population will ultimately reproduce and add more people to that population. As the growth rate increases, the doubling time decreases. Populations from 8000 BCE to 1650 doubled about every 1500 years as evidenced by the relatively slow rate of population increase (Figure 2.10). The population next doubled in 200 years, then 80, 45, and 36 years, when the world reached 4 billion people in 1975, and is now over 6.5 billion people in 2006. Viewing population growth in this way is useful because most everybody can understand what life would be like if the number of people in our particular location were doubled within a generation. A graphical presentation of doubling time vs. the rate of natural increase is shown in Figure 2.13. A rate of 4.0% annual increase will double the population of a country in 18 years. The World Bank estimated Kenya to have such a rate in 1979, dropping to 3.7% in 1988 with a population of 23.3 million.[12] The population in Niger in 2005 was 14 million with an annual rate of increase of 3.4%, and its doubling rate has been increased to 20.5 years.[13] Although this may be viewed as an improvement, such an increase in so short a time will likely surpass the carrying capacity of that country, and it will be incredibly challenged to build roads, houses, schools, and sanitation facilities to accommodate this many people in so short a time.

Demographic Transition

Understanding population trends in the world requires an explanation of the differences in growth rates occurring between the developed countries of Western Europe, North America, and Japan vs. the LDCs in Latin America, Asia, and Africa. The developed countries have exhibited slowly declining death rates over the last century

FIGURE 2.14 Demographic transition in developed countries showing declining death rates and birthrates resulting in a low rate of natural increase. (Adapted from United Nations Population Fund, *The State of the World Population 1999: 6 Billion, A Time for Choices,* 1999.)

with the gradual introduction of sanitary principles (chlorinated water supplies, pasteurized milk), increased agricultural production, and improved medical care (vaccines, antibiotics, etc.). During this time, the industrialization of society and mechanization of agriculture reduced the economic benefit of having more children, whereas those children that were born had a better chance of survival. Therefore, as the death rate declined, so did the birthrate. This has resulted in a diminishing difference between birth and death rates and a very low rate of natural increase, resulting in a stable population with very long doubling times (Figure 2.14).

Incomplete Demographic Transition

The LDCs had a very different experience. Most of these countries are or were agrarian societies without government-sponsored programs for social security. These are subsistence economies, where rural inhabitants extract a living directly from plants and animals, requiring much human labor to accomplish simple tasks without the aid of electricity or tap water. Children in the household are required to cultivate crops, care for livestock, cook food, and produce simple products for the market. Such activities may require up to 6 h a day collecting wood, water, and fodder. Children in India as young as six are expected to perform these duties. As resources diminish, there is a need for even more hands to collect water and firewood, which are located increasingly farther from the home.[14] Such children, especially males, are also viewed as a source of security in old age, because the governments in LDCs do not have the resources to institute social security or have unstable policies that fail to capture the trust of its citizens.[15] The populations in such countries had remained stable for perhaps thousands of years with high birthrates and death rates. It was not uncommon for half of the children in a family to die, thereby maintaining a relatively low growth rate.[14] The sudden introduction of better nutrition, modern

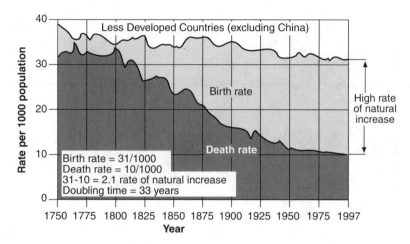

FIGURE 2.15 Incomplete demographic transition for lesser-developed countries showing declining death rates with birthrates remaining high, resulting in a high rate of natural increase. (Adapted from United Nations Population Fund, *The State of the World Population 1999: 6 Billion, A Time for Choices*, 1999.)

medicine, insecticides, improved water quality, and methods to isolate human wastes resulted in a startling decrease in deaths, but without the time for adjustment in birthrates. The number of births has remained high for many of the reasons discussed earlier. The combination of high birthrates and lowered death rates has resulted in very high rates of natural increase (Figure 2.15).

CURRENT POPULATION TRENDS IN DEVELOPED COUNTRIES

The world's population is growing at a rate of 1.2% annually and reached 6.5 billion people in 2005. This growth rate is a remarkable decline from the peak of 2.06% growth rate reported in 1970, and we entered the year 2000 with a declining rate. According to the United Nations, rates are expected to continue to fall gradually for the next 20 years, and more rapidly thereafter.[3] However, we may have already set in motion the potential for destructive impacts to global ecosystems and some sociopolitical structures as the population momentum sweeps around the world. These problems are not equally balanced, nor are the population increases equally distributed. It is expected that the human population will reach 8.9 billion by the year 2050.[11,16] The U.S. media has announced in bold headlines that *population explosion is over*, and that the real crisis ahead is *population implosion* as many countries now face fertility rates that are below the replacement level of 2.01.[17] This may be true in the developed nations, where the United Nations predicts that during the next century industrialized nations will have fewer people than it has today.[2] From a different view, over 95% of the annual increase in the world's population is occurring in the LDCs, and more than 74 of the LDCs are expected to double their population in less than 30 years, adding nearly 80 million people to the earth annually. A clear division has developed between the developed and LDCs that will likely have enormous geopolitical consequences. We need to examine the current population trends based on these differences.

Are you interested in learning more about current population trends and policies around the world? Visit http://www-unix.oit.umass.edu/~envhl565/weblinks.html and click on one or more of the following sites:

- Interactive Game — 6 billion people
- U.S. Census Information
- Negative Population Growth
- Population Institute
- World Health Organization
- World Watch Organization
- Zero Population Growth
- U.S. Population Clock
- Population Links
- Population Reference Bureau
- International Programs

Population Decreases in the Developed Countries

Declines in Fertility

In 1970 there were 19 countries reporting declining fertility rates, whereas in 2005 over 68 countries have reported below-replacement fertility rates.[3,6] Nearly all the European countries have below-replacement fertility rates, as do the Asian countries of China (1.6), Japan (1.3), Thailand (1.7), South Korea (1.2), and Hong Kong (1.0).[6,16] In North America, the countries of Canada (1.5) and America (2.0) are also at below-replacement levels. Those countries with the lowest fertility rates include Macao SAR (0.8), Hong Kong (1.0), Taiwan (1.2), and Slovenia (1.2).[6] The European birthrates of the 1980s were already remarkably low; then declined precipitously by another 20% to 1.4 in the early 1990s, and plummeted to 0.0 on the eve of the year 2000. These rates cannot sustain a population, and even if the European rate were to return to the replacement rate of 2.1 by the year 2060, Europe will have lost almost 25% of its population. Nearly 50 years ago about 32% of the world's population lived in the modern nations of Europe, North America, and Japan. Today, 20% of the world's population lives in these countries, and by 2020 it is projected to be 16%. In 2050 that figure will have dropped to 12% of the world's population.[17] The population in these regions fell so rapidly and so far that it was unprecedented in the world's history. The potential for major economic, geopolitical, and environmental changes remains to be discovered, but are likely to be dramatic. These low-fertility countries will have to look to active older people and immigrants to supply needed services and contribute to the economy.[3] There was such concern that, for the first time ever, the United Nations convened population experts and leaders from around the world in December 1997 to consider aspects of the low and rapidly declining fertility rates.[17]

The global fertility rate in the years 1950 to 1955 was 5.0, and on course to explode the world's population beyond its carrying capacity. The rate remained steady for another 15 years until the period 1975 to 1980, when the global TFR fell to 4 children. In 1995 the rate had fallen to slightly less than 3, and fell below 2.79 in 1997. Demographers had earlier made projections of human populations reaching nearly 10 billion people by the year 2050. Now their projections would show a population that is 650 million short of the original figure, because some projected birth will not occur. Now the United Nations predicts that the population will reach 8.9 billion people by 2050 and 10 billion by the year 2100, following a 21st century medium-fertility scenario.[18]

Concerns about Decline

The industrialized countries are deeply concerned about the static or declining workforce. By the year 2025, Japan will lose 12.2 million people in the age group 15 to 64, whereas Germany will lose 4.5 million, and Italy will see a loss of 4 million. These developed countries are experiencing a rapidly aging population and a significant decline in the ratio of working-age people. In Western Europe there is an average of less than 4.3 workers in the age group of 15 to 64 years for each elderly person. The ratio of workers to elderly is projected to decline further by the year 2025, with Britain at 2.9 to 1, followed by France (2.7), Germany (2.5), and Japan (2.3), where almost a third of the population will be elderly.[15] If the present trend of a tumbling population in Japan continues, its population in 2100 will be 55 million people; down nearly 50% from the 125 million people today.[19] In developed countries, populations are projected to slowly rise until 2025, and then decline, so that the total number in 2050 will be about the same as today.[16] There is concern throughout Europe and Japan that the declining population will result in decreasing real estate prices as the demand declines along with the population. A smaller workforce and declining economy are also viewed as creating a country of lesser importance. Further, a high proportion of elderly nonwage earners in countries where governments support the elderly place an ever-increasing burden on the diminishing ratio of wage earners. There is likely to be economic pressure on a country that exhibits a high dependency ratio of many aged vs. fewer young to alleviate the imbalance by delaying retirements and reducing retirement benefits. We see some of that already occurring in the U.S.[20] The world economies are becoming more globalized, and those countries having a declining population with a high percentage of nonworking people will be unable to compete against nations of young, lower-wage, productive workers such as China and India. Consequently, there is a growing effort among some governments to reverse this population trend.[20]

Vastogirardi, a small village in southern Italy, has experienced a decline in fertility rate for many reasons, including many women choosing to have a career and men enjoying their bachelorhood.[21] The people of Vastogirardi fear that they will lose their culture, traditions, and schools. Although birthrates are low all over Europe, Italy and Spain have the lowest of all — an average of 1.2 bpw (births per woman), slightly lower than Greece and considerably lower than the U.S. rate of 2.1 bpw.[21] One of the main reasons for Italy's low birthrate is that state aid to families with children is lower in Italy than other countries, such as Germany or France. As Italian women who work are often also responsible for domestic chores, it is not uncommon to avoid having a second child in order to fulfill their obligations to work and the home. This village has also witnessed many businesses closures owing to people moving to bigger towns. There was only one child born in Vastogirardi last year.[21]

Officials in Kyokushi, a small town in the southern island of Kyushu, Japan, are offering a gift of $5000 to parents who have a fourth or subsequent child. The former city of Kyokushi became part of the new city of Kikuchi as of March 22, 2005. Other Japanese towns are offering cash rewards, and also considering reduced land prices and cash incentives for newlyweds.[19] It seems that very few couples are impressed and feel that the amounts offered are too small. Many feel that the children are costly to raise and interfere with their free time. Others feel pressured to raise perfect children and do not want the burden of having to be perfect parents. Children raised with such lofty expectations often rebel. The declining fertility may also be attributed to the growing number of women who marry later in life, or not at all. The total fertility rate dropped from 1.57 in 1989 to 1.39 in 1997, and is now 1.3.[6,22] In Japan, the average woman marries at 27, and nearly 17% of women in their early 30s are unmarried.[23] Makoto Atoh, director-general, Institute of Population Problems in Tokyo, points to the advancement of women as the major factor in fertility decline. In Japan, the failure of husbands to participate in child care and housework is also considered a significant factor in influencing women to have fewer children.[19] Higher education for women, new aspirations, and higher incomes are considered to be factors for declining fertility rates in many countries.[17] In fact, as illiteracy among women decreases in a country, the average number of children born to those women declines, and conversely, the more income a women brings into the family, the fewer children she is likely to bear (Figure 2.16).[11,16] Unfortunately, women's rights as well as their voices are often silenced around the world, with over 100 million girls in the next decade to be married before their 18th birthday, and some before they turn 10 years old.[11] Also, pay is often reduced during a woman's maternity leave, making it even less likely that she will have many children.[22] In Italy, unmarried children often continue to live at home with their parents, contributing to one of the lowest birthrates in the world.[20] Another factor that may decrease fertility rates is increased urbanization. As more people move to cities, children become more expensive to maintain and are unnecessary for farm work. Improved methods of contraception, late marriages, more divorces, and substantially reduced infant mortality (giving parents confidence their children will survive) may all play a role in decreasing fertility rates.[17,20] Is the concern about declining birthrates realistic? A large population is not a necessary condition for global power and influence, as illustrated by the most rapidly growing nations in Africa, Asia, and Latin America. These nations are growing most rapidly but are also the most poverty-stricken. Population increases are expected for three reasons: declines in mortality, population momentum, and a fertility rate that is still above the replacement level (despite a dropped rate from 6 bpw to 3 bpw).[16] Certainly, if population size were the yardstick of global power, India and China would be the world's two superpowers, with Israel and Great Britain being essentially without power or influence.[13] Dr. Nafis Sadik illustrates that Brazil and Japan had similar growth in economy between 1960 and 1984. Brazil's per capita GNP (gross national product) doubled in that time, Japan's escalated 7.4 times. The high fertility rate consumed much of the expanding economy to provide for the larger population, whereas Japan's economic and political clout rose in the pressure of a declining fertility.[23] The concerns about declining populations may

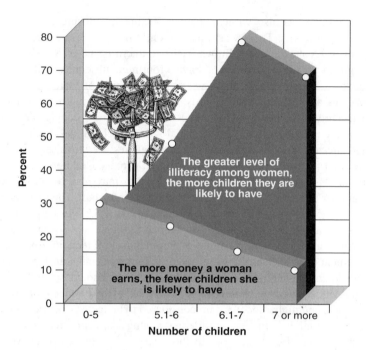

FIGURE 2.16 The number of children a woman is likely to have, based on her level of literacy or the earned income she contributes to the household. (Adapted from Kristof, N.D., Baby Makes 3, but in Japan That's Not Enough, *The New York Times International*, October 6, 1996, p. 1.)

not be warranted, but the concerns are apparent and as countries act on those concerns, the future may see upward shifts in fertility for some of those countries as governments look for successful ways to influence fertility behavior.

Increasing Death Rate

Some countries in Eastern Europe and the former Soviet Union have experienced mortality increases. The trends influencing the rise in adult mortality have been cited as increased poverty, inadequate nutrition, increased stress and overcrowding, and a greater incidence of infectious diseases.[3] These factors most likely have been brought about because of the current political changes and instability in the region, and their persistence could be a source of concern.

Fertility Rates in the U.S.

The U.S. has a demographic history of rapidly changing patterns of fertility with mortality remaining stable (Figure 2.17). Subtle changes in social attitude appeared to produce rather significant changes in fertility rates, and subsequently cause marked changes in the rate of natural increase. It is not surprising that the *Great Depression* resulted in major decline in fertility, reaching a low of 2.2 in 1936, and then showed a rapid increase during and after World War II. The postwar demographers, however, predicted a fertility decrease, but were surprised by the continued and rapid increase in population, which peaked in the early 1960s. The population then rapidly declined

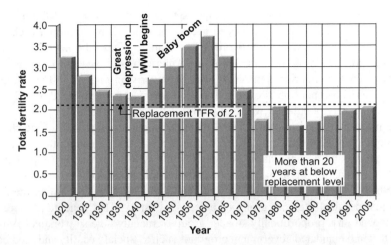

FIGURE 2.17 Total fertility rate (TFR) for U.S. from 1920 to 1997. (From U.S. Census Bureau, U.S. and Word Population Clocks, April 2006, http://www.census.gov/main/www/popclock.html.)

to a below-replacement level in 1971. The demographers had predicted that the birthrates would remain high after the 1960s, but once again were unable to predict the decrease. There are explanations given retrospectively that attribute the decline to a more educated female population seeking greater career opportunities and challenges. The ideal American family size grew smaller as couples realized that large families were an economic disadvantage. The availability of inexpensive and effective birth control methods made it easier for couples to reach this goal. Unpredictable changes in social attitudes had produced dramatic changes in fertility and the rate of natural increase. The TFR continued to drop from 1971 through 1976. It rose again in the 1975 to 1980 period to 2.1 for unexplained reasons, and then fell again in 1986 to a low of 1.7. Since then the TFR has continued to rise to the present level of 2.0 while remaining below the 2.1 replacement level for the past 35 years.

Immigration and the Changing Racial Landscape in the U.S.

Although the TFR has remained below replacement levels, the U.S. has one of the most liberal immigration policies in the world; immigration adds at least another 850,000 to 1.2 million people each year to the county's population, with an estimated 12 million unauthorized foreign-born people in the U.S. in 2006.[13,24] There were 37 million foreign-born residents in the U.S. in 2005, which included both unauthorized and legal immigrants.[11] If federal immigration policies do not change, these numbers will likely go unchecked for the next 50 years. Because TFR has been at or below replacement level for 27 years, immigration is considered to be among the most important demographic factors in the growth rate of the country. About 82 million of the projected 390 million people living in the U.S. in 2050 will either have immigrated since 1991 or will be children of those immigrants. This group will then represent 21% of the nation's population, and total growth due to natural increase (0.8%) and net migration (0.2%) would be expected to double the nation's population in less than 70 years.

For the past 30 years, most immigrants to the U.S. have been from Latin America and Asia. Hispanic women have been averaging more than 3 children each, black women more than 2, and white women less than 2. Within the next 25 years, the under-65 population of Hispanic and Asian Americans is expected to double, as the number of white Americans under 65 will be fewer than in 1995.[17] The number of black Americans under 65 is expected to increase by a third during this period, and the white elderly population will balloon by 67% to over 47 million people, so that in the next 25 years more than 75% of the elderly, but less than half of the children, in America will be white. This divergence in race and age is leading to a situation in which an expanding population of elderly white will be expecting support from a working population of tremendous diversity. Another coinciding event is that the aging baby boomers (those born between 1946 and 1964) are becoming the largest and most politically potent group of elderly in history. This generation of retirees will depend on a proportionally smaller group of racially diverse workers to support them through mandated government programs such as Social Security and Medicare.[17]

The Social Security system was established on a pay-as-you-go basis, such that the money a person pays each year is not saved for their pension, but is used to pay the pension of an older person alive at the time. The system was established assuming the ratio of workers to retirees would remain high, but this is not happening. There were 16 workers per retiree in 1950, and by the year 2020 there will be only 2.2 workers per retiree (Figure 2.18). Consequently, the percentage of worker income going to the Social Security system has increased and will likely increase more drastically, with a probability of one of the supporting workers being white and the other being nonwhite. Such a trend may create a gap between racial generations on issues of school funding and social security. The powerful elderly white electorate is demonstrating its unwillingness to fund an increasingly minority school-age population across America, while consuming greater amounts of limited resources for expensive health care.[25] In the state of Florida, Medicaid nursing home costs multiplied more than fivefold from $200 million to $1.2 billion in the 11 years from 1985 to 1996. It is expected to reach a budget-destroying $8 billion by the year

FIGURE 2.18 Number of workers per retiree in the U.S. from 1950 to 2025.

2000.[25] The education and support of our increasingly racially diverse children is critical if the U.S. is to maintain a position of economic competitiveness in the world and if it expects ungrudging support of the elderly white.

At the time of writing of this section, there are competing bills and proposals in the U.S. Congress with regard to immigration. The U.S. House Bill on Border Protection, Antiterrorism, and Illegal Immigration Control Act (H.R. 4437) would expand the employee verification program, requiring U.S. employers to submit Social Security and immigration numbers of newly hired workers within 3 d to government agencies. Steep fines and imprisonment would be faced by those who support or shield illegal immigrants. The bill would also authorize the construction of a $2 billon, 700-mi fence along the Mexico–U.S. border. The Senate is still grappling with this issue, and there are many who support President Bush's approach, which would provide a mechanism for unauthorized foreign-born people to earn legal immigrant status after several years and U.S. citizenship thereafter. There is division among Democrats and Republicans on this issue.[24] A final conference bill is expected in 2006.

CURRENT POPULATION TRENDS IN LDCs

In the previous 12 years, fertility has fallen in a number of countries. Kenya dropped from a TFR of 7.5 to 4.9, and Cote d'Ivoire from 7.4 to 5.2.[6] The African continent now averages a fertility rate of 5.1 and is likely to double its population within the next 25 years. Bangladesh almost cut its fertility rate by half from 6.2 in 1985 to 3.0 children per woman in 2005. The TFR in Turkey has declined from 4.1 to 2.4, Syria from 7.4 to 3.7, and Pakistan from 6.5 to 5.6. Western Asia now has a TFR of 3.6 and South Central Asia is at 3.2 with population-doubling times of 35 and 38.8 years, respectively.[4,6] Despite a continuing decline in fertility among many of the LDCs, the populations of 74 of these countries is expected to double in less than 30 years because of the forces of population momentum and incomplete demographic transition. More than 80% of the world lives in LDCs, and about 98% of the annual increase in population of over 75 million people has occurred in these countries.[6] Nearly 3 billion people will enter their childbearing ages within the next 20 years whereas 2.8 billion people will die, leaving nearly 1.2 people (600 million couples) to produce 1.8 billion children if the current fertility rate of 3 children per women is maintained. Of these 1.8 billion children born in the next 20 years, 1.76 billion will be born in the LDCs.

India represents an excellent example of the role of population momentum in maintaining a high growth rate in spite of declining fertility. The fertility rate in India is now 3.0, having dropped from 4.3 children per woman in less than 12 years.[6,26] However, its enormous population size and large percentage of women in their reproductive age resulted in a population increase of 25 million newborns in 1997. In general, India's population grows about 16 million a year, which means approximately 72,000 Indians are born every day.[27,28] India's fertility rate remains well above replacement level, and when combined with its current population momentum, the country is expected to overtake China within the next 30 years as the world's most populous nation. The Indian population, which continues to grow by 2% annually, will reach at least 1.5 billion before it stabilizes.[27] India has 4 times as many people as the U.S., but is only 40% as big.[28] Despite these remarkable numbers, the most critical regions for population growth continue to be sub-Saharan Africa, along with Western and South Central Asia as seen in Table 2.1.[4,6] These countries are experiencing environmental degradation, deterioration of urban centers, economic stagnation, increased morbidity and mortality from infectious diseases, malnutrition, and violence.[1,6] The contention by the U.S. news media that the population explosion is over may be premature. There remain a number of reasons to be concerned about the world's expanding population[6]:

- The world's population is expected to increase by 85 to 100 million people per year for many years because of population momentum and continued high fertility rates among the LDCs.
- Although fertility rates have declined globally, 33 countries still have fertility rates in excess of 6 children per woman, accounting for almost a half billion people; rates continue to grow geometrically.
- In the last 27 years, more than 2 billion people were added to the global population despite the declining fertility rates in many countries. Ninety-eight percent of these annual population increases were in LDCs.
- Nearly a quarter of the world's population survives on less than a dollar a day. Nearly a quarter lack drinking water or sanitation, whereas 20% lack adequate food. More than a half billion are homeless or without adequate shelter, and 86 countries are presently classified by the United Nations Food and Agriculture Organization as low income and food deficient because they lack the resources to purchase sufficient food.[6]
- There are 4.8 billion people in developing countries and three fifths lack access to basic sanitation.[3]
- Death rates are increasing because of aging populations in developed countries. In developing countries, a return of infectious and parasitic diseases, AIDS, as well as war and other violence is increasing the death rate. Ninety-seven percent of the deaths from emerging and re-emerging infectious diseases (malaria, dengue fever, tuberculosis, and cholera) have occurred in the LDCs, causing more than 25% of the 68 million deaths in these countries since 1995.
- Ninety-five percent of HIV-infected persons live in LDCs.[29]
- AIDS has killed more than 4.5 million people in sub-Saharan Africa alone since 1985, and HIV infects more than 10% of the adult population in these countries and has left more than 10 million children orphaned. It is now the leading cause of death in Africa.[6,29,30]
- As the world's billions scramble for resources to feed, clothe, and house themselves, they are devastating natural resources, causing: (1) pollution of the oceans, lakes, streams, and air; (2) deforestation; (3) desertification; (4) extinction of numerous plant and animal species; (5) elimination of wetlands; (6) decline in arable land resulting in diminished food supplies; and (7) mass migration to cities, creating the phenomena of urbanization.

TABLE 2.1
Demographic Indicators of World and Regional Population Development

	TOTAL POP. (millions) 2005	PROJECT. POP. (millions) 2050	AVERAGE GROWTH RATE 2005	TOTAL FERTILITY RATE	DOUBLING TIME IN YEARS
WORLD TOTAL	**6,464**	**9,075**	**1.2**	**2.6**	**58.3**
Developed Regions[1]	1,211	1,236	0.3	1.57	233.0
Less Developed Regions[2]	5,253	7,839	1.4	2.82	50
AFRICA	**905.9**	**1,937**	**2.1**	**4.83**	**33**
Eastern	287	678	2.4	5.41	29.1
Middle	109	303	2.7	6.18	25.9
Northern	190	311	1.7	3.04	41.1
Southern	54	56	0.4	2.81	175
Western	263	587	2.3	5.59	30.4
ASIA (excluding China)	**2,617**	**3,888**	**1.6**	**3.0**	**23.3**
China	**1,303**	**1,437**	**0.6**	**1.6**	**116**
EUROPE	**728**	**653**	**0.0**	**1.42**	**-**
Eastern	297	223	-0.5	1.29	-
Northern	96	105	0.3	1.66	233.0
Southern	149	139	0.3	1.37	233.0
Western	186	186	0.2	1.56	350.0
LATIN AMERICA & CARIBBEAN	**561**	**782**	**1.4**	**2.47**	**50**
Caribbean	39	46	0.8	2.42	87.5
Central America	147	210	1.5	2.55	46.6
South America	375	527	1.4	2.44	50.0
NORTH AMERICA	**331**	**438**	**1.0**	**1.98**	**70.0**
Canada	32	43	0.9	1.49	77.7
United States	298	395	0.9	2.04	77.5
OCEANIA	**33**	**47.6**	**1.2**	**2.27**	**58.3**
Australia/New Zealand	24	33	1.0	1.78	70.0

[1]More developed regions comprise North America, Japan, Europe, Australia and New Zealand.
[2]Less developed regions comprise all regions of Africa, Latin America and the Caribbean, Asia (excluding Japan), and Melanesia, Micronesia, and Polynesia.

Source: United Nations Population Fund, *State of World Population 2005: The Promise of Equality, Gender Equity, Reproductive Health and the Millenium Development Goals*, 118 pp, 2005, http://www.unfpa.org/swp/swpmain.htm.[56]

- Many areas with high fertility rates are beginning to see rising death rates due to the spread of HIV, depletion of aquifers, and shrinking of available cropland to support each person. For example, in Botswana life expectancy has fallen from 61 years in 1990 to 44 years in 1999, and is expected to drop to 39 years by 2010 on account of HIV.[31]

Many of these concerns are covered in the following chapter, but as population increases are likely to aggravate the problems in these areas, this seems the appropriate place to discuss future population trends.

PREDICTED FUTURE TRENDS IN POPULATIONS

We have already seen how difficult it is to predict population growth. There is, however, general agreement that the world's population will continue to grow substantially over the next 50 years, with most of the growth occurring in the lesser-developed regions of Latin America, Africa, and parts of Asia. The world's population is predicted to grow from its current 5.84 billion to over 8 billion in the next 25 years.[5,11] The developed regions are expected to grow from 1178 billion to 1220 billion, or 3.5%, whereas the LDCs will increase by more than 47.8% from 4.6 to 6.8 billion people. Current projections show the world's population leveling off at slightly more than 9262 billion people by 2050 and 11.5 billion approaching the year 2150 (Figure 2.19).[6] The population of the developed nations will be stable at nearly 1.5 billion, leaving nearly 98% of the population originating in the LDCs. The population of Africa will almost double in the next 45 years to 1.9 billion (Table

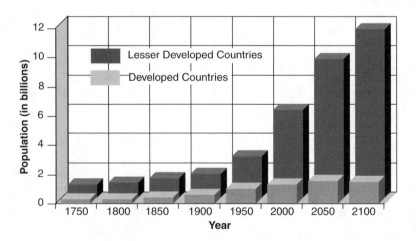

FIGURE 2.19 Population comparisons between developed and lesser-developed countries from 1750 to 2100. (Adapted from Population Reference Bureau, 1997 World Population Data Sheet, 1997 and United Nations Population Fund, The State of the World Population 1997, The right to choose reproductive rights and reproductive health, New York, 1997, http://www.unfpa.org/swp97e/toc.html.)

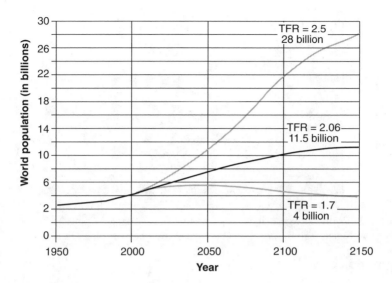

FIGURE 2.20 Projected world populations up to 2150, based on various TFR possibilities. (Adapted from Doyle, R., Global Fertility and Population, *Scientific American*, March 26, 1997; and Montavalli, J., Contents Under Pressure, *E Magazine*, November–December 1996, p. 28.)

2.1). Asia will increase its population to 3.88 billion people and Latin America will grow by 40% to 782 million. Contrast this with Europe, which will show a decline from 729 to 653 million. North America will show a rate of natural increase in population during the next 45 years from 331 to 438 million or a 24% increase in population, excluding the effects of migration.[5,6] These predictions are risky and depend on accurately forecasting trends in mortality rates and the childbearing behavior of millions of couples. The demographers have, therefore, projected several possible outcomes for global population allowing for deviations in fertility rates. The median or best estimate by the United Nations is that the world population will stabilize at 11 billion people around the year 2150 if the world fertility rate drops to 2.06 and life expectancy is 85 years (Figure 2.20). If fertility rates stabilize at 1.7 children, the world population would fall to 4 billion, down from the nearly 6 billion people today. Conversely, if the TFR falls only to 2.5 globally, the world population will escalate to 28 billion people and continue to climb.[32,33] The social, political, and ecological consequences would be greatly aggravated by such a tidal wave of people and would far outstrip the earth's carrying capacity, resulting in unprecedented human calamities. Even if the median prediction of 11.5 billion people is approached in the next 150 years, the major effects will be felt in the developing countries. Because 98% of the expected increase in the world population will occur in the poorest countries, world leaders and demographic experts argue that the population growth rates should be stabilized before they reach these theoretical levels projected by the United Nations.[5]

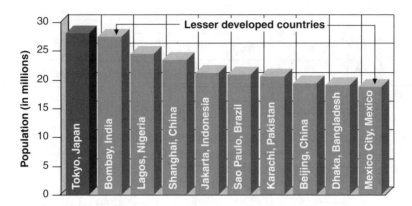

FIGURE 2.21 World's largest cities projected for 2015. (Adapted from The Environment: A New Positive Image, *The Weekly Review*, Kenya, November 10–24, 1995.)

What Is Urbanization?

There is another dynamic to the world's increasing population. The mass migration of people to the cities is known as *urbanization*. In 1950, when the black and white television was beginning to become commonplace in American homes, 30%, or 750 million, of the world's 2.2 billion people lived in cities. There was only 1 city in the developing world that had more than 5 million people. Forty years later, in 1990, there were nearly 2.3 billion living in cities, or nearly 43% of the global population. This number will reach 50% by the year 2000 and is expected to escalate to 61%, or 5.12 billion people, in the year 2020. By this time, 95% of the world's growth will be in the cities, and 80% of the world's urban residents will live in the LDCs by the year 2025.[6] *Megacities*, defined as having a population of more than 10 million, will be commonplace by the year 2015 with 9 of the 10 largest cities being in the developing countries (Figure 2.21).[34] To put this population trend in perspective: megacities such as Lagos, Nigeria, and Dhaka, Bangladesh, are growing at a rate 10 times faster than New York or Los Angeles. In 1975 there were 5 megacities and in 1995 there were 14 megacities; the number of megacities is expected to reach 26 by the year 2015.[35]

The movement of masses of people into urban centers is prompted by the rapid destruction of rural ecosystems as increasing hordes of people attempt to eke out an agricultural life on land of diminishing quality. The depletion of resources, and the lack of jobs and opportunities drive millions of desperately impoverished people to the cities.

In many of these countries, the rate of urbanization exceeds the ability of local and national governments to provide necessary employment and services such as sewage disposal and clean water supplies. Uncontrolled settlements erupt as unwelcome growths around nearly all the major developing cities and continue to grow even faster than the core urban centers. These shanty towns often are dense clusters of corrugated metal shacks or cardboard walls covered with plastic wrap.[1] According

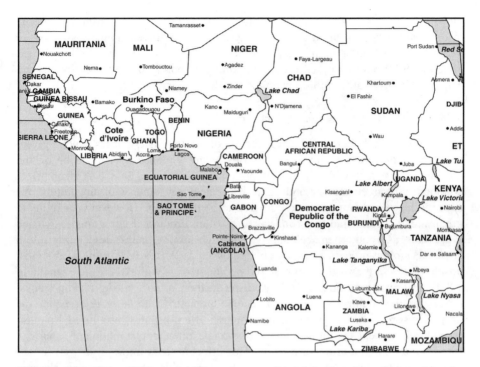

FIGURE 2.22 Areas of Western Africa, where traditional borders have broken down by merging megacities.

to a World Resources Institute report, an estimated 20 to 50% of urban inhabitants in developing countries live in impoverished slums or squatter settlements without access to adequate water, sanitation, and refuse collection.[35]

Few residents have electricity, proper sewage disposal, or clean water supplies. People urinate and defecate in streams that are filled with garbage, and which incubate clouds of swarming mosquitoes.[1] Whether viewed by satellite or from windows of a car, internal African borders are disintegrating into huge megalopolises, as evidenced by the countries of Ghana, Togo, Benin, Nigeria, and Cote d'Ivoire (Figure 2.22). Clear borders no longer exist as shanty towns merge into each other, creating an endless obscene vista of abandoned cars, wire mesh, metal shacks, and stores built of shipping containers facing into puddled streets of floating garbage.[1] These urban growths are breeding grounds for disease, violence, and environmental degradation.

The Role of Urbanization in the Spread of Disease

Three quarters, or 8 million, of the 12 million HIV-positive people worldwide are located in Africa, where the disease is accelerated by war, refugee movements, and uncontrolled settlements linked by improvised road systems. Tuberculosis, cholera, and malaria have returned with a vengeance. Substandard housing, inadequate sanitation, and unsafe water in the densely populated cities account for 10 million deaths globally each year, with 4 million infants and children dying from waterborne diseases alone.[34] Such disease outbreaks produce dramatic economic losses. As an example, a 1994 outbreak of the plague in the Indian city of Surat, killing 54 and affecting 5000 others, caused economic damages approaching $1.5 billion and an exodus of nearly half a million people from the country.[34]

Violence in Developing Urban Centers

Violence in densely populated urban centers of developing countries is not a given. However, there are a number of countries where violence appears to be on the increase, consistent with significant population increases. In the African tribal community, it is customary and natural to eat at any table with any family. An entire village or community is considered to be the family. However, as exodus to urban areas occurs, traditions and customs often erode and communal existence evaporates, as young men seek survival through joining gangs, descending into a criminal, violent process and making cities such as West Africa some of the most unsafe places in the world. Gunmen roamed the streets with AK-47s in Somalia in 1992 and cruised the streets of Mogadishu in their armed rusting pickup trucks, terrorizing the country. Somalia had become a country with no working economy, no police force, and no government.[36] Internal clan disputes over limited natural resources have kept stability a remote prospect despite the intervention of American forces under Operation Restore Hope. This effort erupted into a disastrous armed conflict on October 3, 1993, when U.S. Rangers raided a site near the Olympic hotel where followers of the warlord Mohammed Farrah Aidid were meeting.[37] In 1994, gunfire and mortar attacks triggered terror and stampedes at the Zaire–Rwanda border in one of the world's most vicious war zones. The million refugees of Rwanda's civil war stretched for miles as they accumulated in these death camps and succumbed by the thousands to cholera, dysentery, measles, and other diseases. The prime minister of Rwanda, Faustin Twageramunga, tried in desperation to coax his people back into the capital city, where violence had disrupted all services including phones, stopped the flow of water, electricity, and turned it into an armed camp with military checkpoints.[38]

Currently, there is an ongoing conflict dating from 1993 in the Darfur region of western Sudan, predominantly between the Janjaweed, a militia group recruited from local Arab tribes, and the non-Arab peoples of the region. The Sudanese government publicly denies that it supports the Janjaweed, but allegedly supplies arms and assistance, and has participated in joint attacks with the group.

The Western media describes this conflict as *ethnic cleansing* and *genocide*. A recent British parliamentary report estimates that over 300,000 people have already died, and the United Nations estimates that 180,000 have died in the past 18 months of the conflict. More than 1.8 million people have been displaced from their homes, and 200,000 have fled to neighboring Chad.[39,40]

Urban poverty is a socially destabilizing force, with hordes of impoverished people fleeing decaying rural ecosystems to urban centers only to be threatened by the loss of native customs in expanding cities. Here, their values are undermined, physical necessities are absent, and the spread of diseases is imminent.[1] Many countries in Africa, and others outside, are experiencing the rise of criminal anarchy, unprovoked crime, and the erosion of nation states and international borders as disease, overpopulation, and disappearing resources become prominent.

Environmental Degradation

In countries such as Cote d'Ivoire, Sierra Leone, Guinea, Ghana, and Brazil, the destruction of primary rain forests is proceeding at a blinding speed. Less than 6% of the primary rain forest of Sierra Leone remains, compared to the 60% that was intact more than 40 years ago. Deforestation has led to soil erosion with more flooding and a proliferation of mosquitoes. As the tentacles of the urban population expand outward, ever-greater resources are required. Soil degradation is most pronounced in those areas of the world where the densest populations occur, such as the West African coast, the Middle East, the Indian subcontinent, China, and Central America. The degradation of soils is readily apparent in China, where land has been destroyed by deforestation, loss of topsoil, and salinization. Water supplies have also been heavily contaminated, forcing large numbers of people to the cities and to coastal China.[41] The result has been surges in crime, growing regional disparities, and internal conflict with the looming possibility of government decentralization.[1] Global carbon dioxide levels are 31% higher than at the beginning of the Industrial Revolution, whereas 90% of commercially important fish such as tuna, cod, and swordfish are gone. The world's croplands have been diminished by 20%, and half of the world's forests are gone with another 30% very fragmented or degraded.[11]

The environment is very likely the national security issue of the early 21st century, in which expanding populations, re-emerging diseases, soil erosion, depletion of water, air pollution, hazardous chemicals, and radiation will likely force mass movements of people and incite group conflicts. Environmental disputes may very well merge with ethnic and historical disputes, resulting in a core foreign policy that drives the political issues. There is concern by some that the doubling and tripling of populations in the developing nations over the next 25 years will accelerate the erosion of natural resources, prompting greater likelihood of inter- and intranational conflicts.[1]

THE CONTROL OF POPULATION

EMPOWERMENT OR FORCE?

The International Conference on Population Development (ICPD) was held in Cairo beginning September 5, 1994. The major topic of the conference was women's rights. The draft program of action proposed that a full range of reproductive and health care services, including contraception and sex education, should be made available to all.[42] One of the most crucial points of the conference was the need for participation of women in the global workforce at different levels. Education and employment were discussed as tools to make women more powerful. The stabilizing of human population is no longer a mystery. It is best accomplished by improving women's education, providing them with paid employment, and making effective and inexpensive contraception available.[14,42] It has been estimated that if women had the power to make decisions about sexual activity and its consequences, 80 million unwanted pregnancies, 20 million unsafe abortions, 500,000 maternal deaths, and 333 million sexually transmitted infections could be avoided.[29] Unfortunately, this is not an easy solution in many countries where poverty, cultural traditions, and lack of educational opportunities exist and contribute to illiteracy of nearly 1 billion people worldwide. The largest gathering ever of world leaders reaffirmed the need to maintain gender equality, reproductive health, and HIV/AIDS at the top of the development agenda at the 10-year review of ICPD known as *IPCD+10*.[43] Two thirds of the illiterate population are women who as young girls were denied this education. In much of Asia, a woman's body may not be considered her own, nor the child in her womb. Collective decisions are often made about a woman, and they are accepted because the extended family is the institution that works; the state is often not trusted.[42] Countries attempting to bring population growth under control without first empowering women and providing effective birth control have often resorted to oppressive population control policies.

POPULATION POLICIES IN SOME COUNTRIES

India

India was the first country to introduce family planning, in 1951. The programs focused on the use of the unreliable *rhythm* methods, followed 10 years later by the use of intrauterine devices (IUDs) as the TFR continued to climb. The use of IUDs led to a number of complications in largely undernourished and anemic population that had little access to proper medical care. Consequently, there was little acceptance and populations continued to rise. Efforts shifted to male sterilization in the 1970s with significant government abuses as attempts were made to achieve regional sterilization quotas.[44] More recently, after being exposed to the widely accepted electronic media, the aspirations of the people to lead a better life have led more people to realize the virtue of a small family. In October 1997, the focus shifted to a more broad approach to address the need to limit family size.[23,27] Literacy campaigns, particularly among women, combined with education about the advantages of smaller families are playing a greater role in causing this trend.[26] India conducts a census every 10 years. The most recent reports available suggest that India has succeeded in slowing its rate of growth from 2.1% annually to less than 1.8%. However, this rate is still sufficient to double the population in less than 30 years. Some parts of India continue to have a high fertility rate. Preference for sons has led couples that have girls to continue to have children until a son is born.[38] Girls are often seen as an economic liability. There have also been sex-selective abortions and unreported female infanticide.[38] This continued oppression of females in India has lead to high rates of illiteracy and less health care.[45]

If stabilized population growth is to be achieved, India must improve the status of Indian women by further expanding access to education, increasing opportunity for employment, and supporting programs that encourage later marriage.[29] All states in India have made notable progress in improving education and literacy among women over time, but equality has yet to be achieved.[38,44] The average age of marriage has increased considerably, and there has also been an increased use of effective family planning methods.[45] Government efforts to achieve birth reduction quotas through sterilization, abortion, or heavy socioeconomic penalties may result in regional and/or temporary successes. However, these efforts run counter to the popular will and may lead to abuses and failures. In 1977, Prime Minister Indira Gandhi led her Congress party to a crushing electoral defeat, which was associated with these efforts to force a sterilization program on men during her previous 2 years of emergency rule.[26]

Peru

Critics of a sterilization program that began in Peru in 1995 charge that government health workers are taking advantage of poor rural women who speak only the local dialect. They report that the health workers are rewarded by promotions and money and so vigorously pursue government-imposed sterilization quotas by offering clothes and food to women in exchange for agreeing to become sterilized.[46] The TFR for Peruvian women who are uneducated and live in rural areas is 7.0, whereas the average TFR for Peruvian women with some college education is 1.7. The nation's poor women are, therefore, likely targets for sterilization. Critics further argue that the women are not being told about alternative methods of contraception or that tubal ligation is nearly always reversible. The opportunity for making informed decision by these women is often minimized. An official of the U.S. Congressional Subcommittee on International Operations and Human Rights questioned whether the Peruvian government is more interested in family planning or population control.[46]

China

China continues to enforce a one-child policy in the nation's largest cities such as Beijing and Shanghai. The one-child policy has successfully reduced China's population growth by 250 million.[47] There are severe penalties in much of China for having a second child, which may include job loss and fines equivalent to 3 years' salary for each parent.[41] It was not long ago (1980s) that family planning goals were accomplished through forced abortions and sterilization. The one-child policy, which was announced in 1979, is gradually yielding to a healthier economic growth. The improved economy has permitted couples in some medium-sized cities to have additional children by simply paying the steep fines without concern for retribution. However, economic growth also appears to be reducing the number of births as families find new economic prosperity. Also, there are financial incentives to encourage couples to abide by the one-child policy.[47] The emerging government policy of China is moving away from enforcement to education and encouragement. There is an emphasis on later marriages, longer intervals between children, and smaller families.[47]

Some officials have come to realize that persuasion is more effective than threats in effective long-term family planning goals.[47,48] Improvements in quality and choice of contraceptives in recent years has increased the effectiveness of family planning.

But the one-child policy has led to increased sex discrimination against women. Threats have led to underreported births of both males and females, with the birth of a girl being twice as likely to be ignored.[47] It is believed that one half to one third of the difference in infant sex ratios is due to underreporting. Unreported daughters may be adopted out, left with relatives, or abandoned in orphanages. Sex ratios are further skewed by illegal use of ultrasound to determine fetal gender, which has led to widespread abortion. The girls who survive often have less health care and education.[47]

Mexico

The population of Mexico is currently 107 million; more than 21 million live in Mexico City alone.[6,49] If it were not for the success of sex education and family planning practices, Mexico's population could have been over 134 million today. The population growth rate has dropped from 3.5% in 1974 to 1.9% today.[6,49] Twenty-five years ago, the average number of children in a Mexican family was 7; today it is down to 2.6. The emphasis has been on promoting smaller families and better relationships. Some areas are even offering free birth control. Many families are realizing that large families are not really possible with Mexico's poor economy. Today, more than 1 in 3 women work outside the home, whereas 20 years ago less than 1 in 5 did. Also, educators are handing out leaflets about medical care, offering courses on sexuality, and providing free HIV testing.

Iran

It is estimated that Iran will double its population from 65 million to 115 million by 2050.[2] However, Iran's family planning operation reaches almost the entire population and is one of the most successful operations worldwide.[2] There are advocates that disseminate information, and family planning messages are advertised on television. Thirty-five thousand health volunteers operate mobile clinics, offering free contraceptives. There is, however, one catch: condoms are only offered to married couples as premarital sex is not acknowledged. Iranians who plan on getting married must first have a day of counseling, where family planning is discussed. Many people in Iran are waiting until they are 22 or 23 years of age before they start a family, and most of them only want 2 children.[2] Contraceptive use by Iranian women has increased from 26% to 75% since 1979.[2] Although fertility rates are still high, they continue to drop rapidly.

FAMILY PLANNING VS. POPULATION CONTROL

The government-directed programs that set a policy for establishing an optimum population size are referred to as *population control*. In the countries described earlier, the policies were directed at reducing population. In a number of developed countries, such as in Western Europe, policies may be directed at promoting a higher TFR by providing state financial support for hospital expenses related to childbirth; for monthly support for the parents during the first year or two of the child's life as in Austria; or even tax deductions or special bonuses. Population control is in contrast to family planning programs, which are directed at assisting couples in having the number of children they desire, regardless of number. Family planning is parent centered and places the desires of the individual or parent for children before that of a state or national policy. Some policymakers argue that this style of family planning may not be enough to stem the tide of population growth in these countries, where many children are considered an important resource for gathering wood and water, harvesting crops, or performing other forms of manual labor. The simple act of getting water may take several hours a day, and it takes longer and longer as resources diminish. Children are also viewed as a source of security in old age in those countries without a financial support system for the elderly.[14] The persistent desires for large families in many LDCs is estimated to add nearly 660 million people in the next 50 years.[6]

There is the additional problem of funding needs for family planning services in foreign countries. The ICPD in Cairo (1994) agreed that it would take at least $17 billion annually to provide services for population and reproductive health programs and the prevention of sexually transmitted diseases (STDs) including HIV. Unfortunately, annual world expenditures are much less than half of this amount, with the dollars contributed appearing to have leveled off.[6] One issue contributing to the decrease in funds supporting family planning efforts has been the freezing of much of the promised U.S. aid owing to the antiabortion lobby in the Republican-controlled Congress.[11,50] Such shortfalls in LDCs are expected to result in as many as 220 million unintended pregnancies. As many as 88 million of these will likely be terminated by abortions. The pregnancies and abortions may cause as many as 117,000 additional maternal deaths, with up to 1.5 million women contracting some form of chronic health problem. This unmet demand for family planning services may likely cause an increase of 1.2 billion in global population.[6] A better quality of life for the world's billions may depend on (1) the willingness of governments to set reasonable policies with adequate funding, (2) development of greater motivation for small families through improved economies, and (3) universal access to knowledge and use of family planning services.

In July 1999, delegates from 180 countries approved proposals to slow world population growth, including offering women greater access to abortions as well as sex education to adolescents.[51] This was a significant advance from what was agreed to at the ICPD in Cairo, because it forces governments to increase and improve access to reproductive health programs. The belief is that women who are educated and have access to reproductive health care have fewer children. There are new goals in place to increase access to universal health care, increase education for women, and reduce maternal mortality and HIV infection.[50] These goals have been reaffirmed at world meetings and were at the top of the development agenda at the 10-year review of ICPD, known as *IPCD+10*.[43]

METHODS OF FERTILITY CONTROL

INTRODUCTION

The process of conception requires that a viable sperm have access to a healthy egg. When the sperm enters the egg, the tail is released and the ovum secretes a protective chemical barrier that prohibits other sperm from entering. Methods that prevent fertilization of the egg are called *contraception*, and there are hundreds of different procedures and formulations available today. The enormous variety of contraceptive tools attests to the interest of society in preventing unwanted pregnancies. These methods vary in their risks to health, their efficacy in preventing pregnancies, ease of use, acceptance, and costs. Some of the methods and characteristics are described in the following text.

Contraceptive Methods That Are Reversible

Natural Birth Control and Family Planning

This process is sometimes referred to as the *rhythm method* and relies on voluntary changes in sexual behavior. The method requires that the female predict her fertile periods by examining cervical mucous and/or tracking basal temperature, or charting her menstrual cycle and ovulation times on a calendar. She would then abstain from sexual intercourse during times of the month when she is fertile. This method has proved to be very unreliable because menstrual cycles vary, and the narrow window for sexual opportunity is not always observed.

Abstinence

The term *abstinence* has come to mean the exclusion of sexual intimacy, including oral-genital kissing and mutual masturbation. These latter activities are often referred to as *outercourse*. Either of these methods can be 100% effective in prevention of pregnancies as long as there is no opportunity for sperm to get near the vagina. The problems with outercourse are the increases in the probability of spreading STDs through oral-genital contact and the motivation and opportunity for intercourse is increased.

Hormonal

Oral Contraceptives

Since the introduction of the pill in 1960, it has become widely accepted and is now used by nearly 40% of women who wish to avoid pregnancy. The combined effects of the synthetic hormones estrogen and progesterone in the pill prevents the maturation of ova, prevents the development of the uterine lining, and forms cervical mucous, creating a barrier against sperm. The amount of estrogen supplied in the pill has been decreased to reduce health risks. There remains a limited risk of increased blood pressure, the formation of blood clots, strokes, and heart attacks for some women. This risk is low in most healthy women. The risk is further reduced for women less than 35 years of age and for those who do not smoke. A physician's prescription is required in most Western countries, including a complete gynecological exam. When taken according to directions, the pill approaches a 97 to 99% effectiveness in preventing pregnancy (Figure 2.23). Oral contraception offers no protection against the transmission of STDs.[52]

FIGURE 2.23 Various forms of the oral contraceptive pill designed to be taken daily.

A progestin-only pill, known as the *minipill*, is also available for women who do not want to use estrogen. The minipill is 96% effective and has few side effects, except for the possibility of irregular menstrual bleeding.

Depo Provera®

This is a long-acting synthetic hormone (progesterone) that is injected into muscles of the arm or buttocks every 3 months. It has an efficacy of 99% and is especially desirable for women who do not want to keep track of taking the daily pill or insert a device or chemical into the vagina. Most women lose their menstrual period within a year on this hormone. The adverse side effects are minimal and usually limited to a slight weight gain, occasional dizziness, and headaches.[52]

Norplant®

Norplant consists of five or six silicon rubber capsules (Figure 2.24) containing progestin that are inserted under the skin of the upper arm of the female using minor surgery. Small amounts of progestin are time-released over a 5-year period and work by suppressing ovulation, thickening the cervical mucous, and inhibiting the growth of the uterine lining. Norplant has been successfully tested in more than 40 countries, demonstrating minimal health risks with nearly 100% effectiveness in preventing pregnancies. It has been used more frequently in the U.S. since it entered the market in 1991, but the up-front costs that may exceed $550 are sometimes a deterrent even

FIGURE 2.24 Norplant, showing elongated capsules.

FIGURE 2.25 Samples of spermicidal jellies, creams, and lubricants.

though the cost of the pill over the same 5-year period is more than twice that amount. The distribution of Norplant ended in 2002 and contracts to distribute Norplant in other countries expired in December 2006.

Spermicides

Chemicals that are designed to kill sperm are called *spermicides*. The substance most commonly used is nonoxynol-9, and it comes in a variety of forms including jellies, creams, foams, and waxy suppositories (Figure 2.25). These are all meant to be inserted far enough to cover the cervix, providing a chemical and physical obstacle to the sperm. Spermicides are most useful when used together with a barrier method such as a condom. Together they provide 98% effectiveness.

Barrier Methods of Contraception

Male Condom

Condoms are typically elastic material of latex rubber or polyurethane designed to be rolled onto the erect penis before intercourse (Figure 2.26). They appear in a

FIGURE 2.26 Male condoms appear in a variety of colors, shapes, and textures.

FIGURE 2.27 Female condom is inserted into the vagina with the wider open end extending over the labia.

variety of colors and textures, with and without lubricants, and with and without a reservoir designed to capture the ejaculate. The effectiveness of condoms to prevent pregnancies is reduced to 88% because of improper use or storage, which permits sperm to enter the vagina through tears, rips, or spillage. Formerly, it was believed greater effectiveness could be achieved if the condom was used along with a spermicidal lubricant. Studies have found that the spermicide nonoxynol-9 might actually irritate the vaginal walls, making infection in users more likely. Because of this study the CDC has changed its recommendations, advocating use of a lubricated condom alone until this finding can be researched further.[52] Condoms are one of the top three methods used in the U.S. and provide the additional benefit of protection against the transfer of HIV.

Female Condom, Diaphragm, and Cervical Caps

The female condom is a soft, elastic polyurethane conical bag with an elastic ring (Figure 2.27). The narrow end is inserted into the vagina leaving the wider end to extend out over the labia. It has a reported failure rate of around 12% based on use in the U.S.

The diaphragm is a thin latex rubber cup with a flexible rubber-coated ring intended to cover the cervix. Diaphragms come in a variety of sizes and must be fitted by a trained health professional. Moreover, the user must be familiar with the correct method of inserting and positioning the diaphragm. It is intended to be used with a spermicidal cream or jelly on the concave or inside surface before insertion and then left in place for up to 8 h after intercourse. This permits sufficient contact time to kill any remaining sperm.

The cervical cap is similar in design but smaller than the diaphragm (Figure 2.28), and is also fitted by a health professional. It is designed to be used with a spermicide and left in place for several hours after intercourse. Its effectiveness is somewhat less than a diaphragm because it is smaller and more difficult to insert properly.

FIGURE 2.28 Sponge (left) and cervical cap (right) designed to be placed over the cervix with spermicide.

FIGURE 2.29 IUD showing approximate placement in the uterus.

Intrauterine Devices (IUDs)

IUDs are plastic T-shaped devices that are folded with a special instrument and inserted by a trained health professional into the uterus such that the arms of the T open up and cross the uterus where it is braced. Some brands come with a slow-releasing synthetic progesterone, whereas others are copper wrapped. The specific action of IUDs is unknown, but it is believed they interfere with egg fertilization by sperm. The IUDs must be removed or replaced periodically every 1 to 4 years depending on the type used. Effectiveness in preventing pregnancies can reach 95%. Risks include pelvic inflammatory disease, severe cramps, heavy menstrual flow, uterine perforation, tubal infections, and infertility. These are among some of the reasons for the negative publicity that prompted the manufacturer to remove the Dalkon shield® from the market in the 1970s.

PERMANENT CONTRACEPTIVE METHODS

Sterilization has become one of the most popular methods for contraception in the U.S. among married couples who have achieved their desired level of parenthood. Male sterilization by vasectomy is usually an outpatient procedure, accomplished by making an incision on either side of the scrotum and snipping out a piece of the vas deferens. The two ends of the vas are tied off or cauterized. There is some discomfort in this area for about a week. The risks are minimal and effectiveness in preventing unwanted pregnancies in partners is better than 99%. Women may be sterilized by tubal ligation, which is a slightly more involved surgical procedure requiring small incisions near the navel and just above the pubic area. The abdomen is inflated with a gas so that the physician can easily locate the fallopian tubes with the fiber-optic light on the laparoscope and tie off or cauterize the tubes. This procedure blocks the entry of eggs into uterus, and eggs released from the ovaries dissolve and are reabsorbed into the body. Tubal ligations are largely reversible, whereas vasectomies have generally not met with similar success at being reversed. However, developments in microsurgical techniques have resulted in a reversal approaching 50%, depending on the elapsed time since the vasectomy and the skill of the surgeon.[32]

ABORTION

Abortion refers to the medical means of terminating a pregnancy. In the U.S., this is considered to be a decision that a woman and her physician may make within the first trimester without legal restriction. This right was given under the 1973 *Roe v. Wade* Supreme Court decision.[54] However, individual states were given the right, under later court actions to impose restrictions on abortions. Despite these restrictions, about 1.6 million abortions have been legally performed each year in the U.S. since 1980.[32] Nearly 60 million abortions occur annually on a worldwide basis. Three quarters of the world's population live in countries where it is legal to induce abortion medically. There are different forms of abortion, depending on how many weeks pregnant the woman is at the time of the abortion. The number of weeks of pregnancy is determined from the first day of a woman's last menstrual period. Most pregnancies are terminated in the first trimester, when legal and health consequences are minimal. These are usually referred to as *therapeutic abortions* and are normally accomplished by vacuum aspiration. Abortion can also be safely induced within the first 9 weeks of pregnancy by administering the drug RU-486®. This steroid blocks the action of progesterone and causes the uterine lining to break down, thus expelling the lining and the fetus, which terminates the pregnancy. The drug RU-486 was developed in the late 1970s by a French researcher consulting for Roussel-Uclaf, a French pharmaceutical company. There continues to be fears by many, including antiabortion groups, that RU-486 would popularize abortion by making it available in a doctor's office and therefore remove the need for abortion clinics, which are frequent targets of protestors. Despite many setbacks including halted productions and distribution, RU-486 is now being widely used in France and on September 28, 2000, was approved by the FDA for sale in the U.S. The approval of this drug gives women an alternative to surgical abortion for the cost of a standard abortion.[55]

REFERENCES

1. Kaplan, R., The Coming Anarchy, *Atlantic Monthly*, February 1994, p. 44.
2. Motavalli, J., Now We Are Six, *The Environmental Magazine*, July–August 1999.
3. United Nations Population Fund, *The State of the World Population 1999: 6 Billion, A Time for Choices,* 1999.

4. Population Reference Bureau, 1997 World Population Data Sheet, 1997.
5. United Nations Population Fund, The State of the World Population 1997, The right to choose reproductive rights and reproductive health, New York, 1997, http://www.unfpa.org/swp97e/toc.html.
6. Population Reference Bureau, 2005 World Population Data Sheet, August 2005, http://www.prb.org.
7. Odum, E., *Fundamentals of Ecology*, W.B. Saunders Company, Philadelphia, 1959.
8. Meyers, J. and Krebs, C., Population Cycles in Rodents, *Scientific American*, June 1974, p. 38.
9. Turk, J. and Turk, A., *Environmental Science*, Saunders College Publishing, Philadelphia, PA, 1988, chap. 7.
10. U.S. Census Bureau, U.S. and Word Population Clocks, April 2006, http://www.census.gov/main/www/popclock.html.
11. Nierenberg, D. and MacDonald, M., The Populations Story ... So Far, *World Watch*, September–October 2004, pp. 14–17.
12. Perlez, J., Birth Control Making Inroads in Populous Kenya, *New York Times*, Sunday, September 10, 1989.
13. Zero Population Growth, Frequently Asked Questions, http://www.zpg.org/q-a.htm, February 10, 1998.
14. Dasgupta, P.S., Population, Poverty and the Local Environment, *Scientific American*, February 1995, p. 42.
15. Freed, S.A. and Freed, R.S., One Son Is No Sons, *Natural History*, January 1985.
16. Bongaarts, J., Demographic Consequences of Declining Fertility, *Science*, 282, October 16, 1998, p. 419.
17. Wattenburg, B.J., The Population Explosion Is Over, *New York Times Magazine*, November 23, 1997, Section 6, p. 60.
18. Deen, T., Development: Population Explosion Not Over, Warns U.N., Inter Press Service, March 1998, http://www.oneworld.org/ips2/mar98/population.html.
19. Kristof, N.D., Baby Makes 3, but in Japan That's Not Enough, *The New York Times International*, Sunday, October 6, 1996, p. 1.
20. Ayres, R., The Economic Conundrum of an Aging Population, *World Watch*, September–October 2004, pp. 45–49.
21. Stanley, A., Blissful Bachelorhood and the Shrinking Village, *New York Times*, November 16, 1999, http://www.nytimes.com/news/world.
22. Funabashi, K., Reassessing the Value of Children, *Japan Echo*, 26, February 1999, http://www.japanecho.co.jp/docs/html/260111.html.
23. Sadik, N., United Nations Fund for Population Activities, 1985 Report.
24. Martin, P., The Battle over Unauthorized Immigration to the United States, Population Reference Bureau, April 11, 2006, http://www.prb.org.
25. Tilove, J., Racial Generation Gap Growing, *Sunday Republican*, February 22 1998, p. B2.
26. Kataria, S., India Succeeding in Slowing Population Growth, Reuters, 02, 1997, http://www3.elibrary.com/id/2525/getdoc.c...Q002D029&Form+RL&pubname=Reuters&pubur=0.
27. Visaria, L., Jejeebhoy, S., and Merrick, T., From family planning to reproductive health: challenges India, *Int. Fam. Plann. Perspect.*, 25, 1999.
28. Population Nears 1 Billion in Poverty-Ravaged India, *Daily Hampshire Gazette,* 213 No. 294, 1, August 16, 1999.
29. United Nations Population Fund, The State of the World Population 2000, http://www.unfpa.org.

30. 10 Million Orphans, *Newsweek*, January 17, 2000.
31. Brown, L. and Halweil, B., Breaking Out or Breaking Down, *World Watch*, 13, 5, September–October 1999.
32. Doyle, R., Global Fertility and Population, *Scientific American*, March 26, 1997.
33. Motavalli, J., Contents Under Pressure, *E Magazine*, November–December 1996, p. 28.
34. The Environment: A New Positive Image, *The Weekly Review*, Kenya, November 10–24, 1995.
35. Lynn, W., Megacities: Sweet Dreams or Environmental Nightmares?, *Environmental Science and Technology*, 33, 1999.
36. Nelan, B.W., Taking on the Thugs, *Time*, December 14, 1992, p. 27.
37. Church, G., Anatomy of Disaster, *Time*, October 18, 1993, p. 40.
38. Gibbs, N., Cry the Forsaken Country, *Time*, August 1, 1994, p. 28.
39. Smith, R., How Many Have Died in Darfur, *BBC News*, February 16, 2005 http://bews.bbc.co.uk/world/africa/4268733.stm.
40. Egeland, J., UN's Darfur Death Estimates Soars, *BBC News*, March 14, 2005, http://bews.bbc.co.uk/world/africa/4349063.stm.
41. Xin, W., Population vs. Development: Challenge of the New Century, *Beijing Review*, May 1–7, 1995, p. 12.
42. Elliot, M. and Dickey, C., Population Wars: Body Politics, *Newsweek*, September 12, 1994, p. 22.
43. UNFPA, Consolidating the achievements of the ICPD, May 2006, http://www.unfpa.org/icpd.
44. Conly, S.B. and Camp, S.L., India's Family-Planning Challenge: From Rhetoric to Action, Country Study Series No. 2, The Population Crises Committee, 1992.
45. Arnold, F., Choe, M.K., and Roy, T.K., Son preference, the family-preference process and child in India, *Population Studies*, 52, 301, 1998.
46. Sims, C., Using Gifts as Bait, Peru Sterilizes Poor Women, *New York Times*, February 14, 1998, p. 1.
47. Choi, C.Y. and Kane, P., China's one child family policy, *Br. Med. J.*, 319, 7215, 992, October 9, 1999.
48. Faison, R., China's Strict 1-Baby Policy Giving Way to Economic Changes, *Sunday Republican (Springfield)*, August 17 1997, p. 1.
49. McConahay, M.J., A Smaller but Better Future, *Sierra*, July–August 1999.
50. Roach, J., Family Planning Urged for Developing Nations, October 6, 1998, http://www.enn.com/enn-news-archive/1998/10/1000689/famplan.asp.
51. U.N. Moves to Slow Population Growth, *Weekend Gazette*, A5, July 3, 1999.
52. Donatelle, R. and Davis, L., *Pregnancy, Childbirth, and Birth Control, Access to Health*, 3rd ed., Prentice Hall, Englewood Cliffs, NJ, 1988, chap. 8.
53. http://www.unaids.org/whatsnew/press/eng/durban120700.html.
54. Boston Women's Health Collective, *The New Our Bodies, Ourselves*, Simon and Schuster, New York, 1984, p. 218.
55. Kaufman, M., Abortion Pill Wins FDA OK, *Union News*, September 29, 2000.
56. United Nations Population Fund, State of World Population 2005: The Promise of Equality, Gender Equity, Reproductive Health and the Millenium Development Goals, 118 pp, 2005, http://www.unfpa.org/swp/swpmain.htm.

3 Environmental Degradation and Food Security

CHAPTER OBJECTIVES

A student, after reading this chapter, will be able to:

1. Discuss the impact of population on resources and ecosystems.
2. Define the following terms and explain their response to population growth: retrogression, soil erosion, desertification, deforestation, wetlands destruction, and wildlife destruction.
3. Define the term *food security* and discuss the reasons leading to food insecurity among many nations worldwide.
4. List the suggested steps that might be taken to minimize global food insecurity.
5. Explain the most likely reasons for a growing food insecurity in the U.S.
6. List and discuss the demographics of the populations in the U.S. at risk of food insecurity.

INTRODUCTION: THE DEBATE

The ability of our planet to sustain and feed the dramatic increases in human population growth has been an ongoing debate stretching back over 200 years. An English clergyman and economist by the name of Thomas Malthus published an article in 1789 titled "An Essay on the Principle of Population." He concluded that human populations grow exponentially (2, 4, 8, 16, …) and that plants grow arithmetically (1, 2, 3, 4, …) so that eventually uncontrolled human growth would lead to "misery and death," as growth exceeds the capability of the land to feed the expanding population. The predictions of Malthus were not widely accepted initially, and were even scorned as opportunities for migration relieved population pressures, and technological advances in agriculture increased the available food supply.

It was not until the post–World War II period, when growth rates sharply escalated, that world leaders and demographic scientists pressed the alarm button, proclaiming that overpopulation was the greatest threat to continued habitation of the planet by humans. Government reports and published papers proclaimed that the world in the year 2000 would become more polluted, greatly crowded, ecologically unstable, and vulnerable to economic and political disruption. Those holding these views are often considered to be *neo-Malthusians*. There would likely be more regional water shortages and major reductions in forested land. Significant deterioration in agricultural soils has been reported through the processes of erosion, desertification, waterlogging, and salinization. World climate is predicted to change owing to the influence of greenhouse gases such as carbon dioxide that are produced in ever-greater amounts by human activities such as combustion of fossil fuels. These gases are then thought to absorb greater amounts of heat energy from the sun, producing global warming. The combustion of fossil fuels is also thought to increase the formation of acid compounds in the air, leading to increased acid deposition in lakes, streams, and rivers that impact many forms of aquatic life. There are predictions of increasing numbers of animal and plant species becoming extinct as pollutants increase and habitats are destroyed for agriculture and development. There is no shortage of professionals and lay people who predict worldwide calamities as a direct consequence of unbridled human growth.[1-5]

TECHNOLOGY AND POLICY WILL SAVE THE DAY

There is a real concern among many political leaders and the scientific community that the rapid expansion of human population cannot continue unabated without creating dramatic environmental crises. The human population may have already become so large, and the momentum for further growth so great, that any technologies which might mitigate the pressures cannot be developed or applied soon enough to prevent disasters from occurring.

There are many who counter these views and argue that overpopulation is not the cause of a degraded environment, but develops from poverty and unequal distribution of wealth. It is suggested by some that the earth has resources capable of feeding numbers as great as 16 billion.[1] Still others predict the earth has an enormous productive potential capable of supplying food for 400 to 1000 billion people.[6] Such wide discrepancies in the predicted carrying capacity of the earth highlights the uncertainty in making such predictions. The obverse view to "population is a problem" is that the real threat to global stability is the failure of nations to pursue economic trade and research policies that increase food production, more evenly distribute food and resources, and limit environmental pollution. These views usually recognize that more people are not a problem but an asset that drives the need for greater technology and innovation to sustain larger numbers of people. People who adhere to these views are sometimes called *Cornucopians*, after the mythical horn of plenty. It may have been the pressures of population that resulted in the development of new, genetically superior crops with dramatically higher yields, which became known as the *green revolution*. Preceding the green revolution was the development of more efficient and economical energy sources.

The Green Revolution

It was in the mid-1960s that an interdisciplinary team of scientists from a worldwide network of agricultural centers began to operate under the Consultative Group on International Agricultural Research (CGIAR). Research from these groups has resulted in the development of new varieties of important food crops such as rice, wheat, and corn.[7] The International Rice Research Institute in the Philippines was responsible for developing genetically improved strains that doubled the yields of rice per acre. Strains were developed that resist diseases, pests, drought, and flooding. Corn production per acre has continually risen since the 1960s, with the world average rising from 1 to 1.75 tons per acre. Corn is one crop that has shown increasing yields per acre of 1 to 2% even without the contribution of genetic engineering or other technologies. Using the latest technologies and best seeds, the state of Iowa now averages in excess of 4 tons per acre of corn, or 5 times the world average, and Iowa master farmers now average nearly 8 tons per acre.[7] The new varieties have been shortened, with thicker stalks to support the heavier weights, and the leaves were also made shorter and more upright to reduce the problem of shading and crowding of nearby plants. These plants respond to increased fertilization by producing much greater yields of corn, rice, or wheat. Wheat crops in Mexico exploded by a factor of 4.3 times the yield per acre when genetically enhanced seeds were used compared to yields from previously grown crops.[7] Dramatic increases in yields per acre of these crops ranging from 50 to 200, or even 400%, have been seen in many areas around the globe. The International Rice Research Institute in the Philippines developed genetically improved strains of rice that have doubled yields.[6] There are now varieties being developed that put more energy into developing grain-bearing seed heads, and other characteristics are being added that include resistance to drought, flood, salt, and pests in order to further boost yields. So striking has been the increased production that the incorporation of these new varieties of seeds and processes became known as the green revolution (Figure 3.1).

Historically, the improved yield in crops was a relatively slow process resulting from the selection of seeds that were better performers within existing crops. These seeds varied slightly in genetic makeup from the others and possessed the characteristics necessary to produce somewhat greater yields per acre. The investment by

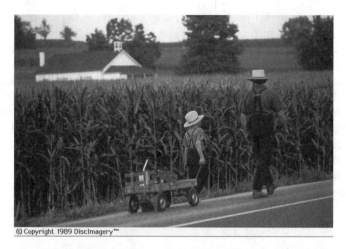

© Copyright 1989 DiscImagery™

FIGURE 3.1 The world markets and the green revolution may promote monoculture technology that could prove to be ecologically unstable. (Photo from Discimagery.)

governments in international agricultural research allowed scientists to rapidly accelerate this process through crossbreeding and mutation by ultraviolet or gamma radiation. The technique of crossbreeding involves the selection of genetic and morphological variants that possess one or more of the desired characteristics and then crossing the plants in the hope that the most desired characteristics of the different plants will appear in the new crossbred plants. As an example, corn with a heavy stalk but few ears might be crossed with corn that has many ears but a thin stalk that won't ordinarily support the additional ears of corn. The desired product would be corn that has a thick stalk and many ears (Figure 3.2). The genetic makeup of plants can be artificially altered by using ultraviolet or gamma radiation and producing mutations. The mutations will produce many variations of the plants, but some of these could be beneficial and produce the desired characteristics of many ears and a thick stalk, as described in this example (Figure 3.2). More recently, these processes have been improved with the advent of genetic engineering, which uses various artificial techniques to alter genetic information and produce new and unique plants. One of these techniques involves gene transfer. Gene transfer techniques were initially limited to bacteria that might be beneficial to certain plants because they were much simpler to treat in this way. Most plants are difficult to manipulate with gene therapy, but newly developed techniques use a "gene gun" that incorporates the desirable DNA fragments into a plastic bullet. This bullet is then fired by a gene gun into a barrier that shatters the bullet, creating a genetic spray that pierces the plant target at over 1000 mph.[7] Leaf cells are then cultured with growth hormones to produce plant shoots in a few weeks that exhibit characteristics associated with the newly incorporated genetic materials (Figure 3.3). Viral-resistant barley has been created this way, and these techniques are being used to develop corn that can play host to nitrogen-fixing bacteria and so require less fertilizer.[6] On March 3, 1988, the USDA and Delta and Pine Land Company announced that they had a patent on an agricultural genetic technology. This technology is referred to as the *Terminator*™

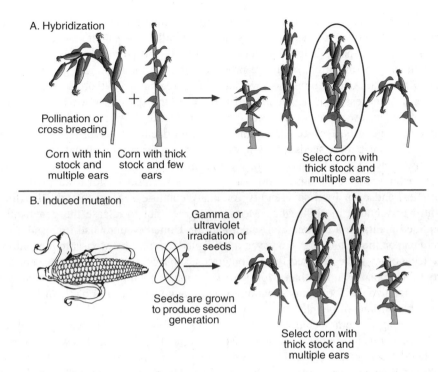

FIGURE 3.2 Historically, plant improvements were achieved with crossbreeding and artificial mutation. (Adapted from Turk, J. and Turk, A., *Environmental Science*, 4th ed., Saunders College Publishing, Philadelphia, PA, 1988.)

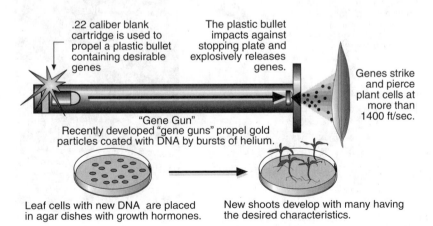

FIGURE 3.3 Methods of gene transfer in plants may use the gene gun, which blasts the desirable genetic fragments into the plant. (Adapted from Burdansky, S., 10 Billion for Dinner Please, *U.S. News and World Report*, September 12, 1997, p. 57.)

seed. A Terminator seed has an embryo-killing toxin genetically spliced into the gene sequence before a gene promoter sequence. To allow the parent plant to produce seeds, a genetic spacer is inserted between the toxin and promoter sequences. This way, in early germination, seeds are produced because the toxin and promoter are not in contact. At this time, however, a protein recombinase that recognizes the spacer as unnecessary is activated, and it removes the spacer sequence. The toxin and promoter now make contact, and the resulting toxic protein kills the embryo in late embryogenesis. The seed sits in the ground producing no offspring, forcing the farmers who use these seeds to buy new ones every year from multinational corporations.[8] Since 1.4 billion people rely on farm-saved seeds for their survival, the Terminator seed technology could lead to a loss of farmers, local adapted crop varieties, and even food sovereignty for many countries.[9] Most thought that this controversy ended in 2000 when Monsanto agreed not to release the genetically modified plants on a commercial scale, and the United Nations had imposed a *de facto* ban on the technology. However, a 2005 pledge report by Monsanto may allow the technology to be released for use on cotton, tobacco, grass, and pharmaceutical crops.[9] Companies other than Monsanto have made it known that they intend to commercialize the Terminator technology and many view this as a battle of life and death for billions of people.[9]

Other agricultural techniques being developed involve the improved application of fertilizers using global positioning satellite (GPS) systems, and new soil-tilling methods that reduce erosion (Figure 3.4). *Precision farming* depends on the use of GPS satellites that exactly match the rates of application of seeds and fertilizer to existing soil conditions. Signals from the satellite are sent to the computers on board the tractors, which determine the coordinates and adjust the amount of fertilizer dispensed in each area of the field. This process reduces waste and increases yields.

Computers on-board tractor receive signal from GPS satellite and determine field coordinates, then adjust fertilizer dispersion.

Global positioning satellite (GPS) sends specific signals...

..on location and local soil conditions to receiving systems on tractors

Some artwork based on Metro Creative Graphics, Inc

FIGURE 3.4 "Precision farming" techniques use satellite maps and onboard computers to adjust control rate of seed and fertilizer spread. (Adapted from Burdansky, S., 10 Billion for Dinner Please, *U.S. News and World Report*, September 12, 1997, p. 57.)

There are now *no-till* planters that cut a narrow trough or slit through the crop residues into the sod, drop the seed in place, and then cover the seed. This substantially reduces soil erosion.[6] New techniques are being developed all the time. However, there remain concerns about the application of new agricultural technologies. The new breeds of plants that are selected or genetically altered to produce higher yields often require productive soils, along with extensive application of fertilizers, water, and insecticides to promote the potential growth of the plants and protect against insect invasion.[10] The higher yields that may be achieved often prompt farmers to plant single crops or monocultures because they may be profitable in the short term, but could result in extensive ecological damage through the depletion of soil nutrients, erosion, and the attraction of pests. The requirement of these high-yield plants for more water, pesticides, fertilizers, and production equipment demands ever-greater needs for energy as these components all require fuels to be produced or delivered. Many areas of the world likely to experience drought or poor water control (necessary for rice production) have not adopted these crops widely. Diffusion of high-yield crops is slow in areas that have inadequate infrastructure or market access, such as sub-Saharan Africa or the hillside systems of Latin America and Asia.[10]

There is little debate that these advances in agricultural technologies have contributed significantly to reducing hunger in millions of people. Many lesser developed countries (LDCs) have adopted the seeds and technologies produced via the green revolution. China's rice and maize crops are almost entirely planted in modern varieties, and although some exceptions exist, most of the growth in production in developing countries is a result of the higher yields generated by the green revolution.[10] Nearly three fourths of rice and wheat crop areas and more than half of all maize croplands were planted in modern varieties.[11] However, growth of the human population in many of these LDCs has exceeded the capacity of even these technological wonders in agricultural production,[6] and the energy to support high-yield cropping techniques is becoming more expensive.[10]

Agricultural researchers claim that we could expect more increases in crop yields provided money continues to flow into agricultural research and into credit and technical assistance for farmers who wish to apply the technologies. However, funding for CGIAR centers has been falling, causing a reduction in more than 100 scientists to leave the organization, and U.S. aid for foreign agricultural programs has sharply fallen by more than 75% since the 1980s.[7] Further, grain harvests have leveled off or seen a reduction in growth rate such that global cereal reserves will continue to remain below minimum safe levels for world food security.[12] Therefore, as global population increases and per capita food supply has leveled off or even decreased in critical areas, it is difficult to share the optimistic view that new technologies will cause another quantum burst of food productivity in the next 20 to 50 years. Decreasing freshwater supplies, reductions in fertile croplands, lesser effectiveness of fertilizers and pesticides, and the increasing likelihood of major climatic shifts in some of the more productive areas of the world all would make it more difficult to meet future food demands at affordable prices. Efforts will need to be expanded in stabilizing world population and in developing sustainable agriculture if we are to avoid catastrophes to human populations.

Energy

Although there have been major technological advances in fuel use, LDCs continue to use wood as a primary fuel for cooking and heat. Wood is being used at such a rapid pace in some LDCs that forested regions have been decimated, and the collection of wood for fuel may require several hours each day or as much as 25% of average income.[13]

On the other hand, the history of fuel use in the developed nations moved from wood to more efficient fuels. As the industrial revolution entered the 17th and 18th centuries, a process for making charcoal from wood was developed, and it became the main fuel for smelting of iron and virtually depleted much of the forested land of the U.S. Technology again entered the picture as wood resources became a limiting factor, and a process was developed that converted coal to a useable coke form necessary for powering "iron refineries." As the need developed for economical fuels to power automobiles, trucks, tractors and, eventually, planes, petroleum was refined into gasoline and oil. The requirements for petroleum increased, and new technologies were developed that permitted the detection and extraction of previously unrecoverable sources, driving the cost of oil down for many years so that it has become an inexpensive fuel for much of the world when compared to total wages. This, however, is no longer the case as prices escalated sharply in 2006 because of strong petroleum demand, limited refining capacity and surplus oil production, and deep concerns about political tensions with oil-producing countries such as Iran.[14] The world price for oil is likely to remain high through 2007 as none of these factors appears likely to change soon.[14] The consequence is that these high oil prices are putting billions of people around the globe to significant hardships. A few of the developed nations appear on the verge of entering yet another energy transition as the use of petroleum is reduced in favor of less polluting alternatives. Humankind continues to respond with the development of new, more efficient, energy sources. As technologies expand, they create the possibility of feeding, clothing, and housing even greater numbers of people. Authors such as Julian Simon argue that human adaptation and change will avert disaster as more technologies are developed in response to growing population pressures.[15,16] Although such authors argue that technology will support ever-increasing populations, there is a practical limit to the maximum number of people this planet can sustain with a given quality of life. The tragedy of the human condition is already evident in many parts of the world. Tens of thousands die each day from malnutrition and diseases in countries where resources have been exhausted and the land laid to waste. This is contrasted by those countries that have brought population under control, have enormous wealth, and a very high standard of living such as the U.S., Western Europe, Japan, Australia, New Zealand, and other developed countries. Poor developing countries are generally blamed for the world crises because of their high growth rate and subsequent environmental damage. However, Paul Ehrlich, professor of biological science at Stanford University, thinks the impact of humanity on the earth's life-supporting systems are not merely decided by the numbers of people, but also by how people behave. The impact of human activity on environments can be summarized by the following relationship[17]:

$$I = P * A * T$$

where
   I is the impact of human energy-related activity on the globe
   P is the population size
   A is the affluence in terms of per capita consumption
   T is the technologies to supply each unit of consumption

According to Ehrlich, it is energy-intensive activities that cause most of the environmental degradation. The average energy consumption per person in the developed "rich" countries is 7.4 kW, whereas the average person in the LDCs use less than 1 kW. There were 1.211 billion people in the "rich" nations in 2005 so their total environmental impact as measured by energy use was 8.96 trillion watts. Some 5.22 billion people lived in poor nations in 2005; hence, their total impact was 5.22 trillion watts. The relatively smaller population of those in the developed countries accounts for nearly two thirds of the environmental destruction as measured by energy use. About 5% of the world's population lives in the U.S., where 35% of all raw materials are consumed. Moreover, the gap between the rich and poor countries is growing. A United Nations report shows a widening economic gulf occurring among the world's poorest countries, driven by wars and political infighting.[18] Such unrest is thought responsible for reductions in per capita gross domestic product in the Congo, Burundi, Somalia, Sudan, and the Central African Republic, for example.

Attitude and Behavior

The question posed by many is whether human society has the capacity to adjust to the dizzying pace of population increases without major political upheavals, with the accompanying wars and environmental destruction resulting in human calamity. We have seen in this chapter that neo-Malthusian arguments predict the inevitable misery of humankind as the population momentum outstrips the carrying capacity of the earth.[19] We have also presented the optimistic Cornucopian outlook that more people will drive the technology to support the world's growing billions. Which of these views will ultimately be realized? Will we progress in a smooth transition to a world of global stability and health through carefully planned and thoughtful processes, or will national and personal interests prevail at the expense of the larger global community, resulting in the continued destruction of our environment and depletion of world resources? What are the attitudes and behaviors that may impact this outcome?

The Tragedy of the Commons

If we are to believe the arguments proposed by Garrett Harden, the individual or nation that follows a healthy environmental protection ethic may be doomed unless everybody else follows the same path.[20] Where there is common land, air, or water, some people will take advantage of this situation and overgraze, overfish, or pollute that environment to their own advantage at the expense of others. The deforestation and soil erosion in ancient Greece,[21] the overgrazing in the Sahel of northern Africa,[13] or the polluting of air in New England by plumes of sulfur dioxide emissions from the Midwest are all examples of this behavior. As one person or segment of society emits their toxic brew into the environment while avoiding costly controls, the profit margin is enhanced in the short term. Others, seeing the advantage are likely to follow suit provided no adverse consequences immediately follow. It is probably axiomatic that many members of any society will likely pass on the consequences of their destructive actions if they will benefit in the short term and receive little or no negative consequences from that action. Harden refers to this as the *tragedy of the commons.*[20]

The Pioneer

The consequences of laying waste to a land in the past were minimized by the ability of the population to emigrate, just as the pioneers in America could move west. There were enormous resources that permitted the early pioneers to cut trees, harvest the land, and trap or shoot wildlife with ecological impunity. The pioneer mentality cannot be continued indefinitely in the presence of massive population increases. The opportunity of mass emigration has been largely eliminated throughout the world as most habitable areas are now populated. The consequences of unrestricted use of resources well beyond their replacement level will eventually result in eco- logical and human disaster. We must seek a sustainable development in which resource use balances replacement and the discharge of pollutants is minimized.

Declining Investment in Technologies

The optimistic attitude of the Cornucopians toward population increases demands an investment in new technologies that will help sustain these populations. Unfortunately, the major gains in food crops experienced as part of the green revolution are unlikely to continue in the absence of investment in research and development. Further, these technologies, along with financial support, must reach the farmers of the developing countries. Just as it would be impossible to feed today's population with 50-year-old technology, it will therefore be impossible to feed the population of the year 2050 with today's technology. However, government funding for such organizations as the CGIAR centers, which were largely responsible for the green revolution, has been falling. Despite contributions from the World Bank to shore up the losses, more than 100 scientists left the organization in 1993–1994 alone.[6] Further, U.S. aid for agri- cultural programs has sharply declined by more than 75% since 1980.[6]

Family Planning Cuts

The 104th Congress of the U.S. reduced foreign assistance overall in 1996 with a 25% decrease in USAID funds and a 35% cut in the family planning/population assistance budget. The allowance was spread over a 15-month period that reduced the USAID's 1996 population funding by nearly 80%.[1] The U.S. has continued to withhold $34 million from the United Nations Population Fund (UNFPA) every year of the Bush administration because of disputes over abortion. The global gag rule originated with the Reagan presidency and continues by banning discussion of abortion in reproductive health clinics even when considered legal in many coun- tries.[22] Much of this reduced spending by the U.S. is intertwined with the debate over abortion. There continues to be severe shortfalls in planning assistance in LDCs resulting in up to (1) 220 million unintended pregnancies; (2) 117,000 additional maternal deaths and 1.5 million women who experience permanent impairment; and (3) 9.3 million additional deaths of infants and young children.[4] Because the U.S. has normally contributed the most to family planning, the effect worldwide is very limiting to programs in LDCs, and it is predicted that unavailability of family planning services will likely increase abortions. This is the very opposite effect that was intended by those leading the effort to cut population assistance funds.[1] The loss of U.S. population funds has scaled back or eliminated reproductive health services in some of the world's poorest countries where fertility rates are the highest, such as Kenya and Ethiopia.[22] The consequence of failing to meet contraceptive needs of women in these countries is that there will likely be an additional 23 million unplanned births, 22 million induced abortions and 142,000 pregnancy-related deaths, and 1.4 million infant deaths.[22]

What Now?

We are now faced with two very different views on the effect of world population on the human condition and global ecosystems. The pessimistic view warns that population momentum will hurl humankind beyond the carrying capacity of the earth, creating enormous human misery and suffering, and dramatic destruction of environmental support systems. Others view people as an asset that will finally prevail with the development of new technologies and policies which will prevent a cataclysm. As with most polarities in life, the reality is probably somewhere in between, and is already being played out in regions around the world. Many of the LDCs are showing a downward trend in birthrates because of policies that empower women through education and economic opportunity. Overall, world growth rates are declining and the public news media boldly broadcasts that the population problem is over. Certainly, these are very substantial reasons to be optimistic about the future. However, just as there are positive examples, so are there dramatic failures that give reason to be apprehensive about the future.[23] Just as we can believe that technology will rescue humankind, we can also point to declining investments in agricultural research, family planning, and the preservation of ecosystems. Social and political will may not be capable of making the necessary adjustments, and ecological damage may already be sufficiently severe that it is irreversible in many parts of the world.[24] The ability of people to make predictions about the consequences of increasing populations depends on too many variables to be accurate. There are no simple answers to these issues. We can, however, examine the current state of human affairs and global ecosystems to capture a glimpse of where we have been and where we are now. This may give students of environmental health enough information to personally develop their own vision of the future world.

IMPACTS ON THE ENVIRONMENT

We have all seen a spider web, sometimes glistening in the early sunlight, still wet from the morning dew. Touch one strand, and the entire web will move and vibrate.

Our environment is so intertwined that each human action ripples across it, causing imperceptible changes throughout the system that become magnified by the billions of people. As the population increases, the need for food increases. Land is cleared or grassland is cut and uprooted. Top soils are lost, nutrients depleted, and balanced ecologies of insects, birds, and mammals are upset as habitats are destroyed. Fertilizers and pesticides are applied that wash into streams and seep into groundwater, causing soil and water contamination that biomagnify up the food chain, with attendant reproductive abnormalities and disease in wildlife and humans. As insects develop resistance and natural soil nutrients are depleted, more chemicals are applied to croplands. These chemicals require energy in their manufacture that is drawn from the combustion of fossil fuels. These combusted fuels release pollutants into the air that produce acidic deposits and climate-warming gases. Climate change and acidification of soil and water may cause further ecological damage, contributing to a downward spiral from a sustainable environment. The circle of influence moves from more people, greater deforestation, poorer soil, more deserts, fewer species of animals and plants, and less food. This is the sequence of events we will explore in the following pages.

DEFORESTATION

Although life is highly adaptable, most species require a source of energy such as sunlight (required by most plants), nutrients, and moisture. Organisms are also affected by the presence of oxygen and the availability of warmth. Life exists in a relatively narrow shell about the globe extending a few kilometers into the air and below the sea. The richest and most diverse forms of life tend to be in the tropical rain forests, whereas harsher conditions at the top of the Himalayan mountains or in the frozen, bleak landscape of the polar ice caps harbor very few life-forms. Just as conditions of temperature, moisture, and nutrients vary in regions around the world, so do the types and diversity of vegetation and animal life change. Several major regions exist around the world that are identified by the major types of vegetation. These are called *biomes*, and include tropical rain forests, temperate forests, prairies, deserts, and arctic tundra. As our present topic is deforestation, we will focus on tropical rain forests because the majority of them occur in areas of the world at risk from overpopulation, and many are being threatened with slash-and-burn techniques to make room for croplands.

RAIN FORESTS

The tropical rain forests (Figure 3.5) are located predominantly in the LDCs in a latitudinal band forming a belt around the globe from 20 1/4° S to 30 1/4° N. Rain forests are located in many countries along this band, including Mexico, Belize, Salvador, Guatemala, Nicaragua, Columbia, Ecuador and Brazil, eastward to Cote d'Ivoire, Nigeria, Congo, the Central African Republic, and further east to Malaysia, Indonesia, Philippines, New Guinea, and the Solomon Islands (Figure 3.6).

The benefits of the tropical rain forests include their being (1) a major producer of oxygen for the global atmosphere, (2) the major carbon dioxide sink, (3) a

FIGURE 3.5 Typical lowland tropical rain forest. (NASA, The Earth Observing System Educator's Visual Materials, 1998, http://eospso.esfc.nasa.gov/cos-edu.pack/p24.html.)

potential source of new pharmaceuticals useful in the treatment of human disease, and (4) an important source of species diversity. Rain forests cover less than 10% of the earth's surface but contain more than a third of the earth's biomass, which is instrumental in absorbing large amounts of carbon dioxide and releasing great quantities of oxygen and so serving as the "lungs of the planet."[25] In 1995, a team of British, Australian, and Brazilian ecologists found proof that the Amazon rain forest absorbs tons of carbon dioxide produced by burning fuels. The globe's 1 billion acres of rain forests may absorb one sixth of the amount of carbon dioxide produced by burning fuels each year.[26]

Some of the most important and historical drugs isolated from plants include those used for used for treating circulatory disorders and high blood pressure. Some of these come from the Indian Smoke Root, *Rauwolfia serpentina*. Caloniea, from the Malaysian tropical rain forest tree, is presently being developed and tested for the treatment of HIV-infected persons in the U.S.[27] Plants from the Peruvian rain forests could help doctors in their fight against deadly tuberculosis. In a study of about 1250 plant extracts from Peru, about 46% showed an ability to inhibit the growth of *Mycobacterium tuberculosis*, the bacterium that causes tuberculosis. There are a myriad of other pharmaceuticals being used and investigated in the treatments of cancer, blood pressure, HIV, and other diseases. There are extensive efforts under way to investigate the possible pharmaceutical benefits from tropic rain forest plants before they disappear as a result of ecosystem destruction.

In spite of the numerous benefits from rain forests, they are disappearing at an alarming rate. This process is leading to *deforestation*, which has been defined as the permanent decline in crown cover of trees to a level that is less than 10% of the original cover. Prior to the human population explosion, the tropical rain forests are estimated to have covered 16 million kilometers, or nearly 11% of the earth's land area. By 1987, tropical rain forests were disappearing at the rate of 42 million acres each year, representing a loss of 115,000 acres each day. WRI reports that half of the world's original forests have been lost and only one fifth of the remaining forests are "frontier forests" — pristine areas that have not been disturbed or degraded by

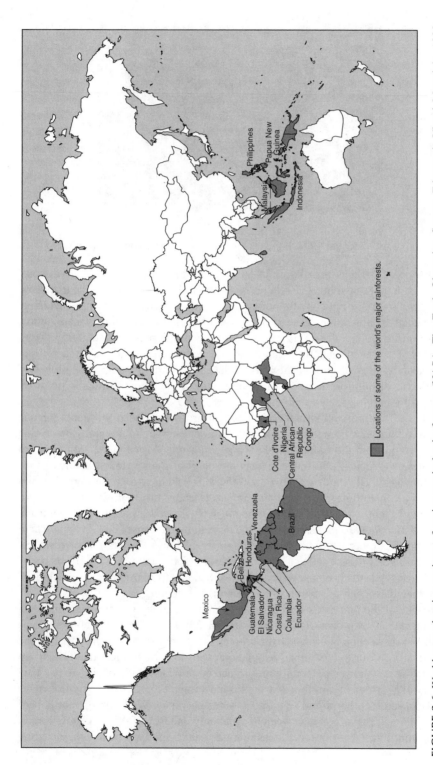

FIGURE 3.6 World map showing regions of major tropical rain forests. (NASA, The Earth Observing System Educator's Visual Materials, 1998, http://eospso.esfc.nasa.gov/cos-edu.pack/p24.html.)

human activities.[28] Today, global rates of deforestation remain the same as 10 years ago. The Food and Agricultural Organization (FAO) estimates the rate of South American tropical deforestation at 21,000 mi^2 annually, or about the size of the U.K.[29]

Tropical deforestation contributes significantly to the net movement of carbon (carbon dioxide) to the atmosphere as the concentration of carbon in forests is higher than in the nonforested areas that replace them. Additionally, tropical rain forests occupy less than 7% of the land surface but contain more than half of all plant and animal species. The greatest problem associated with tropical deforestation may be the massive extinction of species. The diversity of plant and animal species is affected by (1) destruction of habitat, (2) the formation of fragments of forested area less than 100 km^2, and (3) edge effects occurring as far as 1 km into the forest from adjacent deforested areas.[30] This boundary zone has greater exposure to winds, easier access by livestock and humans, and many other effects that result in a loss of plant and animal species at the edge. Edge effects greatly magnify the adverse effects of deforested areas.

The Brazilian Amazon is the largest region of tropical forest in the world, containing 31% of the world's total.[30] NASA's (National Aeronautics and Space Administration) Global Inventory Monitoring and Modeling Study showed that in 1987, fires burned 77,000 mi^2 of the Brazilian part of the Amazon rain forest, or an area greater than New York by one and half times.[31] During the months between July and October 1987, 4% of Brazil's Amazon jungle was destroyed, and most of this destruction followed the route of the World Bank–financed road through Rondonia known as BR-363. Brazil's poor and destitute have flocked to this area in search of arable land, and they clear the land by setting fire to it. NASA's heat-sensitive film taken from satellites show nearly 350,000 fires were started in 1987, and perhaps even contributing to the destruction of the earth's ozone layer. Despite worldwide attention, the deforestation of the Amazon continues. Brazil released figures in May 2005 that showed deforestation in the Amazon reached nearly 10,000 mi^2 each year since 2002 and almost returned to the highest recorded level of 11,000 mi^2 set in 1995.[32] However, the soil is so thin and of poor quality that the settlers have to clear new plots every year or two. With no canopy to break the intense downpours, the thin topsoil is easily washed away. The birds and other natural predators of insects are eliminated through habitat destruction, and mosquitoes often abound, creating greater opportunities for vector-borne disease transmission. Colonization, and subsequent subsistence farming, is among the leading reasons for clearing the rain forest, but clearing for cattle pasture, commercial agriculture, and logging also remain very high.[32]

The consequences of tropical deforestation are well documented and well understood globally. There are, in fact, 219 international groups concerned with saving the Amazon forests.[26] Unfortunately, reforestation efforts have met with dismal failure in the wake of the continued rapid growth of human populations. According to the Brazilian government's Institute for Space Research, weather satellites detected 39,900 fires in July of 1995, a nearly fivefold increase over the 8503 fires reported in July of 1994. These fires are not limited to Brazil, but span a number of countries including Uruguay, Bolivia, and Paraguay, producing a plume of smoke covering over 6 million square kilometers, producing respiratory irritation and closing down airports on many days.[27] Environmental experts contend that Brazil's failure to analyze the extent of deforestation since 1991, along with stepped-up levels of burning in the Amazon region, raise doubts about the claims that the destruction of the Amazon rain forest has slowed as a consequence of steps taken by the government. This situation has worsened, with Landsat images revealing that it was burning at an alarming rate in 1997. Scientists working with the Washington-based Environmental Defense Fund report they "have never seen it so bad."[28] Further, the areas revealed by high-resolution Landsat images only show cleared areas and do not reveal the almost equal areas being impoverished beneath the canopy.[28] The drought-causing effects of El Nino are expected to promote even greater decimation through burning.

It was reported in 2006 that nearly 28% of the Brazilian Amazon was subjected to incipient human pressure associated with fire activity, where fire zones are defined as the 10-km radius around a forest fire. These were detected by satellite between 1996 and 2002.

Human pressure may vary within fire zones. More than 60% of fire zones are concentrated near deforested or urban zones that have been subjected to intensive pressures such as logging. The remaining 30 to 40% is found in more remote locations, suggesting the fires are caused by shifting agriculture carried out by traditional *mestizo* communities or indigenous populations. Forests in these areas may be subjected to less intensive pressures such as hunting, harvesting of nontimber forest products, and selective logging. The number of annual forest fires has nearly tripled from 16,000 per year in 2000 to almost 42,000 per year, showing an acceleration of human activity.[33]

Central America has been experiencing tremendous losses of its forests to industrial felling and the practice of slash-and-burn agriculture by the rural, landless poor. The felling of native forests has been leading to serious soil erosion in these parts. Panama, El Salvador, and Guatemala have been found to have the worst soil erosion in the world, according to a study conducted by the Center of Studies and Social Action of Panama. Nearly 98% of El Salvador's natural forest cover has vanished, with Panama and Guatemala close behind.[34] Dramatic price increases in oil by OPEC in the early 1970s placed kerosene and other petroleum fuels out of reach for millions of the world's poorest, and left them no alternative but wood. Poor families in El Salvador and Guatemala are spending an entire day every week looking for firewood and traveling up to 12 km from their homes to find it.[34] The recent major increases in world oil process are likely to worsen this situation.

The tropical forest of Africa accounts for 18% of the world total and covers 3.6 million square kilometers of land in West, East, and Central Africa. Annual rates of deforestation in the region vary from 38,000 acres in Gabon to nearly 742,000 acres in Cote d'Ivoire. West African forests have been reduced to patches. African forests are the most depleted of all the tropical regions, with only 30% of historical stands remaining.[35] Just as in part of South America, many Africans must travel miles away from home to gather firewood and/or expend 25 to 30% of their income on wood to burn.

A study published by the Worldwatch Institute, a nonprofit research organization dedicated to the analysis of global issues, concludes that nearly three quarters of the middle mountain range in Nepal has been deforested in the last 40 years. India has lost 40% of its forest cover within the last 30 years, and Pakistan has only a fraction of its original forested land. The deforestation contributes to extensive flooding, costing Pakistan about a $1 billion each year in crop, building, and infrastructure losses.[36]

The rate of deforestation in the Pacific and Asia could decimate some 450 million acres before the year 2000. The greatest rates of deforestation are occurring in Indonesia, followed by Thailand, Malaysia, India, and Laos.[36] Hundreds of thousands of acres are also being lost in Nepal and Vietnam.

Although 70 tropical countries have agreed to support the International Tropical Forests Action Program (TFAP) initiated in 1993, the full implementation of this program will require substantial investments of time and resources.[37] According to the Global Environmental Outlook Report released by the United Nations Environment Program (UNEP) at Nairobi, Kenya, "International and national funds and political will remain insufficient to halt further global environmental degradation."[24]

Forests in Developed Countries

The area under forest cover in developed countries has remained fairly unchanged during the last decade.[24] The forested areas in these countries are stable and increasing in the temperate regions, where alternative fuels and forest management programs have sustained timber harvests. The major issues surrounding temperate forests include (1) damage by air pollution, (2) the loss of biological variety, and (3) fragmentation with the loss of forest edge. Many West European countries have actually experienced an increase in forest cover. Since the mid-1960s, Sweden, France, and the U.K. have experienced increases of forested area of 22, 34, and 37%, respectively. The U.S. has seen a significant return of forests compared to the 18th and 19th centuries when farmers cut down all but 2% of the trees in the Northeast.[38] However, in several important ways the forests are fundamentally different from those of previous centuries. At 90 years old, the forest ecosystems are not yet mature. Ironically, the oldest trees are preserved in cities and suburbs. Exotic species from other continents have taken hold. Many native species are threatened by diseases and pests. Expanding suburbs have partitioned or fragmented the forest into patches, creating lots of forest "edges." Maine, Vermont, and New Hampshire, as well as northern New York, contain 26 million acres of forest known as the Northern Forest. This precious resource is in serious danger, according to the Northern Forest Alliance, a coalition of about 25 local and national environmental groups. The danger stems from the process of clear-cutting by lumber companies, by recreational pressures from millions of campers,[38] and from the constant threat of acid deposition due to air pollution.

Nevertheless, developed countries also bear inescapable responsibility for the rapid loss of forest sources. North America, Europe, and Japan, with just 16% of the global population, consume two thirds of the world's paper and paperboard and half of its industrial wood.[38] The World Bank has invested in sustainable forest development and offered incentives to companies to harvest a variety of tree species and plant replacement trees. The Brazilian government has begun talks with representatives of the major industrialized nations on the progress of an ambitious program to save the Amazon rain forest. The project, which was set up in 1991 and is funded mainly by the G-7 countries, aims to reduce deforestation, protect the Amazon's unique biodiversity, and encourage sustainable development. Brazil has lifted a ban on new logging permits in the Amazon rain forest after landowners and loggers agreed to slow their rates of forest destruction. However, Brazil's authorities refused to issue new logging permits in February after they discovered more than 15,500 km^2 — an area the size of Belgium — was cleared in 1998. The total area of selectively logged forests in the Amazon is unknown, although estimates indicate this activity may affect 10,000 to 20,000 km^2 of forest per year in the Brazilian Amazon. Some of these forests are converted to agricultural and pasture land soon after timber is harvested, whereas other areas remain as logged forest.

A large part of the timber harvested in the Brazilian Amazon, or approximately 47% in 2001 and 43% in 2004, is thought to be illegal. Many licensed loggers fail to implement forest management plans or they harvest illegally in public unclaimed lands, causing significant damage to the forest, often from wildfires.[33]

Soil Degradation

What Is Soil?

Do not make the mistake of telling a farmer that his "dirt" must be really good to grow such excellent crops. Normally, he (or she) will acidly remind you that this is "soil," not dirt. What is the difference? The difference is that you can grow things in soil. Soil consists of small particles of rock and minerals mixed with a major proportion of plant and animal matter in various stages of decay. This organic matter is filled with living organisms including bacteria, algae, fungi, protozoa, insects, mites, and worms. A handful of soil is a complex, living ecosystem that provides the nutrients and conditions for plant growth. As plants grow and ultimately die, they serve as the nutrient source for a living mass of microorganisms that effectively recycle the organic materials into fertile soils. This organic material supports the growth of plants, which may in turn become the nutrients for higher forms of animal life.[7] Plants are called *autotrophic* because they synthesize their own food from inorganic substances. Plants derive carbon, oxygen, and hydrogen from carbon dioxide and water. Plants also require nutrients, which they derive from the soil, including nitrogen, phosphorous, calcium, magnesium, and potassium. These are called *macronutrients* because plants use them in significant amounts. There are a number of minerals that plants require in only trace amounts. These are known as *micronutrients* and include substances such as zinc and molybdenum. Soils vary in their physical nature. Some soils are sandy and so contain larger "gritty" particles that allow water to flow through rapidly and carry away or leach nutrients. Clay soils consist of extremely small or "greasy" particles that are bound together very tightly by chemical attraction. These soils tend to be impermeable to water, allowing rain to puddle on the saturated surface. Silt tends to be composed of fine particles that are larger than those producing clay but smaller than sand particles. Soils best suited to agriculture consist of sand, silt, and some clay in a homogeneous mixture referred to as *loam*. Fertile soils contain complex organic matter that has been biologically broken down so that original plant and animal matter is unrecognizable. This organic matter is humus and it serves to (1) retain moisture, much as a sponge; (2) serve as an insulator to heat and cold; and (3) bind and release nutrients to plants in useable forms. Soils that lack humus have low or poor fertility.[7]

Soil Biomes

Not all soils are equally fertile, and soils also vary substantially about the globe in their ability to sustain intensive agriculture. The most fertile soils for the growth of foods such as corn, wheat, and rice are located predominantly in the global temperate zones of northern America, Europe, and Asia. A smaller belt of the temperate zone climate may be found in Argentina, southern Africa, Australia, and New Zealand (Figure 3.7). Such temperate zones of forest, woods, and grassland tend to have fertile and deep soils of 1 to 2 m because the moisture and temperature of these regions permit the growth of substantial plant cover while permitting organic matter to accumulate more rapidly as it decomposes. In higher latitudes the growing seasons become very short, and substantially less organic matter is produced. The boreal forest, or taiga, is an example of this biome and is represented by a great circle of spruce and pine trees stretching from Alaska through Canada and Russia. In these higher latitudes, the growing seasons become very short and substantially less organic matter is produced. Such organic matter is decomposed very slowly resulting in thin soils. In even higher latitudes is the tundra. This is an area where there are no trees, but great expanses of mosses, lichens, sedges, and occasional low-lying bushes. The soils here are very thin because the weather is extremely cold, which restricts plant growth and also inhibits their decomposition. Further, the thin soil lies atop a permanent layer of ice or frozen soil known as *permafrost*. Any disruption to the sensitive tundra takes many years to recover. Southern, or "horse," latitudes at 20 to 30° north and south exhibit some of the earth's driest and largest deserts such as the Sahara because the regions receive a constant supply of warm, high pressure air.

Soil Erosion

Consider that the soils which sustain life around the globe are really extraordinarily thin and vary from 1 to 2 m to just a few centimeters. The critical layer of soil is in constant flux as new organic matter decomposes and is added to the soil, and some of it is carried or swept away by a process called *erosion*. In most natural conditions, the replacement of soil is equal or greater than erosion. This allows soil depth to be maintained or to increase. However, the practice of poorly managed agriculture has caused erosion to proceed more rapidly, resulting in impoverished soils. As woods are cut and fields are plowed to plant crops, soils are lost to the effects of wind and runoff water (Figure 3.8). Techniques to prevent soil erosion have been practiced in many parts of the world for centuries. Traditionally, such practices have included the planting of nitrogen-using (corn) and nitrogen-fixing (beans) crops in alternate years in the same space or planting them in alternating

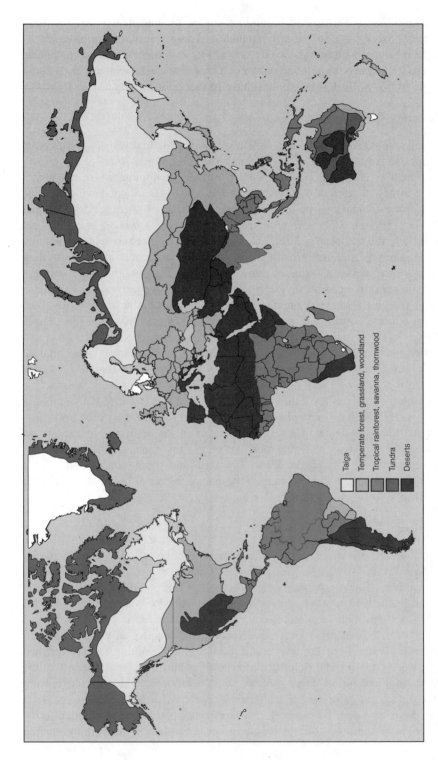

FIGURE 3.7 World map showing regions of major soil biomes including temperate, taiga, tundra, deserts, and tropics.

bands in the same year. This is termed *rotation*. When fields remain unplanted every few years, this time interval permits moisture and nutrients to return in a process called *fallowing*. When farmers are driven to plant on hillsides, they must create level fields along the contour of the slope in a process called *terracing*, else erosion will occur at an enormous rate and wash away the topsoil in a generation or two.[7] However, these techniques are not widely used, and soil erosion looms as one of the greatest potentially devastating problems facing the world today even as most countries fail to recognize it as such.[33] Driven by logarithmically expanding populations, the developing nations of the world are forced to continuously cultivate their best farmland while moving agriculture into areas with decreasing agricultural sustainability. Many of these countries look to the U.S. for food, and that creates a high demand and high commodity prices. Farmlands in the U.S. have little time to lie fallow or be rotated as the land is pressed into service to grow food for the world's billions. North America now exports more than 100 million tons of grain per year, which is a fourfold increase over exports in the 1950s, and three times the combined exports of Australia, New Zealand, and Western Europe.[7] According to the U.S. Department of Agriculture (USDA), soil surveys show that more than 30% of natural top soils in North America have been lost since farming began in the U.S.[39] The average rate of topsoil erosion in the U.S. in the early 1980s was 4.7 tons per acre, with many of the agriculturally intensive states showing levels of 8.7 tons per acre. At this higher rate, and assuming an average topsoil of 12 in., there would be a complete loss of topsoil in less than 150 years.[39] Nearly half of the cropland in the U.S. is losing soil to erosion, with about 3 million acres being destroyed every year. In areas of moderate rainfall and temperate climate, it takes 30 years to form an inch of topsoil, which can be lost in 10 years with excessive grazing, monocropping, and ground cover destruction.[40] The devastation of croplands in the Midwest in the 1930s by wind erosion led to the term *dust bowl* and, subsequently, the creation of the Soil Conservation Service, which was renamed in the 1990s to the Natural Resources Conservation Service (NRCS). The NRCS has inventoried soil erosion in the major croplands of the U.S. and shown that through crop rotation, improved tilling methods, and the Conservation Reserve Program (CRP) there has been either an improvement or no increase in average erosion rates since the 1980s.[40] The CRP was started under the Food Security Act of 1985 and offers farmers financial incentives to remove eroding land from production for periods of 10 years so they may recover. Despite these improvements, excess erosion still occurs in southern Iowa, northern Missouri, parts of western and southern Texas, and eastern Tennessee. A large section of the Great Plains stretching from Nebraska to northeastern Texas is unstable, and a prolonged drought could cause the entire area to become a Sahara-like desert.[40] A significant reduction in the U.S. food production would have a devastating effect on world food supply.

When viewed globally, soil erosion claims over a billion acres every year, and 1.2 billion acres of global cropland is losing topsoil so rapidly that these acres are expected to become unproductive in the next few decades.[41] India is averaging topsoil losses at the rate of 9 tons per acre and is losing about 5 billion tons of topsoil a year at a rate double that of the U.S. Food security is a major issue in Africa, West Asia, and Latin America. The major concern in these areas is the availability of land

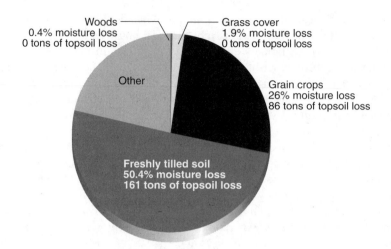

FIGURE 3.8 Loss of soil and moisture in a season per acre under various ground cover conditions. (Adapted from Turk, J. and Turk, A., *Environmental Science*, 4th ed., Saunders College Publishing, Philadelphia, PA, 1988, chap. 9.)

and the prevention of degradation. More than 1 billion acres of land in Africa is moderately or severely degraded. Nearly 47% of the grazing lands in Latin America have lost their fertility because of erosion, overgrazing, and salinization.[24] As the lesser-developed countries move toward another 5 billion people in the next decades, continued soil erosion in these countries could result in harvest reductions of 30% or more during the same period. Upwardly spiraling populations may make the political will and funds to establish soil conservation programs unlikely.[24]

THE PROCESS OF DESERTIFICATION

Fences pieced together by plywood, corrugated cardboard, and rusted scrap metal are the futile efforts of the residents in Komsomolky to slow down the inevitable advancing desert from finally consuming their town. The Kalmyks, who inhabited the region for nearly 400 years, understood the fragility of the Black lands in Southern Russia, which were once the bed of a larger Caspian sea. They never tilled the soil. They raised sheep with flat-bottomed feet and did not let cattle graze in the area. These careful custodians of the grassy steppes of central Kalmykia were expelled by Josef Stalin in 1943, and in their place were brought irrigation projects to develop cropland and sharp-hooved sheep. Because the topsoil lay over the former extensive bed of the Caspian sea, the irrigated land became a salt marsh and the sheep destroyed the thin topsoil causing the desert to begin advancing in the 1950s.[42] The encroaching sands bury everything in sight and forced 25 towns to be abandoned while threatening 40 others. State-financed plans to reverse the trend have been halted as money ran out in the mid-1990s.[42] The agricultural mismanagement of this land has promoted the desertification of this delicate eco-system and it will take generations for the land to be restored even if the practices of mismanagement are halted.

What Is Desertification?

The preceding scenario is being repeated in a number of areas throughout the world, but it was not until the late 1960s that desertification became a matter of international concern. Worldwide, with more than 10 million acres of farm land becoming unproductive each year, dust bowls are multiplying and raising legitimate concern about our planet's capacity to feed its rapidly growing population. During a period of severe drought in the Sahelian region of Africa from 1968 to 1973, the people suffered enormous tragedy through starvation. This drew worldwide attention to the problems of trying to survive on arid and semiarid land, especially on the perimeters of deserts. As a consequence there have been a number of initiatives, including a United Nations Conference on Environment and Development that was held in Rio de Janeiro in 1992 (Rio Earth Summit). Here, more than 100 countries began the process of developing a legally binding international contract to combat desertification. This resulted in the United Nations Convention to combat desertification (known as *the Convention*). According to the convention, desertification was defined as "land degradation in arid, semiarid, and dry subhumid areas resulting from various factors, including climatic variations and human activities."[43] The definition was adopted by the United Nations Environment Programme (UNEP), and areas susceptible to desertification were defined as those subjected to arid climates (arid, semiarid, and dry subhumid). About 15 billion acres, or one third of the earth, is dry land, and 2.5 billion of these dryland acres, or 16% of the earth's surface, are hyperarid deserts where there is little or no growth. This leaves 12.5 billion acres, or one third of the earth's surface, that falls into the categories of arid, semiarid, and dry subhumid climates (Table 3.1). These areas include 51,720,000 km² in nearly 100 countries that may be directly or indirectly affected by desertification. Some 900 million people live in these areas that are susceptible to drought and desertification, and 200 million

TABLE 3.1
Types of Land Areas at Risk of Desertification by World Region [in square kilometers × 1000]

Climate Type	Africa	Asia	Australia	Europe	North America	South America	TOTAL
Hyperarid	6,720	2,777	0	0	30	260	9,780
Arid	5,040	6,260	3,030	110	820	450	15,710
Semiarid	5,140	6,930	3,090	1,050	4,190	2,650	23,050
Dry Subhumid	2,690	3,530	510	1,840	2,320	2,070	61,500
TOTAL AREA	30,335	43,508	8,923	10,498	25,349	17,611	136,224

Source: Adapted from Brazil Embassy, What is desertification?, Science and Technology Section, National Research Network's Brazilian Regional Center, Feb. 30, 1997, http://brasil.emb. nw.dc.us/nuna/ deserts/ingles/se/~edesert/desertmi.html#dm.

FIGURE 3.9 Land degradation in arid, semiarid, and dry subhumid areas resulting from various factors, including climatic variations and human activities, defines the process of desertification. This was once completely forested land in Africa.

of these people, or about 22%, are presently affected by this process.[43] Even more disturbing from a socioeconomic view is that the majority of areas at high risk for desertification are also home to some of the world's most impoverished people in LDCs. Poverty is being globally recognized as among the major factors associated with the degradation of arid, semiarid, and dry subhumid regions.[43] The expanding population forces people to cultivate and overuse more and more marginal lands, which are increasingly less resilient to human and climatic disturbances. Growth conditions fluctuate from season to season and even within a season. Dryland areas may receive an entire season's precipitation in a few days, or experience prolonged droughts, although average rainfall over a decade may be normal. The people who inhabit these regions must adjust to these conditions. Drought years may require land to remain fallow or be used in another way. However, poverty and the need for food is an enormous pressure that defies a flexible land use response and leads to desertification[44] (Figure 3.9).

The Costs of Desertification

The process of desertification is ongoing, and about 14.8 million acres (60,000 km²) are being removed from production annually because of poor land management, including overgrazing, improper irrigation, and intensive use. Economic losses are calculated to be $40 billion, and the cost of recovering these lands worldwide is estimated at $10 billion annually.[43] These $40 billion in losses from land degradation are considered to be on-site, or immediate, losses from reduction in productivity and are distributed according to Table 3.2. Costs may be many times higher if you were to include off-site, or indirect, costs involving the loss of wells

TABLE 3.2
Estimated On-Site or Immediate Losses from Land Degradation with Loss in Productivity

	Estimated Direct Loss from Land Degradation (in Billion US Dollars)
Africa	9.3
Asia	21.0
Australia	3.1
North America	4.8
South America	2.7

Source: Brazil Embassy, What is desertification?, Science and Technology Section, National Research Network's Brazilian Regional Center, Feb. 30, 1997, http://brasil.emb.nw.dc.us/nuna/deserts/ingles/se/~edesert/desertmi.html#dm.

from silt formation, illiteracy among the young as they are forced to scavenge, and the cost of refugees and emergency aid.[43] There are also costs in human misery as the world's largest population increases are occurring in areas of the world at greatest risk for desertification.

Another loss difficult to measure is the reduction in biodiversity of important food crops. Many of our important food crops such as wheat, barley, and millet originate in drylands. The species that have adapted to drylands are few in number, and their loss to nonsustainable agriculture methods is a great threat to biodiversity. This is further exaggerated by the green revolution, which has developed high-yield crops but at the cost of fewer and fewer varieties. Overall, we are seeing a reduction in the genetic pool of important food crops. Consequently, the lack of genetic variability in the important food crops places us all at food security risks from changes in climate, emerging or reemerging pests, changing soil conditions, or other unpredictable events.

Another inestimable cost derives from the plight of environmental refugees. There is no accurate way to estimate the numbers of environmental refugees as they are not officially protected through the United Nations, although the figure is thought to greatly exceed 10 million.[43] The flight of refugees from degraded land has caused extensive disruption in developing countries, resulting in economic, social, and political difficulties. These problems may worsen as populations increase in LDCs, causing greater poverty, increased desertification, and reduced food production.

WETLANDS

What Are They?

Wetlands are found everywhere, ranging from tropical forests to the furthest reaches of the tundra. Wetlands appear virtually in every biome except the frozen expanses of Antarctica. Wetlands are those areas of land where water saturation is the major factor influencing the nature of soil development and the communities of plants and animals that live in the soil and on the surface.[45] The National Research Council defines a wetland as "an ecosystem that depends on constant or recurrent shallow inundation or saturation at or near the surface of the substrate."[46] There is a somewhat different definition of wetlands used for regulatory purposes and written under the Clean Water Act (40 CFR 230,3[t]): wetlands are "those areas that are inundated or saturated by surface or ground water at a frequency and duration sufficient to support, and that under normal circumstances do support, a prevalence of vegetation typically adapted for life in saturated soil conditions." Because acreage that is classified as wetlands falls within the guidelines of a protected environment under federal law, the ultimate decision in this matter depends on EPA's definition of a wetland. Some ambiguity in delineating wetlands still exists. For example, the number of consecutive days that are required for land area to have surface water can vary from 14 to 21 d, depending on the scientific or legislative source. Other factors such as growing season, temperature changes, or the existence of records of environmental activity for the specified land site all must be taken into account. The key wetland features that are in question include the hydrology, anaerobic substrate, and biota. Policymakers and scientists dispute the number of wetland features that must be proved to be present.[47,48]

Wetlands are zones that fall between areas of land and water and both separate and join the land and open water. Wetlands are covered or soaked by water for long periods during the year. They provide habitat for numerous species of plants and animals, including some that have specifically adapted to wetlands and appear nowhere else such as cattails, mangrove trees, or wood ducks.[47] Wetlands appear in a variety of forms including swamps, bogs, prairie potholes, bottomland hardwood forests, and estuaries. Swamps are often transitional environments between large open water and drier land. They are home to species from both extremes and are marked by flood-tolerant trees and shrubs.[45] Bogs are usually a stagnant, acidic, nutrient-poor ecosystem filled with peat and sphagnum moss. Its abundance of decaying material consumes enormous amounts of carbon dioxide. Prairie potholes are freshwater marshes found primarily in the upper Midwestern U.S. This glacially formed landscape is pockmarked with potholes that fill with snowmelt and spring rains, and are among the most important wetlands in the world, supporting more than 50% of North American migratory waterfowl and wide varieties of invertebrates and plants (Figure 3.10).[46] River swamps or bottomland hardwood forests are found along streams and rivers in the Southeast and South Central flood plains of the U.S. These are deciduous forested wetlands consisting of gum and oak trees adapted to areas of seasonal flooding or covered with water most of the year. The trees in these wetland systems often have fluted or flaring trunks and knees or aerial roots as seen in the bald cypress.[45] Estuaries are inland marine waterways found worldwide. They are dominated by grass or grasslike plants and lie mostly between barrier islands and beaches. Nutrient-rich tides make estuaries among the most naturally fertile habitats in the world.[46] The EPA lists a number of wetland categories, including bogs, bottomland hardwoods, fens, mangrove swamps, marshes (tidal, saltwater, freshwater), playa lakes, prairie potholes, swamps, vernal ponds, and wet meadows.[45]

FIGURE 3.10 Various types of wetlands (From USEPA, Office of Wetlands, Oceans, and Watersheds, Facts about Wetlands, August 25, 1997, http://www.epa.gov/owow/wetlands/facts/types.html.)

Benefits of Wetlands

Wetlands, because they are often inaccessible and boring landscapes of gray and muddy brown identified with insects and snakes, have not historically received much positive press. However, wetland areas are very beneficial. For example[45–47]:

• Wetlands purify and replenish water supplies. They serve as natural recharge areas to groundwater. They act as natural water filters because the rich biomass of plants and microscopic animals strain pollutants from water flowing toward streams or soaking into the ground.
• Wetlands are extremely rich in biomass (the amount of plant and animal life). They are equivalent in biomass acre for acre to a tropical rain forest. Wetlands in the U.S. are habitat for 5000 species of plants, 190 kinds of amphibians, and one third of all bird species.
• Wetlands are an important source of food. Many species of fish and shellfish such as shrimp, crabs, and crayfish spawn in salt marsh wetlands.
• Wetlands absorb large amounts of carbon dioxide from the air taken in by plants or stored in peat moss.
• Wetlands control flooding in low-lying areas, because they work like sponges, soaking up floodwaters from overflowing rivers. Vegetation slows the water flow, thereby sparing farms and communities downstream. The mammoth flooding of the Mississippi in 1993 worsened because the farm-filled states of Iowa, Indiana, and Missouri have drained all but a small percentage of farmlands.
• Wetlands protect coastal areas from storms. Fierce coastal storms can create huge waves that wetlands are able to absorb and so buffer communities further inland.

- Wetlands provide recreation and beauty. Hunters and sport fisherman find enjoyment, and often dinner, in wetlands.

Wetland Losses

The U.S. has lost more than half of its wetlands since the early days of colonization. Wetlands were considered a nuisance that made farming difficult and as settlements moved westward, these areas were filled in. The great farm areas of Ohio, Iowa, and California have seen 90% of the wetlands disappear. There were no federal laws to protect wetlands until 1972, when Congress enacted amendments to the Clean Water Act that prohibited discharge of fill material into "waters" of the U.S. without a permit from the U.S. Army Corps of Engineers. A federal district court ruled that this included wetlands and unleashed a bitter controversy over issues of property rights and the role of government.[47] Unfortunately, the Army Corps of Engineers has been granting quick building permits in environmentally sensitive areas under the Clean Water Act. Permit 26, one of the 30 different quick permits under this act, allows alteration of up to 10 acres of wetlands if they are isolated from other bodies of water. The regulations allow for small lots to be developed without requiring extensive documentation about the environmental impact of the project. Acquiring such permits may take as little time as 1 month.[49] Florida has lost an estimated 2000 acres in a 4-year period from this process, and nationally, the estimate is tens of thousands of wetland acres lost since the program started. From the mid-1970s to the mid-1980s, 2.6 million acres (1.05 million hectares) of U.S. wetlands were destroyed. The most recent and greatest losses have been in Louisiana, Mississippi, Arkansas, North Dakota, South Dakota, Nebraska, and Florida. An estimated 300,000 acres (120,000 hectares) of wetlands are drained or filled every year. About 87% is lost to agriculture, with urban development taking the rest.[47] More than three quarters of U.S. wetlands are privately owned, and so preservation attempts are difficult to mandate. One successful approach incorporated a provision in the 1985 Farm Bill by removing farm program benefits from farmers who drain wetlands for agriculture. Enacted under the same legislation was the CRP, which provides financial assistance to farmers who remove lands from cultivation and so promote the restoration of wetlands habitat. With wetlands in U.S. rapidly vanishing, ecologists are seeking methods of restoring or replacing them. A symposium at the annual meeting of the American Association for the Advancement of Science in February 2000 offered evidence that human-made wetlands may be just as good at water filtration as untouched wetlands, but they may not offer the same rich resources for wildlife. These efforts have seen a return of thousands of acres to wetlands along with significant increases in waterfowl populations.[50] A report released by the U.S. Fish and Wildlife revealed that between 1998 and 2004, there was a net gain of 191,750 acres of wetlands and other waters for a net gain of 32,000 acres annually. This is the first time ever that a study has reported a net gain in water resources. The problem, however, is that the increase is mostly in new pond acreage (700,000 acres, an 11% increase) A closer evaluation of the report reveals a net loss of more than half a million acres of naturally occurring wetlands such as estuarine emergent wetlands (salt marshes), freshwater emergent, and freshwater shrub wetlands. Natural wetlands provide for waterfowl and shorebird habitat, floodwater reduction, shoreline stabilization, fish habitat, and other functions, whereas special-purpose ponds and ornamental lakes offer little vegetation and provide limited wetland functions, with the most use as a stormwater retention device.[51]

Wetlands are being threatened in many other parts of the world. As an example, despite concern for the destruction of the Amazon rain forest, another part of the Amazon is also suffering from rapid population and development. Brazil's Panatanal, the largest wetlands in the world, is being threatened by human development, thus jeopardizing many of the rare 2000 species of animals and birds.[52]

THE LOSS OF BIODIVERSITY AND EXTINCTION OF SPECIES

Background

The extinction of species is not a new phenomenon. There have been many catastrophes on earth, but none so destructive and overwhelming as the "mother of mass extinctions" some 250 million years ago at the end the Permian period. This mass extinction was far more severe than those events that occurred 65 million years ago and caused the extinction of the dinosaurs and nearly half of all species.[53] The Permian extinction caused 90% of all species in the oceans to disappear; two thirds of reptiles and amphibian families perished, and up to 30% of insect orders were lost. The events leading to this mass extinction are hypothesized to have involved a huge and prolonged series of volcanic eruptions that released large volumes of carbon dioxide into the air and increased climatic instability.[53] The return of life may have taken 5 million years and resulted in dramatic changes in the forms of life that emerged. Very likely, most of the species (perhaps 99%) that have ever existed on this planet are now extinct. The diversity of species has ultimately recovered from each of these cataclysms as the evolvement of new species through evolution has been somewhat greater than the extinction rate. Today, more than 1.8 million currently living species have been identified. There is general consensus, however, that the planet is now losing more species than are being created, and that the activities of humans are the reason for a rapidly growing species extinction and loss of biodiversity. *Biodiversity* refers to the range of animal and plant species and the genetic variability among those species. The greater the genetic variation, the more likely it is that there will be survivor species in the event of major catastrophes.

As the planet has experienced a series of cataclysms over several million years and has recovered each time with abundant and diverse life, what is the major concern about the loss of a relatively few species? In order to respond to this question, we need to remember that the forms of life that emerged after cataclysms were very different, and that humans are a relatively recent phenomenon in the history of the earth. A reduction of species diversity will often result in an unstable ecosystem that is in danger of collapse if key species in that ecosystem disappear. Since genetic variation is the basis for evolution, the diversity of genes and species "affects the ability of ecological communities to resist or recover from environmental change, including long-term climatic change."[54] A human-induced cataclysm resulting in major losses of species could lead to a retrogression in the global ecosystem incapable of supporting large populations of complex creatures such as humans.

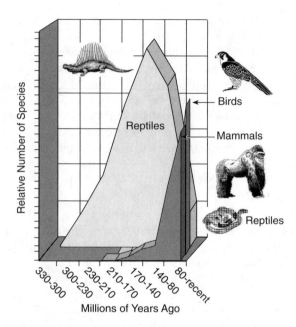

FIGURE 3.11 Comparative evolution of birds, mammals and reptiles over geologic time. (Adapted from Turk, J. and Turk, A., *Environmental Science*, 4th ed., Saunders College Publishing, Philadelphia, PA, 1988, chap. 9.)

Previous extinctions have been associated with major climatic changes arising from such forces as volcanism, or geologic changes that forced continents to clash and seas to rise or fall. Records of fossils show that entire groups of organisms including fish, reptiles, birds, and mammals have replaced one another over long periods of time. Birds and mammals, better able to respond to changes, began to proliferate as the numbers and species of reptiles declined (Figure 3.11). These changes are measured in millions of years. Even the cataclysm of volcanic eruptions in the Permian period is thought to have occurred over several hundred thousand years. This is in sharp contrast to the sudden and precipitous global changes being effected by humans. The massive increase in human population with corresponding pollution of land, air, and water is resulting in such rapid ecosystem changes that many species are unable to adapt. The extinction of species is thought to be occurring more rapidly than at any other time in recorded history.

More recently, loss of biodiversity has also become a public health concern. One of the better-known examples is the potential loss of medicines derived from plants. It has been estimated that on average 1 useful pharmaceutical can be derived per 125 plant species. Unfortunately, 3000 of the world's 300,000 cataloged plant species may become extinct in the next 25 years.[48] Disease epidemics may also be associated with a loss of biodiversity as ecosystems are disturbed and disease-producing organisms or conditions emerge. As an example, the deforestation of tropical rain forests eliminates the natural predators of mosquitoes, which may then proliferate along with many vectorborne diseases, including malaria (see Chapter 6).

Loss of Biodiversity

A report by the United Nations Environmental Programme titled "The Global Biodiversity Assessment" was among the first comprehensive, peer-reviewed reports ever to examine the earth's biodiversity. The report concludes that (1) 4000 plants and 5400 animal species are threatened with extinction, (2) species have recently become extinct at 50 to 100 times the average expected natural rate, and (3) 1.75 million, or about 13% of the 13 to 14 million species on earth, have been scientifically identified.[55] Background extinction rates through geological time have been roughly estimated at the rate of 1 mammal and 2 bird species every 400 years. Documented extinctions for mammals and birds during the last 400 years are already at 50 times that rate, and this figure is probably underestimated. The highest levels of biodiversity are in the tropical forests, where current rates of species extinction have been estimated at 1 to 11% per decade.[56]

A study by the World Conservation Union in October 1996 showed that the world is entering a period of major species extinction rivaling five other periods in the past half billion years. Of the 4327 known mammal species, 1096 are at risk, and 169 are at extremely high risk of extinction in the wild in the immediate future.[57] Besides mammals, 11% of birds, 20% of reptiles, 25% of amphibians, and 34% of fish are threatened. Of the 26 orders of mammals, 24 are threatened, with the most at risk being certain species of elephants, primates, rhinoceros, and tapirs. The pattern of species extinction and threatened species follows the greatest population densities closely. The countries with the most threatened mammals are Indonesia (128 species), China (75 species), and India (75 species), where 43% of the world's population lives, and which are the most densely populated countries (Figure 3.12). A United Nations report called the "Global Biodiversity Outlook 2" was released Monday, March 20, 2006, and claims that we must make unprecedented efforts to attain the goal of slowing down the enormous losses of species if the UN goals set in 1992 regarding the problem of decreasing biodiversity are to be achieved by 2010.[58,59] The report revealed that the current estimated pace of extinctions is 1000 times faster that at any time in the past and that the major causes such as habitat change, overexploitation, invasive alien species, climate change, and nutrient loading are not likely to change in the near future.[58]

The U.S. is the country that has the potential for losing more plant variety than any other, with 4669 of all existing plant species being threatened with extinction. Australia follows with approximately 2245 species threatened, followed by South Africa with 2215.[5]

Threats to Biodiversity

Some of the threats to biodiversity include (1) the destruction and fragmentation of habitat; (2) the harvesting of wild species for food, lumber, and other products; (3) the introduction of nonnative animals and plants; (4) the pollution of air, water, and soil with toxic chemicals and waste products; and (5) the increased ultraviolet radiation and global warming resulting from atmospheric changes and possible human activities.[54]

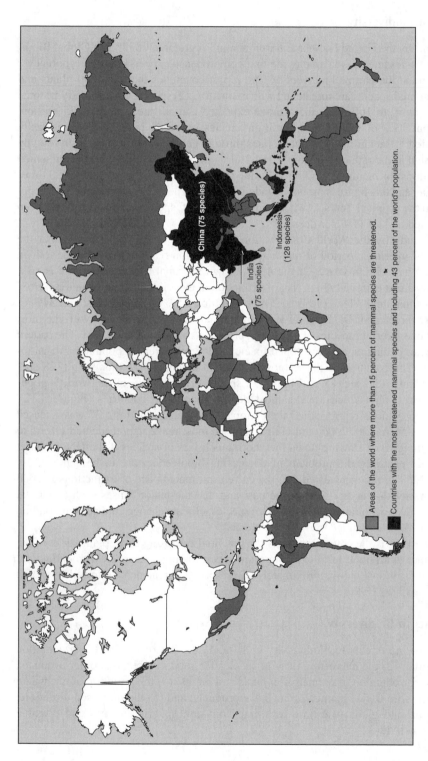

FIGURE 3.12 The majority of threatened mammal species occur in lesser-developed countries. (Adapted from Doyle, R., Threatened Mammals, *Scientific American*, January 1997, p. 32.)

Loss of Habitat

The most significant threat to biodiversity today is the elimination of habitat for agricultural and habitat purposes. As expanding populations assimilate geography by cutting down forests, filling in wetlands, and plowing fields, most species are unable to adjust to these rapid changes and perish. The greatest biological diversity occurs in the rain forests, where much of the habitat destruction is occurring. But habitat destruction occurs on many levels and even threatens the mass extinction of freshwater mussels. More than half of the 300 species of mussels in the U.S. are in trouble to some degree, and most of this is due to human activities such as the building of dams, gravel mining, sewage, and polluted runoff.[60] For a lesson in how quickly ecosystems fragment across international borders, a report of December 2000 suggests looking no further than Lake Victoria. Lake Victoria is the world's second-largest freshwater lake, covering an area of 67,850 km². This vast expanse, about the size of the Republic of Ireland, forms the headwaters of the River Nile. Prior to 1970, Lake Victoria, the world's largest tropical lake, had more than 350 species of fish from the Cichlid family. Ninety percent of these were unique to the lake. Fishing for tilapia and Nile perch provides a living for many of the Luo people who live along the lakeside, but also caused a collapse in the lake's biodiversity and deforestation in the three countries that border the lake: Uganda, Tanzania, and Kenya. Wood is needed to dry the oily perch, but the resulting deforestation is jeopardizing the lake and its fish. Forest clearing has increased siltation and eutrophication (often the result of human intervention, eutrophication is the process by which a water body is overwhelmed by plant life and nutrients at the expense of oxygen) in the lake, jeopardizing the Nile perch and tilapia fishery — the cause of the problem to begin with. Lake Victoria illustrates how profound and unpredictable trade-off can be when management decisions do not take into account how the ecosystem will react.[4]

Overharvesting

Many species have been harvested to the point of extinction. For example, many fish stocks such as cod in the North Sea off New England are heavily exploited, with as much 60% of the fishable stock being removed annually. Despite warnings by the International Council for the Exploration of the Sea (ICES), and government action to reduce allowable catches or closing fisheries completely, the data suggest that management efforts have not been successful.[61] The Northern "right" whales are estimated to be fewer than 300 in number. Their decline is attributed to the devastation by turn-of-century whalers, who preferred to hunt these whales because of their slow-moving and docile nature. Hence they were called the "right" whale to hunt. Deaths are attributed today to the slowness of the beasts, due to which collision with ships results in one third of the whale deaths per year.[62] The small number of the remaining whales leads to inbreeding, resulting in a loss of genetic variability that further jeopardizes the species. People along the coast of the Philippines use dynamite and cyanide to stun fish and then pound the reefs with crowbars to scare them out into waiting nets. Some fish are eaten, and others are sold as exotic aquarium fish. The natives see this as necessary for day-to-day survival. The process, however, is devastating the coastline coral reefs. Although efforts are under way through the Coastal Resources Management Project to empower local communities to act in their own self-interest, it is difficult to protect the environment from impoverished people seeking survival.[63]

Organized crime in countries such as the U.S., Japan, Russia, and many other countries has discovered a new high-paying commodity: wildlife. The species they trade include parrots, elephants, whales, bear, tigers, snakes, and rhino. Body parts from many of these animals are used in some cultures to improve sexual prowess. Earlier, criminals used boa constrictors to carry cocaine across borders. Now they sell the snakes for faster and easier money.[64]

Nonnative Species

In the aftermath of World War II, no one paid much attention to the invasion of the tiny Pacific island of Guam. The opportunistic brown tree snake, which grows up to 2 m in length, hitched rides from the island of Manus to Guam by slithering into the machinery used by the U.S. military. Without natural enemies, the population of brown snakes grew quickly within 30 years and eliminated 8 of the island's natural bird species and sharply reduced several others.[65] The freshwater mussels, already under attack from changing habitat and pollution, are also being displaced by Zebra mussels inadvertently transferred from the Black, Caspian, and Azov seas. Many nonnative species of fish including large-mouth bass and carp have been implicated in the extinction of as many as two thirds of the native species, and now threaten many others.[66] Rainbow trout never encountered "whirling disease" before the parasite was unknowingly transplanted here from Europe. In some reaches of the Colorado river, the disease has wiped out all but 1% of the population of young rainbows and has infected trout in many streams in the Rocky Mountains on both sides of the Continental Divide.[67]

Pollution

Acid deposition, promoted by the release of nitrogen oxides and sulfur oxides from combustive processes, have caused dramatic ecologic changes in the U.S., Europe, and in many cities of the lesser-developed countries. The acidification of lakes and streams has led to juvenile recruitment failure among fish, resulting in the disappearance of many species in a number of industrialized countries. The acidification may also be leading to forest decline in many parts of the world, and it is thought to contribute to the loss of 40 to 50% of the mycorrhizal fungi in parts of Western Europe. These fungi are considered "key" to many ecosystems because they live on plant roots symbiotically and facilitate the uptake of nutrients to the plants.[68]

A significant decrease in some species of fish and birds on the Southern California coast has been attributed to the 80% decline in the population of zooplankton. This decline is explained by some researchers as a consequence of rising ocean surface temperatures of 1.2 to 1.6°C. Some scientists suggest this warming is a result of the "greenhouse" effect. A steeper density gradient, resulting from ocean surface warming, has slowed the upwelling to the surface of nitrates, phosphates, and other nutrients used by phytoplankton. The phytoplankton have diminished sharply from lack of these nutrients. Because zooplankton feed on the phytoplankton, their numbers have also been markedly reduced and, consequently, this effect ripples up the food chain to larger fish. If the ocean surfaces continue to warm over the next few decades, the biological impacts could be devastating.[69]

Deformed frogs have been reported in Minnesota, Wisconsin, South Dakota, Vermont, and even Quebec. Frogs have been found to have limb abnormalities. Some have stumps, others have four front limbs, sunken or missing eyes, and small or missing sex organs. Aquatic frogs appear to have the most severe abnormalities. Amphibians without these deformities are difficult to find.[70] The cause of the abnormalities is uncertain, but has been hypothesized to be increased ultraviolet radiation resulting from stratospheric ozone depletion, heavy metals, pesticides, or a combination of these factors. Studies are under way to determine the cause, and there is concern that these effects may signal potential health effects for human populations.[70] Based on key findings of the 2006 Global Amphibian Assessment (GAA), there are 5918 known amphibian species. About 32% of 1896 species are threatened compared to 23% of all mammal species.[71] The GAA reports that at least 34 amphibian species are known to be extinct and that as many as 165 species may be extinct, and up to 43% are declining. Habitat loss and fungal diseases are major contributors to losses while many species are declining for unknown reasons.[71]

Protecting Endangered and Threatened Species

Legislation first aimed at protecting wildlife in the U.S. was introduced as a bill in 1926. It had little authority to protect. In 1969, the Department of the Interior added species to the list that were threatened with extinction worldwide. This international initiative helped to form the Convention on International Trade in Endangered Species (CITES), which attempts to reduce the trafficking of rare and exotic animals and plants.

In 1973, the Endangered Species Act (ESA) was promulgated in the U.S. Species are constantly evaluated by the Fish and Wildlife Service (FWS) and the National Marine Fisheries (NMF) to determine if they should be listed as threatened or endangered based on scientific data alone. Consideration is not given to economic or political concerns, and no branch of the federal government may proceed with a project that might threaten an endangered species or its habitat without first contacting the FWS.[72] Because projects may be stalled or blocked when threatened species are encountered, the ESA has many critics. However, 99.5% of the 100,000 projects going to consultation proceeded, usually with some modification. Less than 0.5% of the projects were prohibited. Even so, some isolated cases received national attention and thrust the ESA into the spotlight. The snail darter, a 3-in. fish, became famous in 1978 when it was listed under ESA as endangered and so temporarily halted the construction of the multimillion dollar Telico dam project on the Little Tennessee River. Thousands of jobs were expected to be lost in the Pacific Northwest when the spotted owl was listed as threatened in 1990, causing a fury among the loggers in the area.[63] The ESA seeks to protect the presently listed 1000 species and subspecies of mammals, mollusks, and plants classified as endangered in the 50 states.[73] The ESA also attempts to protect some 200 additional species classified as threatened, and using trade sanctions, intends to protect 523 foreign and oceanic species outside the U.S. Many environmentalists praise the ESA for reducing the

FIGURE 3.13 Endangered or threatened species whose numbers have increased include the whooping crane, peregrine falcon, American bald eagle, grizzly bear, and the red wolf. (From U.S. Fish and Wildlife Service (USFWS) http://fws.gov/. Whooping crane photo by Steve Hillenbrand; grizzly bear photo by Don Redfern; bald eagle photo by Robert Fileds; gray wolf by USFWS.)

extinction rate of some animal species in the U.S., and even increasing numbers in as many as 65 species. Numbers have increased for the American bald eagle, whooping cranes, grizzly bears, the red wolf, peregrine falcons, and many other species (Figure 3.13). More than 29 species have been delisted from the government's endangered species list, including the bald eagle, the peregrine falcon, and the little-known Pahrump poolfish.[74] This marks the first time in the 30+ years of ESA's existence that such a large number of species have been targeted for removal from the list. Some groups, such as the Environmental Defense Fund (EDF), hope that politics does not drive the process, as efforts are made to show Congress that the ESA is working and deserves less criticism.[74] However, many opponents of ESA argue its powers are excessive and override the property rights of the individual, threaten jobs, and interfere with economic progress. ESA is the only federal law that addresses biodiversity. It expired in 1992 and is being considered for reauthorization in the spring of 2006. Many organizations are calling for Congress to considers reforming ESA to include protections for private property owners. Bills submitted before Congress propose major changes in ESA that include (1) compensation to landowners if federal actions lower property values, (2) better scientific justification for listing species, (3) tax incentives to property owners for habitat conservation planning, and (4) removal of protection of endangered species habitat.[54,75] The idea of providing property owners incentives as part of the ESA reauthorization has gained new support based on a recent study that found populations of most endangered species in the U.S. live in a relatively few critical "hot spots" covering less than 2% of the land area, and most of the critical tracts are on private land.[76] These hot spots contain endemic species, in that they are limited to a small area, and they are being disturbed by urban development and agriculture. These hot spots are concentrated mostly in Hawaii, California, and the Southeast, mostly Florida. Spending money on these limited areas would be very cost-effective.[76]

Scientists are constantly working on new ways of protecting and rebuilding populations of endangered species. However, these techniques have greatly varying degrees of success. The cloning of Dolly the sheep presented scientists with new options for the repopulating of endangered species. At China's National Academy of Sciences, zoologist Chen Dayuan and his team has already used adult panda cells to produce a cloned embryo. Some teams are already even working on cloning cells taken decades ago from species that are now extinct.[77]

Experts now generally agree that it is important to protect large areas of landscape to protect species and keep them off the endangered species list to begin with, while permitting greater flexibility in balancing development and species protection.[76] Scientists and policy-makers are also learning that absolute bans on wildlife destruction has led to disaster in LDCs because (1) large areas of land have been destroyed by the uncontrolled growth of wildlife, (2) the wildlife is eliminated as a natural resource for the native human populations, and (3) these policies require very costly enforcement. The solution, as seen in some wildlife markets in the U.S., is to allow and regulate markets for use of these species in ways that will protect the native peoples and balance the use of natural resources.[78] In some programs, such as the one called "campfire" in Zimbabwe, the villagers actually are the owners of the wildlife and are allowed to sell permits, under some governmental control, to hunters that allow the villagers to profit while maintaining a sustainable population.[79]

FOOD SECURITY

One of the biggest debates of the 21st century concerns whether or not the world can produce enough food to feed another few billion people. In the American Association for the Advancement of Science (AAAS) meetings, the experts discussed how the world's growing population would be fed in the 21st century. About two thirds of preschool children in Southeast Asia do not get enough nourishment.[80] By 2010, about 680 million people in the developing countries will not be able to have enough food, and by 2020, about 135 million preschool children will be malnourished. Population activists, as well as some social scientists, believe the answer is no, and that the only solution is to limit the numbers of mouths to feed. However, many agronomists and other scientists say yes, provided there are ample funds for research on improved plant varieties and agricultural technologies that will increase yields.[80] Depending on whether you live in an impoverished nation or a wealthy one, there is ample room for pessimism or optimism.

Are you interested in learning more about environmental degradation and food security around the world? Visit http://www-unix.oit.umass.edu/~envhl565/weblinks.html and click on one or more of the following sites under Chapter 3:

- General Links to Environmental Degradation
  - United Nations Environment Programme UNEP

- • UNEP's Global Environment Outlook
- • Earthwatch home page
- • SD-Dimensions: Environmental Information
- • United Nations Development Programme
- • Rachel Carlson web site
- • Rainforest
  - • Spacelink — Land
  - • NASA Landsat Pathfinder Humid Tropical Forest Inventory Project
  - • Search.spacelink.nasa.gov
  - • The Amazon as Seen by NASA
  - • Jungle Canopy Crane
- • Climate
  - • World Conservation Monitoring Centre — Home Page
  - • World Resources Institute
  - • Earthwatch Institute International
  - • NASA Spacelink — An Aeronautics & Space Resource for Educators
- • Soil erosion
  - • GEO-3: The Complete Report:
  - • Food and Agriculture Organization of the United Nations
  - • Control of Soil Erosion
- • Wetlands
  - • Office of Wetlands, Oceans, and Watersheds
  - • Lost Wetlands
  - • Facts About Wetlands
- • Biodiversity
  - • U.S Fish & Wildlife: The Endangered Species Act and What We Do
  - • Endangered Species Program, U.S. Fish & Wildlife Service
- • Food Security
  - • The HungerWeb
  - • United Nations Feed the Hungry Site
  - • Second Harvest

FOOD PRODUCTION

For nearly 40 years, the world production of grain rose by more than 2% a year, but declined to scarcely 1% a year in the 1990s. This is a level that is markedly below the 1.5% annual increase in the world's population.[4] The 1997 world carryover of grain stocks remained at a low level of 55 d of consumption, which is only 4 d more than the record low of 51 d experienced in 1996. A 70-d supply in carryover stocks of grain is thought important to maintain a minimal level of food security in the world.[4] Food security is said to occur when all people have physical and economic access to the basic food they need to work and function normally.[70] This 1996 low surplus in grain drove prices of wheat and corn up, resulting in 2.5% increase in grain harvest to 1.71 billion acres and a record 1997 grain harvest of 1.84 billion tons. Even so, the world was unable to rebuild depleted stocks to a secure level.[4] The 1998 cereal production is forecast to be just slightly below this record and would be sufficient to meet consumption requirements but would remain below minimum safe levels for world security. Malnourishment effects 1 in every 3 people on earth, and about 18 million people, mostly children, die each year from starvation.[4] Thirty-seven countries are expected to have food emergencies, with Africa standing out as the continent with the most serious food shortages.[12] Countries with critical or low food security are shown in Figure 3.14. Despite these concerns, trends in recent cereal yields are expected to meet the demand the much of the world except sub-Saharan Africa based on projections through 2025.[81] Cereals are central to the human diet and represent more than 63% of calorie intake when indirect consumption of grain fed livestock is considered. Cereal production has been lagging behind the growth rate of the human population since 1984, but some areas of the world are faring better than others.[81] If one looks at regional trends, Africa and the Middle East, including West Asia and North Africa, are doing poorly in terms of food production. Africa continues to experience widespread political instability, government neglect of agriculture, an AIDS epidemic, and rapid population growth.[81] South Asia (India, Pakistan, and Bangladesh) are experiencing a rise in per capita cereal output to around 225 kg, and Southeast Asia (China, Japan, and Indonesia) has seen a marked upward shift following the late 1980s. Europe (including the former Soviet Union, and North America, including Australia, showed a marked decline in per capita cereal output. The Soviet Union experienced massive economic and political collapse that caused a major reduction in cereal output by 1995. The U.S., Canada, and Australia provided more than 20% of the global cereal harvest although they collectively represented less than 6% of the world's population. The main reason for the decline in production in these countries is linked to the European Union (EU), which flooded the world market with their government-subsidized cereal surpluses and drove the market price down.[81]

Reasons for Regional Food Shortages

Distribution of Population and Food

Today there is a huge food gap between the developed countries and most developing nations. Already this gap poses moral and political as well as economic problems of major dimensions, and if it widens, these problems will become increasingly serious.

The majority of the global population already lives in developing countries, including China, which faces the greatest food problem. Approximately 1 billion live in the poorest 30 to 40 countries — roughly the same number as in the well-fed developed nations. Moreover, the proportion of the world total living in the developing nations will unquestionably continue to grow; 98% of the annual increase in the world's population is occurring in the developing countries.

The growing food gap between rich and poor nations has been — and will continue to remain — in large part due to the differences in rates of population growth, as food output has actually been increasing somewhat more rapidly in the latter than in the former. However, because of a more rapid rate of population increase in the poor countries, their average rate of per capita increase in food output was less that one third the rate in the developed countries. Although there has generally been steady growth in the world production of most crops from the 1960s onward,

FIGURE 3.14 Countries with low and critical food security. (Adapted from Brown, L.R. and Kane, H., *Reassessing the Earth's Population Carrying Capacity*, W.W. Norton, New York, 1994.)

when viewed as per capita production, there are unequal results among countries. Food production in Africa has shown a steady increase beginning in the 1960s, with a doubling in food production, but the per capita trend has declined because of the enormous population increases. Food production fell behind population growth in 64 of 105 developing countries between 1985 and 1995. If countries are to feed the greater than 8 billion population expected by the year 2050, Africa would have to increase production by 300%, Latin America by 80%, Asia by 70%, and North America by 30%.[4] Asia and Latin America have shown significant increases in both absolute and per capita food increases, but the former Soviet Union has experienced a decline of 40% in agricultural production since 1989 (Figure 3.15).[12]

In a fundamental sense, the world food problem, viewed globally, is less than a problem of absolute shortages than a problem of unequal distribution between the rich and poor countries. Moreover, the inequality of food distribution is a basic problem within many developing countries.[81]

On the other hand, the production trends do not support a completely optimistic view. The growth rate of agricultural production worldwide has slowed from a 3.0% growth rate in the 1960s to 2.3% in the 1970s, and less than 2.0% in the mid-1990s.[81] Growth rates in cereal production have been even more disturbing, declining from 2.8% in the 1960s to nearly 2.1% in 1992 (Figure 3.16).[10] So, though actual yields in agriculture and cereal may have increased in the 93 developing countries studied, the rate of increase has slowed.

Pessimistic View

Some authors such as Lester Brown, President of the Worldwatch Institute, predict dire consequences for future food production. He maintains that agricultural technologies are going unused, fisheries and rangelands are approaching the limits of biological productivity, freshwater is becoming increasingly scarce, fertilizer applications are becoming less effective, and social disintegration is occurring in many LDCs.[82] There is also concern that the tropical and subtropical environments are so fragile in many LDCs that they will be unable to sustain significant food production increases. About 45% of the potential cropland in sub-Saharan Africa and Latin America is under forest or in protected areas.[83] Nearly three quarters of the potential cropland is so fragile that development would result in dramatic losses in biodiversity, increased carbon dioxide emissions, decreased carbon storage capacity, and high financial and ecological costs.[84] Many experts agree that the potential to expand cropland area is quickly disappearing in most areas of the world.[85] China represents a dramatic example of this problem. More than 21% of the world's population lives in China, and its arable land covers less than 7%, resulting in a per-person share of 0.086 hectares, or only 25% of the world average.[86] The acreage for agricultural development is diminishing while existing farmland is being overcultivated, polluted, and nearly one third of the land is threatened by soil erosion.[73] Overgrazing and reclamation of farmland has turned a quarter of the country's grassland into desert, with more than 300 cities or towns having a shortage of water.[86] In many of the LDCs, agriculture is hampered by unfavorable growing conditions. There is a continuing threat of civil strife, the impoverished cannot afford to pay for food, and

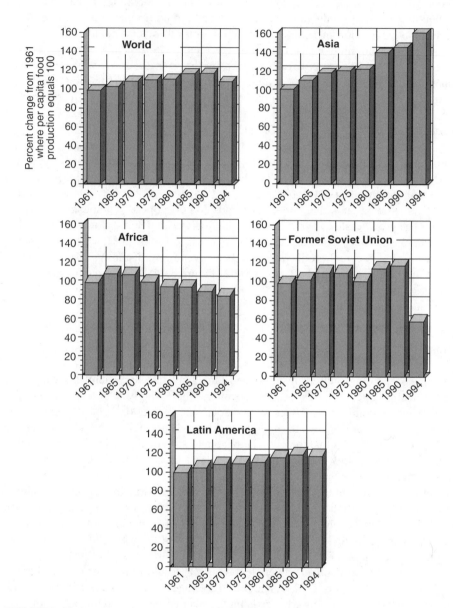

FIGURE 3.15 Some trends in food production on a per capita basis. (Adapted from FAO, http://www.fao.org [WAICENT/faoinfo/econ-tie/giews/English/fo/fo9804/httoc.htm].)

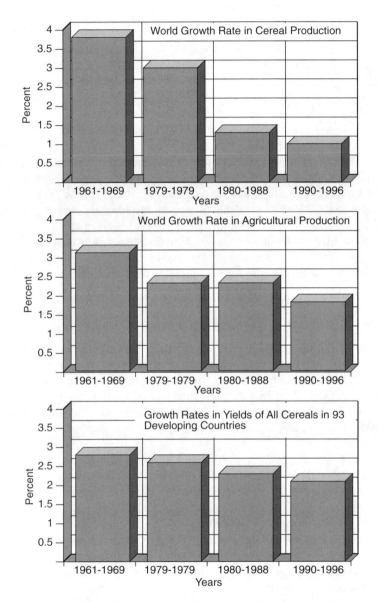

FIGURE 3.16 Growth rates of world agriculture and world cereal production compared to trends in actual cereal yields in 93 developing countries. (Adapted from FAO, http://www.fao.org [WAICENT/faoinfo/econ-tie/giews/English/fo/fo9804/httoc.htm].)

governments cannot provide the policies or infrastructure to the rural poor that promote domestic agriculture production.

Optimistic View

A number of organizations, including FAO of the United Nations, the World Bank, and the International Food Policy Research Institute (IFPRA), which is a component of the CGIAR, have published studies looking at food production estimates from 2010 through 2030.[81,87–89] The studies and projections point to a number of hurdles with respect to worldwide food production, but are similar to the FAO prediction that expected production increases will be able to meet the effective demands of a rising world population providing there are substantial investments in agricultural research.[12,87] Effective demand refers to that "demand level" corresponding to purchasing power without regard to food requirements. Unfortunately, the sharp rise in number of food-related crises has shifted priorities. In 1986, the World Food Programme directed 75% of its available monies to development projects. By the mid-1990s, more than 85% of its resources were going to meet emergency and refugee needs as those affected by natural and political disasters rose from 44 to 175 million people in the same years.[90] Admittedly, this last revelation is not optimistic, but must be included to realistically balance the view that we can meet the world's effective demand for food.

SOURCES

Where will the food come from to feed the world's billions? Increases in food supply must come from one or more of the following sources: (1) increases in yield (tons per acre), (2) increases in arable land placed under cultivation, and (3) cropping intensity (fewer fallow periods or more than one crop per year or field). The FAO projects that 66% of food production will come from increased yields, about 21% from arable land expansion, and the remaining 13% from increasing cropping intensity (Figure 3.17).[12]

Worldwide gross agricultural production is expected to increase in the next 20 years, but its growth will likely be at a slower rate of 1.8% per year compared to the previous decades because population growth rates and food demand have slowed in developing countries, and there is inadequate income growth in many countries to purchase large supplies of food.[91]

There are scientists who believe that ability to expand cropland is limited, and that it is disappearing in many areas of the world. The potential for increasing agricultural land is limited by (1) the significant costs of developing an infrastructure in remote areas, (2) the lesser productivity of these alternative areas, and (3) the trade-offs in environmental destruction of sensitive ecosystems.[85]

The FAO projection of increased yields is based on their assessments that there is room for improvement by farmers who are not achieving peak yields. Many countries are showing only half the yield (3.7 metric tons) being experienced in the best-performing countries (6.7 metric tons).[85] The FAO projects that South Asia will make the strongest contribution of increased yields in crops of about 82%, whereas Latin America and Africa will experience less than 53% yield increases.[12]

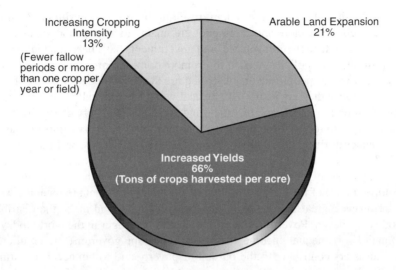

FIGURE 3.17 Projected sources of food increases. (Adapted from FAO, http://www.fao.org [WAICENT/faoinfo/econ-tie/giews/English/fo/fo9804/httoc.htm].)

This optimism is not shared uniformly. It is argued that increasing numbers of farmers in developing countries are already using crops with the highest genetic yield potential, and that rice crops at experiment stations in Asia have leveled off for a number of years.[85] Further, the high-input approach of using genetically programmed, high-yielding varieties has required large amounts of water, pesticides, and synthetic fertilizers. This approach can increase yields, but may also cause the loss of genetic stocks and reduction in soil quality. Further, many LDCs do not have the funds necessary to support the increased costs of fertilizers, pesticides, and seeds. Alternative models are being evaluated and promoted that are more friendly to the environment. Integrated farming systems are being proposed that employ sustainable crop intensification strategies. These strategies include (1) improved irrigation systems, (2) structured water pricing to reduce overuse, (3) alternative rotation of crops, (4) selective pesticide use, (5) use of pest-resistant varieties, (6) improved soil testing and fertilizer application, (7) regional crop-breeding programs, and (8) more education of farmers.[12] High-input and crop intensification strategies may not be viable in fragile ecosystems because of drought risk, inadequate support structures, and lower yield response. In these areas, alternative strategies might involve (1) erosion and moisture control, (2) recycling nutrients, (3) and integrating annual and perennial crops in what is known as *mixed farming*.

FOOD SECURITY

Worldwide

Temporary deficiencies in food and famine may occur in the presence of droughts, floods, political strife wars, or poor harvests. These events are on the increase and produce significant suffering in the short term. However, chronic undernutrition is a more difficult and pervasive problem, resulting in a food security crisis in many LDCs. In such areas, more and more poor are unable to afford food, conditions for agriculture become more unfavorable, and governments have been unable to provide the infrastructure or policies needed to promote domestic production.[12] The lack of food security means people are unable to acquire the basic food they need to function normally because they are restricted physically and economically. Civil strife, inadequate marketing, poor storage facilities, and inadequate delivery capabilities reduce physical access to food.[12] Protein-energy malnutrition (PEM) is the most devastating form of undernutrition and is the one most referred to when world hunger is considered.[92] The world production of food calories is 17% greater than 1975, even with an increase of population by 70%. This means that there is at least 2720 kilocalories (kcal) of food per person per day, which is sufficient to meet the world's needs. However, great disparities in income and arable land make the distribution of food very uneven. Poverty is the primary cause of hunger in the world today. The combination of unequal income distribution, corrupt governments, conflict, and natural disasters contribute to the burden of poverty and to hunger. It is estimated that 1.08 billion people in developing countries live on less than $1 day, with nearly 800 million people suffering from chronic hunger.[92]

Food insecurity is most pronounced in sub-Saharan Africa, where more than 300 million people may be undernourished by the year 2010 with a per capita food supply of 2170 calories per day, which is about 60% of the per-person calories available to people in developed countries. In other areas of the developing world, there is reason to be more optimistic. The growth rates and undernutrition are declining in East Asia, which is rapidly becoming industrialized. It is estimated that by the year 2010 the number of undernourished in South Asia will be 4%. Latin America is seeing progress in that its undernourished population is projected to decline by more than half since 1988 to a low of 6% by 2010.[12] Overall, chronic undernutrition is expected to decline in the 93 developing countries studied from 20% in 1988–1990 to 11% in 2010.[12] However, many countries will experience shortfalls in food production. The World Bank estimates that they will be importing 15% of their grain by 2101, and net imports to LDCs are expected to increase from 90 to 160 million tons in the years from 1990 to 2010.[12]

Although there is a hypothetical potential to meet the worldwide demand for food, the reality is that food crises in Africa, South Asia, and other countries are likely to persist. The IFPRI has made recommendations that is hoped will lead to sustainable land use, efficient agricultural systems to produce low-cost food, and sufficient food that is physically and economically accessible to every person.[93] For these goals to be met, IFPRI recommended that (1) governments of LDCs need to be strengthened; (2) agricultural and extension services need expansion; (3) agricultural yields need to be intensified with methods that sustain natural resources; (4) low-cost, efficient agricultural input and output markets need to be developed; and (5) international assistance needs to be expanded.[93] The target set at the 1996 World Food Summit was to reduce the number of undernourished people estimated in 1990–1992 at 824 million to half this number by 2015. That number has only slightly declined to 815 million as of 2002, with the only real progress being made in Asia, Latin America, and the Caribbean.[92] The trends are moving in the wrong direction for many other developing countries. The goal of halving hunger by 2015 appears very unlikely with new technologies and significantly greater investment.[94] It is estimated that 10 preschool children continue to die every 60 sec from malnutrition and that this number has not changed since 1980 even though summits are held and programs are promised.[94]

Hunger in America

It seems unlikely, even absurd, but the American economy continues to do well just as a hidden crisis in poverty and hunger threatens millions of Americans. The rate of hunger in America grows steadily worse in the face of this prosperity, creating our own food insecurity problems as the gap widens between the rich and the poor, the secure and the threatened.[95] Second Harvest, a nationwide network of food banks, released a study in 2006 that revealed more than 27 million Americans, almost 36% of them under 17 years of age, resort to using food distribution programs such as soup kitchens and food pantries. This represents nearly one of every ten Americans who at least once in a while make use of these services.[96] In a report from the Tufts University Center on Hunger and Poverty and Nutrition Policy declared that nearly 35 million Americans live in hungry or food-insecure households.[95] Food insecurity has increased by 3 million households, with slightly over 12% of all U.S. households being food insecure.[97] About 4.4 million suffered from food insecurity that was sufficiently severe to classify them as "hungry" under the USDA conservative measure.[97] Seventy-three percent of 25 million people served annually by Second Harvest are food insecure as listed in their 2006 report.[96]

The U.S. Conference of Mayors (USCM) is expecting the demand for food aid to rise in 92% of American cities, with request for emergency assistance rising 16% in 1997 alone. Nearly 20% of these requests were unmet.[95] The problem of feeding the rising number of hungry Americans is being revealed by a number of charitable and government organizations. The USCM reported a 16% increase in food emergency requests based on a survey of 29 cities. In December 2004, the USCM reported that requests for emergency food assistance increased in 26 of the 27 cities surveyed and that the requests increased by an average of 14%.[95] Catholic Charities reported a 14% rise in people fed from 1995 to 1996 as they served 5.6 million people. The Congressional Hunger Center suggests the problem is being underestimated and that some areas of the country are reporting greater than 42% increases in food aid demand.[95]

What is the profile of a person who makes use of food banks? Do you imagine a single man? Perhaps you think of an uneducated person, or a substance abuser? The face of hunger has changed to reveal that the food-insecure are children, mothers, the newly unemployed, the underemployed, and the working poor.[95,96] The Second Harvest report shows that 26 million Americans received food from food banks in 1997. The majority were female (62.4%), with 38% being under 17 years of age, and 16% over 65 years of age (Figure 3.18). Most were white (47.1%) and the majority had a high school diploma (60%), with nearly 19% having had some college education.[95] Two out of the 5 households using charity have at least one person employed, and in 50% of these cases, the employment is full-time.[96] The numbers of working families requiring food aid is increasing as a percentage, but the majority

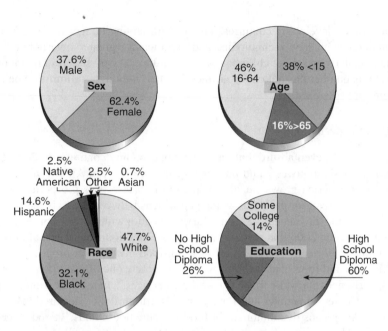

FIGURE 3.18 Demographic profile of Americans using food banks and pantries. (Adapted from Roberts, L. and Roberts, S.V., Hunger: A Startling Crisis, *USA Weekend*, March 27–29, 1998, p. 4.)

of people requiring aid still include those with disabilities, single mothers, and seniors on fixed incomes.

There is probably no single reason why more and more people are seeking food aid in the presence of a strong American economy. Policy analysts in the Heritage Foundation, a conservative think tank, argue that the more programs you have to hand out food, the more people will use them. They also maintain that there is no evidence for widespread hunger or undernutrition in America.[97] This view is in contrast to that of J. Larry Brown, who in 1994 as director of the Center on Hunger, Poverty, and Nutrition at Tufts University in Medford, MA, claimed that hunger in America could be ended in 6 to 8 months by expanding the food stamp program and increasing funding for targeted programs in WIC (Special Supplemental Food Program for Women, Infants, and Children).[97] Unfortunately, the study conducted for Second Harvest in the summer of 1997 by Larry J. Brown and Dr. J.T. Cook from the Tufts University National Center on Hunger, Poverty, and Nutrition revealed that food stamp cuts in the welfare bill from 1997 through 2020 will amount to nearly $28 billion, or approximately 24 billion pounds of food.[99] In order to meet this shortfall, programs such as Second Harvest would have to increase capacity by 425%. Their most optimistic scenario projects a 10% growth rate per year over the next 6 years, and would meet less than a third of the food aid required. Food insecurity in America is already a growing problem, and additional government cuts are not likely to result in increased personal giving. Christine Vlademiroff, President and CEO of Second Harvest, states that private donations are highest in areas where state and local governments give generously to the same issues that charities address.[99]

Obtaining enough food for its citizens to pursue a healthy and active lifestyle should be a primary objective of any government, especially the richest and most powerful in the world. The millions of food-insecure households in the U.S. cannot reach this goal and must often make choices between paying for other necessities such as housing, medical care, or fuel for heating, or transportation.[98] Forty-two percent of the clients served by Second Harvest report having to choose between paying for food and paying for utilities or heating fuel.[96] The recent sharp rise in fuel costs will likely worsen this problem and force even greater numbers to seek assistance.

REFERENCES

1. Motava, W.J., Contents Under Pressure, *E. Magazine*, November–December 27, 1996.
2. United Nations Environment Program, Global environment outlook, Environmental Health, April 22, 1997.
3. Kaplan, R., The Coming Anarchy, *Atlantic Monthly*, February 1994, p. 44.
4. Fornos, W., 1997 World population overview, The Population Institute, December 30, 1997, www.population institute.orgloverview97.html.
5. Barney, G.O., *The Global 2000 Report,* New York, Pergamon Press, 1980 (Vol. 1, Summary Report; Vol. 2, Technical Report; Vol. 3, Government Global Model), 360 pp.
6. Budiansky, S., 10 Billion for Dinner Please, *U.S. News and World Report*, September 12, 1997, p. 57.
7. Turk, J. and Turk, A., *Environmental Science*, 4th ed., Saunders College Publishing, Philadelphia, PA, 1988, chap. 9.
8. The Terminator Technology: New Genetic Technology Aims to Prevent Farmers from Saving Seed, RAFI (Rural Advancement Foundation International) Communiqué, March–April 1998.
9. Gordon, B., Terminating the "Terminator" Seed: Broken Promises and Sterile Seeds Could Alter the Lives of Food Producers, *Utne*, March 2, 2006, http://www.utne.com.
10. World Resources Institute (WRI), Chapter 10: Food and Agriculture, A Guide to the Global Environment, World Resources 1996–1997, http://www.wri.org/wri/wr-96-97/fa-txt2.html.
11. Alexandratos, N., Ed., *World Agriculture: Towards 2010, An FAO Study*, John Wiley and Sons, Chichester, U.K. and Food and Agriculture Organization of the U.S., Rome, 1995, p. 38.
12. Food and Agricultural Organization (FAO) of the United Nations, Food Outlook: Global Information and Early Warning System (GIEWS) on Food and Agriculture: Rome, April, No. 2, 1998, http://www.fao.org(WAICENT/faoinfo/econon-tie/giews/english/fo/fo9804/httoc.htm.
13. Dasgupta, P.S., Population, Poverty and the Local Environment, *Scientific American*, February 1995, p. 42.
14. MSNBC.com, High global oil prices seen through 2007: U.S. government makes prediction on demand outlook, geopolitical worries, Reuters, May 3, 2006, http://www.msnbc.msn.com/id/.
15. Simon, J., *The Ultimate Resource*, Princeton University Press, Princeton, NJ, 1981, 495 pp.
16. Simon, J. and Kahn, H., *The Resourceful Earth*, Basil Blackwell, Oxford, 1984, p. 585.
17. Ehrlich, P., Too Many Rich People: Weighting Relative Burdens on the Planet, *Our Planet*, 6, 1994, p. 31.

18. AP (Associated Press): Geneva, U.N. gap widening between countries, *Daily Hampshire Gazette,* September 25, 1, 1997.
19. Glantz, M.H., Is the Stork Outrunning the Plow?, United Nations Environment Programme: Network Newsletter, 9, 2, 1994.
20. Harden, G., The Tragedy of the Commons, *Science*, 162, 1968, p. 243.
21. Runnels, C., Environmental Degradation in Ancient Greece, *Scientific American*, March 1995, p. 96.
22. Nierenberg, D. and MacDonald, M., The Population Story ... So Far, *World Watch*, September–October 2004, pp. 14–17.
23. Repetto, R., Ed., *The Global Possible Resources. Development and the New Century,* Yale University Press, New Haven, CT, 1985, p. 538.
24. United Nations Environmental Program (UNEP), Global environmental outlook, Environmental Health, April 22, 1997.
25. Greenpeace USA, Satellite provides dim view of amazon, Briefings, November–December 1988.
26. AP, Scientists Hopes are Borne Out: Rain Forests and Carbon Dioxide, *New York Times*, November 8, 1995, p. A4.
27. Farnsworth, N.R., New Drugs from the Rainforest, *World Health*, 2, March–April 1996, p. 49.
28. Earth's Green Lungs Begin to Fade, *International Wildlife*, 25, September–October 1999.
29. Iwokrama, International Centre for Rain Forest Conservation and Development, Global Rain Forests, May 9, 2006, http://www.iwokrama.org/forest/globrainforests.htm.
30. National Aeronautics and Space Administration (NASA), Tropical deforestation and habitat degradation in the Brazilian Amazon Basin, The Earth Observing System Educator's Visual Materials, 1998, http://eospso.esfc.nasa.gov/cos-edu.pack/p24.httnl.
31. Moeller, D.W., *Environmental Health*, Harvard University Press, Cambridge, MA, 1997, chap. 19.
32. Butler, R., Deforestation in the Amazon, 2004, Mongabay, May 5, 2006, http://www.mongabay.com/brazil.html.
33. Baretto, P., Jouza, Jr., C., Mogueron, R., Anderson, A.M., and Salomao, R., Humand pressure on the Brazilian Rainforests, March 2006, World resources Institute: Imazon, Amazon Institute of People and the Environment, March 2006, http://pdf.wri.org/amazon_glossy_overview.pdf.
34. Hernandez, S., Environment: Afro-Asian trees to the rescue in Central America, InterPress Service English Newswire, 01, 08, 1996, http://www2/efi-brary.com/id/2525/getdoc.cgi?id=53502173XOX634&Oqoo2D60&fores=R L.
35. Brevik, J.J., Global Environmental Outlook, *Environmental Health*, April 22, 1997.
36. Haniffa, A., Degradation Hits Human Security, India Abroad, 12, 106, 1996, http://www2.elibrary.com/id/2525/getdoc.c.cos&form=ri.&pubname=lndia-Abroad&puburl=O.
37. Depth News Asia, Action for Tropical Rainforests, *World Press Review*, August 13, 1993.
38. Brown, L.R. and Woff, E.C., Soil Erosion: A Quiet Crisis in the World Economy, Worldwatch Paper, Worldwatch Institute, Washington, D.C., 1984, p. 49.
39. U.S. Dept. of Agriculture (USDA): Analysis of Policies to Conserve Soil and Reduce Surplus Crop Production, Washington, D.C., Agricultural Economic Report No. 534, 21, 1985.
40. Doyle, R., Soil Erosion of Cropland in the U.S., 1982–1992, *Scientific American*, October 1996, p. 34.
41. Stutz, B., The Landscape of Hunger, *Audubon*, 2, 95, March–April 1993.

42. Filipov, D., Desert's drift buries Russian region's dreams, *Boston Globe*, 9, 27, A26, 1997.

43. Brazil Embassy, What is desertification?, Science and Technology Section, National Research Network's Brazilian Regional Center, February 30, 1997, http://brasil.emb.nw,dc.us./nuna/deserts/ingles/se/~edesert/desertmi.html#dm.

44. National Aeronautics and Space Administration (NASA), Desertification Atlas (Africa) Maps I — 1, Global Change Master Directory, October 24, 1997, http://gcmd.gsfc.gov/cgibin/detdy?=display+dif&entry=NBld0270/101.

45. EPA-OWOW (Environmental Protection Agency, Office of Wetlands, Oceans, and Watesheds), Facts about Wetlands, August, 25, 1997, http://www.epa.gov/owow/wetlands/facts/types.html.

46. Secraig, B., What is a Wetland?, *Sierra*, May–June 1996, p. 44.

47. Enteractive, inc., Disappearing wetlands, Earth Explorer, February 1, 1995, http://www3.elibrary.com/id/2525/getdoe.c.:. lorer&puburl=http*c*//www.enteractive.com.

48. Adler, T., Scientist Dispute Legislator's Take on Wetlands, *Science News*, 148, 56, 1995.

49. Cushman, J.H., From Wetlands to Asphalt, a Parcel at a Time, *New York Times*, October 27, 22, 1996.

50. Stevens, W.K., Prairie Ducks Return in Record Numbers, *New York Times*, October 11, 1, 1994.

51. Dahl, T.E., Status and Trends of Wetlands in the Conterminous United States 1998 to 2004, U.S. Department of the Interior; Fish and Wildlife Service, Washington, D.C., 2006, 112 pp.

52. Marina, M., Progress Threatens Brazil's Vast Wetlands: Foes Say Canal Would Drain Pantanal, CNN Interactive, online Internet, April 23, 1997, http://www.cnn.com/earth/9704/23/brazil.wetlands.

53. Erwin, D.H., The Mother of Mass Extinctions, *Scientific American*, July 1996, p. 72.

54. Callahan, J.R., Vanishing biodiversity, *Environ. Health Perspect.*, 104, 4, 386, 1996.

55. Hilchey, T., Biodiversity Study Sees More Species in Danger, *The New York Times*, November 14, Science Section 1, 1995.

56. Worldwide Fund for Nature, Climate Change and Biodiversity Conservation: Worldwide Loss of Biodiversity, Executive Summary, 1996, chap. 3, http://www.panda.org/ida/climate/cIimate.3/Pagel.htmI.

57. Doyle, R., Threatened Mammals, *Scientific American*, January1997, p. 32.

58. Doyle, A., Humans Spur Worst Extinction since Dinosaurs, Reuters, Know Now, Monday, March 20, 2006, http://go.reuters.com.

59. Ananthaswamy, A., Earth Faces Sixth Mass Extinction, NewScientist.com News Service, March 18, 2004, http://www.newscientist.com/article.ns?id=dn4797.

60. Cushman, J., Freshwater Mussels Facing Mass Extinction, *New York Times*, October 3, 1995, p. C1.

61. Cook, R.M., Sinclair, A., and Stefarmson, O., Potential collapse of North Sea cod stocks, *Nature*, 358, February 6, 1997.

62. Allen, S., Trying To Do the Right Thing, *The Boston Globe*, September 26, 27, 1995.

63. Butler, S., Rod? Reel? Dynamite?, A Tough-Love Program Takes Aim at the Devastation of the Coral Reefs, *U.S. News and World Report*, November 25, 1996, p. 56.

64. Lemonick, M.D., Animal Genocide, Mob-Style, *Time Magazine*, November 14, 1995, p. 77.

65. Jones, A., Stowaway Snakes Deplete Bird Populations on Guam, CNN Interactive, 1, November 9, 1996.

66. Luoma, J., Boom to Anglers Turns Into Disaster for Lakes and Streams, *New York Times*, November 17, 1992, p. 3.

67. Stevens, W., Bizarre Parasite Invades Trout Streams, Devastating Young Rainbows, *The New York Times*, September 19, 1995, p. C9.
68. Wilson, E.O., *The Diversity of Life*, Harvard University Press, Cambridge, chap. 1.
69. Kaiser, J., Vanishing Zooplankton, *Science News*, March 11, 1995, p. 157.
70. AP, Many Reports of Deformities among Frogs are Puzzling, *The New York Times*, 33, October 13, 1996.
71. International Union for Conservation of Nature and Natural Resources (IUCN), Global Amphibian Assessment, May 2006, http://www.globalamphibians.org.
72. Chadwick, D.H., Dead or Alive: The Endangered Species Act, *National Geographic*, March 21, 1995.
73. U.S. Fish and Wildlife Service, Threatened and Endangered Species System, Summary of Listed Species, May 8, 2006, http://ecos.fws.gov/tess_public/Boxscore.do.
74. AP (Associated Press), Animals to come off list, *Daily Hampshire Gazette,* May 1998, p. 1.
75. Biddison, J., De-Prioritizing people, Townhall.com, March 28, 2006, http://www.townhall.com/.
76. Stevens, W.K., "Hot spots" for American endangered species cover surprisingly little land, *New York Times*, January 1, 1997, p. 1.
77. Mzurek, R., Back from the Dead, *New Scientist*, October 9, 1999, http://www.new-scientist.com/article/ng6422071.800.html.
78. Budiansky, S., Killing with Kindness, *U.S. News and World Report*, November 25, 1996, p. 48.
79. Satchell, M., Save the Elephants: Start Shooting Them, *U.S. New and World Report*, November 25, 1996, p. 52.
80. AP (Associated Press), Experts Debate Limits of Earth's Resources, CNN Interactive, October 27, 1996.
81. Dyson, T., World food trends and prospects to 2025, *Proc. Natl. Acad. Sci. Colloq.*, 96, 11, May 25, 1999, pp. 5929–5936.
82. Brown, L.R. and Kane, H., *Reassessing the Earth's Population Carrying Capacity*, W.W. Norton, New York, 1994.
83. The World Bank, World Development Report 1992, Washington D.C., 1992, p.135.
84. Ehrlich, P.R. et al., Food Security, Population and Environment, *Popul. Dev. Rev.*, 19, 1, 7, 1993.
85. Pinstrup-Andersen, P. and Pandya-Lorch, R., Alternating Poverty, Intensifying Agriculture, and Effectively Managing Natural Resources, Food Agriculture and the Environment Discussion Paper No. 1, International Food Policy Research Institute, Washington D.C., 1994, p. 6.
86. Xin, W., Population vs. Development: Challenge of the New Century, *Beijing Review*, May 1–7, 1995, p. 12.
87. Food and Agriculture Organization of the U.S. (FAO), The State of Food and Agriculture 1994, FAO, Rome, 1994, p. 11.
88. Mitchell, D.O. and Ingco, M.D., 'Ibe World Food Outlook, The World Bank, Washington D.C., 1993.
89. Rosegrant, M.W., Agcaoili-Sombilla, M. and Perez, N.D., Global Food Projections to 2020: Implications for Investment, International Food Policy Research Institute (IFPRI) Food, Agriculture and the Environment, Draft Discussion Paper, IFPRI, Washington, D.C., 1.
90. Webb, P., A Time of Plenty, A World of Need: The Role of Food Aid in 2020, International Food Policy Research Institute (IFPRI) 2020, Brief No. 10, Washington D.C., 1995, p. 2.

91. Pimental, D. et al., Environmental and Economic Costs of Soil Erosion and Conservation Benefits, *Science*, 267, 5201, February 24, 1995, p. 1117.
92. Hunger Notes, World Hunger Facts 2006, World Hunger Education Service, updated January 21, 2006, http://www.worldhunger.org/articles/Learn/world%20hunger%20facts%202002.htm.
93. International Food Policy Research Institute (IFPRI), A 2020 Vision for Food, Agriculture, and the Environment, Draft Paper (117PRI), Washington D.C., June 1, 1995.
94. Lichtarowicz, A., Millions More Starving By 2015, *BBC News*, St. Louis, MO, Friday, February 17, 2006.
95. Roberts, L. and Roberts, S.V., Hunger: A Startling Crisis, *USA Weekend*, March 27–29, 1998, p. 4.
96. America's Second Harvest, Hunger in America 2006, 2006, http://www.hungerinamerica.org.
97. Shapiro, L. and Rosenberg, D., How Hungry is America?, *Newsweek*, March 14, 1994, p. 59.
98. Food research and Action Center (FRAC), Hunger and food insecurity in the U.S., September 10, 2005, http://www.frac.org/html/hunger_in_the_us/hunger_index.html.
99. Brown, L.J. and Cook, J.T., Food stamp cuts bring grim forecast, Second Harvest Update, Winter, 1997. http://www.secondharvest.org/websecha/e-uw97tuft.htm.

4 Environmental Disease

CHAPTER OBJECTIVES

A student, after reading this chapter, should be able to:

1. Discuss and define the concepts of environment and disease.
2. List and explain the factors influencing environmental disease, including infectious disease, physical and chemical injury, ionizing radiation, developmental disease, neoplastic disease, and nutritional disease.
3. Explain the process of genetic replication in mammalian cells. List the major genetic components and discuss their primary actions and features.
4. Describe and discuss the major mechanisms of developmental and genetic disease.
5. Describe and provide a schematic of the new processes in recombinant-DNA and genetic screening techniques.
6. Discuss and describe the new concepts on the origins of cancer, risk factors, and current trends.

INTRODUCTION

Hardly a day passes without a news media report on the hazards of chemicals in our environment producing a possible cornucopia of diseases ranging from cancer, reproductive disorders, failed pregnancies, birth defects, asthma, emphysema, and any number of other diseases causing premature death or chronic illness. Some scientists report that chlorinated organic and heavy metals already abound in our bodies, and are so ubiquitous that even the most remote life-forms in the Arctic oceans demonstrate measurable levels in their bodies. Some of these chemicals have been implicated in developmental diseases in amphibians. Chemicals such as certain types of pesticides can mimic natural regulatory chemicals in animals. Mimics of the chemical retinoic acid have been linked to deformities in amphibians and fish, resulting in an increasing incidence of extra and missing limbs.[1] We are, according to some, risking our health, jeopardizing our lives, and threatening the life-supporting ecosystems on which humanity depends by the careless use and disposal of toxic chemicals.[1]

We do all of this, it is reported, because we continue to manufacture and disperse billions of tons of over 70,000 potentially toxic chemicals into the biosphere without regard for nature or an appreciation of the danger that these chemicals can bring to our earth or to us. Exposure to chemicals in pesticides such as MPTP have been associated with the development of Parkinson's disease. Parkinson's disease had been previously thought of as a purely genetic disease.[2] This new association with environmental pollution may be the first in a series of diseases previously thought of as genetic, but which may have an environmental component. Diseases such as heart disease, coronary artery disease, certain types of cancer, obsessive-compulsive disorder, and Tourette's syndrome may all be associated with an environmental cause.[3]

The possibility that manufactured and discarded chemicals in our environment could potentially produce harm erupted into the public consciousness in 1978 when New York State Commissioner of Health, Robert Whalan, declared Love Canal in New York a health emergency with his document, "Love Canal: Public Health Time Bomb."[4] A firestorm of activity lasting 2 years revolved around this revelation, including EPA (Environmental Protection Agency) reports, nongovernment studies and reports, and news media events. The resulting emotional intensity accelerated the condition to a point where the federal government was forced in 1980 to relocate more than 700 Love Canal homeowners at a cost exceeding $30 million. A sensitized public, now critical of government and environmentally astute, wanted answers and immediate resolution in December 1982, when local flooding spread dioxin throughout Times Beach, MI. The dioxin was a contaminant in waste oil sprayed as duct control at a horse arena and onto unpaved roads in the mid-1970s. Soil levels, after flooding, were up to 1000 times higher than federally recommended safe levels, and in an effort to protect the public health and calm angry, frightened citizens the federal government ordered an evacuation of Times Beach at a cost of $33 million. Alarmed, frightened, and on guard, the public became responsive to revelations in the news media about chemicals and radionuclides in our midst that could cause potential harm.

Since the 1980s, popular magazines such as *Time*, major newspapers led by the *New York Times*, a myriad of television shows, and numerous books have warned against the evils of polluting our land. Well-known people such as Ralph Nader, Lester Brown, and Sam Epstein have authored an array of books vigorously challenging humanity to stop polluting our environment with synthetic chemicals. These and other writers speak of "chemical violence ... approaching epidemic proportions,"[5] "the merchandising of poison ... a huge global growth industry,"[6] and "toxic chemicals (that) threaten every man, woman, and child."[7] Environmental advocacy has found its way into the television airwaves in documentaries such as "The Poisoning of America" narrated by Hugh Downs.[8] Even cartoons have taken up the environmental protection flag, and children watch as corporate villains threaten the health of innocent citizens with flagrant disregard for the proper disposal of festering and foul-smelling ooze in their bid for higher profits.

What are the messages given through the public media about toxic chemicals? Among the more prominent messages are that (1) exposure to toxic chemicals has dramatically increased the risk of cancer, (2) common household and agricultural chemicals are causing many human diseases and death, (3) polluted air and water are major sources of disease risk, and (4) environmental chemicals are interfering with the reproductive process in humans and producing harmful effects in the fetus and young children.[4] The major health concerns among many Americans revolve around what they hear about on television or radio talk shows, or read in newspapers and in magazines. What Americans have read is that cancer is on the increase,[9,10] that cancer-causing chemicals or carcinogens pervade our food and drinking water,[11] and that these are even spread onto lawns in the form of pesticides and fertilizers.[12] Sometimes, chemicals introduced to help the environment have unforeseen detrimental effects on the environment. MTBE (methyl tertiary-butyl ether) was mandated to be added to automobile fuel to aid in more efficient burning. This was meant to reduce the amount of toxic chemicals released, as well as lessen the amount of fuel burned altogether. Recently, MBTE has been found in increasing amounts in drinking water supplies throughout the country, including Long Island, NY. MBTE has been classified as a possible carcinogen by the EPA.[13]

Since the 1970s, both the popular media and the professional, peer-reviewed articles claim that nearly 90% of cancers can be traced to environmental causes and so may be largely preventable. Genetic factors by themselves (heredity) are said to explain only about 5% of cancers, whereas the remainder can be attributed to external or environmental factors that act in conjunction with acquired and genetic susceptibility.[14] The message is said to be optimistic in that exposure to environmental carcinogens is theoretically preventable, so that cancer risk can be modifiable and subsequently reduced. Based on what the public hears and reads in the news media and in some professional writing, it is easy to understand why many people associate the major cancer risks with synthetic chemicals released in the environment. This, however, would be largely incorrect. To explain why this is so, we need to begin with definitions of the terms *environment* and *disease* — the subject of this chapter.

DEFINING THE TERM ENVIRONMENT IN RELATION TO DISEASE (CANCER)

Koren defines environment as the "sum of all surrounding physical conditions in which an organism or cell lives, including the available energy in living and nonliving materials,"[15] though it is defined elsewhere as "the surrounding conditions or influences that affect an organism or the cells within an organism."[16] Based on such definitions and the pervasive use of the words "environmental toxins or chemicals" in the media, it is little wonder that most people interpret this to mean cancer-causing chemicals in our air, food, water, and workplace. However, the use of the word "environment" in relation to cancer causation was originally intended to be much broader and is thought to have originated with Dr. John Higginson, who was the founder and director of the World Health Organization's (WHO) International Agency of Research on Cancer. When he reported that nearly two thirds of cancers had an environmental cause, he was "… considering the total environment, cultural as well chemical … mode of life … what surrounds people and impinges on them."[17] Epidemiologists and other cancer researchers use the word "environment" to include almost everything that is not inherited. The scientific community has come to learn that many factors contribute to the risks of cancer, including personal habits and cultural behaviors. The term "environment," in its broadest definition, encompasses lifestyle-associated factors, including smoking, diet, alcohol consumption, sexual and reproductive patterns, workplace, infections, food additives, and pollution, along with the strictly physical environment.[1]

Some of the major factors associated with cancer risk are presented in Figure 4.1. Diet represents one of the highest risks associated with cancer. Food-eating patterns high in fat and low in fruits, vegetables, and grains increase risks of colon, uterus, and breast cancers. An unhealthy diet may contribute up to 35% of all cancers. Diet and other risk factors are discussed in more detail later in this chapter under the subtitle Cancer. The use of tobacco products is thought to contribute to 30% of cancers (Figure 4.1).1 High risks of cervical cancer are associated with certain sexual behaviors such as having sexual intercourse early in life, having multiple sexual partners, or contracting specific sexually transmitted diseases.[18] Estimates of cancer risk associated with reproductive behavior suggest that such risks account for 1 to 13% of all cancer deaths, with the best estimate being 7%. Medicines and some medical procedures are also included as cancer risks; for example, diethylstilbestrol (DES) caused reproductive cancers in the offspring of women who used the drug

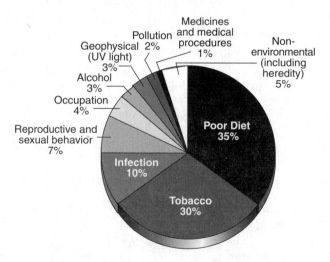

FIGURE 4.1 Proportions of cancer deaths associated with various environmental factors. These are best-estimate figures. (Adapted from *Morbidity and Mortality Weekly Report*, June 19, 1992, p. 47.)

during pregnancy to prevent premature births. DES is a nonsteroid estrogen that was used as a substitute for natural estrogenic hormones.

The best estimate for the contribution of pollution to cancer risk is around 2% of the total risks. Pollution may be defined as "the presence of a foreign substance — organic, inorganic, radiological, or biological — that tends to degrade the quality of the environment so as to create a health hazard."[15] The vision presented to us in the public media and in a number of published books is that chemical pollutants contribute substantially to cancer risk. This term "environmental" has been corrupted to mean "chemical" and, consequently, many people believe that when they read the words "90% of cancers are environmentally caused," these cancers are necessarily caused from exposure to industrially produced toxic chemicals. In fact, such chemicals contribute to a very small percentage of cancer deaths, whereas personal habits, including cigarette smoke, alcohol abuse, and unhealthy diet, contribute to a combined risk of 70% or more to cancer deaths.[19] These behaviors are well within the control of the individual to modify, but translating belief or knowledge to changes in behavior is not always easy. Although 82% of Americans believed that improving their diet can lower cancer risk, and 61% were aware that obesity was a risk for cancer, more than half of those surveyed in an American Cancer Society study believed there was little they could do to reduce their risk. The less than 15% that made dietary changes within the last year did so for weight loss or to reduce risk from heart disease, and not to reduce cancer risk.[20]

DEFINING DISEASE

Disease may be defined as "a definite pathological process having a characteristic set of signs and symptoms which are detrimental to the well-being of the individual."[15] There is no lack of diseases among the human population; they take on an array of symptoms, are produced by a variety of sources, and are resolved in a number of ways and at differing times. Diseases that already exist in a community where they are maintained in a low but constant incidence are said to be *endemic*. When there is marked increase in the incidence of disease within a limited area affecting ever-increasing numbers of people, this is referred to as an *epidemic*. Should the epidemic spread throughout the world, as in the case of various outbreaks of influenza, it is called a *pandemic*. Many diseases such as measles are acute and have a rapid onset, are usually self-limiting, and are of relatively short duration. Diseases having a slow onset and lasting for extended periods are said to be chronic and may include cancer, emphysema, some forms of heart disease, or AIDS. Diseases may be further classified by the form of the disease (Table 4.1). These diseases may be infectious in nature, or the result of chemical or physical injury. They may be developmental diseases involving defects occurring during embryogenesis or the consequence of genetic errors. Neoplastic diseases such as cancer develop from multiple genetic defects acting in concert with a number of endogenous or environmental factors producing susceptibility. Nutritional diseases are often the consequence of poverty and a weakened society, but these characteristics may also be intimately bound with environmental conditions.

TABLE 4.1

HUMAN DISEASE FORMS

1. Infectious Diseases
2. Physical and Chemical Injury
    Mechanical
    Thermal
    Ionizing Radiation
    Chemical
3. Developmental Diseases
4. Neoplastic Diseases
5. Nutritional Diseases

INFECTIOUS DISEASE

Most of humanity suffers from infectious diseases due to the pathological process that occurs when a microbial agent invades the body. The agents of such disease that are always inside the cell, or intracellular, include viruses (HIV, measles, mumps, human papilloma virus, etc.), chlamydia (often associated with nongonococcal urethritis), and rickettsia (epidemic typhus and Rocky Mountain spotted fever). Organisms that may be intracellular or extracellular include bacteria (strep throat, botulism, staph infections, gonorrhea, etc.), mycobacteria (tuberculosis), fungi (histoplasmosis, ringworm, jock itch [candidiasis], and protozoa [Giardiasis, amebiasis]). Metazoans are multiple-celled animals that are always extracellular because of their significant size, and would include parasitic nematodes and round-worms (ascarids, trichinae), the cestodes or segmented flatworms (tapeworms), and the trematodes including the liver, lung, and intestinal flukes. Insects may also be included in the category of extracellular parasites and would include mites, ticks, and lice (Table 4.2).

For a disease to perpetuate itself, the disease organism must be in constant supply. Such sources can be living organisms or inanimate objects that provide the conditions for survival, multiplication, and transmission of organisms. These sources are called *reservoirs*. Reservoirs may include humans, animals, and nonliving objects. As human populations grow and spread into previously unsettled regions, it is likely that they will be exposed to new reservoirs and potentially new diseases never experienced before. Diseases such as yellow fever, Ebola fever, and possibly HIV, may have been introduced into human populations in this manner.[21] Most human diseases are harbored within the living bodies of humans and are transmitted directly or indirectly to others. Diseases may be transmitted from persons who exhibit signs of the disease, or by persons known as *carriers*, who exhibit no symptoms but harbor and transmit the disease organisms. Some people may never develop symptoms, whereas others may pass the disease during a latent or symptom-free stage. Humans serve as reservoirs for such diseases as AIDS, hepatitis, and tuberculosis.

A number of wild (sylvatic) and domestic animals serve as reservoirs of disease agents that infect humans. Diseases that are spread mostly from animals to humans are called *zoonoses*, and may include Lyme disease, Eastern equine encephalitis,

TABLE 4.2
Agents of Infectious Disease, Reservoirs of Infection, and Mechanisms by which Diseases are Spread

AGENTS OF INFECTIOUS DISEASE		RESERVOIRS OF INFECTION	THE SPREAD OF INFECTION
Virus Chlamydia Rickettsia Bacteria Mycobacteria Fungi Protozoa Metazoa	Always intracellular → Intracellular → Always extrrcellular	**Human** 1. Exhibit signs of disease. 2. Does not exhibit signs of diseases. This person is known as a carrier. **Animal** 1. Domestic. 2. Wild or sylvatic. diseases transmitted by animals to humans are called zoonoses. **Nonliving** 1. Soil 2. Water	**Contact** 1. Direct - no intermediate object involved in transmission; also known as person-to-person such as kissing or sexual intercourse. 2. Indirect - Transmitted from reservoir to host by nonliving object sometimes called a fomite; may include tissues, cups, utensils, etc. 3. Droplet - Organisms are in droplets produced by sneezing, coughing, and talking, and travel less than a meter. **Vehicles** Transmission of disease agents by a medium such as water, food, or air. 1. Waterborne 2. Airborne - travel more than a meter in air **Vectors** Animals that carry pathogens from one host to another; arthropods most important group. 1. Mechanical - passive transport on feet or other body parts. 2. Biological - active process in which agent reproduces in vector and then is transmitted by bite of insect, or in feces or vomit.

Rocky Mountain spotted fever, rabies, bubonic plague, salmonellosis, toxoplasmosis, trichinosis, and more than 140 other known zoonotic diseases. Some diseases that have animal reservoirs mutate in these animals such that they are able to change their characteristics enough that they continue to cause epidemics year after year. One such disease is influenza. In 1997, a strain of influenza that had been previously thought to only infect birds, was linked to deaths in Hong Kong. If this was confirmed, and the strain was particularly virulent, it was feared that a pandemic similar to the large influenza pandemic of 1918 might ensue. The 1918 pandemic infected nearly every person on earth and was linked to an estimated 30 million deaths.[22] The major nonliving reservoirs of most human diseases are soil and water, where disease-producing fungi, and bacteria such as *Clostridium perfringens* or *Cl. botulinum* occur. Many organisms can survive in water, including parasites and bacteria

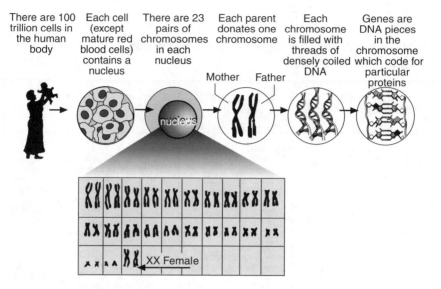

There are 100 trillion cells in the human body

Each cell (except mature red blood cells) contains a nucleus

There are 23 pairs of chromosomes in each nucleus

Each parent donates one chromosome

Mother   Father

Each chromosome is filled with threads of densely coiled DNA

Genes are DNA pieces in the chromosome which code for particular proteins

nucleus

XX Female

FIGURE 4.2 A perspective of DNA in the human cell. (Adapted from Elmer-Dewitt, P., The *Genetic Revolution, Time,* January 17, 1994, p. 46.)

that produce gastrointestinal disease and that may be passed to the water from human feces or animal wastes.

The major routes of infections for most infectious disease agents include (1) contact, (2) vehicles, and (3) vectors (Figure 4.2). The types of contact transmission involved in transmission include (a) direct, (b) indirect, and (c) droplet contact. Direct contact transmission occurs when the disease is passed directly from a source to the host by physical contact without an intervening object. Also known as person-to-person contact, this type of transmission may occur through kissing, touching, or sexual intercourse. AIDS, infectious mononucleosis, influenza, the common cold, and measles are examples of diseases transmitted in this way. When the disease agent is transmitted from a reservoir to a host by a nonliving object, the process is called *indirect contact transmission.* The object on which the disease is transmitted is often referred to as a *fomite* and may commonly include eating utensils, drinking cups, shared toothbrushes, tissues, toys, syringes, and bedding. As examples, hepatitis and AIDS may be transmitted by contaminated syringes. Organisms such as the viruses involved in the common cold may be transmitted by droplet infections, whereby the organisms are contained within mucous droplets which are expelled to the air in sneezing, coughing, or even talking, and they travel short distances to be inhaled or ingested by persons very near by (usually less than a meter).

When disease agents are transmitted by various media such as air, water, food, intravenous fluids or blood, and drugs, this is referred to as *vehicle transmission* (Table 4.2). Airborne transmission refers to disease agents that travel some distance greater than 1 m in the air and are usually carried by dust particles or extremely small droplets. Fungal spores, staphylococci, and measles virus can be transmitted by particles in the air. Diseases spread by water have usually been contaminated by human sewage or by animal waste that contains disease organisms. Foodborne disease usually occurs when food is improperly refrigerated or insufficiently heated and permits the disease organism to grow to numbers capable of producing toxins or invasive numbers of bacteria. In many cases, the disease organisms are carried in the flesh of a food product that has been previously infected (i.e., tapeworm) or has come into contact with the offending organism and serves to simply transport the organism (some forms of hepatitis). Diseases transmitted by air, water, and food are discussed in subsequent chapters.

Vectors are animals that carry pathogens from one host to another, either from another human who is infected or from an infected animal. Most vectors of significance to humans are arthropods (joint-legged invertebrates) including insects, spiders, and crabs. Insects represent the largest and most significant vectors of human disease. When the arthropods transmit the disease by simply carrying the organisms on the feet or other body part to a person's food or an open sore on the person, then the transmission is passive, and the process is referred to as *mechanical transmission*. Most vectorborne diseases have worldwide importance in that they produce disease in tens of millions of people, besides influencing the social, economic, and political climates of many countries. In such diseases, the disease agent goes through some reproductive stage in the vector and is transmitted by a bite, fecal matter, or vomit of the vector in an active process referred to as *biological transmission*. Lyme disease and an increasing number of other diseases such as ehrlichiosis are spread by deer ticks and Lone Star ticks, depending on the region of the U.S. Some of these new diseases are not easily detectable, and therefore more difficult to treat than Lyme disease.[23] Malaria is spread by the female anophelene mosquito in which the malarial parasite (*Plasmodium* sp.) undergoes a sexual cycle before it can reinfect another human by the bite of the mosquito. Because the sexual cycle occurs in the mosquito, the mosquito is known as the *definitive host*, and the asexual reproduction takes place in the human, who is the *intermediate host*.

PHYSICAL AND CHEMICAL INJURY

Physical injury includes mechanical, thermal, and radiation injuries. Mechanical injury results from a transfer of a damaging excess of kinetic injury to tissues resulting in abrasions, lacerations, punctures, contusions, broken bones, and projectile wounds.[24] Automobile accidents, falls, shootings, stabbings, and other forms of trauma to the human body continue to be among the most important form of death and injury in the world. The leading cause of injury-related deaths in the U.S. is motor vehicles, accounting for nearly 40,000 deaths per year, with more than 40% of those deaths in the 16- to 19-year age group, with alcohol involved in nearly half of all motor vehicle deaths.[25] Falls are the leading causes of accidental deaths in the home. Nearly 40,000 children are injured each year, after falling from cribs, high chairs, and bunk beds. Another 27,000 are injured in infant walkers, especially from falling downstairs. More than 750,000 falls and injuries occur from accidents on stairs, whereas falls from ladders account for another 100,000 injuries each year in the U.S. The elderly above 75 years are 12 times as likely to die from falls as from all other groups combined.[24] Violence within the home and the community has become a significant public health problem. Firearms-related deaths in the U.S., which include homicides, suicides, and unintentional shootings, escalated sharply in the 30 years from 1968, when there were 23,875 such deaths. Deaths from firearms now approach 38,000 per year, with most deaths involving males aged 15 to 19 years.

Thermal injury results when tissues are exposed to extremely low or extremely high temperatures. Low temperatures can produce hypothermia or frostbite. Frostbite occurs when extreme cold freezes body water and forms intracellular crystals that burst cell membranes, killing affected cells. Very high temperatures can produce hypothermia or burn injuries. Burn injuries claim about 500 people a year in the U.S. These deaths are generally from residential fires and most commonly occur among the very young or the very old, and the black or hispanic populations. Major reasons for the fires include children playing with matches, faults in electrical devices including heating sources, and careless smoking.[25]

When radiant energy interacts with matter to form charged particles, that energy is referred to as *ionizing radiation* and may include electromagnetic radiation (gamma and x-radiation) or particle radiation (alpha, beta, and neutron). Exposure to radiation may come from radon, diagnostic use of x-rays, use of radionuclides in medicine, and emissions form catastrophically failed nuclear power plants such as Chernobyl. Injury results from the radiolysis of water molecules in tissue that form free radicals (ion pairs), which react immediately with components of the cell, causing cross-linking of macromolecules, thereby changing the tertiary or three-dimensional structure of the molecule. These changes may impair the function of proteins, deoxyribonucleic acid (DNA), and other molecules, resulting in reactions from subtle interference to cellular death.[24]

Chemical injuries result from exposure to chemicals that produce a deleterious effect on organ systems by stimulating or depressing metabolic functions. Some chemicals may function at trace levels to promote or stimulate organ response, whereas larger doses may retard activity or destroy it entirely. The major mechanisms of chemical injury may be categorized as (1) interfering with enzyme activity, (2) directly combining with some cell component other than enzymes, or (3) producing a secondary action in which a chemical causes the release or formation of a more harmful substance. In this last case, allergens represent a good example, in that otherwise harmless substances such as pollen, animal dander, or even certain foods can cause life-threatening allergic reactions in some people. Some chemicals produce a systemic disease. Systemic refers to conditions or a process that affects the whole body. Many environmental pollutants such as lead, mercury, cadmium, and chlorinated hydrocarbons are systemic poisons in that they affect a broad array of organ systems. Systemic pollutants are normally ingested, though inhalation and skin absorption may also occur. Once in the body, they are distributed through the bloodstream to various organs, where common manifestations include central nervous system disorders, stomach and bowel disorders, and disruptions to the blood-forming tissues.

DEVELOPMENTAL DISEASE

Every individual within each species is very much a representation of the information contained in the genome together with the expression of those genes in the development of that individual. Developmental disease occurs when faults or mistakes occur within the genes (or chromosomes), or when stages in development of the fetus are disturbed. Faults or mistakes occur within the genes when the genetic information is disturbed by chemical or radioactive insult, inherited errors, or random (spontaneous) errors in transcription. Major categories of genetic disease include (1) single-gene defects or (2) cytogenetic defects (abnormalities in the number or structure of chromosomes).[24] When defects arise during the embryonic period of development, these are said to be *developmental* or *teratologic* defects.

Some teratologic diseases are inherited, and some are produced by cytogenetic defects, whereas most occur from interference with embryologic development in an individual whose genetic makeup would ordinarily produce a normal individual. The causative factors in this case are extragenetic or extrinsic factors, and the agents producing these factors are referred to as *teratogens*. This is most often an environmental disease in that external factors are involved, including maternal behaviors (tobacco and alcohol use, inadequate nutrition, use of various medications such as accutane or DES, stress, etc.), infectious diseases including toxoplasmosis or German measles, and exposure to chemical and physical agents in the home or workplace environment. The defects occurring during the embryonic stages of development are said to be *congenital defects*. More detailed information on developmental disorders are given later in this chapter.

Neoplastic Disease

The new and uncontrolled growth of abnormal tissue from the transformation of normal body cells is referred to as *neoplasia*. Such growths are referred to as *tumors*, or *neoplasms*. Tumors may be benign or malignant and vary dramatically in form, rates of growth, potential injury to host, and appearance. Malignancies may derive from a number of exogenous factors (arising from the environmental) and endogenous factors (originating from within body). The exogenous factors may include (1) habits (tobacco use, poor nutrition, alcohol use, sexual and reproductive activities), (2) ionizing radiation, (3) chemical exposure, (4) environment (socioeconomic, geographical, and occupational differences), and (5) oncogenic viruses. Endogenous factors contributing to cancer risk may include (1) gender, (2) age, (3) hormonal imbalance, (4) impaired immune system, and (5) genetic predisposition.[24] This topic is covered in more detail later in this chapter.

Nutritional Disease

The consequences of overpopulation, including an inadequate food supply in many parts of the world have been discussed, in Chapter 3. Inadequate diets are also seen to occur in developed nations such as the U.S. Fast food represents an increasing percentage of the American diet. Critics of this trend point to the overprocessing of the food for the purpose of homogeneity. This processing removes much of the nutritional value from the foods, often increasing the sugar and fat content of the food. Obesity in the U.S. has risen from 8% of the population in 1980 to 20% currently.[26] In lesser-developed countries (LDCs), most nutritional diseases are coupled with political, cultural, or emotional problems that make solutions more difficult than simply increasing caloric intake. Malnutrition, as a consequence of inadequate diet, is often associated with ignorance and poverty, which are manifestations of a weakened society.

In the broadest definition of environment, malnutrition may be considered an environmental disease, but it must be known that in most nutritional diseases there is a social or emotional factor that is contributing to the problem. The consequences of an inadequate diet are far more disturbing to the child than to the adult because the child needs food for growing in addition to functioning. Deprivation of nutrients at vital stages in development may result in delayed maturation, permanent damage, or even death. In the LDCs such as certain African nations, many 1- to 3-years-old children display a disease of dietary insufficiency known as *kwashiorkor*. These children are severely protein deprived and may exhibit anemia, diarrhea, graying or reddening of hair, edema, ascites (distended belly), fatty degeneration of the liver, failure to grow, and apathy with extreme weakness. Nutritional marasmus may appear in children 1 year of age who lack sufficient food and show signs of wasting, failure to grow, and complications of infections with multiple vitamin deficiencies. Infantile scurvy develops from a lack of vitamin C (ascorbic acid) in the diet and, consequently, the synthesis of collagen, and mucopolysaccharides is disturbed. This results in widespread hemorrhages, anemia, failure to heal, tenderness in limbs, and possible cardiac failure with sudden death.[24]

ENVIRONMENTAL DISEASE

We now have a better understanding of the terms "environment" and "disease," and it is appropriate here to combine these terms and derive a more concise definition of the term "environmental disease." Environmental disease refers to any pathological process having a characteristic set of signs and symptoms that are detrimental to the well-being of the individual and are the consequence of external factors, including exposure to physical or chemical agents, poor nutrition, and social or cultural behaviors. Factors involved in the development of environmental disease include polluted air and water, excess noise, sunshine (UV radiation), ionizing radiation, food contamination, occupational exposure, toxic wastes, poor nutrition or diet, stress, and tobacco use. There is a rising public awareness concerning environmental disease, and these concerns have driven extensive research efforts into understanding the mechanism producing cancers and developmental abnormalities. The current scientific efforts are directed at the cellular level, where injury to the genetic makeup of the cell can have ominous implications for the individual and possibly for several generations of progeny that will carry the genetic errors as inherited traits. A study of mutations, birth defects, or cancer must start with a knowledge of the genetic code.

Are you interested in learning more about environmental disease? Visit http://www-unix.oit.umass.edu/~envhl565/weblinks.html and click on one or more of the following general links to environmental disease under Chapter 4:

- OncoLink: A University of Pennsylvania Cancer Center Resource
- www.organogenesis.com

- Center for Childrens Environmental Health and Disease Prevention Research
- Welcome to the March of Dimes Home Page
- Your American Cancer Society
- e.hormone: Your gateway to the environment and hormones
- NCI's CancerNet Cancer Information
- LBNL ELSI Home Page
- Gene Gateway: Exploring Genes and Genetic Disorders
- Human Mutation
- CDC
- Welcome to PubMed

THE ROLE OF GENETICS IN DISEASE

STRUCTURE AND FUNCTION

There is a thread of life that courses through all living things that defines and programs the commonality and differences among and between species. This thread of life is deoxyribonucleic acid (DNA), a spiral, staircase-shaped molecule over 3 billion steps long compressed in the nucleus of living cells. DNA is the basic informational macromolecule, the basis of heredity, whose chemical structure has been known for more than 45 years. The storage and transcription of biological information occurs within DNA, which has the power to express that information in the form of proteins. When viewed in a concentrated and pure form outside the cell, it is a clear, thick, viscous liquid that looks like water but has the consistency of mucus and is threadlike.

DNA is located within the cell nucleus of each of the human body's 100 trillion cells (except mature red blood cells; Figure 4.2). Each nucleus contains 46 chromosomes arranged in 23 pairs. Each parent contributes one of the chromosomes in every pair. The chromosomes in eukaryotic cells are tightly coiled (condensed) around a group of proteins known as *histones*. Genes are segments of DNA that code for a particular protein. The DNA of each species or cell type is specific for the species and reproduces that species precisely, with the exception of mutations. When unfolded, DNA is a very long, threadlike molecule that is formed by four recurring subunits called *mononucleotides*. The mononucleotides are made up of one purine or pyrimidine base, one molecule of phosphoric acid, and one molecule of a sugar called 2-deoxy-D-ribose. Except for the base component, the nucleotides are identical and attached to each other in a long chain by phosphodiester links between carbon atoms in each molecule (Figure 4.3). The base components include adenine and guanine (purines), plus thymine and cytosine (pyrimidines). These four different recurring mononucleotides are the base elements of the DNA coding system from which all proteins are ultimately formed, including over 1000 enzymes, numerous structural proteins, lens protein, myosin in striated muscle, hemoglobin in blood, nucleoproteins, and mucus. The proteins are essential for growth, the building of new tissue, and the repair of injured or broken-down tissue. The genetic message for the production of protein is imparted initially by the specific sequence

FIGURE 4.3 The chemical backbone structure of DNA. (Adapted from Lehninger, A.L., in *The Molecular Basis of Biological Energy Transformations*, W.A. Benjamin, New York, 1965, chap. 11.)

of the four basic mononucleotides forming the long DNA strand. As an example, starting along one section of the DNA strand you might find a sequence that reads A_C_G_A_A_C_G_T_. Chromosomes are, then, gene sequences along strands of DNA compressed and floating in the fluid of the cell nucleus. A thousand genes may be carried in a single chromosome, and it is with the precise replication (copying) of this genetic information that genes are duplicated and transferred to daughter cells.

The chromosomes do not exist as single structures in higher plants and animals but are homologous pairs such that every chromosome has a near twin counterpart, and the two strands intertwine to form a double-helical structure so that the two molecules run in opposite directions (Figure 4.4). The pair nucleotide bases fit together in a precise way, held together by hydrogen bonds between the base pairs and by hydrophobic bonding (Figure 4.4). Each strand of DNA contains the same information because the base pairs are complementary, in that adenine is always paired with thymine, and guanine with cytosine. Dividing cells provide each daughter cell with one complementary strand of the parent and then synthesize a complementary strand of the parent to produce double-stranded DNA that is a duplicate of the parent DNA. This process is referred to as *replication*.

Double helical DNA

FIGURE 4.4 The double-helical structure of DNA and the mechanism of hydrogen bonding.

Consider the human genome of 23 chromosomes pairs as 23 chapters in a book that describes the complete set of instructions for making a person. Within each chapter are sentences (genes) composed of words that are each 3 letters in length. There are only 4 letters available to make these words, and they are A (adenine), G (guanine), C (cytosine), and T (thymine). There about 100,000 genes (sentences) in this book and so tens of thousands of different proteins are described by the sequences stored in the genetic code. In this genetic code, a group of three nucleotide bases (three letters) is called a *codon* and specifies the position of one amino acid in a protein molecule that will ultimately accumulate with all other proteins to produce a living individual. Proteins are composed of specific sequences of amino acids linked by peptide bonds, and the sequence of amino acids ultimately determines how the completed protein folds or layers into specific three-dimensional shapes. The shape of the completed proteins is usually critical to its function. Enzymes are proteins that interact with a substrate much as a key in a lock to catalyze the formation of a new molecule or product. Changes in the three-dimensional structure of the enzyme could reduce or eliminate this ability and cause outcomes that might range from barely perceptible to severe, including death.

PROTEIN BIOSYNTHESIS

The DNA molecule is locked in the nucleus and cannot reach the ribosomes outside the nucleus in the cell cytoplasm, where protein synthesis occurs. The DNA must first be transcribed to another molecule called messenger RNA or mRNA. Messenger RNA is quite similar to its DNA counterpart except that it is single-stranded, contains the nucleotide base uracil instead of thymine, and has the sugar D-ribose instead of

FIGURE 4.5 A schematic representation of protein synthesis on the ribosome. (Adapted from Lehninger, A.L., in *The Molecular Basis of Biological Energy Transformations*, W.A. Benjamin, New York, 1965, chap. 11.)

2-deoxyribose in the four mononucelotides. The complementary mRNA, once formed, diffuses through the nuclear membrane into the cytoplasm and attaches to the surface of cellular ribosomes where protein synthesis occurs. The mRNA now carries the code that will direct the specific sequence of amino acid molecules to be inserted into the peptide chains that will form the protein.

The coding is accomplished through the sequence of three consecutive (triplet) mononucletoide units (codons) in the mRNA that each represents a single, specific amino acid molecule (Figure 4.5). The consecutive nucleotide codon triplets (i.e., A-U-G) are read out by the protein-synthesizing machinery of the ribosome to form peptide bonds among a specific sequence of amino acids, to form a complete protein. Amino acids are first prepared for this process by an activation enzyme for each amino acid that forms a high-energy link between the amino acid and an amino acid carrier molecule called transfer or soluble RNA (sRNA), which is specific to each of the 20 amino acids. The sRNA molecule has a sequence of nucleotides that recognize and attach to a specific codon triplet. The recognition site for nucleotides on the sRNA is called an *anticodon* and is thought to be a complementary triplet of bases to each mRNA codon. The activated amino acid and its anticodon forms a specific bond with the corresponding codon of mRNA. Thus, the amino acid tyrosine with its specific anticodon will attach only to the triplet codon sequence of A-U-U appearing as a sequence on the mRNA attached to the surface of the ribosome.

Once fixed in position, enzymes form peptide bonds between this amino acid and the end of the peptide under construction, releasing sRNA while other enzymes peel the peptide chain from the ribosome. A great deal of energy is input into this reaction through the formation and breaking of peptide bonds via adenosine triphosphate (ATP). The accuracy of replication is therefore ensured thermodynamically by the creation of very stable attachments between the triplet codon and its anticodon. However, mistakes do occur, and if a single amino acid is out of sequence, the coding is disrupted and an amino acid may be eliminated or replaced; the corresponding protein will be altered.

DISEASES OF GENETICS AND DEVELOPMENT

Genetic Abnormalities

A change in the nucleotide base sequence of DNA is called a *mutation*. When such changes occur, there is often a resulting change in the protein encoded by that gene. If this protein is an enzyme, that enzyme may become less active or completely inactive as its amino acid sequence is changed, causing alterations in the active site of the enzyme. Most genotypic changes are harmful or damaging, but some mutations are beneficial. That is how species change and compete in an ever-changing environment. Many simple mutations are neutral or silent and cause no change in the product of the encoded gene. This can occur when the amino acid that is changed is in a nonvital portion of the protein molecule, so that protein function remains. However, many changes cause variations in protein function resulting in a spectrum of decreased or increased activity. These are genetic polymorphisms reflecting the existence of several variations of the same gene. As an example, the P450 genes produce enzymes that detoxify chemicals such as certain polyaromatic hydrocarbons (PAHs) that may occur in tobacco smoke. Variations in P450 genes result in decreased activity of the resulting enzymes and are associated with an increased risk of lung cancer in smokers in about 10% of the Caucasian population.[14] Other genes, such as GSTM1, also have been shown to have a natural function of inactivating some of the carcinogens in tobacco smoke. This gene is deleted in 50% of Caucasians, and this may increase the risk of lung cancer from secondhand smoke by 2.6 to 6 times.[27]

Unfortunately, many human diseases are associated with genetic defects. These defects can occur as (1) single-gene defects known as a *point mutations* or *base substitutions* and (2) cytogenetic defects in which there are abnormalities in the number or structure of chromosomes. A point mutation occurs when a single nucleotide base is replaced at one point in the DNA molecule with a different base. The

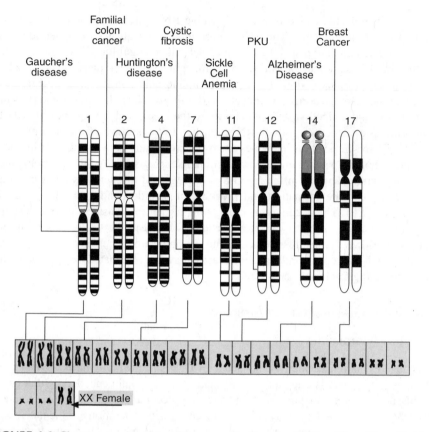

FIGURE 4.6 Chromosome locations of some genetic diseases. (Adapted from Elmer-Dewitt, P., The Genetic Revolution, *Time*, January 17, 1994, p. 46.)

DNA replicates with a substituted base pair such that cytosine might be substituted for Guanine. This base substitution could result in an incorrect amino acid in the synthesized protein. The effect may be imperceptible or neutral, or may result in one of many serious human ailments such as cystic fibrosis, phenylketonuria (PKU), hereditary spherocytosis, alpha-1-antitrypsin deficiency, Gaucher's disease, achondroplastic dwarfism, hemophilia, or certain forms of diabetes (Figure 4.6). Point mutations may occur spontaneously because of occasional mistakes created in the process of DNA replication, resulting in a spontaneous mutation without the involvement of external factors. There are also agents in the environment including chemicals or radiation that promote such mutations and are said to be *mutagens*. When such mutations occur, they become part of all cells of next-generation individuals. Defective genes may be inherited according to Mendelian laws, just as are normal genes.

Some genes are inherited as autosomal dominant genes in that they are always expressed phenotypically (physical appearance) even when appearing on just one of the paired chromosomes. These are usually insignificant defects because they are always expressed and have occurred over many generations. The hereditary disease spherocytosis is caused by a point mutation that is autosomal dominant. Dominant genes that are lethal before reproductive age would clearly not survive in a population. A lethal dominant gene produces Huntington's disease, but this disease is normally expressed in the fourth and fifth decades.[24]

Codominant genes are partly expressed if present as a single allele (an *allele* is a pair of genes situated at the same site on paired chromosomes) and fully expressed if present on both. If the gene is defective on one chromosome, with a normal gene on its counterpart chromosome, the expression of the gene may be subdued. When defective on both chromosomes, the disease may be fully expressed and possibly even fatal before adulthood. An example of a codominant gene is that which produces Sickle cell disease. Sickle cell describes the distinctive crescent or "sickle" shaped erythrocytes (red blood cells) that develop in deoxygenated erythrocytes carrying hemoglobin S. It is a disease seen primarily in the black race. Hemoglobin S differs from normal hemoglobin A by the substitution of a single amino acid in one of the three polypeptide chains that make up the protein portion of the molecule. The misshapen erythrocytes are quite fragile and become "sickled" and accumulate in the tiny capillaries, where oxygen is released to the tissues. This causes vascular obstruction, tiny infarcts, and fibrosis. When the defective gene responsible for sickle cell disease is present on one chromosome, about 40% of the hemoglobin will be hemoglobin S. When defective in both genes, the sickle cell disease will be fully expressed, leading to severe anemia and probable death in early adulthood.

Some genetic diseases are inherited as autosomal recessive, and this means that a defective gene must occur in both chromosomes at complementary sites for the disease to be expressed. A person with this genetic disease is deprived of an essential protein, resulting in a potentially severe disease such as phenylketonuria (PKU) or cystic fibrosis.[24] There may also be dominant and recessive defective genes in the sex chromosomes as well, producing a variety of sex-linked diseases including hemophilia-A, color blindness, and forms of muscular dystrophy.

Some genetic diseases are produced by abnormalities in the number or structure of the chromosomes and may be referred to as *cytogenetic defects*. Such defects tend not to be inherited, but result from disorders occurring during mitosis when eggs or sperm are formed in the parent or when the embryo is developing. These defects involve large parts of the chromosome and most often appear as deletions or additions of chromosome material. Most such defects are fatal, and the fetus fails to survive embryonic development. About 20% of fetuses that spontaneously die while in the uterus have such genetic defects. There are survivable cytogenetic disorders, however, and one of the more recognizable is trisomy 21. Trisomy 21, or Down's disease, is characterized by the addition of an extra chromosome to chromosome number 21, which is expressed phenotypically as various degrees of mental retardation, predisposition to leukemia, congenital heart disease, and characteristic facial features including epicanthal folds of the eyelids, giving an Mongoloid appearance to the eyes, flat nose or absent bridge, and low-set ears.

Teratologic Diseases

A number of physical defects arise during the period of embryonic development and may be the result of inherited or cytogenetic defects in the genome, though most result from factors that cause problems in embryonic development in a person otherwise genetically normal. Such diseases are *teratogenic* and the external factors are *teratogens*. Teratogens may include ionizing radiation; chemicals such as accutane, dioxin, diethylstilbestrol (DES), thalidomide, organic mercury, and alcohol; pathogenic infections including toxoplasmosis, German measles (rubella), cytomegalovirus, and syphilis; and maternal conditions such as age, stress, nutritional adequacy, smoking, drinking, hormone levels, and previous number of births (Table 4.3).

Most teratogens exert their effect during certain critical time windows in the early stages of tissue and organ formation known as *organogenesis*. This is a period occurring from about the 18th to 60th day after conception, with the most sensitive period around the 30th day. Exposure prior to this time is very often fatal to the developing fetus, whereas any exposure after 8 to 9 weeks occurs only after the organs are developed and so will not produce major physical defects, but it can produce blindness, mental retardation, or damage to sexual organs. As an example, the sedative thalidomide, synthesized and marketed in Western Europe in the 1950s, was eventually linked to an an unusual birth defect in which the long bones of the arms and legs were missing so that hands and feet were attached directly to the torso. This condition was known as *phocomelia* or *seal limb* disease.[22] Most of the women giving birth to children with this defect had taken the mild sedative thalidomide during the first trimester of pregnancy, but more than 10,000 such births occurred before the drug was removed from the market. A similar critical window of exposure was demonstrated for German measles (rubella). It was discovered in the early 1940s that women who contracted this disease during the first trimester were at high risk of having babies with congenital cataracts, deafness, and heart abnormalities.

Just as chemicals taken into the body during critical windows of development can cause physical abnormalities, so can the absence of certain important nutritional compounds. The absence of folic acid, a B vitamin, is associated with neural tube defects including spina bifida and anencephaly. Spina bifida is characterized by the protrusion of a portion of the spinal cord from the spinal column, whereas anencephaly is a disorder in which most of the brain is absent. Consequently, the simple addition of this vitamin to the diet of a woman about to conceive can prevent a seriously debilitating and often-fatal disease. There is a growing list of physical and chemical agents that have the ability to produce birth defects. It should be remembered, however, that 600 to 800 of these agents are teratogenic in laboratory animals, and not necessarily humans. About 30 chemicals are currently known to be responsible for human birth defects.[4,28,29] However, the science of teratology is relatively new, and the little that is known rests on those substances and conditions that produce clear abnormalities. Very little is known about teratogenic substances that may produce mild deficiencies, lack of vigor, or slight losses in mental acuity.

New studies have suggested a much stronger link between fetal development and many adult-onset diseases. A study in England found that women who were pregnant during times of severe food shortage gave birth to more children who became obese as adults. It is thought that factors such as food shortages may "program" the fetus to retain food, predisposing them to obesity. Other possible diseases that may have a fetal implications may include heart disease and breast

TABLE 4.3
Some Selected Teratogens, Sources, and Effects

TERATOGENS	COMMENTS
Chemicals (including pharmaceuticals)	
Accutane (isotretinoine)	A derivative of vitamin A used in the treatment of cystic acne; when used during pregnancy, 25% of births result in defects
Alcohol	May cause fetal alcohol syndrome (FAS); associated with microcephaly, low IQ, memory problems, abnormalities of upper lip, teeth, hyperactivity
Anesthesia	Various anesthetics associated with miscarriages and structural deformities
Cigarette smoke	Associated with low birth weight, stillbirths, miscarriages
Dilantin	Dilantin is a drug used to control seizures and produces a wide spectrum of birth defects
DBPC (1,2, dibromo-3-chloropropoane)	Widely used as a nematocide between 1955-1977; produced infertility in factory workers
Dioxin	An unwanted byproduct in some forms of paper production, mass burn incineration, and production of defoliants; produces miscarriages and stillbirths
Methyl mercury	Often produced by soil microorganisms from organic mercury wastes that reach the bottom of streams and bays. Concentrated in fish flesh, eaten by humans, produces gross physical and mental abnormalities
Tegison	A pharmaceutical producing an effect similar to Accutane
Thalidomide	Used in the 1960s in Western Europe as a mild sedative; resulted in phocomelia or absence of long bones in legs and arms
Pathogens	
Cytomegalovirus	A virus in the herpes family transferred across the placenta from mothers with latent infection; about 1/4 of infants develop symptoms that result ultimately in death
German measles (rubella)	Virus crosses placental barrier to reach fetus, most serious in first trimester causing congenital cataracts, deafness, and heart defects
Syphilis	Sexually transmitted disease producing microcephaly and mental retardation
Toxoplasmosis	Disease caused by protozoan parasite, *Toxoplasma gondii* found in many mammals; the congenital form causes lesions in central nervous system, jaundice, and anemia
Ionizing Radiation	
Gamma or x-rays nuclear fallout	Causes eye problems, microcephaly, and mental retardation

Source: Adapted from Whelan, E.M., How to reduce your risk of cancer, *Priorities,* Summer, 36, 1990, and Nader, R., Brownstein, R., and Richards, J., Eds., *Who's Poisoning America: Corporate Polluters and their Victims in the Chemical Age,* Sierra Club Books, San Francisco, 1981.

cancer, as well as other physiological characteristics such as body size and metabolism.[30] Expectant mothers wishing to protect their developing fetus from teratogenic effects should be under a physician's care, pursue a healthy diet (including folic acid), avoid smoking and alcohol consumption, reduce stress, and limit exposure to known chemical teratogens and infections known to produce teratogenic effects.

New Approaches in Genetics

Methods of Studying Genes

Artificial techniques for creating recombinant DNA developed rapidly in the 1970s and 1980s, permitting the transfer of genes between unrelated species and creating a powerful new science called *genetic engineering*. These techniques relied on the discovery of a number of critical tools and techniques, including the discovery of restriction enzymes. These are microbial enzymes that are used to cut DNA into several predictable pieces. These fragments come together to form genetic "maps" that are unique to each individual, as we all contain many slightly different forms of some genes. Most of these polymorphisms are, in effect, harmless, but invaluable in using DNA as forensic evidence in courtrooms, as well as genetic screening tests.[31]

These fragments of DNA from different sources can also be spliced together through the enzymatic action of DNA ligase, thus creating a recombinant-DNA molecule (Figure 4.7).[32] Once created in a test tube, the recombinant-DNA molecules can be inserted in cells where copies of the gene may be produced, or protein products may be created from the newly inserted gene. The protein product can then be harvested and used for many different purposes. One of the original uses of this process was in the production of insulin. The two, small, polypeptide chains of insulin are created by inserting the gene for each chain into two different bacteria, which are then cultured in large amounts to produce the polypeptide chains. The polypeptides are linked to another compound normally produced by bacteria and so must be separated from the compound, and then the insulin polypeptide chains are joined together to form a complete insulin molecule. The genetic manipulation in insulin production require methods for inserting the recombinant-DNA molecules into cells. The method developed relies on a fragment of self-replicating DNA molecule called *cloning vector*, to which the target gene is attached and transmits the gene from one organism to another (Figure 4.8). Once inside the bacterial or yeast cell, the cloning vector will replicate itself, including the newly inserted target DNA sequence. The most commonly used vectors are plasmids, which are circular, self-replicating, gene-containing DNA fragments found in most bacteria and some yeasts such as *Saccharomyces cerevisiae*. Restriction enzymes can be used to cut out a section of the plasmid and insert the desired DNA sequence into it. The plasmid must then be inserted into bacterial or yeast cells. One method involves placing bacterial cells in a solution of calcium chloride, which makes them competent to take up the plasmids when placed together in a process called *transformation*. In addition to human insulin, many useful proteins are developed in this way including erythropoetin, growth hormone, epidermal growth factor, bone mor-

The double stranded DNA is cut with specific restriction enzymes at particular recognition sites.

Once cut, the DNA fragment with "sticky ends" is released to join with a fragment from another DNA source cut by the same restriction enzymes.

The fragments from different sources join to form a linear or circular molecule (as in a plasmid). The final step is to join the backbones of the two DNA fragments with DNA ligase, producing a molecule of recombinant DNA.

FIGURE 4.7 Restriction enzymes help create recombinant DNA. (Adapted from Tortura, G., Funke, B.R., and Case, C., *Microbiology: An Introduction*, 5th ed., Cummings Publishing, Redwood City, CA, 1995.)

FIGURE 4.8 The genetic engineering of insulin. (Adapted from Tortura, G., Funke, B.R., and Case, C., *Microbiology: An Introduction*, 5th ed., Cummings Publishing, Redwood City, CA, 1995.)

phogenic proteins, taxol, monoclonal antibodies, factors useful in treating hemophilia, and many others.

Another application of gene manipulation is already becoming commonplace in the food that we eat. New techniques are being employed that insert DNA from one species into crops. These genes encode proteins that reduce reliance on fertilizers, make crops more palatable in taste and appearance, and encode natural defenses against pests. The U.S. in particular has been at the forefront of this new technology. However, in the face of rising opposition from consumer groups, the increase in transgenic crop-planted acreage has leveled off. China, facing a burgeoning population, has embraced the new technologies to a larger extent.[33] These new technologies are not without controversy. One possible consequence is the decline of the insects and other fauna that are dependent on them, including the Monarch butterfly.[34]

In addition to making products, recombinant-DNA technology can produce millions of exact copies (cloning) useful in analytic techniques. This technology enables researchers and clinicians to determine the exact sequence of nucleotide bases in DNA, using techniques that are now largely automated. Such sequencing permits the identification and cloning of mutant genes such as those responsible for cystic fibrosis (CF). This cloned gene can then be used to diagnose CF and other genetic diseases employing techniques based on hybridization. Once a gene is located and its sequence of nucleotide bases determined, a hybridization test can be devised to detect the presence of the gene even before the disease is manifested, in a process called *genetic screening*. Such diagnostic tests exist for cystic fibrosis, Huntington's disease, breast cancer, Duchenne muscular dystrophy, retinoblastoma, neurofibromatosis, and many others.

One such hybridization technique is called Southern blotting, and a variation on this has been automated by manufacturers into a probe array (gene chip) that permits rapid genetic screening of a wide array of genetic disorders. In the original Southern blotting technique, DNA is extracted from human cells and fragmented by restriction enzymes (Figure 4.9). The fragments are separated by gel electrophoresis and then transferred to nitrocellulose filter by blotting. The DNA is transferred to the filter as an exact copy of the gel and is exposed to a radioactive probe. If you are screening for cystic fibrosis, the probe consists of cloned DNA fragments with the defective gene, and these are radioactively labeled. Once exposed, the probe will pair up with complementary bases (hybridize) on the patient's separated DNA on the filter if a close match exists. When exposed to x-ray film, the gene fragment containing the hybridized pair will show up as a band in the developed film and confirm the presence of the abnormal CF disease gene fragment in the patient.[35]

The introduction of gene chip technology has greatly simplified and automated this process. In a typical assay, the user extracts and purifies the patient nucleic acid, prepares a fluorescent-labeled target including the genes of interest, and fragments the DNA for hybridization. The material is applied to a single-use probe array (gene chip) containing hundreds of thousands of different oligonucleotide probes (Figure 4.10). Computer-controlled equipment incubates, washes, and processes this material automatically. The probe arrays are then laser-scanned to detect any fluorescent-tagged probes indicating a complementary match with the patient's DNA. In this way, genetic disease can be rapidly screened. Such devices are useful in sequence analysis, genotyping, and gene expression monitoring. Such technology is being

FIGURE 4.9 The hybridization technique called Southern blotting. (Adapted from Tortura, G., Funke, B.R., and Case, C., *Microbiology: An Introduction*, 5th ed., Cummings Publishing, Redwood City, CA, 1995.)

used in the Human Genome Project. The project leaders completed the full human genome sequence in the year 2003 by developing a detailed map of every one of the 60,000 genes in human DNA and the sequences of the 3 billion nucleotide bases positioned along those genes. Now these researchers are moving forward with research aimed at improving human health and fighting disease.

The Hunt for Environmental Genes

Similar techniques are also being proposed by director of the National Institutes of Environmental Health (NIEHS) in the Environmental Genome Project. This effort is based on experiences of researchers who have implicated certain genes in enhancing environmental risk. There is, for example, the gene coding for the enzyme paraoxonase, which produces various versions of the enzyme. These versions differ in detoxifying capability. Ten percent of Caucasians and 25% of Asians express a version of the enzyme paraoxonase that converts organophosphates such as the nerve

FIGURE 4.10 A probe array (gene chip) used for genetic screening developed by Affymetrix Inc., Santa Clara, CA.

gas sarin into a less toxic form nearly ten times faster than those who do not produce this version.[35] This became evident during the sarin nerve gas attack in Tokyo subway by Aum Shinrikyo cult members in 1995. Variations in gene coding for cytochrome P450 and NAT families can increase bladder cancer risk by sixfold in smokers compared to nonsmokers. The HLA-DPbet1 genetic marker is found in people who are much more susceptible to beryllium disease (lung disorder caused by exposure to beryllium) than those who do not have the marker. Based on decades of research demonstrating variations in genes that detoxify or activate offending chemicals, every person is suspected of having unique drug-metabolizing enzymes, causing differences in how they are handled.[35]

These variations in susceptibility may have broad implications for setting environmental policy and regulations. Current standard-setting methods set permissible exposure at some fraction (i.e., 1/10th) determined to be acceptable for the general population, without regard for members of the population who may be at higher risk. The coming trend in environmental risk assessment is to include genetic susceptibility and so determine both distribution and ranges for risk. The Environmental Genome Project is proposed to assist in this effort. Substantial technical obstacles exist, including reliability of the gene sequencing process, determining the relevance of polymorphisms to disease, and demonstrating how genotypes relate to phenotypes (physical manifestations).[35] Francis Collins, director of the National Human Genome Research Institute, has urged NIEHS to collaborate because of the substantial complexity of the task. The collaboration is expected to take place and so begin a project that will attempt to identify genetic predisposition to environmental pollutants, which may ultimately be reflected in setting environmental policy and developing regulations. Although the technology is exciting, legal and ethical questions abound. For instance, will workers be obligated to submit to genetic screening before working with toxic chemicals? Will the results of the screening be a basis for job denial?

The Promise of Gene Therapy

Genetic screening will rapidly advance the discovery of genetic diseases while offering improved capabilities in gene therapy. Gene therapy may involve removing cells with defective genes from a person, substituting them with normal or healthy genes, and then placing the cells back into the patient, where the cells will begin to produce normal product and "cure" the disease. As fantastic as this sounds, these procedures are being attempted by such people as Dr. W. French Anderson, director of gene-therapy programs at the University of Southern California/Norris Comprehensive Cancer Center in Los Angeles. Dr. French and colleagues performed the first federally approved gene therapy procedure in 1990 when they replaced a faulty gene in Ashanthi DeSilva, a 4-year-old girl with severe combined immune deficiency (SCID, or "bubble-boy" disease). The disease results from a deficiency in the enzyme adenosine deaminase because of a rare and deadly mutation. Dr. French et al. extracted some white blood cells from the girl, and exposed them to viruses that were genetically engineered to carry the normal gene for the production of adenosine deaminase. When the white blood cells were returned to her bloodstream, her cells began producing the necessary enzyme, restoring immunity. The white blood cells eventually die, requiring follow-up treatments.

Although very promising, the success of this gene therapy story has not been easy to duplicate. Many technological problems in subsequent human trials have prompted the National Institutes of Health (NIH) Committee on Gene Therapy to suggest that NIH funding for such activities be confined to laboratory animals.[36] Ten years of gene therapy experimentation has produced a number of adverse reactions and at least one death. It is unclear if these adverse reactions were related to the gene therapy or not.[37] Technical problems center around methods of introducing DNA into the right place in human cells, and then having the genes express themselves. Ideally, inserting the "corrected" genes into undifferentiated human stem would provide long-term or permanent cures as the stem cells would serve as an unending source for the differentiated cells throughout the body. Unfortunately, stem cells are few in number and divide infrequently, making the introduction of vectorborne genes difficult. Even so, gene therapy sits on our horizon as a promising technology and the ultimate solution to curing gene-based diseases. Even infectious diseases such as AIDS may be cured via gene therapy. A new treatment under development is to introduce a gene into an HIV-infected person's white blood cells, the very cells that HIV targets. This introduced gene would produce a protein that prevents the HIV virus from being able to replicate, thus denying the virus the reservoir that it requires to reproduce.[38]

In the meantime, current practical approaches to using recombinant-DNA technology is in genetic screening, which can detect errors in DNA. In some cases, the patients may be advised to follow certain activities and diets that will reduce their risk from the disease. In other cases, it may be possible to counteract the effect of faulty genes by replacing missing or abnormal proteins.

The Ethical Dilemma

The discovery of defective genes is progressing at a much faster pace than the development of other treatments and gives rise to many ethical and legal dilemmas. One of these dilemmas revolves around whether to inform patients that they have a difficult or impossible-to-treat disease. For example, Huntington's disease can now be identified through genetic screening. However, this disease is autosomal dominant and so is invariably expressed. The disease normally shows itself in the fourth and fifth decades of life as a relentlessly progressive disease leading to mental retardation, emotional disturbances, loss of self-control, debilitation, and death. There is presently no known cure for the disease. Another gene, BRCA1, results in a 50 to 85% risk of breast cancer when present.[39] The knowledge of the presence or absence of this gene may cause some women to decide to prophylactically have one or both of her breasts removed before cancer develops. Other women have decided to have their breasts removed at the first indications of cancer.[40] Knowledge that one has the defective gene for this disease is not liberating and could lead to more immediate unwelcome effects, including severe depression and even suicide. However, the genetic knowledge gained by a person may be important to other family members. In the event that severe disease or early death is likely, some argue there must be an ethical obligation to communicate it to a spouse, or a significant other because it influences their lives, such as the choices of whether or not to have children.[41]

The other ethical questions being asked center on whether employers or insurance companies could use the information to identify people who may be insurance or occupational risks. What would a potential employer do if they knew in advance a person were at increased risk to developing cancer or Alzheimer's disease? Prenatal testing could ultimately reveal a comprehensive disease profile of an individual, potentially leading to a type of eugenics through abortion, or perhaps stigmatism by society, because of revealed inborn genetic deficiencies. The ethical and legal questions are enormous and only beginning to be realized as the weight of the technology settles upon the social consciousness. Perhaps there is no more pressing need for genetic information than in the area of cancer, and yet the information that comes with it often generates fear.

CANCER

WHAT IS IT?

The clinicians and scientists treating and researching cancer now understand that it is a genetic disease caused by multiple mutations in genes responsible for controlling cell division.[42] Most cancers develop from the interaction of genetics with environment and perhaps less than 5% can be attributed to hereditary factors alone. Most scientists attribute the majority of cancers to a combination of exogenous (external environmental) factors in combination with endogenous (internal) factors that facilitate the accumulation of genetic errors on the prolonged pathway to cancer.[14] Many endogenous and exogenous factors enhance susceptibility in this process (Figure 4.11). Endogenous factors originate from within the body and include gender and ethnicity, health status, age, and heredity (genetic polymorphism). External factors include cigarette smoking, inappropriate nutrition, exposure to cancer-causing chemicals and physical agents (ionizing radiation), and oncogenic viruses. It has been hypothesized that frequent cellular telephone use may increase the risk of certain types of brain cancer. Studies have been inconclusive in establishing this relationship, however.[43]

Eventually, a single abnormal cell develops that has accumulated enough mutations to bypass all the regulatory mechanisms that normally control cell growth. It divides again and again, accumulating mutations that discard or silence protective

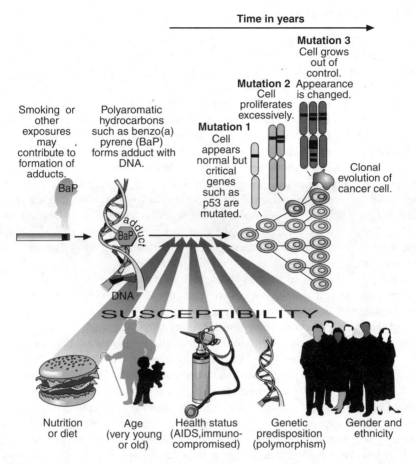

FIGURE 4.11 The evolution of cancer cells and role of susceptibility. (Adapted from Perera, F.P., Environment and Cancer: Who is Susceptible?, *Science*, 278, November 7, 1997, p. 106.)

genes that ordinarily monitor replicating DNA for errors. The malignant cells grow into a small mass that develops nutrient-bearing blood vessels in a process called *angiogenesis*. The blood vessels stimulate further growth of the tumor and establish a pathway for the slippery cancer cells to metastasize to other body sites.[44] Cancer produces its damaging effects through (1) local damaging effects, (2) metastasis, and (3) systemic effects. Local damaging effects include the obstruction of airways, urinary tract, and digestive tract, resulting in infections and stasis. They can erode blood vessels, causing acute or chronic hemorrhaging or the perforation of intestines or the stomach, causing peritoneal infections. Adhesions may develop in the abdominal or thoracic organs. When bone marrow is affected, critical bone marrow is replaced, leading to anemia and thrombocytopenia. Very often, malignancies travel, producing tumors at distant sites in a process called *metastasis*. Tumors may also produce systemic effects, causing (1) endocrine tissues to produce either too much or too little of essential hormones, (2) alteration of intrinsic blood coagulation process, (3) autoimmune inflammation of the skin, and (4) wasting away.[24]

How Does Cancer Develop?

In a typical pathway leading to cancer, a chemical agent such as benzo(a)pyrene (BP) may be associated with smoking activity, and it combines with human DNA to form an adduct (carcinogenic residues bound to DNA), which leads to increased mutations that may eventually accumulate and lead to cancer. The smoke from cigarettes contains as many as 4000 compounds, with more than 40 of these known to be carcinogenic. Included among these are benzo(a)pyrene, benzene, and arsenic. The act of smoking (or being exposed to passive smoke) brings these substances into intimate contact with the delicate tissue of the respiratory tract. Cancer develops from multiple mutations in the genes producing proteins that control cell division. Some of these genes are tumor suppressor genes based on their ability to produce proteins that can repair defective cells or prevent cells from growing that are grossly defective. Other defective genes (such as ras) produce proteins that inappropriately stimulate cell division. Collectively, these cancer-causing genes are called *oncogenes*. When behaving normally, these genes cooperate in repairing damaged cells, destroying greatly damaged cells, and replacing dead cells.

Cancers are thought to emerge through a process known as *clonal evolution* whereby a single cell develops a genetic mutation that enables it to divide even when normal cells do not replicate (Figure 4.11).[45] These cells copy their defective DNA to daughter cells that show similar unrestricted growth. At some point, a descendant of this line undergoes additional mutations that remove the mutated cell further from regulation, allowing the progeny to pass through tissue, enter the bloodstream, and travel to other body locations and colonize distant organs (metastasize). In support of this mechanism, a brain cancer known as *astrocytoma* follows a defined sequence of genetic changes, involving at least (1) inactivation of the p53 gene, (2) loss of a gene on chromosome 9, (3) enhancement of epidermal growth receptor by another altered gene, and (4) loss of one copy of chromosome 10.[45] Certain colon cancers often result when defects develop (1) in both alleles of a gene called APC (adenomatosis polyposis coli), (2) in certain ras genes that promote cell growth, and (3) in the p53 gene. The genetic changes occur early in APC, whereas subsequent changes in the other genes appear much later. These cumulative genetic errors may be facilitated by defective genes such as hMLH1 and hMSH2, which code for proteins that correct errors in the order of nucleotide basis during DNA replication by scanning for errors, detecting them, and repairing them much like a spellchecker on a computer (Figure 4.12). Defects in these genes would result in incomplete or nonfunctioning proteins that would permit errors to accumulate prior to becoming cancerous.

Eventually, the most important step in the development of a cancer cell is its failure to die when genetic mutations render them cancerous. This process of cellular self-destruction is called *apoptosis* and occurs when cell damage accumulates to the point where cell integrity is compromised.[46] At this point the cell produces enzymes that cause the cell's DNA to disintegrate in an orderly way, causing cell death. One of the most important genes coding for enzymes involved in apoptosis is p53, a tumor suppressor gene. More than half of all human cancers involve a p53 gene either mutated by free radicals or inherited as a defect. The normally functioning

Genes such as hMLH1 and hMSH2 code for proteins that correct errors in the order of nucleotide bases during DNA replication. These proteins are known as "spell-checker" proteins, and defects in the gene will produce defective "spell-checker" proteins that fail to correct errors. When such errors occur in genes controlling cell growth, a major step forward in the development of a cancer cell occurs.

FIGURE 4.12 The action of "spellchecking" proteins created by hMLH1 AND hMSH2 genes.

enzyme halts cell division until DNA can be repaired, or turns on the mechanism for cell destruction if the DNA is severely damaged. Mutations occurring along the p53 gene appear at characteristic sites and with unique base sequences known as *biomarkers*, depending on the damaging environmental agent at fault. Cigarette smoke causes damage to specific locations along p53, whereas UV light produces damage at different sites. Such information may eventually be useful in risk assessment, or even legal proceedings.[47] Certain p53 mutations may also reduce the ability of cells to limit blood vessel formation.

Tumors may stimulate their own growth by creating its own blood supply in a process known as angiogenesis. It is this nourishing process that is believed to give cancer cells their enhanced ability to multiply and spread. Dr. Judah Folkman and his colleagues have been testing anti-angiogenesis compounds for years as a possible mechanism for cancer treatment. They have discovered that endostatin and angiostatin prevent and reverse the formation of blood vessels in a broad array of tumors growing in rodents.[44]

The news was received in 1998 with enthusiasm by many, and some skepticism by experienced researchers and clinicians. There are now more than 50 anti-angiogenesis drugs in clinical trials. Check with the American Cancer Society for more information at www.cancer.org. Many cancer cells are unstable and discard both copies of the p53 gene, whereas other cells retain mutant copies. It turns out that cells with normal p53 genes are quite sensitive to chemotherapy, whereas those with mutant p53 genes tend to be very resistant. This sensitivity of cells to p53 actually is the cause of many of the difficult side effects of chemo and radiation therapies. Cells that are only slightly mutated are destroyed, which causes the common side effects of nausea, hair and weight loss, and a weakened immune system.

Some researchers have found that suppressing p53 during these therapies may lessen the side effects.[48] New therapies are being sought that may prove more effective against cancers with mutant p53 genes. Cancer is a complicated disease involving different cells, often expressing many different genes, influenced by multiple growth factors in different ways, and variably responsive to pharmaceuticals. Consequently, combination treatments of chemotherapy with multiple drugs, gamma radiation, and other interventions in combination are proving more successful.[49] With the use of DNA genechips to screen cancer cells by the proteins that they produce, cancers will be named by the genes they express and not by the site, which is the current practice (e.g., colon or stomach cancer).[50] Eventually, treatment in the 21st century will rely on determining the specific DNA damage and custom-fitting the medicinal cocktail to each tumor.[44] Some of the genes associated with cancers are listed in Table 4.4.

Routine testing for genetic errors is very likely 10 years off, although a long-range project called the Cancer Genome Anatomy Project is under way to create a DNA library of tumor types. Individual screening for cancer risk has been associated with technological problems to date. This difficulty combined with the potential abuse of genetic information and the limited use of such information in preventing cancer has prompted the National Advisory Council for Human Genome Research to declare such testing premature in general medical practice.[14] However, advances in the knowledge of genetic and acquired susceptibility in the general population will likely emerge in future environmental health and regulatory policies as high-risk groups are defined by the genetic information that they carry.[14] The risk of cancer can be largely reduced by modifying our behaviors (Table 4.5). The science of epidemiology and genetics (molecular epidemiology) is producing more knowledge on the common factors that influence genetic susceptibility to cancer. Planners responsible for drawing up strategies to protect public health will be challenged to keep abreast of this information.

MAJOR CANCER RISKS

Smoking

Smoking cigarettes is recognized as one of the leading causes of cancer in the U.S., causing more than 30% of all cancers. In excess of 88% of the 120,000 people newly diagnosed with lung cancer each year are smokers. Lung cancer used to be a disease primarily of men, but the popularity of smoking among women has driven the number of cancer cases upward so that lung cancer surged past breast cancer as the leading cause of cancer deaths among women beginning in the 1980s.[29] Since the 1960s there has been a decline in smoking of more than 15% among American adults, which is being reflected in a decline in lung cancer.[51] Smoking is showing some increase among teenage girls and blacks. There is also a disturbing trend of increased cigarette smoking outside the U.S., with dramatic increases in countries of the former Soviet Union, China, South Korea, and Spain.[52] The disturbing pro-

TABLE 4.4
Some Known Cancer Genes

NAME	ACTION OF DEFECTIVE GENE	COMMENTS
ras (rat sacrcoma)	Causes cell to divide again and again	Plays role in 90 % of pancreatic cancers, 50% of colon cancers, 25% of lung cancers
EGF (epidermal growth factor)	Takes in growth signals, relays them to this protein; abnormally high; high conecentrations in tumors	Found in high concentrations in 40% of tumor cells, some prostate tumors
HER-2/neu	Helps relay signal telling cells to divide; excess produced with defective gene	Rapidly fatal, hard-to-treat breast cancer
BRCA-1	Uncertain	Isolated on chromosome 17; appears in 5% of 182,000 breast cancers; 85% of women with altered BRCA-1 develop breast cancer
p53	Fails to block growth or destroy damaged cells	Found in 50% of all cancers;; these cancers tend to be more aggressive and often fatal
p65	Linked to overproduction of hormones asscociated with breast and prostate cancers	p65 protein marker detectable in blood and decreases as tumors are destroyed during therapy
hMSH2/hMLH1	Fails to correct errors in DNA replication	Proteins normally correct order of nucleotide bases during DNA replication; errors linked to colorectal cancers; gene errors in 1 of every 200 people

Source: Adapted from Wallis, C., Molecular revolution, cancer: a special report, *Time,* 48, May 18, 1988; and ELSI Project, All About Breast Cancer Genes, LBL, June 1988, http//www.LBL.gov. education/ELSI/screening-main.html.

liferation of smoking abroad developed in concert with some American cigarette manufacturers comes at a time when cigarette smoking in the U.S. is under attack by public health officials. This battle was given fresh impetus by EPA's designation of environmental tobacco smoke as a class A carcinogen. While the federal and state governments battle with the major cigarette manufacturers in court, many health departments and agencies are passing health regulations banning smoking in public places.

Diet

Dietary factors also play an important role in cancer risk. Diets that are limited to 20% fat or less of total calories have shown dramatic decreases in prostate specific

TABLE 4.5
Modifiable Behaviors for Reducing Cancer Risk

MODIFIABLE BEHAVIORS
1. Quit smoking and use of other tobacco forms
2. Avoid excess alcohol consumption
3. Adopt a diet low in fat, high in vitamins A, C, and E. Include cruciferous vegetables such as broccoli, cabbage, brussel sprouts, and cauliflower
4. Adopt a high-fiber diet such as whole grain cereals, breads and pasta
5. Reduce total fat intake
6. Reduce intake of smoked, charred, and salt-cured foods
7. Avoid obesity through diet and exercise; maintain a sensible weight
8. Avoid psychosocial stress
9. Avoid known chemical carcinogens and overexposure to UV radiation (sunlight)

Source: Adapted from Perera, F.P., Environment and Cancer: Who is Susceptible?, *Science,* 278, 1068, Nov. 7, 1997; and Nadakavukaren, A., *Our Global Environment,* 4th ed, Waveland Press, Inc., 1995, Chapter 6.

antigen (PSA) circulating in the blood from the high twenties (indicating dangerous cancer activity) to levels of 0.4 to 2.5 (suggesting the tumors were no longer growing).[46] High fat intake has been linked with increased risks of colon and prostate cancer, whereas being excessively overweight encourages the growth of endometrial cancer.[53]

The precise mechanism for how low fat diets may lower cancer risk is unknown, but studies on mice suggest that calorie restriction boosts apoptosis and gene repair.[46] Apoptosis may also be triggered by butryric acid produced when bacteria ferment fiber in the gut. Consumption of high-fiber foods instead of refined flour and processed foods also appears to reduce the risk of cancer.[53] Cruciferous vegetables including broccoli, cabbage, cauliflower, kale, and Brussels sprouts all produce a powerful isothiocyanate anticarcinogen known as sulforaphane. Sulforaphane appears to increase the production of detoxifying enzymes that quickly eliminate cancer-promoting substances from the body.[54] Fruits and vegetables rich in antioxidants such as vitamins C and E protect against a variety of cancers by reducing free radical damage to DNA or to membranes, thereby reducing the access of environmental toxins. Garlic and onions have a long history of medicinal use throughout the world. Both garlic and onions have strong antimicrobial activity and also increase production of detoxifying enzymes through the allylic sulfides (allium) in the plant tissues. Many fruits, vegetables, and nuts contain ellagic acid, which binds to DNA, preventing the powerful carcinogen aflatoxin from forming a DNA adduct. Research in Sweden has shown that human breast milk contains a protein with anticancer activity. This protein seems to confer natural protection in infants from cancer and other diseases such as bacterial infections. In laboratory tests this protein seems effective against many different types of cancer cells.[55]

Although not a food, aspirin has been linked to reduced cancer risk. Persons who use aspirin regularly experience half the rate of colon cancer and lower rates of esophageal cancers than do persons who do not use it.[46] An enzyme called cox-2 prevents DNA damaged cells from dying, and aspirin inactivates this enzyme, allowing colon cancer cells *in vitro* (in the test tube) to die. Aspirin has been associated with reduced risk of colorectal cancer in humans, and Dr. Aaron Marcus of the New York Veterans Affairs Medical Center and Cornell University Medical College has recommended that people at risk of colorectal cancer take a single aspirin tablet (325 mg) every other day provided there are no contraindications.[46]

TRENDS IN CANCER

For the first time in nearly two decades, the rates of Americans diagnosed with cancer and dying from it have begun to decline. The results published in 1998 represent a collaboration of several agencies, including the American Cancer Society (ACS), The National Cancer Institute (NCI), The Centers for Disease Control (CDC), including the National Center for Health Statistics (NCHS). The data on cancer incidence were generated from the NCI's Surveillance Epidemiology and End Results (SEER) program representing 9 cancer registries around the country and 9.5% of the U.S. population.[51] The data on cancer deaths come from the NCHS, which collects death certificates from every state and so represents of the entire U.S. population. The data show that after increasing for 1.2% annually from 1973 to 1990, the rate of new cases (incidence) of 23 major cancers sites combined fell an average of 0.7% a year from 1990 to 1995. The overall cancer death rate, which had risen 0.4% annually from 1973 to 1990, fell by an average of 0.5% per year in the 1990s (Figure 4.13). A decline in adult smoking from 42% in 1965 to 25% in 1995 has played a major role in the declining incidence of cancer and overall cancer death rates.

Another reason for the decline may be the increasing use of mammography to detect breast cancer early with follow-up treatment. The decline in death rates for many other cancers is not easily explained. The overall cancer death rate for males declined 0.9% per year from 1990 to 1995, and women showed a decline of 0.1%. Blacks were more likely to be diagnosed with cancer between 1990 to 1995 than any other racial group, for all categories except breast cancer. Black women diagnosed with the disease were more likely to die from it than white women.[51]

Although the cancer experts are elated by the evidence that some progress is being made in the struggle against cancer, there remains the cold statistic that more than 550,000 people will die from cancer this year in the U.S. Further, the goal for reduction in death rates for most common cancers (causing most of the deaths) has not been met since President Nixon declared war against cancer with a "campaign" that began in 1971 with the National Cancer Act.[56] The National Cancer Institute stated that its goal was to reduce cancer mortality by 50% by the year 2000. In the years 1973 to 1992, overall cancer mortality actually rose.[57] Carcinomas of the lung, breast, prostate, pancreas, and ovary still claim the lives of hundreds of thousands people and the prognosis for most diagnosed cancers that are advanced, invasive, or metastatic has not improved much in the last 25 years.[56]

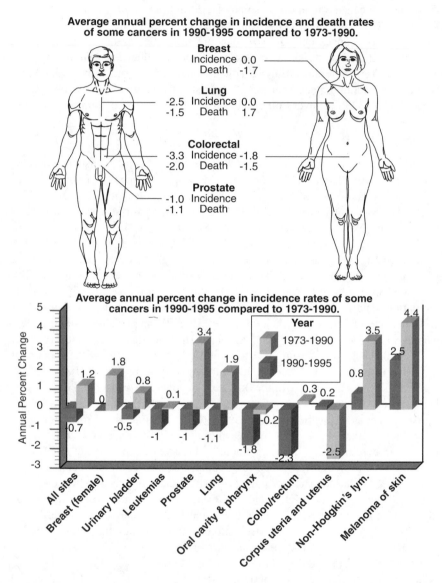

FIGURE 4.13 Average annual percentage change in incidence and death rates for various cancer sites in the U.S. in 1990–1995 compared to 1973–1990. (Adapted from Wingo et al., Cancer Incidence and Mortality, 1973–1995: A Report Card for the U.S., *Cancer*, 1197, March 15, 1998.)

Additionally, the NCI has reported that the rate of cancer among American children has been steadily rising by nearly 1% per year. With an increase in incidence from the early 1980s to the early 1990s of 36.6% in children under 10 years. The incidence rate in infants aged 0 to 1 years has increased by 53.6%.[58] Acute lymphoblastic leukemia in boys and girls increased by 27% between 1973 and 1990.[42] This has resulted in double-digit increases over the last few decades. Cancer is still rare in children when compared to adults and is diagnosed each year in an estimated 8000 children below the age of 15. It is, however, the most common form of fatal childhood disease, accounting for nearly 10% of all childhood deaths, and 1 of every 600 newborns is likely to contract cancer by the age of 10. Acute lymphoblastic leukemia rose by 27% in boys and girls from 1973 to 1990, but the rate has declined in boys since then, though still rising in girls. Glioma, a form of brain cancer, increased by 40% in children from 1973 to 1994. Recent and spectacular advances in the treatment of these childhood diseases has produced a decrease in the percentage of deaths in spite of the increased number of children diagnosed with cancer.

The causes of the cancers are uncertain, but some experts point to toxins in air, food, dust, and drinking water as probable suspects. Carol Browning, the former director of EPA, would like to see increased federal support to investigate the effects of environmental pollutants on children, and to begin looking at children's total cumulative risk from all exposures to toxic chemicals.[58] Children tend to be more susceptible to environmental toxins because (1) they take in more food, water, and air and accompanying carcinogens relative to body weight compared to adults; (2) nursing infants get higher exposures of substances such as dioxin in breast milk than adults exposed to background levels; (3) children have a higher internal dose of toxins and greater genetic damage than adults who have similar exposures to tobacco smoke and polycyclic aromatic hydrocarbons (PAH); (4) children have less efficient detoxification and repair systems; (5) they have a higher rate of cell proliferation during early developmental stages; and (6) children have increased absorption and retention of toxins.[14] Just as health and regulatory policies are likely to target more specific at-risk groups because of enhanced genetic knowledge, children may well become another more specific target for such policies and regulations. It may take a number of years to confirm that children require lower doses of carcinogens to promote disease, but if this is confirmed, it is expected that more rigid controls on pesticides, toxic wastes, and other chemicals will occur. Congress has already started to move in this direction with the approval of new laws that incorporate potential exposures to children when standards are established for pesticide residues in food and contaminants in drinking water.[58]

REFERENCES

1. Sessions, S.K. et al., Morphological Clues From Multi-legged Frogs: Are Retinoids to Blame?, *Science*, 284, April 1999, pp. 800–802.
2. Tanner, C., Parkinson's disease in twins: an etiologic study, *J. Am. Med. Assoc.*, January 27, 1999.
3. Cowley, G., The Real "Hot Zone," *Newsweek*, 136, November 27, 2000, pp. 66–67.

4. Whelan, E.M., *Toxic Terror: The Truth behind the Cancer Scares,* 2nd ed., Prometheus Books, Buffalo, New York, 1993, chap. 1.
5. Nader, R., Brownstein, R., and Richards, J., Eds., *Who's Poisoning America: Corporate Polluters and Their Victims in the Chemical Age*, Sierra Club Books, San Francisco, CA, 1981, chap. 1.
6. Grossman, K., *The Poison Conspiracy,* The Permanent Press, Sag Harbor, New York, 1983, chap. 1.
7. Naar, J., *Design for a Liveable Planet,* Harper and Row, New York, 1990.
8. Downs, H., The Poisoning of America, ABC-TV, New York, September 8, 1988.
9. Waldholz, M., War on cancer: new study disputes claims of progress, *Wall Street Journal,* 1, May 6, 1986.
10. Brown, D., Nonsmoking Cancer is found to Increase in Baby Boomers, *Boston Globe*, February 9, 1994, p. 1.
11. Epstein, B., *The Politics of Cancer,* Sierra Club Books, San Francisco, CA, 1987, chap. 1.
12. Skow, J., Can Lawns be Justified?, *Time,* 63, June 3, 1991.
13. Rather, J., Contaminant From Gas Is Found In Water, *New York Times*, August 29, 1999.
14. Perera, F.P., Environment and Cancer: Who Are Susceptible?, *Science,* 278, November 7, 1997, p. 1068.
15. Koren, H., *Illustrated Dictionary of Environmental Health and Occupational Safety,* CRC/Lewis Publishers, Boca Raton, FL, 1996, p. 407.
16. Thomas, C.L., Ed., *Taber's Cyclopedic Medical Dictionary,* 13th ed., F.A. Davis Co., Philadelphia, PA, 1977.
17. Interview with John Higginson, *Science,* September 28, 1979, p. 1363.
18. Whelan, E.M., How to reduce your risk of cancer, *Priorities,* Summer, 36, 1990.
19. Trichopoulis, D. and Hunter, D.J., What causes cancer? Cancer Causes, Screening and Prevention, April 22, 1998, http://oncolink.upenn.ed/cuaseprevent.html.
20. Associated Press (AP), Most link cancer to diet, but eating habits unchanged, *Daily Hampshire Gazette,* Northampton, MA, 2, February 24, 1987.
21. Preston, R., What New Things Are Going To Kill Me?, *Time,* 69, November, 1999.
22. Gadsby, P., Fear of Flu, *Discover,* January 1999, pp. 82–89.
23. Gorman, C., Beyond Lyme, *Time,* July 26, 1999.
24. Bickley, H.C., *Practical Concepts in Human Disease*, The Williams and Wilkins Co., Baltimore, MD, 1975, chap. 7.
25. Morgan, T.M., Gordon, L., Walker, B. et al., *Environmental Health,* 2nd ed., Morton Publishing Co., Englewood, CO, 1997, chap. 15.
26. Cummings, C.H., Entertainment Foods, *The Ecologist,* 29, January–February 1999, pp. 16–19.
27. Associated Press (AP), Mutation May Raise Cancer Risk from Secondhand Smoke, Cable News Network, December 1, 1999, http://www.cnn.com/health/aids/.
28. Sheperd, T.H., Detection of human teratogenic agents, *J. Pediatr.,* 101, 810, 1982.
29. Nadakavukaren, A., *Our Global Environment,* 4th ed., Waveland Press, Long Grove, IL, 1995, chap. 6.
30. Underhill, W., Shaped By Life in the Womb, *Newsweek,* September 27, 1999.
31. Shreeve, J., Secrets of the Gene, *National Geographic,* October 1999, pp. 43–75.
32. Tortora, G., Funke, B.R., and Case, C., *Microbiology: An Introduction,* 5th ed., The Benjamin Cummings Publishing Company, Redwood City, CA, 1995, chap. 9.
33. Smith, C.S., China Rushes to Adopt Genetically Modified Crops, *New York Times*, October 7, 2000.

34. Brower, L.P., Will Biotechnology Doom the Monarch?, *Defenders of the Wildlife,* Fall, 1999.
35. Kaiser, J., Environmental Institute Lays Plans for Gene Hunt, *Science,* 278, October 24, 1997, p. 569.
36. Jaroff, L., Keys to the Kingdom, *Time, Special Issue: Frontiers of Medicine,* 124, Fall, 1996.
37. The Washington Post, Belated Gene Therapy Records Reported Deaths, *Daily Hampshire Gazette,* 214, 126, January 31, 2000.
38. BBC Online, Gene Therapy Advance For Kids, *British Broadcasting Company,* February 20, 2000, http://news2.thls.bbc.co.uk/hi/english/health/newsid%5f647000/6472820.stm.
39. Couzin, J., Quandaries in the Genes, *U.S. News and World Report,* November 1, 1999, pp. 64–66.
40. Hyman, D. and Huang J., The Courage to Heal Cancer Survivor Doesn't Regret Her Difficult Decision, *Sunday Republican (Springfield),* October 24, 1999.
41. Berg, K., Ethical Aspects of Early Diagnosis of Genetic Diseases, *World Health,* 5, September–October 1996, p. 20.
42. Glasser, R.L., *The Greatest Battle,* Random House, 1976, chap. 1.
43. Center for Devices and Radiological Health, Consumer Update on Mobile Phones, October 20, 1999, http://www.fda.gov/cdrh/ocd/mobilephone.html.
44. Gorman, C., The Hope and the Hype, *Time, Cancer: A Special Report,* May 18, 1998, p. 38.
45. Cavenee, W.K. and White, R.I., The Genetic Basis of Cancer, *Scientific American,* March 1995, p. 72.
46. Kohitak, R., Cancer's Nemesis: The Suicide Gene, *Chicago Tribune,* 257, September 14, 1997, p. 1.
47. Harris, C.C., At the Crossroads of Molecular Carcinogens and Risk Assessment, *Science,* 62, December, 24, 1993.
48. Recer, P., Drug Said to Aid Cancer Therapy Toleration, *Boston Globe*, September 10, 1999.
49. Nash, J.M., The Enemy Within, *Time: Special Issue: The Frontiers of Medicine,* Fall, 1996, p. 14.
50. Saltus, R., Cancer Cells' Gene Signature May Aid in Diagnosis, *Boston Globe,* October 15, 1999.
51. Wingo, P.A., Reis, A.G., Rosenberg, H.M., Miller, M.D., and Edwards, B.K., Cancer incidence and mortality, 1973–1995: a report card for the U.S., *Cancer,* 1197, March 15, 1998.
52. Brown, L.R., Kane, H., and Ayres, E., *Vital Signs,* Worldwatch Institute, Norton, W.W., 1993.
53. Willett, W.C., Diet and Health: What We Should Eat, *Science,* 264, April 22, 1994.
54. Davis, D.L., Natural Anticarcinogens: Can Diet Protect Against Cancer?, *Health and Environment Digest,* 1, 4, February, 1990.
55. Radetsky, P., Got Cancer Killers?, *Discover,* June 1999.
56. Sporn, M.B., The war on cancer, *Lancet,* 347, 177, May 18, 1996.
57. Kosary, S.L., Rees, I.A.G., Miller, B.A., Hankey, B.F., Harras, A., and Edwards, B.K., Eds., *SEER Cancer Statistics Review, 1973–1992, Tables and Graphs,* National Cancer Institute NIH, Bethesda, MD, Pub. No. 95, 2789, 1995.
58. Cushman, J.H., U.S. Reshaping Cancer Strategy as Incidence in Children Rises, *New York Times,* 147,

5 Toxicity and Toxins

CHAPTER OBJECTIVES

A student, after reading this chapter, should be able to:

1. Discuss and define the concepts of the toxic triangle: poison, hazardous material, and hazardous waste.
2. List and explain the various methods of absorption, including diffusion, facilitated diffusion, active transport, and special processes.
3. Explain the processes of endocytosis, including phagocytosis, pinocytosis, and receptor-mediated endocytosis.
4. Describe and discuss the major mechanisms by which toxic materials produce their adverse effects, including: (1) inactivation of enzymes, (2) direct effect on cells and tissues, and (3) production of intermediate compounds or secondary action.
5. Describe and provide an overview of the immune system, the cellular and humoral immune system, and allergic mechanisms.
6. Discuss and describe the adverse health effects associated with endocrine disruptors, PCBs, dioxin, lead, mercury, asbestos, and organic solvents.

INTRODUCTION

In the U.S., there are currently more than 70,000 synthetic chemicals currently in commercial use, and with most of them, their toxicity is not widely known or understood. More than 60,000 are thought to be potentially harmful, and their effects could be devastating, causing cancer, neurological and reproductive disorders, or other injuries.[1] Some of these chemicals are so widely distributed and of such potential harm to human health and ecology that they serve as sentinel models for understanding the potential harm for the indiscriminate use and disposal of toxic agents. The compounds to be discussed in this chapter include: (1) dioxin, polychlorinated biphenyls (PCBs), and other endocrine disruptors; the heavy metals lead and mercury; asbestos; and the organic solvents. These represent a wide divergence in chemical make-up, mechanisms of toxicity, distribution, and patterns of exposure. Some substances, such as lead, are recognized to have existed as a hazard since early civilization when lead, for example, was used to store wine, to pipe water, and even as vessels in which to cook food.[2] Dioxin, PCBs, and other organochlorines have only recently been synthesized as products, or unwanted by-products, and have been implicated as agents in reproductive disorders or other diseases. Exposure to many of these chemicals at critical points in human development may produce lifelong adverse effects or even effects that will be passed on to subsequent generations.

Toxicological information is incomplete because animal testing requires significant time and money, and deals with one chemical at a time. Therefore, such testing provides insufficient information on the interactive effects that may occur during or after multiple chemical exposures. Additionally, animal studies often fail to detect future or delayed conditions. Human exposure studies are seldom performed because of the ethical problems associated with exposure of humans to various chemicals and metals. In addition to these insufficiencies, the federal government is currently reducing support for further research, while private corporations are filling the gap, which may potentially lead to bias in both study designs and interpretations.[1]

Similar to the legal premise that a person is innocent until proved guilty, a chemical is often considered safe until proven otherwise by the appropriate federal agency. Conferring such a privilege on chemical substances is not the best approach to protecting the public health. As 1000 to 2000 new chemicals are introduced each year into our society, there is a significant opportunity for untested materials to enter our environment and expose humans, wildlife, and plants to toxic effects. Sometimes, decisions on what substances are allowed to enter the environment are political and economic, with little regard for the effects on human health. In other instances, the health effects are simply unstudied or not recognized.[1] A potentially toxic substance produces its adverse effect by interacting with humans (or organisms) and the environment in a relationship referred to as the *toxic triangle* (Figure 5.1). Consequently, a poison or toxic substance does not constitute a hazard unless it makes contact with the organism in a form and quantity that can cause harm. Furthermore, the environmental conditions of heat, humidity, or even noise may aggravate the response to the poison.[3] Hazardous substances or wastes may be explicitly defined under federal law, and many of them are listed by name and regulated under the Clean Air Act (CAA), the Toxic Substances Control Act (TSCA), and the Clean Water Act (CWA). A

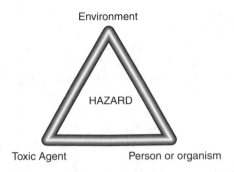

FIGURE 5.1 The toxic triangle. (Adapted from Schroeder, H.-A., *Environment,* 13, 18, 1971.)

hazardous substance is defined in the Comprehensive Environmental Response Compensation and Liability Act (CERCLA) as any chemical regulated under the previously mentioned acts. Hazardous materials are defined and regulated under the U.S. Department of Transportation's (DOT) regulations (49 CFR). The definition is quite broad and pertains to "a substance or material, including a hazardous substance, which has been determined by the Secretary of Transportation to be capable of posing an unreasonable risk to health, safety, and property when transported in commerce." Hazardous wastes are solid wastes defined under 40 CFR.261 as "any discarded material, garbage, refuse, sludge or other discarded material, that is solid, liquid, or confined gas which is abandoned, inherently waste-like, or recycled." Listed wastes appear under 40 CFR 261.30. Wastes not listed may still constitute a hazard if they possess defined characteristics of toxicity, corrosiveness, reactivity, and flammability. Toxic substances are those that (1) can produce reversible or irreversible bodily injury, (2) have the capacity to cause tumors, neoplastic effects, or cancer, (3) can cause reproductive errors, including mutations and teratogenic effects, (4) produce irritation or sensitization of mucous membranes, (5) cause a reduction in motivation, mental alertness, or capability, and(6) alter behavior; or cause death of the organism.[3]

People continue to be exposed to a wide spectrum of toxic and hazardous substances that potentially threaten their health. In most cases, the consequences of such exposures are unknown. The following sections in this chapter outline some of the principle mechanisms of toxicity and spotlight a few of the more widely distributed toxic and hazardous chemicals in our society.

EXPOSURE AND ENTRY ROUTES

EXPOSURE

For a toxic substance to produce its harmful effects on the human body, a person must first be exposed to the chemical. The harmful effect may occur upon contact, or the substance may be absorbed, after which the adverse effect is produced. Some toxic materials, such as strong bases or acids, can express their harmful effects by causing chemical lesions upon contact with the skin. Although no absorption of toxic substances takes place, the lesions produced are local and often result in killing areas of tissue in a process known as *necrosis*. Absorption refers to the passage of substances across the membranes through some body surfaces into body fluids and tissues by any of a variety of processes that may include diffusion, facilitated diffusion, active transport, or special processes.[3,4] Diffusion is a passive process that occurs when molecules move from areas of high concentration to those of low concentration. When this process involves water molecules across a semipermeable membrane, it is referred to as *osmosis*. The chemical nature (basic or acidic), electrical charge, and the size and shape of the molecules determine their ability to passively diffuse across the membrane.

Some molecules such as amino acids and sugars require specialized carrier proteins to be transported across a membrane. No high-energy phosphate bonds, such as adenosine triphosphate (ATP), are required in this process, and so it is referred to as *facilitated diffusion*. When ATP is required, in conjunction with special carrier proteins to move molecules through a membrane against a concentration gradient (i.e., high concentration to low), this process is called *active transport*. Typically, about 40% of the ATP in a human body cell is used in this process of actively transporting molecules across the cell membrane. This process is important in nerve impulse conduction, in which sodium and potassium are actively transported across the nerve cell membrane by the sodium-potassium ATPase pump. The energy from ATP is required because three sodium ions are actively transported out for every two potassium ions actively transported in.[3,4]

Particles and large molecules that might otherwise be restricted from crossing a plasma membrane can be brought in or removed by the process of endocytosis. There are three major types of endocytosis including phagocytosis, pinocytosis, and receptor-mediated endocytosis. Phagocytosis (*phag* = "eat"; *cyto* = "cell") occurs when specialized cells such as phagocytes extend out streams of cytoplasm, much like an amoeba, and engulf solid particles to form a sac around the particles. The sac detaches from the plasma membrane inside the cell, and enzymes attack the particles. When the materials are liquid, a small drop adheres to the cell membrane surface, where an indentation forms to surround the drop and form a vesicle, which in turn detaches from the membrane in a process called *pinocytosis*.[3]

Waste products, poisons, nutrients, and hormones may be located in interstitial fluids (between cells) at lower concentrations than in the cells. These substances bind to receptor sites on the cell surface, causing a vesicle to form around the substance in a process called *receptor-mediated endocytosis*. In a series of steps, the receptor is separated, and the substance is enzymatically digested.[3,4]

ROUTES OF ENTRY

Toxic substances may be gases, liquids, solids, or vapors. There are several ways in which toxic substances can enter the body. They may enter through the lungs by inhalation, through the skin, mucous membranes, or eyes by absorption, or through the gastrointestinal tract by ingestion.

Respiratory System

The respiratory system is composed of the nose, pharynx, larynx, trachea, bronchi, and lungs (Figure 5.2). Its function is to supply oxygen to the body's cells and to

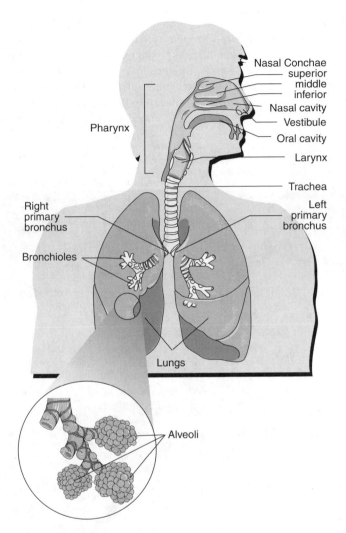

FIGURE 5.2 The respiratory system.

expel carbon dioxide from the body in a process called *respiration*. The act of breathing, or ventilation, draws air into and expels it out of the lungs. The exchange of oxygen and carbon dioxide between the atmosphere and blood is known as *external respiration*, whereas the exchange of gases between blood and individual cells is called *internal respiration*. Air is drawn into the nasopharyngeal area through nasal hairs at the vestibule and then over mucous-membrane-covered bony plates called the *nasal conchae*. These structures combine to filter, warm, and moisten inhaled air. Additionally, nerve endings in this nasal area may be stimulated to cause a sneeze reflex, helping to eliminate mucus and trapped particles. The inhaled air then enters the trachea–bronchial area, where the bronchi branch into numerous bronchioles with even smaller diameters until they terminate into thin-walled, delicate alveoli where gas exchange takes place. The trachea, bronchi, and bronchioles are lined by a velvety layer of cilia with mucous cells throughout. Particles such as dust and pollen of 10 mm or larger are removed by the constant streaming of mucus propelled from the bronchial and tracheal passages by the cilia beating at over 1300 times per minute in a process known as *mucociliary streaming*. The tubelike structures of the bronchi and trachea are also largely surrounded by smooth muscle layers that contract in response to irritating substances and allergens (Figure 5.3). This bronchoconstriction narrows the lumen and restricts the flow of air, other gases, and

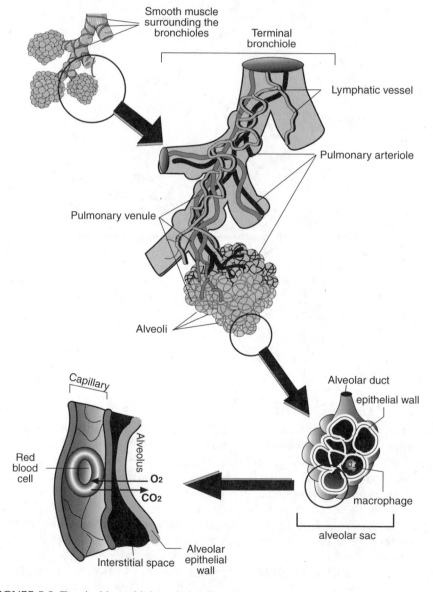

FIGURE 5.3 Terminal bronchiole and alveoli.

particles from reaching more delicate tissues deeper in the lung. This process may also be combined with excess mucous secretions or other fluids that make breathing difficult. When prolonged and severe, this process can be life-threatening, as in asthmatic attacks. A cough reflex may be initiated at the level of the bronchi to eliminate accumulated mucous in which foreign substances have become lodged.[2,3,5]

Inhaled air ultimately reaches the delicate, elastic and extremely thin alveolar sac at the terminus of the bronchioles (Figure 5.3). The alveoli are surrounded by tiny capillaries in which red blood cells give up their carbon dioxide and receive oxygen in exchange. Within the lumen of the alveoli are macrophages that serve as the final line of defense against microbes and a variety of organic and inorganic materials.[2]

Particulates in the air stream greater than 5 μm are mostly removed within the nasopharyngeal or bronchial areas by mucociliary streaming and either swallowed, sneezed, or coughed up. Smaller particles in the range of 1 μm or less reach the alveoli, where they may be (1) engulfed by the macrophages, (2) moved up the respiratory tree in a mucous sheet, (3) lodged in the interstitial tissues of the lung, or (4) found in the lymphatic spaces and lymph glands from which they are carried to the bloodstream.[3] Some inorganic particles cannot be digested by the macrophages and may prove to be lethal to them, resulting in the formation of scar tissue and loss of elasticity in the associated alveolar tissue. The size of particles largely determines where in the lungs they will be deposited, whereas the solubility of gases determines their fate in the lungs. Highly soluble gases such as sulfur oxides and ammonia tend to act in the upper moist airways of the respiratory tree, whereas less soluble nitrogen dioxide tends to reach deep into lung tissue, where it can cause tissue destruction and be more easily absorbed into the bloodstream.

The Skin

The skin is normally thought of as a relatively impervious barrier to chemicals and biological agents, and for the most part this is true. The skin is the body's largest organ, consisting of many interconnected tissues covering an area of nearly 2 m^2 in the average adult. The skin (1) regulates body temperature through sweat glands, (2) provides a physical barrier to dehydration, microbial invasion, and some chemical insults, (3) excretes salts, water, and organic compounds, (4) serves as a sensory organ for touch, temperature, pressure, and pain, and (5) provides some important components of immunity.[6,7] The skin consists of an outer thinner layer called the *epidermis*, which is attached to an inner and much thicker layer called the *dermis* (Figure 5.4). Beneath this layer is a subcutaneous layer of glandular and adipose tissues. The skin is effective in keeping out water and water-soluble (hydrophilic) materials, although small amounts of substances may be absorbed through hair follicles or sweat glands. Materials may also enter breaks in the skin from cuts, injections, insect or snake bites, and high-pressure steam or liquids.[7] Absorption of hydrophilic materials can be increased as sweating increases, or in the presence of moisturizers or oily materials. Substances that are fat soluble (lipophilic), such as organic solvents, petroleum compounds, and organochlorine compounds, can be absorbed by passive diffusion through the skin. A variety of substances, such as hexane, trichloroethylene, many insecticides, and military nerve gases fall into this category.

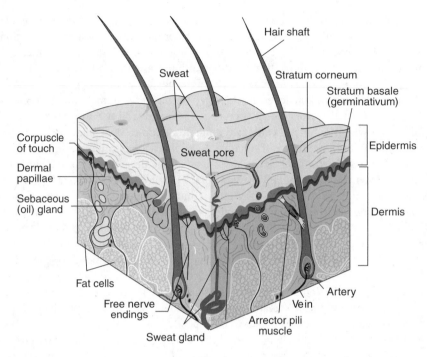

FIGURE 5.4 Cross section of human skin.

The eye is also an important route of absorption, especially by way of the moist epithelial tissue surrounding the eye in the socket that normally facilitates movement.

The Gastrointestinal Tract

The gastrointestinal tract is a continuous tube nearly 30 ft in length extending from the mouth to the anus (Figure 5.5). The components of the digestive tract include the mouth, pharynx, esophagus, stomach, small intestine, large intestine, and anus. The purpose of the digestive system is to ingest and move food for digestion and nutrient absorption. The system also functions to eliminate indigestible substances by defecation.[7,8] The gastrointestinal tract is a major route of absorption for many toxic agents, including mercury, lead, and cadmium, that appear in food and water. Most of this absorption occurs between the stomach and the upper portion of the intestine. Toxic substances can also be transferred from hands to mouth.

The inner layer of the small intestine or mucosa consists of a velvety layer of finger-like projections of 0.5 to 1.0 μm long called *villi*. There are up to 40 of these villi per square millimeter, giving a tremendous absorptive capacity to this section of the intestinal tract. Within each villus is a venule, a capillary network, an arteriole, and a lymphatic vessel. Nutrients, as well as toxic agents, can penetrate through the epithelial cells of the villus, enter the blood and lymph vessels, and be carried to various parts of the body (Figure 5.6). Nutrients, including carbohydrates, proteins, sodium, other electrolytes, and calcium, are absorbed using a combination of facilitated and active transport. Toxic materials that are similar in configuration, electrical

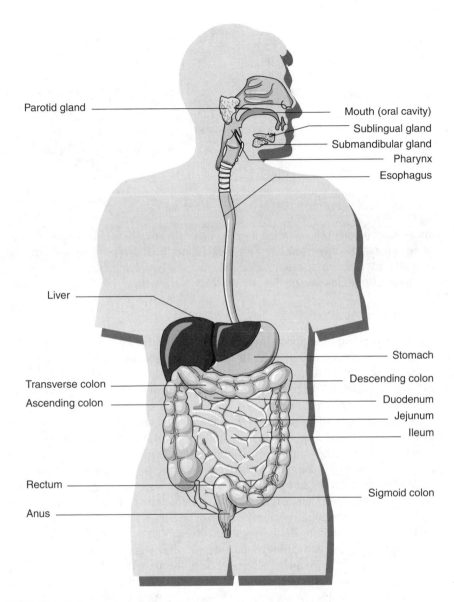

Parotid gland
Mouth (oral cavity)
Sublingual gland
Submandibular gland
Pharynx
Esophagus
Liver
Stomach
Descending colon
Transverse colon
Ascending colon
Duodenum
Jejunum
Ileum
Rectum
Anus
Sigmoid colon

FIGURE 5.5 The gastrointestinal tract.

charge, and molecular size may also be transported in this way. Such substances include lead, mercury, cadmium, and fluorides. The absorptive process can be influenced by physical health such as the presence of parasites, viruses, infectious bacteria, ulcers, cancer, or other digestive disorders. Absorption is also affected by nutritional status, the chemical and physical nature of the substances in the intestinal tract, level of physical activity, and age. Both spastic movements and mucous secretion are defensive capabilities of the intestinal tract that serve to remove many noxious agents either through diarrhea, vomiting, or both. However, soluble toxic

FIGURE 5.6 A cross section of villi in the gastrointestinal tract.

chemicals may be quickly absorbed into the bloodstream whereas less-soluble ones are often carried into the lower bowels for excretion. It is also possible for particles of up to 50 mm to penetrate the mucosa and enter the lymphatic channels and blood vessels to reach the internal organs.

MECHANISMS OF ACTION

Nearly 2500 years ago, Hippocrates described toxic effects of numerous poisons including lead, mercury, and arsenic. In the Middle Ages, the art of poisoning and the study of toxic effects became a formidable science in the hands of Catherine de Medici, who reported on the use of poisons, their clinical signs and symptoms, their potency, site of action, and how rapidly they produced their effects.[3] Today, we know that the toxic actions of pollutants vary dramatically, based on dose, length of exposure, and individual response. In fact, the symptoms that develop among a group of people exposed to the same toxic substance may vary to such a degree as to appear to be caused by completely different substances. A toxin can produce a harmful effect upon an organ by either stimulating or depressing the normal meta-bolic actions of that particular organ. Very small amounts of toxic agents have the ability to stimulate function in a variety of organs, whereas larger doses may impede or even destroy the activity of the organ involved. The body is able to detoxify foreign substances principally through the liver and other organs, such as the kidneys, intestines, and lungs. Sometimes, adverse effects occur when the quantity of the toxic substance in the body overwhelms the body's detoxification mechanisms or when injury or illness has compromised the body's ability to detoxify substances. The harmful effects of environmental toxins are dominated by three principal mech-anisms, which include (1) the toxin's influence on enzymes, (2) direct chemical combination of the toxin with a cell constituent, and (3) secondary action as a result of the toxin's presence in the system.[12]

EFFECTS OF TOXIC AGENTS ON ENZYMES

Enzymes are complex, three-dimensional proteins that have the unique ability of catalyzing (promoting or speeding up) metabolic activities in the body without

Substrate

Products

Apoenzyme
(Protein
component)

Cofactor
(Nonprotein
component)

Holoenzyme
(Whole
enzyme)

FIGURE 5.7 Components of an enzyme and mechanism of action.

undergoing any changes themselves. Most enzymes are highly specific and selective. The enzyme "prototype" may be envisioned in the following manner: an enzyme has the capacity to hold molecules together by binding to them, keeping them stabilized for a reaction to occur between them, and then moving on when the reaction has occurred. Many enzymes require a protein component (apoenzyme) and a nonprotein component (cofactor) to become active. The apoenzyme and its cofactor combine to form a complete and active enzyme called a *holoenzyme*. Cofactors are usually metal ions such as manganese, magnesium, copper, or iron. They may also be an organic molecule called a *coenzyme*, including vitamins such as nicotinamide adenine dinucleotide (NAD+) and nicotinamide adenine dinucleotide phosphate (NADP+). Enzymes act on substrates to add or remove molecules of water, oxygen, or hydrogen, or amino or other functional groups. Enzymes may also rearrange atoms within a molecule or join molecules (Figure 5.7). Enzymes are important in nearly all life functions including most aspects of energy production and biosynthesis. Many toxic substances have the ability to (1) interfere with or block the active sites of the enzyme, (2) inactivate or remove the cofactor, (3) compete with the cofactor for a site on the enzyme, or (4) alter the enzyme structure directly, thereby changing the specific three-dimensional nature of the active site (Figure 5.8). Hydrogen cyanide and hydrogen sulfide inactivate enzyme cofactors by binding to the iron present in the cytochrome oxidase system and inhibiting enzymes critical to the final steps in glycolysis and ATP production.[2,9] Similarly, dithiocarbamate will bind to copper and prevent the activity of copper-dependent enzymes important in the breakdown of ethanol. Consequently, alcohol consumption results in severe nausea and headaches. Substances similar to dithiocarbamate are used in alcohol antiabuse drugs.[9] Cadmium and beryllium are believed to inactivate enzymes by blocking the sites on the enzyme where cofactors such as magnesium or zinc normally attach.[10]

The inhibition of enzymes may be reversible or irreversible. Usually, reversible enzyme inhibition occurs when the enzyme acts upon a substance similar to its normal substrate and so is unavailable to carry out its normal function. When a toxic

| Blocking the active site of the holoenzyme | Inactivate or remove cofactor | Compete with cofactor for the site on the enzyme | Altering active site of enzyme |

FIGURE 5.8 Actions of enzyme inhibitors.

chemical such as lead, arsenic, parathion, ozone, or fluorine covalently bonds to an enzyme, the bonding tends to be permanent for that particular enzyme molecule and is therefore considered an irreversible reaction. In most of these cases, the active group of the enzyme is blocked or destroyed.[9]

The Direct Action of Pollutants on Cell Components

Many substances affect cells directly. Strong acids, bases, and phenol will combine directly with cellular materials and also kill tissue, producing areas of necrosis. Nitrous and sulfuric oxides and ozone are powerful oxidizers that irritate mucous membranes. Inhalation of phosgene or chlorine gases can blister tissue and produce edema. Inhaled carbon monoxide reacts directly with hemoglobin to produce carboxyhemoglobin and prevents the attachment of oxygen, which has a much lower affinity to hemoglobin than carbon monoxide. The greater the level of carboxyhemoglobin, the more severe the symptoms. Many sulfur-containing compounds including sulfonamides produce sulfhemoglobin in blood which is unable to carry oxygen. A number of aromatic amines, nitrites, and nitrocompounds can oxidize hemoglobin to methemoglobin, which is unable to carry oxygen, resulting in oxygen-poor blood and a syndrome in infants called "blue baby."

Pollutants That Cause Secondary Actions

Many substances may themselves be harmless but can cause the formation of other chemicals in the body that are harmful or potentially lethal. As an example, fluoro-acetate (rodenticide 1080) is enzymatically treated as acetic acid in the respiratory metabolic chain, causing the formation of fluorocitric acid. This substance prevents the formation of NADP and ATP by inhibiting aconitase. The consequence is often lethal in relatively small doses.[3,9] Methanol is degraded to formaldehyde, which attacks the optic nerve and causes blindness. In most instances, people are aware of the danger of these chemicals, but many are threatened by such innocuous substances such as animal dander, pollen, milk products, shellfish, peanuts, and even sunlight. These exposures all trigger an exaggerated immune response in some people, resulting in mild allergies from substances associated with hay fever to rapidly fatal anaphylactic shock. The intermediary substances produced include histamine, serotonin, and similar products. Many pollutants also have an allergic component. Toluene diisocyanate (TDI) is a component of polyurethane foam and is reported to produce allergic reactions.[5] The air pollutants nitrogen dioxide and sulfur dioxide are also reported to have an allergic component. In fact, there is an allergic or immune component to many environmental diseases. Many parasites, bacteria, viruses, and toxic chemicals may have an allergic component. What are allergies? How do otherwise harmless substances cause the release or production and final expression of intermediate substances in the body which can result in significant discomfort and even death? To understand this mechanism, a discussion of specific immunity and allergies is in order.

Immunity and Allergies

The immune system is critical to the defense of the human organism against invading microbes and abnormal or cancerous cells. An immune system debilitated by disease, chemotherapy, or inherited defects will permit the rapid onset of overwhelming infections or possibly cancer. The movie titled *The Boy in the Plastic Bubble* starring John Travolta was based on the true story about a boy who had inherited severe combined immunodeficiency syndrome (SCIDS). The absence of immune responses in this boy required him to live within an antiseptic, plastic-lined room in his home. Attempts to build an immune system in the boy through bone marrow transplants failed, and he died soon after. Just as threatening is an exaggerated response of the immune system, as seen in anaphylactic shock from insect bites or food allergies, which can lead to a painful death within minutes. Chemical pollutants such as ozone can depress the immune response by inactivating alveolar macrophages. Others, such as nitrogen dioxide or sulfur dioxide, have an allergic component that makes the reactions to exposure more severe. Immunity is a double-edged sword.

Origins of the Immune System

Immunity is based on the premise that certain immune cells in the body can recognize microbes, tissues, and other substances that are "nonself" or foreign, and so destroy, encapsulate, or remove them. There are two separate but cooperating components of the immune system known as *humoral* (antibody-mediated) *immunity* and *cellular* (cell-mediated) *immunity*. Cell-mediated immunity (T cells) respond primarily to viruses, fungi, foreign tissue transplants, and cancer cells. In humoral immunity, antibodies produced by B cells are directed against bacterial and viral infections, and soluble antigens, while also being involved in anaphylactic or "Type I" allergic responses. The responses of cellular and humoral immunity are quite different. A rabbit injected with bovine serum albumin (BSA) and a purified protein derivative (PPD) from a modified tuberculosis bacterium will develop an amplified immune response to these materials within 2 weeks. When these substances are injected separately just under the skin (subcutaneously) in two separate areas, the immune

This reaction developed within 72 hours of subcutaneous injection of PPD to which the rabbit had been previously exposed. The reaction is characterized by influx of macrophages and lymphocytes producing a raised, hard area.

This reaction developed within minutes of subcutaneous injection of BSA to which the rabbit had been previously exposed. The reaction is characterized by subcutaneous bleeding.

Cellular immune response to purified protein derivative (PPD) from Mycobacterium tuberculosis.

Humoral immune response to bovine serum albumin (BSA).

FIGURE 5.9 Demonstration in the rabbit of humoral and cellular components of the immune system.

response is quite different (Figure 5.9). The raised, hardened, warm area is caused by the cellular immune system to the tuberculin PPD. This area is filled with macrophages and lymphocytes and takes 2 to 3 d to develop. The subcutaneous bleeding is caused by humoral (antibody) response to subcutaneously injected BSA, which causes localized vasodilation and leaking of blood components into the area. The mechanisms of these two components of the immune system are discussed in the following sections. Each component of the immune system is formed in the embryonic stages from lymphocytic stem cells that appear in bone marrow. During this development, some stem cells migrate to the thymus gland where they are induced to develop into T cells and then populate specific areas in lymph glands, spleen, bone marrow, and other lymphoid tissue in the gastrointestinal tract. The remaining lymphocytic stem cells come under the influence of tissues (possibly gut-associated lymphoid tissue, bone marrow) to become antibody-producing cells or B cells that also occupy lymphoid tissue but in compartments other than T cells (Figure 5.10).

The Initial Immune Response

The immune system responds to agents, cells, or substances that are foreign or nonself, which are collectively called *antigens*. An antigen is a chemical, cell, or agent that causes the immune system to respond with the production of antibodies (humoral immunity), or T cells (cellular immunity), or both and then react with the antibodies or T cells. Most antigens are high molecular weight proteins often associated with lipids, carbohydrates, or nucleic acids. Some large polysaccharides may also be antigenic. Many molecules are too small to promote antibody formation; however, they may link to a larger carrier molecule such as a serum protein and function together as an antigen. The attached molecule is then called a *hapten*. Under most circumstances, an antigen, such as a bacterial cell, reaches one of the numerous

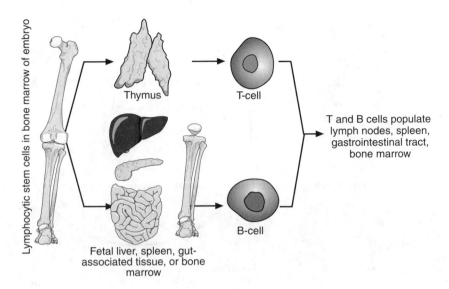

FIGURE 5.10 The origin of T cells and B cells.

lymph nodes in the body, where a large white blood cell known as a *macrophage* (large eating cell) ingests and processes the antigen by partially digesting it and displaying it on the surface of the macrophage along with human leukocyte-associated (HLA) antigens to B cells and T cells. The macrophage also produces lymphokines and interleukins, which promote the growth and activity of T cells and B cells.

Cellular Immunity

T cells respond to a particular antigen and then enlarge, divide, and give rise to clones of several subpopulations of T cells including cytotoxic (killer) T cells, helper T cells, memory T cells, suppressor T cells, delayed-hypersensitivity T cells, and amplifier T cells (Figure 5.11).[11,12] Killer T cells are sensitized to recognize particular antigenic groups on the surfaces of cancers cells, foreign transplanted tissues, viruses, and fungi. As the cells migrate through the body and recognize such chemical groups, they will respond by attaching to the cell and (1) secrete lymphotoxins that can kill the cell directly, (2) release a series of chemicals that attract, hold, and activate macrophages to assist in the destruction of the foreign cell, (3) secrete a transfer factor that sensitizes nearby lymphocytes and recruits them to the battle, and (4) produce interferon, which inhibits multiplication of viruses. Helper T cells stimulate other immune cells to activity. These helper cells are sometimes termed *CD4 cells* because of certain cell receptors on their surface. The CD4 or helper T cells are among the primary targets of the AIDS virus, thereby depressing the immune response. There also appear to be memory cells that are long-lived and will recognize the reentry of a particular foreign cell or material and cause a rapid immune response.

Humoral Immunity

B cells do not normally leave the protection of their locations in the lymph nodes, spleen, gastrointestinal tract, or bone marrow. Their response to foreign invaders is

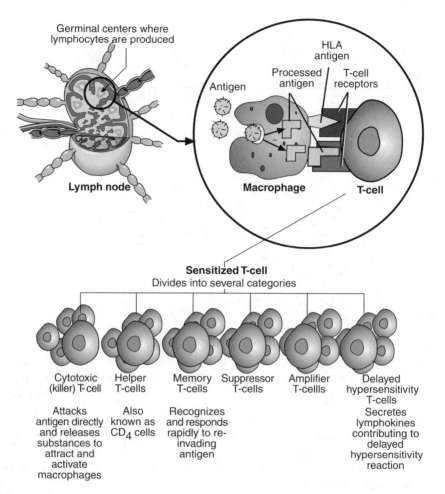

FIGURE 5.11 Action of killer T cells. (Adapted from Tortora G.J. and Anagnostakos, N.P., *Principles of Anatomy and Physiology*, 6th ed., Harper and Row, New York, 1990, chap. 22; Tortora, G.J., Funke, B.R., and Case, C.L., *Microbiology: An Introduction*, 5th ed., Benjamin Cummings, Redwood City, CA, 1995, chap. 17.)

to produce liquid proteins (humoral) known as antibodies and secrete them into the bloodstream, where they can travel to the affected site and carry out their destructive action. A foreign substance or cell is processed by a B cell with antigenic receptors that recognize specific chemical groups on the foreign material. Each B cell is specific to one particular antigenic group. An antigen selects a specific B cell and attaches to its antibody receptor on the B cell. The antigen is processed and presented with its HLA antigens on the B cell surface to helper T cells (Figure 5.12). The helper T cells produce substances that promote B cell proliferation into a clone of identical plasma cells, which produce antibodies to this specific antigen. Other B cells are recruited as memory B cells that are long-lived and will direct a rapid and aggressive response if the identical foreign material should reappear at a later time.

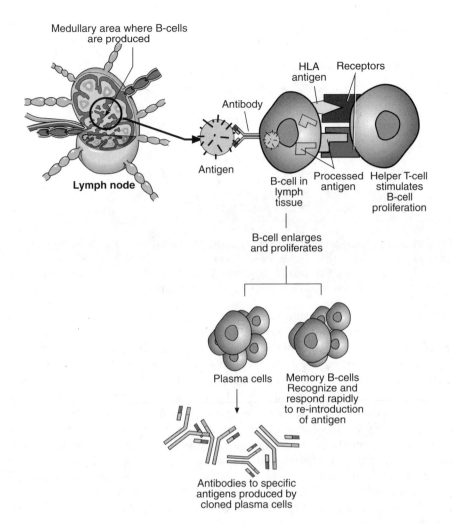

FIGURE 5.12 How B cells respond to antigens in humoral immunity. (Adapted from Tortora G.J. and Anagnostakos, N.P., *Principles of Anatomy and Physiology*, 6th ed., Harper and Row, New York, 1990, chap. 22; Tortora, G.J., Funke, B.R., and Case, C.L., *Microbiology: An Introduction*, 5th ed., Benjamin Cummings, Redwood City, CA, 1995, chap. 17.)

Antibodies to a particular antigen are then released from the B cells into the bloodstream, where they are carried to the affected site and act. The antibodies will attach to the antigenic groups on the foreign substance causing a number of possible actions, including (1) neutralization, (2) precipitation or agglutination, or (3) activation of the complement system. Prior to explaining these activities, it is necessary to describe a typical antibody molecule and its mode of action.

The Antibody Molecule

Antibodies (also called *immunoglobulins*) are proteins with molecular weights of 150,000 to 900,000 kDa. The simplest form of an antibody appears as a bivalent or

FIGURE 5.13 Typical bivalent antibody molecule. (Adapted from Tortora G.J. and Anagnostakos, N.P., *Principles of Anatomy and Physiology*, 6th ed., Harper and Row, New York, 1990, chap. 22; Tortora, G.J., Funke, B.R., and Case, C.L., *Microbiology: An Introduction*, 5th ed., Benjamin Cummings, Redwood City, CA, 1995, chap. 17.)

"Y"-shaped structure consisting of two identical light peptide chains and two heavy peptide chains joined together by disulfide linkages (Figure 5.13). Each light and heavy chain has a constant region composed of specific sequences of amino acids. These regions remain largely unchanged and differ only among the five major classes of antibodies known as immunoglobulins IgG, IgA, IgM, IgD, and IgE, plus 4 subtypes of IgG (IgG1 to 4) and 2 of IgA (IgA1 and IgA2). Light chains exist in two classes, lambda and kappa. Each antibody molecule has either lambda or kappa light chains, not both and each of these classes has a separate function. The stem of the constant region of the heavy chain is also known as the Fc region and is capable of attaching to a number of host cells including mast cells. The opposite end of the antibody molecule, the Fab portion consists of variable regions in the heavy and light chains. The combination of the variable regions on a single heavy chain and its attached light chain form a single antigen-binding site that will attach tightly to one very specific antigenic group.

As there are two such sites on the simplest "Y" form antibody, it is called *bivalent*. Furthermore, the peptide arms of the "Y" are flexible and may form a "T" shape as well. Each possible antigenic group on a foreign substance must have its own unique antibody. Consequently, the variable regions of the antibody are created in a specific three-dimensional form that is preconfigured to only one antigenic group in the B cell clone (Figure 5.14). The largest percentage of antibodies belongs to the IgG class. These bivalent antibodies have a molecular weight of 150,000 MW and can cross the placenta to provide passive immunity in newborns, enhance phagocytosis, neutralize toxins and viruses, and activate the complement system. The role of IgE in allergic reactions will be discussed later.

Antibody Activities

Antibodies can neutralize viruses and organic toxins (as found in snake bites or bacterial toxins) by attaching to receptors and preventing their attachment to host

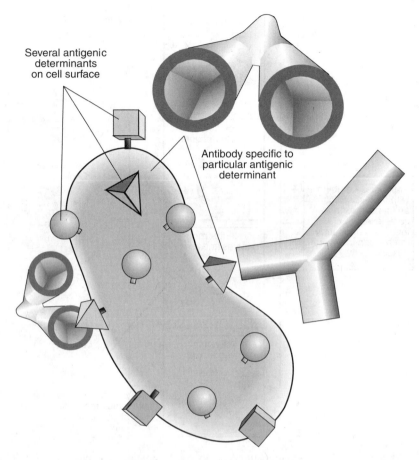

FIGURE 5.14 Specificity of antibody and antigenic determinants.

cells, or neutralizing toxins by blocking active sites. Antibodies will combine with cells or soluble proteins or haptens to promote agglutination or precipitation. This effectively enlarges the particle sizes and enhances the process of phagocytosis by macrophages. Yet another function of antibodies is opsonization in which the bacteria is coated by the Fab region, thereby facilitating subsequent phagocytosis by cells possessing an Fc receptor. The binding of an antibody with its specific antigen can activate the complement system. The complement system is composed of several (at least 20) serum proteins that follow a sequence or cascade of actions beginning with the Ag-Ab binding. This sequence of action enhances phagocytosis, inflammation, and cell lysis (Figure 5.15). When exposed to an antigen, there are two types of responses, namely, the primary response, which is slow in onset, low in magnitude, and short lived, and the secondary response, which is rapid in onset, high in magnitude, and long lived. There is a lag of several days before a specific antibody, IgM, becomes detectable. After a short time, the antibody level declines. This is the primary response. If at a later date we are exposed to the same antigen, there is a far more rapid appearance of antibody in a greater amount. It is of IgG

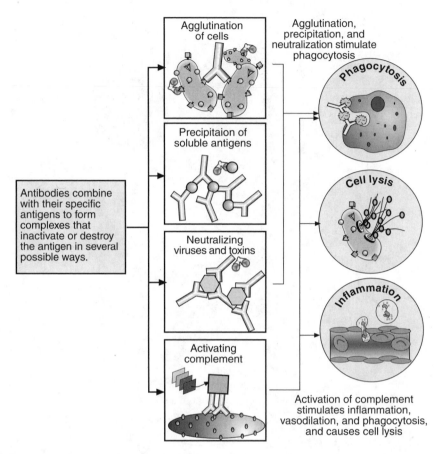

FIGURE 5.15 Antibody activities stimulated by Ag-Ab binding. (Adapted from Tortora G.J. and Anagnostakos, N.P., *Principles of Anatomy and Physiology*, 6th ed., Harper and Row, New York, 1990, chap. 22; Tortora, G.J., Funke, B.R., and Case, C.L., *Microbiology: An Introduction*, 5th ed., Benjamin Cummings, Redwood City, CA, 1995, chap. 17.)

class and remains detectable for months or years. These are the features of the secondary response.

Hypersensitivity

An exaggerated immune response to the presence of an antigen is termed *hypersensitivity* or *allergy*. Allergies are the other side of the two-edged sword of immunity. These immune reactions are usually unnecessary, harmful, and sometimes fatal. People have developed allergies to almost everything, including insect bites, pollen and other plant products, environmental toxins, a variety of foods, sunlight, human semen, and much more. Some people have become so sensitized to chemical substances that they are unable to physically cope in routine surroundings. Such people are said to have multiple chemical sensitivities (MCS), which is a highly controversial subject. There is also a very significant rise in the numbers and severity of asthma cases. The prevailing scientific thought is that environmental pollutants such as ozone and fine particulates contribute to this phenomenon.[13]

There are four major types of hypersensitivity reactions: cytotoxic, cell-mediated, immune complex, and anaphylactic. Cytotoxic hypersensitivity involves complement activation and is most often seen in transfusion reactions when incompatible ABO and Rh antigens on blood cells are transfused to a person, resulting in the donor cells being lysed. Cell-mediated reactions involve T cells that respond to certain antigens on transplanted tissues and organs, and to haptens associated with poison ivy or poison oak, cosmetics, latex gloves and condoms, nickel in coins and zippers, and bacterial extracts, such as PPD from *Mycobacterium tuberculosis*. Immune-complex reactions involve the coupling of antibodies with soluble antigens circulating in the serum in specific ratios that permit the complex to become trapped in the basement membrane of blood vessel cells where they can activate complement and cause inflammatory responses. Anaphylactic, or Type I hypersensitivity, is associated with hay fever, asthma, and anaphylactic shock from insect bites, certain foods such as shellfish, and a variety of chemicals. The reactions are usually immediate, occurring within moments in persons previously sensitized by exposure to the offending antigen. Type I hypersensitivity occurs when the antigen binds in specific ways to IgE antibodies attached to mast cells and basophils. Mast cells occur in the connective tissue of the respiratory tract, skin, and blood vessels. Basophils are a form of white blood cells circulating in the bloodstream. The IgE antibodies attach to the mast or basophil cells by their Fc fragment. Persons who have developed an excess of IgE antibodies run a higher risk of developing an anaphylactic type of allergic response because most of the sites on the mast cell or basophil cells will be covered by IgE antibodies. When an antigen such as animal dander or pollen attaches across two closely spaced IgE molecules specific to that antigen, this causes the mast or basophil cell to degranulate and release histamines that stimulate vasodilation and bronchoconstriction (Figure 5.16). Vasodilation increases the permeability of blood vessels, allowing blood components to seep into the affected area (Figure 5.7). Swelling and redness of mucous membranes in the eyes, nose, and throat are also common symptoms of this type of allergy when the allergens are confined to the respiratory tract. The histamines also cause smooth muscle contraction, which occurs in the bronchi and bronchioles of the lung. This bronchoconstriction decreases the lumens of the respiratory tract and makes breathing difficult. The mast or basophil cells, once triggered by an antigen, cause the synthesis of leukotrienes and prostaglandins that cause prolonged contractions of smooth muscle and increased mucous production, which further aggravates respiratory distress. Localized reaction may lead to hay fever, hives, and asthma. These allergic reactions can also be systemic (involving the whole body), producing breathing difficulties, a precipitous decline in blood pressure, cyanosis, and possibly death.[11,12]

FACTORS GOVERNING TOXICITY

The outcome of exposure to a toxin depends on a number of factors that may include (1) the chemical properties of the substance, (2) the concentration of the substance, (3) the length of time for which the person is exposed, (4) the route of exposure,

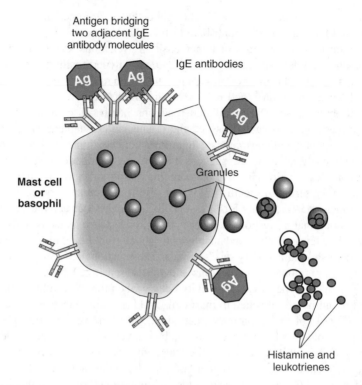

Antigen bridging
two adjacent IgE
antibody molecules

IgE antibodies

Ag   Ag

Ag

Mast cell
or
basophil

Granules

Ag

Ag

Histamine and
leukotrienes

FIGURE 5.16 Mechanism of anaphylactic allergic responses involving IgE. (Adapted from Tortora G.J. and Anagnostakos, N.P., *Principles of Anatomy and Physiology*, 6th ed., Harper and Row, New York, 1990, chap. 22; Tortora, G.J., Funke, B.R., and Case, C.L., *Microbiology: An Introduction*, 5th ed., Benjamin Cummings, Redwood City, CA, 1995, chap. 17.)

(5) whether or not it is biotransformed in the body, (5) the interaction with other pollutants, and (6) a number of personal or internal factors, including a person's age, level of activity or stress, and health status.

Chemical Properties

Chemicals vary in their toxicity to mammals. An oral dose of sodium chloride would require 3000 mg/kg to achieve an LD_{50} in rats, whereas the LD_{50} for strychnine is less than 2.5 mg/kg. The LD_{50} is defined as the dose of a chemical that will cause death in 50% of a defined animal population.

Concentration

The concentration of the substance to which one is exposed is also important because the concentration and the length of time of the exposure determine the dose. Each toxin and health effect has its own dose–response relationship. Adverse effects from exposure to toxic agents may range from a slight irritation, such as itchy, red eyes, to rapid death. Large doses may cause more harm than exposure to the same toxin in small amounts over a long period of time. In acute illness, effects are often witnessed immediately, with characteristic symptoms occurring over a limited time. Chronic effects are usually the result of long-term exposure to small amounts of toxins, with illness developing slowly and insidiously (without obvious identifiable symptoms).[2] A sudden exposure to large amounts of carbon monoxide will lead to dangerous concentrations of carboxyhemoglobin in the blood with resulting unconsciousness and death. Persons exposed to the same toxin in smaller doses over longer periods may slowly develop headaches, dizziness, and chest pains, with the possibility of damage to the heart and blood vessels.[14]

Usually, only very small amounts of an absorbed chemical reach target tissue, which becomes the site of action for the chemical when it attaches to molecular receptors. The effective dose is the amount that reaches general circulation and, ultimately, the location where a particular effect is seen or felt. This dose, along with the speed with which it happens, largely determines the pharmacological effect. Drinking 5 beers in an hour has a noticeably different effect from drinking 5 beers in 4 h. Once in general circulation, the toxic chemicals may be absorbed, stored, biotransformed by other tissues, and/or excreted. The route of entry is also critical to determining the toxic consequences of the agent. Some substances, such as lead and mercury, are more efficiently absorbed through the lungs than the intestines, making airborne forms of these substances of more concern.[15] When materials containing lead or mercury are incinerated or otherwise combusted, the potential consequences of even small amounts should be of greater concern.

The mechanisms of tissue absorption have been previously discussed and include passive and active processes. Some substances may be localized or stored in various body compartments characterized by having similar structures and functions. Lead tends to be stored in long bones, whereas organochlorines tend to be stored in fatty tissue. If the substance is stored at levels above what would be normally expected, this is termed *bioaccumulation*. When metabolic processes alter the structure and characteristics of a chemical, this is called *biotransformation*. A wide number of enzymatic pathways are capable of biotransforming toxic substances into nontoxic substances, and these pathways exist in the liver, kidneys, lungs, intestines, and other organs. Pesticides such as malathion are detoxified in humans and other mammals by the enzyme carboxyesterase. However, such biotransformations are not always beneficial as discussed previously when sodium fluoroacetate is readily biotransformed into fluorocitrate by mammalian cells, resulting in chemical asphyxia.[2,9] Absorbed toxic substances may also be excreted. The major excretory organs are the kidney, liver, and lungs. Chemicals that are not readily soluble in blood and that may have high vapor pressures will likely be exhaled via the lungs. Urine and bile are often the recipients of water-soluble toxic materials. Substances not eliminated in this way are generally biotransformed or stored in some body compartment.

Interactions

Most of the information on toxicity of chemicals is based on studies that attempt to limit all variables but the effects of the single chemical under study. However, most exposures come with multiple chemicals, and these additional chemicals influence the outcome of the exposure. The effects of some substances are additive in that the two substances taken together are the sum of the two substances taken separately. If you are exposed to chemical A, and it produces a measurable response of a 5% decrease in red blood cells (RBCs), and chemical B produces a reduction of 3% in RBCs, the two together (A + B) would be expected to produce an 8% reduction, if their effect was additive. Generally, substances behaving in this way employ a similar set of toxic mechanisms. In our example, if the two substances produce a 20% reduction in RBCs the effect would be much greater than the sum of the two substances, producing a synergistic effect. This has great significance in environmental health because many toxic agents act in this way. Sulfur dioxide is far more toxic in the presence of small particulates or sulfuric acid mists, which carry the pollutant deeper into lung tissue.[16] Again referring to our example, if substance A and substance B are combined and the effect is to produce a reduction of only 2% in RBCs, the effect is less than the sum of the two acting independently, and this is referred to as *antagonistic*. Although particulates enhance the toxicity of sulfur dioxide, they have quite the opposite effect when inhaled simultaneously with nitrogen dioxide by decreasing the toxicity.[17] Stimulants such as caffeine in some teas and coffee will tend to counteract the effects of barbiturates.[9]

Age

There are a number of personal factors that influence the effects of toxic substances. The age of the person is a major factor. The very young and the elderly tend to be at increased risk from toxic substances. The young are more susceptible for a number of reasons including: (1) they inhale or ingest more toxic substance per pound of body weight than adults; (2) they are more likely to put dangerous substances into their mouths; (3) children often absorb materials into the tissues at a higher rate than adults; (4) they will tend to retain larger proportions of toxic substances than adults; and (5) children's nervous systems and detoxifying systems are not fully developed, causing disproportionate effects from toxic materials.[18] The adverse effects of ozone, lead, mercury, and sulfur dioxide are more pronounced in children than in adults.[18,19] The elderly are also at greater risk from the effects of toxic substances as they tend to have preexisting respiratory and cardiovascular diseases,[2] depressed immune systems, and higher levels of stored toxic agents in their bodies than younger people.

Exercise and Physical Stress

Most athletes are aware that you should not run in hot, humid, heavily polluted air. The consequences are reduced performance and even pain in the area of the lungs. There is a tendency when exercising heavily to take in air through the mouth in large amounts. Consequently, higher levels of pollutants bypass the nasopharyngeal area and enter the sensitive tissues of the lungs more directly. It has been known for some time that rats exposed to 1.0 ppm of ozone while being forced to exercise periodically in a rotating cage will die, whereas nonexercising rats will experience little discomfort at the same exposure.[19]

Health Status

The health status of a person is also important in the outcome of toxic chemical exposure. Preexisting cardiovascular disease will render a person more susceptible to a variety of air pollutants. Persons who have asthma or who smoke are also increasingly sensitive to the many pollutants. Alpha1-antitrypsin deficiency is a genetic defect in humans that markedly increases the risk of emphysema from a number of chemical exposures. Genetic polymorphisms, or variations of the same gene, can produce subtle or dramatic differences in response to toxic chemicals. The P450 genes produce enzymes that detoxify some agents in tobacco smoke. Variations in P450 genes result in enzymes of varying activity that may increase the risk of lung cancer for many people.[20] Unfortunately, many human diseases are associated with genetic defects. Debilitating diseases, poor nutrition, and lack of exercise all contribute to increasing the risk of adverse effects from toxic substances.[5]

SOME SPECIFIC EXAMPLES OF TOXIC AGENTS

ENDOCRINE DISRUPTORS AND REPRODUCTIVE HEALTH

Hormone Function

Hormones are a type of chemical messenger that originate in an organ, gland, or other body part, and are then conveyed through the blood to a target organ or tissue which is stimulated by the hormone to increase functional activity or secretion. Hormones do not provide energy and are active in very small quantities. They are critical in the regulation of many life processes, including sexual development, metabolic functions, development of the brain, human growth, and stress response. As examples, androgens regulate the development and maintenance of male sexual characteristics, whereas estrogens stimulate the development of female sexual characteristics. The predominant and most active androgen is testosterone, which is produced by the male testes. The other androgens that support the functions of testosterone are produced mainly by the adrenal cortex (the outer substance of the adrenal glands) and only in relatively small quantities. Hormones function by attaching to specific receptors on the inside or outside of cells. These receptors have a three-dimensional complementary fit to the active site of the hormone. Once the hormone links with its specific receptor site, a series of events is initiated that may involve the production of other chemical products or specific biochemical reactions.[1,21] When the activity of such hormones are disrupted, a number of unwelcome outcomes may occur.

Adverse Effects of Endocrine Disruption

In a report issued in 1997, the EPA concluded that endocrine disruptors can produce cancer, developmental problems, and sterility in animals.[22] There are many examples of adverse effects on reproduction in wildlife and fish in many geographic areas including Florida, the Great Lakes, and California.[23,24] In those areas where fish and wildlife have been exposed to endocrine-disrupting hormones, the effects on reproduction and development have been marked and include the mixing of male and female characteristics, or indeterminate sex; the feminization of male species, and the masculinization of female species; unsuccessful reproduction; thyroid dysfunction in birds and fish; decreased fertility in birds, shellfish, and mammals; gross birth deformities in turtle, fish, and birds (i.e., twisted beaks in birds); and behavioral abnormalities.[25,26] The effects of endocrine disruptors on humans is much more problematic. Significant differences of opinion have arisen concerning the potential threat. Some authorities emphasize that there is no strong evidence that environmental disruptors cause adverse health effects in humans at the current ambient exposure levels, and that there is no causal link between such chemicals and perceived human illnesses.[21] In fact, the science of endocrine disruptors is highly controversial, and there is little knowledge or agreement on the prevalence of the problem.[27] Many people have pointed to endocrine disruptors as the cause of a number of emerging human ailments including (1) reduced sperm counts, (2) precocious puberty, (3) increase in non-Hodgkin's lymphoma, (4) marked increase in males having undescended testicles (cryptochordism), and (5) testicular cancer.[27–29] Studies have shown a reduction in sperm counts to levels that suggests increased infertility within the general population of many different countries.[30] Other studies suggest there may be regional problems of sperm decline, but challenge the assertion that sperm counts are declining worldwide.[31] The frequency of undescended testicles has more than doubled in England and Wales,[32] and has increased in other countries as well.[29] More recent studies have not demonstrated an association between breast cancer and the endocrine disruptors DDE (a metabolite of DDT) and polychlorinated biphenyls (PCBs). In one study, blood levels of PCBs and DDE were measured in 600 pregnant women. When their babies were born, the researchers then measured levels in the mothers' breast milk. Finally, the team monitored the children as they grew and entered puberty. The most prominent effect reported was that boys exposed to DDE and girls exposed to PCBs were heavier than their unexposed peers at age 14. The study also noted that girls with high prenatal PCB exposure tended to hit the first stages of puberty a bit earlier than others.[33] The role of these chemicals in promoting testicular cancer has also not been demonstrated.[27] However, numerous other endocrine disruptors exist, and the level of concern among public health professionals is high. The combination of increased testicular cancer, undescended testicles, and falling sperm counts is thought to be caused by fetal exposures to endocrine-disrupting chemicals.[21] Studies of rats exposed *in utero* have shown that alkylphenols disrupt Sertoli cell development in the fetuses, causing decreased testes size and a decrease in daily sperm production as the rats grow to adults.[34,35] Because humans are exposed from many pathways, at levels similar to those in rats, there is a significant concern that these and other endocrine disruptors may be the reason for increasing reproductive disorders seen in males.[21] Efforts are under way by the EPA, working groups, and an advisory committee, together with industry to develop screening strategies for environmental disruptors and to implement a program by August 1999 with a report to Congress in the year 2000.[27]

TABLE 5.1
Endocrine-Disrupting Chemicals

CHEMICAL	USES
Phthalates	Used as plasticizers; found in medical and household products, clothing, toys, and packaging including plastic wraps, and liners of metal food and beverage cans; phthalates leach from food wraps and lining of containers into food and beverages where they are ingested; phthalates may also be absorbed through the skin and have been associated with testicular toxicity
Dioxin	An unwanted byproduct in the production of some chlorinated hydrocarbons, paper production, and incineration of plastics
PCBs	Used in electrical transformers and capacitors, hydraulic fluids, plasticizers, and adhesives; production was banned in the U.S. in 1977 because of bioaccumulation, persistence, and toxicity
Alkylphenols	Present in detergents, paints, pesticides, plastics, food wraps,and plastic laboratory ware; they are known to be estrogens
Pesticides	Methoxychlor, lindane, dicofol, atrazine, and DDT are pesticides that exhibit weak to moderate estrogenic activity, but mixtures of two or more pesticides have been shown to produce potent activities sometimes 1000 times that of individual pesticides
Natural sources	Grains, some vegetables, nuts, berries, grasses, and soybeans may contain phytoestrogens such as flavinoids and liganans with hormone-like properties

What Are Endocrine Disruptors and How Do They Work?

Endocrine disruptors have been referred to as xenoestrogens, estrogenic, hormone mimicking, and endocrine-disrupting chemicals. The synthetic chemicals presently identified include: a number of pesticides such as DDT, lindane, and atrazine plasticizers, such as phthalates and alkylphenols; bisphenol A; PCBs; dioxin; the food antioxidant, butylated hydroxyanisol (BHA); and several fungicides. A variety of naturally occurring plant compounds or phytoestrogens are also known to have estrogen compounds (Table 5.1). There are at least four different mechanisms by which endocrine disruptors can exert their adverse effects (Figure 5.17):

1. Some endocrine disruptors are similar enough to the native hormone that they can occupy its site on the receptor tissue, causing a cascade of events as if the native hormone had bound to the site. Such a chemical is called a *hormone mimic*. Some chemicals may bind to the site but are not similar enough to activate a response but simply prevent the hormonal action through a hormone block.
2. Endocrine disruptors may cause artificial and unwelcome changes in the numbers of hormone receptor sites on tissues and organs.

FIGURE 5.17 Mechanisms of endocrine disruption.

3. The baseline production of hormones may be diminished or increased.
4. Endocrine disruptors may replace hormones on the carrier proteins that transport hormones via the blood stream to various parts of the body. This makes some proportion of the native functional hormone population unavailable and produces adverse health consequences that may be barely noticeable to very severe.[21]

Pollutants, originally described as estrogens or androgens, were based on their ability to bind to two different types of steroid receptors, and not much thought was given to the possibility of one binding to both sites. However, recent studies *in vitro* reveal that some environmental pollutants can bind to estrogenic, androgenic receptors, and binding proteins. If this phenomenon occurs in living organisms, then the effects of environmental disruptors may occur across several different receptor sites and magnify the effects, making even very low concentrations quite potent.[36]

Reducing Exposure

Although no definitive evidence exists on the role of endocrine disruptors in adversely effecting human health, there is sufficient concern so that federal agencies, industry, and private organizations are attempting to respond to this complex dilemma with more answers through research and recommendations for risk reduction. Many of the endocrine-disrupting chemicals are chlorine based and persistent in the environment. Furthermore, they are ubiquitous and find their way into the food supply through (1) ingestion of contaminated grains and grasses by livestock, which then store the lipophilic chemicals in their fatty tissues, (2) contamination of fruits and vegetables by spraying with pesticides, and (3) leaching of endocrine-disrupting chemicals from plastic wrappers, plastic liners of cans, and polystyrene containers. The majority of exposures may come from such contaminated food. Individuals who are concerned about the exposure to endocrine disruptors may reduce that exposure by:

1. Reducing or limiting ingestion of dairy products and meat high in fat where organochlorines tend to accumulate.
2. Avoiding synthetic pesticides by purchasing foods low in pesticide residues and switching to herbal or scent-based repellents.
3. Keeping children from using vinyl toys or teething rings.
4. Using detergents and shampoos that do not contain alkylphenols such as nonoxynol and octoxynol.[22]

DIOXIN

Dioxins are not manufactured for direct use in any product. Dioxin is actually an unwanted byproduct obtained from heating mixtures of chlorine and organic compounds in industrial processes, such as the bleaching of paper pulp, the incineration of chlorine-containing materials, or in the production of pesticides, herbicides, and chlorophenol wood preservatives. They are formed naturally by the incomplete

FIGURE 5.18 Chemical structure of dioxin.

combustion of wood products and industrial and municipal wastes. Unfortunately, dioxin does not easily break down, therefore dioxin tends to bioaccumulate (the accumulation of toxic chemicals in living things through the consumption of food or water) in the fatty tissue of humans and animals. This substantially increases the exposure of nursing infants whose mothers may have concentrated dioxin in the milk and fatty tissues of the breast.[37] There are 75 separate chemical compounds that make up the group referred to as *dioxins*. The most acutely toxic chemical of this group is known as 2,3,7,8-tetrachlorodibenzo-p-dioxin, or TCDD. The International Agency for Research on Cancer (IARC), part of the WHO, announced on February 14, 1997, that TCDD is now considered a class 1 carcinogen, meaning a "known human carcinogen."[38] Dioxin slowly breaks down in the environment when it is exposed to the ultraviolet rays of the sun, otherwise, dioxin is a stable compound. As it is highly insoluble in water, it is not readily able to migrate through the soil, and once in the soil, it persists.[39]

Dioxin is considered to be one of the most toxic man-made chemicals (Figure 5.18). It is not an estrogen, however, but it can block the action of estrogens.[40] Dioxin is capable of lowering androgens and affecting the levels of thyroid hormones. It affects both the endocrine and the immune systems.[29] Dioxin does not attach to the estrogen receptor; rather, it attaches to a receptor called the Ah-receptor, whose function is unknown. Even though it does not attach to the estrogen receptor, it still exerts both estrogenic and anti-estrogenic activity, indirectly influencing estrogen activity. Its antiestrogenic effects may result from (1) causing the DNA, which is attached via the receptor, to produce an enzyme that is capable of breaking down the body's normal estrogen, or (2) simply by decreasing the number of estrogen receptors available for naturally occurring estrogen.[41]

Although dioxin is considered to be among the most potent synthetic toxic chemicals known, its effects are highly variable among species and even differ among age groups within the same species. Animal studies show wide variations in reproductive and developmental effects from dioxin in different species at different levels of exposures. Adult hamsters are several thousand times more resistant to the toxic effects of dioxin than adult guinea pigs, whereas there is a much smaller difference among the fetuses.[42]

Most of the effects of dioxin are based on animal studies, but there have been unintended or accidental human exposures exceeding normal ambient levels. The largest known dioxin contamination occurred between 1963 and 1971, when 12 million gallons of Agent Orange, a defoliant mixture consisting of 2,4,5,-T and 2,4-D contaminated with dioxin, was sprayed over parts of Vietnam. Anecdotal reports followed by media attention thrust the specter of Agent Orange into the public eye as a possible cause for veterans' health complaints. Most studies to date show little or no adverse health effects from such exposures. A study by the Centers for Disease Control showed Vietnam veterans have slight reductions in sperm concentrations and somewhat greater hearing loss than other veterans.[43] The health of the Air Force unit responsible for aerial spraying of the herbicides (Ranch Hands) was compared with units handling other cargo. The two groups were similar in health, except that Ranch Hands had slightly more basal cell carcinomas.[44] There were also no differences seen in major, minor, or suspected defects among children between Vietnam and non-Vietnam veterans.[45]

In July 1976, pressure built up in the reaction chamber of the ICMESA chemical company that was producing trichlorophenol, the parent compound of hexachlorophene and 2,4,5-T. At noon the pressure became too great, and up to 4 lb of dioxin was released into the air, descending on the town of Seveso and the patios where meals were being served. Many of the animals in the town, including birds, rabbits, and chickens, died within a few hours of exposure. People developed chloracne (a disfiguring acne involving the sebaceous glands), nausea, and temporary nerve damage. Since that time, scientists have been monitoring the health of about 2000 families in Seveso and have documented increases in cardiovascular disease and questionable increases in certain types of cancers. Research carried out by the same group of scientists has strengthened the evidence that dioxin may be a carcinogen in humans. They have found that people living in the second most contaminated area, referred to as zone B, were nearly three times more likely to acquire liver cancer than the general population. In the same zone, the occurrence of a form of myeloma was noted 5.3 times more often among women, whereas among men, cancers of the blood were 5.7 times more likely. The scientists did not find a greater number of cancers in the most heavily polluted area because there was only a small group of people who were most affected and who moved immediately, thereby shortening the duration of their exposure. Those people in zone B had a lower and more prolonged exposure.[46]

Although these findings were not the first to associate dioxin with cancer in humans, the Seveso study has significance. This population was well monitored because new techniques had made blood levels of dioxin easy to measure, which enables researchers to more accurately determine exposure levels. The investigators also based their findings on extrapolations from soil data, which corresponded to the analyses of the blood samples.[46]

In the 1970s, a waste hauler sprayed dioxin-contaminated oil on several towns in Missouri and at the Shenandoah Stables near Moscow Hills in Missouri. More than 50 quarter horses died, in addition to chickens, dogs, cats, and birds. The stable owner's young daughter was hospitalized with aching joints, inflammation, headaches, and diarrhea.[47] In 1983, the EPA announced that the riverside community of Times Beach, MO, should abandon their homes and property and evacuate. During the early 1970s, the roads were oiled to control the dust. Soil analyses revealed high levels of dioxin contamination; the oil had been scavenged from a trichlorophenol factory by a waste hauler.[39] Subsequent studies revealed no related health problems associated with the former residents of Times Beach.[39]

The EPA released a draft dioxin reassessment document in 1994, based on an extensive review of dioxin's toxicity. The document asserts that levels of dioxin present in the environment may be at levels very close to those that can produce health consequences in humans. The report further states that dioxin poses an increased risk for cancers, adverse reproductive and developmental effects, neurological damage (including both cognitive and behavioral effects) from *in utero* exposure, endocrine disruption, and reproductive and development effects.[48]

In a continuing effort to provide a balanced view, it should be reported that some authors report that:

1. There are no formal studies to date that any person has died from environmental exposure to dioxin in the U.S.
2. Dioxin has been reported to cause chloracne, and short-term reversible nerve dysfunction, and may be toxic to both the liver and the kidneys.
3. Dioxin is highly toxic to some species of animals and minimally toxic to others.
4. The evidence that dioxin causes birth defects is inconclusive.
5. Occupational exposures to dioxin may increase the risk of soft-tissue sarcomas.[39]

POLYCHLORINATED BIPHENYLS (PCBS)

PCBs are relatively chemically inert, nonflammable fluids with high plasticizing ability and a high dielectric constant (Figure 5.19). They were commercially synthesized in the U.S. from 1929 to 1977 and used in transformers, capacitors, hydraulic and heat transfer fluids, and solvents in adhesives and sealants. They were also used to coat electrical wires, and as a protective coating on lumber, concrete, and metal surfaces. PCBs were used as an antioxidant in paints, varnishes, and rubber compounds.[48] PCBs were found not only around us in air, food, soil, water, and in human and wildlife fat tissue, but also in plastics, food wraps, cosmetics, detergents, pesticides, and even baby bottles.[1] PCBs are a family of approximately 209 chemical compounds, whose characteristics vary considerably. Some of the PCBs are light and oily fluids, whereas others are heavy, greasy, or wax like. The properties that made PCBs desirable in electric equipment, such as their resistance to thermal degradation and their stability, are the very same properties that have led to their bioaccumulation in the environment.[29] PCBs were banned from most uses in the U.S. in the 1970s due to their persistence in the environment and their ability to bioaccumulate in animal tissue. There are various chemical alternatives to PCBs such as silicone fluids, fluorocarbons, high-molecular-weight hydrocarbons, low-

FIGURE 5.19 Chemical structure of polychlorinated biphenyl (PCBs).

molecular-weight chlorinated hydrocarbons, and high-boiling oils and esters, which work well in electrical equipment. PCBs[49,50] remain widely spread in the environment, including in the fat tissue of humans and animals. More than 94% of fish collected in the U.S. show PCB residues at an average concentration of 0.53 ppm.[51,52]

Sources of PCB Exposure[52]

1. Using old fluorescent lighting fixtures and old appliances such as television sets and refrigerators; these may leak small amounts of PCBs into the air when they get hot during operation
2. Eating food, including fish, meat, and dairy products, containing PCBs
3. Breathing air near hazardous waste sites that contain PCBs
4. Drinking PCB-contaminated well water
5. Repairing or maintaining PCB transformers

Although the U.S. stopped producing many of these PCBs, they still continue to be made in other parts of the world. These PCBs bind to dust particles and are transported through the air, becoming deposited on plants that enter the food chain, or they land on water where they enter aquatic life and bioaccumulate.[29]

A few of the PCBs are dioxinlike, exerting similar biological activities. Some PCBs, or their metabolites, are estrogenic, whereas others can alter the levels of thyroid hormones in the bloodstream.[53] Some PCBs have the ability of binding to proteins in the blood that would otherwise normally bind to thyroid hormones, interfering with the transport of the thyroid hormone and thereby altering normal growth and metabolism.[41] PCBs may also be altered within the body, so one portion of the PCB molecule resembles a portion of an estrogen molecule, allowing for the PCB molecule to bind to the estrogen receptor. This activity will cause either an estrogen or antiestrogen response.

Researchers are consistently finding delayed psychomotor development of children who were exposed to PCBs in the uterus. There is evidence of a connection between paternal dioxin exposure to birth defects in offspring. There are worldwide increases in diseases and conditions of the reproductive system in infants, children, and adults, which may be linked to environmental chemical exposure.[1]

There have been two major PCB poisoning incidents in which the adverse effects have been clearly observed in children born to women exposed to high levels of PCBs. The PCBs were also contaminated with dioxin analogues, known as polychlorinated dibenzofurans (PCDFs). In Japan, in 1968, a large outbreak of what was thought to be PCB poisoning occurred, when 1300 Japanese developed chloracne, eye discharge, swelling in the joints, weakness, and other symptoms after consuming rice oil. Apparently, the rice oil had been contaminated when a PCB-containing heat exchanger leaked fluid from a pipe during processing. This incident was referred to as "Yusho (rice oil) disease."[54] A similar incident occurred in Taiwan in 1979, dubbed the "Yu-cheng disease."[55]

Children were studied over time after the Yu-Cheng incident. Children at birth were observed to have abnormalities in teeth, nails, and pigmentation. Many of the children exhibited developmental delays in addition to being smaller than average size. The symptoms of a decreased body size and a 5-point IQ deficit was observed even after the children continued to mature.[55] Behavioral problems and cognitive deficiencies have also been evidenced in children born to mothers who have eaten PCB-contaminated fish.[55] Studies have also shown that *in utero* exposure to PCBs and related contaminants is associated with poorer short-term memory functioning in early childhood.[56]

Although the environmental levels of PCBs have declined over recent years, the risk of exposure from industrial toxic waste continues because of the current use of PCBs in older electric equipment or its deposition in landfills. The PCBs in such landfills exceed the quantity that has escaped into the environment.[56] PCBs have been banned in this country since 1977, but the persistence and ubiquity of these compounds in the environment has continued to affect children, causing deficiencies in cognitive, behavioral, and motor functions of the brain.[1]

LEAD

Lead has been documented in use since the days of the Egyptian pharaohs. The Romans used it for piping water in parts of the viaduct, in cooking vessels, and as a sweetener and preservative in wines. Its dangers in affecting the human mind have been known for more than 2000 years when it was reported to produce lead-induced anemia, colic, and gout.[57] Despite this awareness of its effect on human health, it continues to be widely used throughout the world, especially in North America, where it has had a history of use in agriculture in the form of lead arsenate for pesticidal use, as solder in pipes, as solder in food containers, and as an antiknock compound in gasoline. It has also been used in paints, inks, and glazes for ceramics. Lead has been used so extensively for such a long time that it has been widely dispersed throughout our environment in substantial quantities and appears in the flesh of every human, nearly all living creatures, and virtually everywhere in the environment, including the most remote locations of Antarctica.[58] Lead has been placed on the forefront of environmental health concerns due to compelling evidence of a wide range of adverse effects. In the 1970s, both federal regulatory and legislative efforts were begun to reduce lead hazards, including the limitation of lead in paint and gasoline (Figure 5.20).[59] Lead was also reduced in evaporated milk and juices, and by 1989, less than 4% of food cans manufactured in the U.S. contained lead solder.[57] Since 1980, there have been intensive actions to further reduce lead exposure from gasoline, paint, solder, and other sources. The EPA regulations requiring the reduction of leaded gasoline were highly successful in reducing the blood lead levels in people.[60] Secondary prevention programs, such as screening for early detection and lead education programs, have also been implemented.

The most significant sources of exposure to inorganic lead include food, water, soil, and lead-based paint.[57] The most current environmental exposures to lead in the U.S. are from lead paint in older homes, although people are exposed to lead in their drinking water supplies, due to the leaching of lead from pipes. Additional sources of exposure may include lead-glazed pottery, medicinal and cosmetic preparations, and food grown in lead-contaminated soil. The fact that lead solder is no longer used in commercial food canning and the phaseout of lead in our gasoline

FIGURE 5.20 Comparison of lead in gasoline and average blood lead levels. (From Schwartz, H. et al., *Costs and Benefits of Reducing Lead in Gasoline: Final Regulatory Impact Analysis,* EPA-230-05-006, Washington, D.C., 1985. With permission.)

since 1983 have reduced lead exposures in the U.S. However, leaded gasoline is currently being used in other parts of the world and will continue to expose pregnant women and children for many more years to come.[58]

Lead may enter the body through ingestion and inhalation, whereas organic lead may also be absorbed through the skin. Lead does not have any normal function in living organisms, but it does produce toxic effects. After lead enters the body, it is able to distribute to various organs, cross the blood–brain barrier, and can cross the placenta. Blood lead levels in the fetus are up to 90% as high as those in the mother.[61] Lead is also excreted into breast milk during lactation.

Whether lead enters the body via ingestion or inhalation, it first enters the bloodstream where its half-life is estimated at 36 d. Therefore, measurements of blood lead levels are calculated as micrograms of lead per deciliter of blood (µg/dl), which gives the most accurate indication of short-term lead exposure. Whereas some lead will make its way to soft tissues such as the brain and kidneys, 50 to 60% of the lead that enters the body is excreted relatively quickly. Over time, the lead will be distributed, deposited, and stored in bones, especially the long bones of the arms and legs.[57] Lead can remain in the long bones for years, and the lead stored here can be released back into the bloodstream secondary to conditions such as high fever, osteoporosis, or pregnancy, which will result in acute lead poisoning.

Lead may cause numerous toxic effects in the body. Lead affects the formation of blood in two distinct mechanisms: (1) by slowing the normal maturation of red blood cells in the bone marrow, decreasing the number of red blood cells and possibly causing anemia, and (2) lead inhibits the synthesis of hemoglobin, by interfering with two important substances that are necessary for the formation of hemoglobin: delta-aminolevulinic acid and coproporphyrin III. In lead poisoning, both of these

FIGURE 5.21 History of blood lead levels considered acceptable by various agencies over time. (Adapted from U.S. Congress, Office of technology Assessment, *Neurotoxicity: Identifying and Controlling Poisons of the Nervous System,* OTA-BA-436, Washington, D.C., April 1990, chap. 10.)

substances are subsequently excreted in the urine in excessive amounts. The inhibition of the activity of delta-aminolevulinic dehydrogenase in red blood cells (an enzyme involved with the synthesis of hemoglobin) serves as a diagnostic marker in the early stages of lead poisoning before the symptoms are pronounced.[61,62]

Over the past 10 years, there has been more and more evidence that lead may have serious health effects at lower exposure levels that were previously thought not to be harmful (Figure 5.21). In the U.S., women of childbearing age have an average blood lead level of about 2.0 µg/dl, whereas the average blood lead level for men of the same age is 4.2 µg/dl.[63] The EPA has listed 10 µg/dl as the maximum acceptable blood lead level for fetuses and young children. The CDC recommends treatment in children found to have blood lead levels over 10 µg/dl.[58] Nearly 22% of black children 1 to 2 years old currently have blood lead levels over 10 µg/dl.[63] In the workplace, employers are not required to remove workers from lead exposure until their blood lead levels reach 50 µg/dl. This occupational health standard will only protect against acute lead poisoning.[58]

Lead impairs fertility in both men and women when blood lead levels approach 50 µg/dl. Lead may act directly on the testes, lowering the sperm count. At high blood lead levels, lead may cause spontaneous abortions and stillbirths. It is interesting to note that in the past, lead was actually used to induce abortions.[64]

The public health threat posed by blood lead levels is worrisome. Young children are at a greater risk for elevated lead levels due to (1) their increased oral activity, (2) increased ability to absorb lead, (3) higher retention of absorbed lead and, (4) the risk of an incompletely developed nervous system.[57] Blood lead levels as low as 10 μg/dl, which was previously thought to be safe, have been associated with developmental delays and deficits in intellectual performance and neurobehavioral functioning, as well as decreased stature and diminished hearing activity.[59]

Over the past 30 years, we have been able to justify the declining levels of blood lead that define a case of lead poisoning. The results of numerous studies carried out over the last several decades continue to show that at levels thought to have been safe, chronic low-level exposure to lead can inhibit the normal development of a child's intellectual ability. A 1979 study in the Boston suburbs of Chelsea and Somerville performed on second-grade students concluded that those whose baby teeth had a higher lead level also exhibited a greater incidence of unruly classroom behavior, lower IQ scores, and a diminished capacity to follow instructions than did classmates with lower lead levels in their teeth (Figure 5.22).[65] A follow-up study was carried out in 1990;[66] all students in the 1979 study were now 18 years of age. In comparison to classmates whose baby teeth had minimal lead levels, the students with higher lead levels had a greater school drop-out rate, higher incidence of reading difficulties, and a lower class rank, all in part due to childhood lead exposure. Several studies conclude that lead exposure at levels even below 10 mg/dl is associated with aggressive and delinquent behavior, impaired neurobehavioral development, and low birth weight.[66,67] Studies on rats confirmed that lead exposure can produce long-term cognitive deficits. Additionally, rats exposed to lead postweaning did not show impairment in performance despite high body burdens of lead, whereas exposures during gestation and lactation (maternal exposure) resulted in significant impairment. The authors concluded that a window of vulnerability to lead levels exists that results in long-term cognitive impairment.[68]

The prevention of lead poisoning should be one of society's top priorities. The reduction of human exposure to lead had been achieved in recent years by the near phaseout of lead in gasoline, paint, food containers, and home plumbing. Nevertheless, preschoolers are continually exhibiting elevated lead levels, possibly secondary to lead paint and dust still present in their homes. Other children are exposed by family members involved in furniture refinishing, pottery, or stained glass businesses who unintentionally carry lead dust home with them on their clothes. Because blood lead levels as low as 10 μg/dl are a cause for concern, there is a continuing need to identify and eliminate the source of lead exposure for every child.

ORGANIC SOLVENTS

Organic solvents are a group of simple organic liquids which have the capacity to evaporate at atmospheric pressure. Organic solvents contaminate our drinking water. Not only can solvents enter the body by ingestion from contaminated drinking water supplies, but they may enter the body via skin absorption and inhalation in the

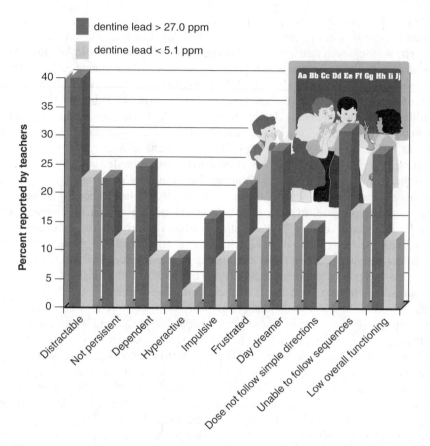

FIGURE 5.22 The negative effect of classroom behaviors associated with dentine lead levels. (Adapted from Needleman, H.L., Rabinowitz, M. et al., *JAMA,* 251, 2956, 1984.)

shower. The total exposure from taking a 10-min shower in contaminated water is actually greater than the exposure after drinking 2 qt of the same water.[69] Organic solvents are widely used in the U.S., both in the home and in industry. Organic solvents are components of many products including paints, varnishes, paint removers, adhesives, glues, degreasing and cleaning agents, pharmaceuticals, plastics, and pesticides. Nearly 50 million tons of industrial organic solvents are manufactured yearly in the U.S.[57]

Individuals may be exposed to organic solvents if they work with electronics or are in the health care field, in auto repair, dry cleaning, laboratory work, or painting businesses. Solvents may enter the body through a variety of ways such as inhalation, skin contact, or ingestion. A major route of exposure is inhalation, as the chemicals are volatile. The amount of solvent entering the body strictly depends on factors such as the concentration of the solvent in the air and the route of exposure.[57] Organs such as the brain, because of its blood supply and high fat content, tend to accumulate high levels of solvents more quickly than organs with lower fat contents. All solvents are fat soluble and will produce effects on the central nervous system.[70]

Researchers have examined the harmful effects of solvents on reproduction, as reported in many human studies. The evidence remains consistent that maternal exposure to solvents during pregnancy increases the risk of spontaneous abortions by two- to fourfold.[71] Birth defects, including cleft palate and lip, cardiovascular malformations, and central nervous system (CNS) defects, have also been reported in association with organic solvents.[72]

Studies of glycol ethers, in both animals and humans, conclude that they damage testicular function, cause testicular atrophy, lower sperm counts, and may lead to infertility.[73] Solvents may also lead to acute leukemias in children of exposed parents, as well as an increased risk of neurological and urinary cancers.

There is disturbing evidence that organic solvents, most notably the glycol ethers, cause spontaneous abortions, birth defects, and childhood cancers. It is estimated that nearly 10 million workers come in contact with organic solvents on a daily basis.[57] Workers and their families need protection through intensive education, informing workers about the health risks involved, and mandatory training sessions to teach employees how to minimize exposure to hazardous substances.

ASBESTOS

Asbestos is a collective term for a group of six fibrous silicate materials: asmolite, chrysotile, tremolite, actinolite, anthophyllite, and crocidolite, which are found throughout most of the world. Nearly 95% of all asbestos in the U.S. comes from just one of the six types of commercially used asbestos: chrysotile. "White asbestos" is used in wall insulation, ceilings, around pipes and vents, and for heat resistance in roofing shingles. The chrysotile-based asbestos is also widely used to reduce friction in automotive brake shoe linings. Asbestos has been used as an essential component in various products and processes, including building materials, brake linings, textiles, and insulation, as well as floor tiles, cement, and potholders (Figure 5.23). There are approximately 3000 uses of asbestos in various industries.[74] Unfortunately, during 1950 to 1972, asbestos was used in spray insulation in buildings, which poses a significant environmental health hazard as they are being demolished. Asbestos can enter our water supplies by airborne settling, dumping of asbestos waste, and by the dumping of effluent from mining operations. Asbestos may also leach from asbestos–cement pipes in municipal water systems. In cities where asbestos–cement pipes are leached heavily by the water, long fibers of asbestos have been known to clog household appliances.[75] Asbestos is a human carcinogen, which can be found in air, water, and food throughout the U.S. Asbestos is naturally released into the environment, but there are also artificial emissions, from mining asbestos ores to the manufacture of asbestos products. Exposure to these fibers may occur throughout the environment, especially in our drinking water supply.

Asbestos, although extremely useful, represents an enormous health risk. It is estimated that of the 8 to 11 million current and retired workers who were exposed to vast amounts of asbestos, between 30 to 50% can be expected to die from asbestos-related cancers.[75] Roughly 5 to 10% of all workers who are employed in asbestos manufacturing or mining will die from mesothelioma (an asbestos-caused cancer of the lining of the lung or abdomen). Asbestos workers who smoke have a 90 times

FIGURE 5.23 Asbestos used in insulating pipes.

greater increase in the incidence of lung cancer over that in nonsmokers.[76] Their families are also at risk, as asbestos workers presumably carry dust on their clothes and in their hair into their homes.

Exposure to asbestos is primarily achieved through the inhalation of these tiny fibers that are suspended in the air, often getting trapped deep within the lungs. Most fibers are carried through the gastrointestinal tract and excreted within a few days, but the few that remain embedded deep in the lungs lead to irreversible disease processes, including asbestosis, lung cancer, and mesothelioma. Certain fibers, such as crocidolite, are more likely to produce disease than other forms of asbestos, but it is generally believed that the size of the fiber may be the more important characteristic in disease production than the fiber type. Asbestosis is a slowly developing disease normally occurring over 20 to 30 years, resulting in gradually worsening breathlessness on exertion and the development of a productive cough. The impoverished respiration leads to a barrel chest, bluish discoloration of the skin, and a restricted chest expansion.[74] The lungs become typically scarred with thickening of the lining of the lungs which is often replaced by calcium deposits. Mesothelioma is a rare, malignant tumor, involving the pleura or the abdominal wall and abdominal organs, which is caused only by exposure to asbestos. The disease is a diffuse cancerous tumor that spreads rapidly, is usually fatal, and results in death within 1 to 2 years after diagnosis.[74] Cancer can also develop in the digestive tract as a result of ingesting asbestos in food, beverages, or swallowing the fibers in contaminated saliva.

Asbestos is durable, flexible, strong, inflammable, and an excellent insulator. Therefore, it is of no surprise that it was widely used in construction. Buildings constructed between 1945 and 1978 are likely to contain asbestos material, whereas pipe insulation manufactured from 1920 to 1972 contained asbestos to prevent heat loss from pipes. In schools, the EPA estimated that 15 million students and 1.4 million teachers and other employees are in buildings that contain asbestos.[77]

Asbestos was determined by the EPA in 1971 to be a hazardous air pollutant as defined by the Clean Air Act of 1970. Although this regulation has helped to control asbestos emissions from manufacturing, roadway surfacing, and demolition projects, the main concern and target from asbestos regulations has been school systems. Congress passed the Asbestos School Hazard Abatement Act (ASHAA) in 1984 to provide financial assistance to schools having significant asbestos problems. The Act was reauthorized in 1990 and required additional training requirements for abatement workers along with revisions to accreditation plans. President Reagan authorized the Asbestos Hazard Emergency Response Act (AHERA) on October 22, 1986, which required schools to inspect for asbestos, develop asbestos-management plans, and implement actions involving repair, encapsulation, or removal of the asbestos. This was all to take place prior to 1988 and to be conducted by accredited asbestos removal personnel. Nearly $300 million has been distributed to schools under these various acts for abatement training, and to conduct other activities related to asbestos.[78] Despite this enormous cost and effort, the average level of asbestos in schools was less than 1/3000 the level (0.0007 fibers per ml) than the federal limit permitted by OSHA in the workplace environment. Work on the removal of asbestos from schools continues today at very significant costs. The EPA estimated that the cost of removing all asbestos from public and commercial buildings throughout the U.S. may exceed $51 billion.[79]

MERCURY

There are three different forms in which mercury is found in the environment: elemental mercury vapor, inorganic mercury compounds, and organic (usually methyl) mercury. The three different forms are significantly different from one another: they are each absorbed by the body differently, they are used and manufactured for different purposes, and they have different effects on reproduction and development.[69]

Examples of the uses of mercury are in the production of chlorine, and for use in thermometers, batteries, and fluorescent light bulbs. For centuries, mercury was an essential part of many different medicines such as diuretics, antibacterial agents, antiseptics, and laxatives. More recently, these drugs have been substituted with others and drug-induced signs of mercury toxicity are rare.[80] Mercury is also a constituent of dental amalgams. Concerns regarding mercury stem from its volatile nature, resulting in global dispersal throughout the air and water. To make matters even worse, inorganic mercury may be methylated by microorganisms that are present naturally in the environment into an even more toxic form, methyl mercury, which tends to bioaccumulate in marine life.[81]

In the 1940s and 1950s, mercury became known as the product that caused acrodynia, also known as *pink disease*. Manifestations of acrodynia included pain and erythema of the palms and soles, irritability, insomnia, anorexia, diaphoresis, photophobia, and skin rashes.[80] Debate continues about the lowest levels at which mercury causes human health effects. Studies have found that the health effects of organic mercury are similar in both animal and human populations. When exposed to organic mercury, cells in the developing brain of mice, according to studies, stop in the middle of cell division.[69] Mercury poisoning usually is misdiagnosed due to the insidious onset, nonspecific signs and symptoms, and lack of knowledge within the medical profession.

Organic mercury is the most toxic form of mercury owing to its ease of oral absorption while circulating in the bloodstream, crossing the blood–brain barrier and destroying the very cells that control coordination. Symptoms of poisoning appear 1 to 2 months after exposure, showing as numbness of the tongue, lips, and fingertips, for example. Speech gradually becomes slurred, and difficulty in swallowing soon follows. The most infamous episode of mass poisoning occurred during the years 1953 to 1961 in the town of Minamata on the Japanese island of Kyushu. A local vinyl chloride manufacturing facility had been dumping inorganic mercury waste into the Minamata Bay, where bacteria in the sediment converted it into organic mercury that then moved through the aquatic food chain, contaminating the fish. The fish showed concentrations of mercury 100,000 times that of the water levels. The population of Minamata relied heavily on a fish diet and vividly recall the day when the cats started acting weirdly and hurling themselves into the ocean to die. Then, in 1956, people started showing the symptoms of CNS damage including tremors, convulsions, numbness of hands and feet, losses in speech, hearing, and vision.[82] In many cases of mercury poisoning, a pregnant mother will remain relatively free of symptoms whereas children born to her will demonstrate gross physical abnormalities and severe mental retardation.[81,82] More than 1000 Minamata Bay sufferers have since died. Minamata's population has shrunk since the time of the poisoning from 50,000 to less than 33,000. The chemical company (Chisso Corporation) implicated in this disaster did not stop dumping until 1968.[82] More than 40 years later, the prefectural government has declared that the fish in Minamata Bay now meet safety standards and the net that confined the fish to the bay has been removed. However, the social stigma and discrimination for victims of Minamata will likely continue.[82] By 1991, more than 2000 cases of Minamata disease had been confirmed.[81] A similar epidemic occurred again in 1965, but this time in riverside villages along the Agano River, Niigata, Japan.[81]

Elemental mercury is the type of mercury which is found in dental amalgam fillings, thermometers, and batteries. It is hazardous only when inhaled. Due to its low vapor pressure, it can be inhaled at room temperature. Elemental mercury can easily pass the blood–brain barrier and cross the placenta to the fetus once it is in the body. Individuals can be exposed to mercury in the air from waste incinerators that burn medical waste or batteries, and also from oil and coal burning, as mercury is a contaminant of these fuels.[83] It is far more dangerous to inhale mercury fumes than to swallow inorganic mercury, such as the mercury which is found in an oral thermometer. If swallowed, the mercury passes through the gastrointestinal tract and is excreted in a few short days.

Inorganic mercury vaporizes at room temperature, and its volatility increases markedly with each degree increase in temperature. This makes it especially hazardous when spilled on floors or surfaces where dirt or grease may exist and combine with the mercury to form tiny drops that expose a larger surface area to air. Vaporized inorganic mercury can enter the body either by inhalation or absorption through the skin following prolonged contact. Symptoms may include damage to kidney tubules, liver damage, interference with coordination, generalized tremors, and possible hallucinations. These symptoms are described among hat makers in 17th century France, who soaked animal hides in a solution of mercuric nitrate in order to soften the hairs. This practice allowed for frequent contact of the hat makers' arms and hands with a solution of inorganic mercury, and they developed the typical symptoms of inorganic mercury poisoning, referred to as *Mad Hatters' Disease* (featured in the children's story, *Alice in Wonderland*).

Organic mercury is a developmental toxin, implicated in some of the largest epidemics of environmental toxicity in recent history. Some of these outbreaks have occurred in Sweden, the Middle East, and Pakistan, where grain seeds coated with the organic mercury fungicide Panogen or Granosan M were mistakenly fed to food animals or used to make bread that was then consumed.[62] Infants studied in these outbreaks exhibited mental retardation, lack of coordination, muscle weakness, seizures, delayed development, and visual loss.[84]

Chronic and intense acute exposure causes cutaneous and neurological symptoms. The classic triad found in chronic toxicity are tremors, gingivitis, and erethism (i.e., a constellation of neuropsychiatric findings that includes insomnia, shyness, memory loss, emotional instability, depression, anorexia, vasomotor disturbance, uncontrolled perspiration, and blushing).

ARSENIC

Since the Middle Ages, arsenic has been a popular way of poisoning people. Arsenic itself is a gray semimetal that is not very poisonous. Arsenic and related compounds occur naturally in granite and mountains as arsenopyrites or arsenic sulfide (*arsenikon* is the Latin word for yellow pigment). Other sources include microbiological activity, pesticides, herbicides, fertilizers, treatments of human waste, and even volcanic eruptions, and industrial sources.[85] Another form is white arsenic or arsenic oxide. This is extremely poisonous, although the symptoms of arsenic poisoning could be confused with those of many other illnesses, and it was also very difficult to detect arsenic after death. All in all, using arsenic was a handy way to get rid of someone. Politicians, popes, and parents were all victims at different times; indeed white arsenic became known as "inheritance powder."

Arsenic poisoning results because arsenic binds very strongly to sulfur groups. Many proteins contain one or more sulfur atoms, and enzymes (also proteins) are responsible for regulating the body's internal chemistry. An enzyme with an arsenic atom bound to sulfur may not be able to function.

Bangladesh and parts of the Indian state of West Bengal are facing a disaster of astronomical proportions. Beginning in the 1970s, tube wells were sunk in Bangladesh and West Bengal to avoid cholera and other bacterial infections present in above-ground water supplies. However arsenic slowly crept into the groundwater and over 6 million people in West Bengal and 30 million Bangladeshis are drinking water containing arsenic at levels over 5 times the WHO limit of 10 ppb.[86] Lowering of the water table resulted in the exposure of arsenic to air in the zone of aeration. This exposure resulted in the oxidation of arsenic minerals, previously present below the water table, in the Bengal sediments. The arsenic oxides migrated to the groundwater and were reduced to the poisonous forms in the water-reduced environment below the water table.

Arsenic poisoning causes a variety of systemic problems like diaphoresis, muscle spasms, nausea, vomiting, abdominal pain, garlic odor to the breath, diarrhea, anuria, dehydration, hypotension, cardiovascular collapse, aplastic anemia, and death.[87]

Do you want to learn more about toxicity and toxins? Point your browser to http://www-unix.oit.umass.edu/~envhl565/weblinks.html and click on any of the Chapter 5 Web links associated with toxicity and toxins including:

- e.hormone: Your gateway to the environment and hormones
- Environmental Performance Index
- Envirofacts Querying Capability

REFERENCES

1. Schettler, T., Solomom, G., Burns, P., and Valenti, M., Generations at Risk: How Environmental Toxins May Affect Reproductive Health in Massachusetts, Greater Boston Physicians for Social Responsibility, and Massachusetts Public Interest Research Group (MASSPIRG) Education Fund, July 1996, chap. 1.
2. Waldbott, G.L., *Health Effects of Environmental Pollutants,* 2nd ed., The CX Mosby Co., St. Louis, MO, 1978, chap. 5.
3. Malachowski, M.J., *Health Effects of Toxic Substances,* Govt. Institute, Rockville, MD, 1995, chap. 1.
4. Tortora, G.J. and Anagnostakos, N.P., *Principles of Anatomy and Physiology,* 6th ed., Harper and Row, New York, 1990, chap. 3.
5. Tortora, G.J. and Anagnostakos, N.P., *Principles of Anatomy and Physiology,* 6th ed., Harper and Row, New York, 1990, chap. 23.
6. Tortora, G.J. and Anagnostakos, N.P., *Principles of Anatomy and Physiology,* 6th ed., Harper and Row, New York, 1990, chap. 5.
7. Malachowski, M.J., *Health Effects of Toxic Substances,* Govt. Institute, Rockville, MD, 1995, chap. 2.
8. Tortora, G.J. and Anagnostakos, N.P., *Principles of Anatomy and Physiology,* 6th ed., Harper and Row, New York, 1990, chap. 24.
9. Malachowski, M.J., *Health Effects of Toxic Substances,* Govt. Institute, Rockville, MD, 1995, chap. 6.
10. Simon, F.P., Potts, A.M. and Gerard, R.W., Action of cadmium and thiols on tissue enzymes, *Arch. Biochem. Biophys.,* 12, 283, 1947.
11. Tortora, G.J. and Anagnostakos, N.P., *Principles of Anatomy and Physiology,* 6th ed., Harper and Row, New York, 1990, chap. 22.
12. Tortora, G.J., Funke, B.R. and Case, C.L., *Microbiology: An Introduction,* Benjamin Cummings, Redwood City, CA, 5th ed., 1995, chap. 17.
13. Friebele, E., The attack of asthma, *Environ. Health Perspect.,* 104, 1, 22, January, 1996.
14. Astrop, P., Kjeldsen, K., and Wanstrup, J., Enhancing influence of carbon monoxide on the development of atheromatosis in cholesterol-fed rabbits, *J. Atheroscler Res.,* 7, 343, 1967.
15. Schroeder, H.-A., Metals in the air, *Environment,* 13, 18, 1971.
16. Amdur, M.O., The physiologic response of guinea pigs to atmospheric pollutants, *Int. J. Air Pollut.,* 1, 170, 1969.
17. Labelle, C.W., Long, J.E., and Christifano, E.E., Synergistic effects of aerosols, *Arch. Ind. Health,* 11, 297, 1955.
18. U.S. Department of Health and Human Services, The Nature and Extent of Lead Poisoning in Children in the U.S.: A Report to Congress, Washington, D.C. Public Health Service, Agency for Toxic Substances and Disease Registry, 1988.

19. Stokinger, H.E., Ozone toxicology: a review of research and industrial experience, 1954–1964, *Arch. Environ. Health*, 10, 719, 1965.
20. Perera, F.P., Environment and Cancer: Who Are Susceptible?, *Science*, 278, November 7, 1997, p. 1068.
21. Crapo, L., *Hormones: Messengers of Life*, W.H. Freeman, New York, 1985, chap. 1.
22. Biondo, B., Are Common Chemicals Scrambling Your Hormones?, *USA Weekend*, February 13–15, 18,1998.
23. Fry, D.M., Speich, S.M., and Peard, R.J., Sex ratio skew and breeding patterns of gulls: demographic and toxicologic considerations, *Stud. Avian Biol.*, 10, 26, 1987.
24. Colburn, T., von Saal, F.S., and Soto, P.M., Developmental effects of endocrine-disrupting chemicals in wildlife and humans, *Environ. Health Perspect.*, 101, 378, 1993.
25. Guillette, L.J., Gross, T.S. et al., Developmental abnormalities of the gonad and abnormal sex hormone concentrations in juvenile alligators from contaminated and control lakes in Florida, *Environ. Health Perspect.*, 102, 680, 1994.
26. Colburn, T. and Clement, C., *Advances in Modern Environmental Toxicology, Vol. XXI Chemically Induced Alterations in Serial and Functional Development: The Wildlife/Human Connection*, Princeton Scientific Publishing, 1992, chap. 1.
27. Ziegler, J., Environmental endocrine disruptors get a global look, *J. Natl. Cancer Inst.*, 88, 16, 1185, August 20, 1997.
28. McLachlan, J.A., Functional toxicology: A new approach to detect biologically active xenobiotics, *Environ. Health Perspect.*, 101, 386, 1993.
29. Birnbaum, L., Endocrine effects of prenatal exposure to PCBs, dioxins, and other xenobiotics: implications for policy and future research, *Environ. Health Perspect.*, 102, 8, 676, 1994.
30. Kimmee, C.A., Approaches to evaluating reproductive hazards and risks, *Environ. Health Perspect.*, 101(Suppl. 2), 137, 1993.
31. EHPnet Forum, New evidence in sperm counts, *Environ. Health Perspect.*, 104, 9, 919, 1996.
32. Jackson, M.B. et al., Cryptochordism: an apparent substantial increase since 1960, *BMJ*, 293, 1401, 1986.
33. Teens before Their Time, *Time Magazine, Health*, 156, 18, October 30, 2000.
34. Sharpe, R.M., Fisher, J.S., Miller, M.M. et al., Gestational and lactational exposure of rats to xenoestrogens results in decreased testicular size and sperm production, *EHP*, 103,12, 1136,1995.
35. Warhurst, M., Introduction to hormone disrupting chemicals, 1998, http://easyweb.easynet.uk/-mwarhurst/estrogenic.html.
36. Raloff, J., Hormone Mimics Get Harder to Pigeonhole, *Science News*, 151, April 26, 1997, p. 254.
37. Smith, A.H., Infant exposure assessment for breast milk dioxins and fumas derived from waste incineration emissions, *Risk Anal.*, 7, 3, 347, 1987.
38. World Health Organization, The International Agency for Research on Cancer, Polychlorinated Dibenzo-para-Dioxins and Polychlroinated Dibenzofurans, Summary of Data Reported and Evaluation, Vol. 69, August 12, 1997, www.iarc.fr.
39. Whelan, E.M., *Toxic Terror*, Prometheus Books, New York, 1992, chap. 9.
40. Safe, S., Astroff, B., Harris, M. et al., 2,3,7,8-Tetrachlorodibenzo-p-dioxin (TCDD) and related compounds as antiestrogens: characterization and mechanism of action, *Pharmacol. Toxicol.*, 69, 400, 1991.
41. McKinney, J.D. and Waller, C.L., PCBs as hormonally active structural analogues, *Environ. Health Perspect.*, 102, 3, 290, 1994.

42. Olson, J.R., Holscher, M.A., and Neal, R.A., Toxicity of 2,3,7,8-tetrachloro-dibenzo-pdioxin in the Golden Syrian hamster, *Toxicol. Appl. Pharmacol.*, 55, 67, 1980.

43. Centers for Disease Control Vietnam Experience Study, Health status of Vietnam veterans, Part 11, physical health, *JAMA*, 259, 2708, 1988.

44. Wolfe, W.H., Michalek, J.E., Miner, J.C. et al., Health status of Air Force veterans occupationally exposed to herbicides in Vietnam, *JAMA*, 264, 14, 1824, 1990.

45. Centers for Disease Control Vietnam Experience Study, Health status of Vietnam veterans, Part 111, reproductive outcomes and child health, *JAMA*, 259, 2715, 1988.

46. Holloway, M., Dioxin Indictment: A Growing Body of Research Links the Compound to Cancer, *Scientific American*, January 25, 1994.

47. Monks, V., The Truth about Dioxin, *National Wildlife*, 4, August–September 1994.

48. Schecter, A., Dai, L.C., Bich, L.T. et al., Agent orange and the Vietnamese: The persistence of elevated dioxin levels in human tissues, *Am. J. Public Health*, 85, 4, 516, April 1995.

49. World Health Organization, The International Agency for Research on Cancer, Poly-chlorinated Dibenzo-para-Dioxins and Polychlroinated Dibenzofurans, Summary of Data Reported and Evaluation, Vol. 69, August 12, 1997, www.iarc.fr.

50. Jensen, S., The PCB story, *Ambio*, 1, 123, 1972.

51. Schmitt, C.J., Zajicek, J.L., and Ribick, M., National pesticide monitoring program, Residues of organochlorine chemicals in freshwater fish, 1980–1981, *Arch. Environ. Contam. Toxicol.*, 14, 225, 1985.

52. U.S. Department of Health and Human Resources, Agency for Toxic Substances and Disease Registry, (ATSDR), Toxicological Profile for Selected PCBs (update), Atlanta, ATSDR/TP-92/16, 1993.

53. Safe, S.H., Polychlorinated biphenyls (PCBs): environmental impact, biochemical and toxic responses, and implications for risk assessment, *Crit. Rev. Toxicol.*, 24, 1, 1994.

54. Kuratsume, M., Yusho, a poisoning caused by rice oil contaminated with polychlo-rinated biphenyls, *HSMSA Health Rep.*, 86, 1083, 1971.

55. Gun, Y.L., Ju, S.H., Chen, Y.C. and Elsu, C.C., Sexual developments and biological findings in Yucheng children, *Chemosphere*, 14, 235, 1993.

56. Jacobson, J.L., Jacobson, S.W., and Humphrey, H.E.B., Effects of in utero exposure to polychlorinated biphenyls and related contaminants on cognitive functioning in young children, *J. Pediatr.*, 116, January 1990.

57. U.S. Congress, Office of Technology Assessment, *Neurotoxicity: Identifying and Controlling Poisons of the Nervous System*, OTA-BA-436, Washington, D.C., April 1990, chap. 10.

58. Schettler, T., Solomom, G., Burns, P., and Valenti, M., Generations at Risk: How Environmental Toxins May Affect Reproductive Health in Massachusetts, Greater Boston Physicians for Social Responsibility, and Massachusetts Public Interest Research Group (MASSPIRG) Education Fund, July 1996, chap. 4.

59. Pirkle, J.L., Brody, D.J., Gunter, E.W., Kramer, R.A., Pashal, D.C., Flegal, M., and Marte, T.D., The decline in blood lead levels in the U.S.: The National Health and Nutrition Surveys (NHANES), *JAMA*, 272, July 1994.

60. Schwartz, J., Leggett, J., Ostro, B., Pitcher, M., and Levin, R., *Costs and Benefits of Reducing Lead in Gasoline: Final Regulatory Impact Analysis*, EPA-230-05-006, Washington, D.C., 1985.

61. Clarkson, T.W., Nordberg, G.E., and Safar, P.R., Reproductive and developmental toxicity of metals, *Scand. J. Work Environ. Health*, 11, 145, 1985.

62. Waldbott, G.L., *Health Effects of Environmental Pollutants,* 2nd ed., The CX Mosby Co., St. Louis, MO, 1978, chap. 12.

63. Brody, D.J., Pirkle, J.L., Krmaer, R.A., Hegal, K.M., Matte, T.D., Gunter, E.W., and Paschal, D.C., Blood lead levels in the U.S. population: phase I of the Third National Health and Nutrition Examination Survey (NHANES III), 1988–1991, *JAMA,* 272, 277, 1994.

64. Agency for Toxic Substances and Disease Registry (ATSDR), Toxicological Profile for Lead, U.S. Department of Health and Human Services, April 1993.

65. Needleman, H.L., Rabinowitz, M., Leviton, A., Linn, S., and Schoenhaum, S., The relationship between prenatal exposure to lead and congenital anomalies, *JAMA,* 251, 2956, 1984.

66. Needleman, H.L., Schell, A., Bellinger, A., Leviton, A., and Allred, E.N., The long-term effects of exposure in low doses of lead in childhood: an 11-year follow-up, *N. Engl. J. Med.,* 322, 83, 1990.

67. Needleman, H.L., Reiss, J.A., Tobin, M.J., Biesecker, G.E., and Greehouse, J.B., Bone lead levels and delinquent behavior, *JAMA,* 275, 363,1996.

68. Kuhhnann, A., MeGlothan, L., and Guilarte, T.R., Developmental lead exposure causes spatial learning deficits in adult rats, *Neurosci. Lett.,* 223, 1010, 1997.

69. Schettler, T., Solomom, G., Bums, P., and Valenti, M., Generations at Risk How Environmental Toxins May Affect Reproductive Health in Massachusetts, Greater Boston Physicians for Social Responsibility, and Massachusetts Public Interest Research Group (MASSPIRG) Education Fund, July 1996, chap. 5.

70. Dick, R.B., Short duration exposures to organic solvents: The relationship between neurobehavioral test results and other indicators, *Neurotoxicol. Teratol.,* 10, 39, 1988.

71. Taskinen, H., Kyyronen, P. et al., Laboratory work and pregnancy outcome, *JOM,* 36, 311, 1994.

72. Holmberg, P.C., Hernberg, S., Kurrpa, K., Rantal, K., and Riala, R., Oral clefts and organic solvent exposure during pregnancy, *Int. Arch. Occup. Environ. Health,* 50, 371, 1982.

73. Wess, J.A., Reproductive toxicity of ethylene glycol monomethyl ether, ethylene glycol monoethyl ether and their acetates, *Scand. J. Work Environ. Health,* Suppl. 2, 43, 1992.

74. Waldbott, G.L., *Health Effects of Environmental Pollutants,* 2nd ed., The CX Mosby Co., St. Louis, MO, 1978, chap. 16.

75. Archer, S.R. and Blackwood, T.R., *Status Assessment of Toxic Chemicals: Asbestos,* U.S. Environmental Protection Agency, Monsanto Research Corporation, Dayton, OH, December 1979.

76. Selikoff, Q., Churg, J., and Hammond, E.C., Asbestos exposure and neoplasia, *JAMA,* 188, 22, 1964.

77. Koren, H. and Bisese, M., *Handbook of Environmental Health and Safety: Principles and Practices,* Vol. 1, 3rd ed., CRC, Lewis Pub., Boca Raton, FL, 1995, chap. 7.

78. Whelan, E.M., *Toxic Terror,* Prometheus Books, New York, 1992, chap. 8.

79. EPA, Study of Asbestos — Containing Materials in Public Buildings: A Report to Congress, U.S. Environmental Protection Agency, Washington, D.C., February 1988.

80. U.S. Department of Health and Human Resources, Agency for Toxic Substances and Disease Registry, (ATSDR), ToxFAQs: Chemical Agent Briefing Sheets (CABS), Arsenic, January, 2006, http://www.atsdr.cdc.gov/cabs/arsenic/arseniccabs_info.html.

81. Ratcliffe, H.E., Swanson, G.M., and Fischer, L.J., Human exposure to mercury: a critical assessment to the evidence of adverse health effects, *J. Toxicol. Environ. Health,* 49, 221, 1996.

82. Pollack, A., Mercury, mostly gone from bay in Japan, still poisons town's life, *Minamata J.*, 1, August 23, 1997.

83. Agency for Toxic Substances and Disease Registry (ATSDR), Draft Toxicological Profile for Mercury, U.S. Department of Health and Human Services, October 1992.

84. Cox, C., Clarkson, T.W., Marsh, D.O., Amin-Zaki, L., Tikriti, S., and Myero, G.G., Dose-response analysis of infants prenatally exposed to methyl mercury: an application of a single compartment model to single strand hair analysis, *Environ. Res.*, 49, 318, 1989.

85. Dhar, R.K., Biswas, B.K., Samanta, G., Mandal, B.K., Chakraborti, D., Roy, S., Jafar, A., Islam, A., Ara, G., Kabir, S., Khan, A.W., Ahmed, S.A., and Hadi, S.A., Groundwater arsenic calamity in Bangladesh, *Current Science*, 73, 1, 48, 1997.

86. Mandal, B.K., Roy Chowdhury, T., Samanta, G., Basu, G.K., Chowdhury, P.P., Chanda, C.R., Lodh, D., Karan, N.K., Dhar, R.K., Tamili, D.T., Das, D., Saha, K.C. and Chakraborti, D., Chronic arsenic toxicity in West Bengal, *Current Science*, 72, 2, 114–117, 1997.

87. Guha Mazumder, N., Das Gupta, J., Santra, A., Pal, A., Ghose, A., Sarkar, S., Cattopadhaya, N., and Chakraborti, D., Non-cancer effects of chronic arsenicosis with special reference to liver damage, *Arsenic: Exposure and Health Effects*, Abernathy, C.O., Calderon, R.L. and Chappell, W.R., Eds., Chapman & Hall, New York, 1997, chap. 10, pp. 112–123.

6 The Trouble with Pests

Brown recluse spider and
resulting bites on human arm

CHAPTER OBJECTIVES

A student, after reading this chapter, should be able to:

1. Discuss and define the term *pest*.
2. List, differentiate between, and classify the major arthropod pests to the "order" level.
3. Discuss and describe the general structure and development of insects and arthropod pests.
4. List and describe the major arthropod and mammalian vectors of disease including kissing bugs, flies, mosquitoes, fleas, sucking lice, ticks, mites, rats, and mice.
5. Describe and provide specific methods for rodent control.
6. Discuss and describe the purpose and function of pesticides and the issues of ecological damage, pesticide resistance, and adverse health effects.
7. List and describe the major classes of arthropod and rodent pesticides, their mechanism of toxicity, and some alternatives to chemical pesticides.

INTRODUCTION

According to a report prepared by Bechtel, Parsons and Brinckerfoff, comanagers of the massive Central Artery/Third Harbor Tunnel project in Boston, MA, rats were sighted at the rate of 500 per hour shortly after 9:00 p.m. on November 12, 1993, in the open-air market known as Haymarket in downtown Boston.[1] Fears were so great that construction would displace the rats into nearby neighborhoods that the construction managers were required to develop plans and countermeasures for exterminating the rats. They were not the first to try and exterminate rats. The blunt-nosed, muscular rat *Rattus norvegicus* arrived from Norway as stowaways on boats in the 1600s and numbered in the tens of thousands by the mid 1700s. Nearly 10,000 rats were killed in the last 4 months of 1742 as the town of Boston paid 4 pence for each rat killed. An all-out effort in 1917 to rid Boston of the more than 700,000 rats failed dismally, and all efforts throughout the following years have failed to eradicate them.[1] Boston is not alone in these efforts. In one 60-ft-deep pit in Mumbai, India, there were 442,186 dead rats of the variety *Rattus rattus*. After more than half a century of relief from the plague, the return of the rat in huge numbers has promoted the return of the plague to many of India's cities.[2] In most areas of the world, rats are feared and detested, and periodically become the targets of military-style campaigns to destroy them as pests. This is not so everywhere. In Mexico's impoverished northern desert region, rats are hunted down as a favorite food and boiled in soup or fried golden brown, though the government views the selling of rats for meat (about $1 each) as illegal.[3] The prospect of hunger and starvation can change attitudes toward what is food and what is a pest.

Also on the comeback trail after a pause in years of effort to destroy them are the mosquitoes. *Aedes aegypti* has returned in enormous numbers to Latin America after funds for mosquito spraying were diverted elsewhere, human populations exploded, sewage and waste were improperly handled, and sanitation suffered. The result has been a dramatic increase in dengue fever.[4] For similar reasons, increasingly larger epidemics of dengue fever and other vectorborne diseases have returned to India, Sri Lanka, Taiwan, China, and Africa.[4] Despite heroic attacks against the female anophelene mosquito to control malaria, these programs have also proved ineffective, and once more malaria affects 500 million people worldwide and causes 2 million deaths per year.[5] Deer mice have proliferated in New Mexico and spread hantavirus to the human population in that area, and on the east coast of the U.S. the deer tick, *Ixodes scapularis*, carries the disease agent for Lyme disease in a complicated life cycle involving the tick, deer, and mice. Rats, mice, mosquitoes, and ticks all have something in common: humanity perceives them as undesirable because they transmit diseases, because of which they have been labeled pests and are targeted for destruction.

WHAT ARE PESTS?

Pests are unwanted plants and animals.[6] Pests are not limited to organisms that transmit disease, but could be any living thing that negatively affects human interests. The interests usually include: (1) a loss of resources such as agricultural crops, food and property damage, and damage to lawns and gardens; (2) agents of disease; and (3) sources of annoyance and discomfort. Pests may be weeds that compete for agricultural space; freshwater mussels that clog intake pipes; rabbits that decimate field crops; moles that burrow through lawns; starlings that threaten collisions with aircraft; wasps that sting; moths that eat through clothes; cockroaches and flies that carry disease organisms onto food; or insects, fungi, and viruses that attack crops or people. Pests cover the entire spectrum of the plant and animal kingdoms, ranging from small viruses to large mammals. Although well represented across the animal and plant kingdoms, the actual number of pests is only a small percentage of the total number. The vast majority of plants and animals are viewed by most as beneficial, worthwhile, or at least neutral. The most unwelcome pests from a public health perspective tend to include arthropods and rodents. The arthropods (phylum Arthropoda) are invertebrate animals with jointed and paired appendages, a chitinous exoskeleton, and segmented bodies. It is the largest animal phylum, with over 700,000 species, including insects, arachnids, myriapods, and crustaceans (Figure 6.1).[7] Within the arthropods are included the insects and arachnids that are medically important as vectors of disease, disease agents themselves, or noxious pests. Certain arthropods and rodents have been associated with numerous disease outbreaks, crop damage, annoyance from bites, and contamination of food sources. Insects, ticks, and mites are involved in the majority of important human vectorborne diseases, and most of these cannot be prevented by vaccines or chemotherapy. Consequently, control of vectorborne diseases often relies on reducing the number of vectors.[8] Early success in the control of insect vectors has been met with reverses as: (1) insects have developed resistance to insecticides; (2) insect control programs have been halted or underfunded; (3) the use of less expensive pesticides such as DDT have been reduced because of environmental concerns and political pressures; (4) the combination of poverty and overpopulation has lead to poor sanitation with greater opportunity for insect proliferation; (5) destruction of forested areas has eliminated natural insect predators; and (6) climate changes, including warming trends, have promoted increases in some insect populations.[4] For these reasons, many insects are rebounding in large numbers and there have been significant increases in vectorborne diseases such as malaria, yellow fever, dengue fever, Chagas' disease, typhus, relapsing fever, the encephalitides, schistosomiasis, and plague. These diseases affect hundreds of millions of people worldwide and reduce the vitality and economies of many nations. Understanding the role of insects in this process is important to the development of public health policy and programs. In order to develop workable biological, physical, or chemical controls, it is critical to understand the biology of and mechanisms for the spread of disease by these organisms.

Do you want to learn more about pests? Point your browser to http://www-unix. oit.umass.edu/~envhl565/weblinks.html and click on any of the Chapter 6 Web links associated with pests including:

- Entomology Index
- Insects on the WWW
- CDVIK
- Crab Louse or Pubic Louse
- Rats

INSECTS AND OTHER ARTHROPODS

GENERAL STRUCTURE AND DEVELOPMENT

Most insects start their life cycle with the fertilization of an egg, and they pass through either a complete or incomplete metamorphosis. Incomplete metamorpho-

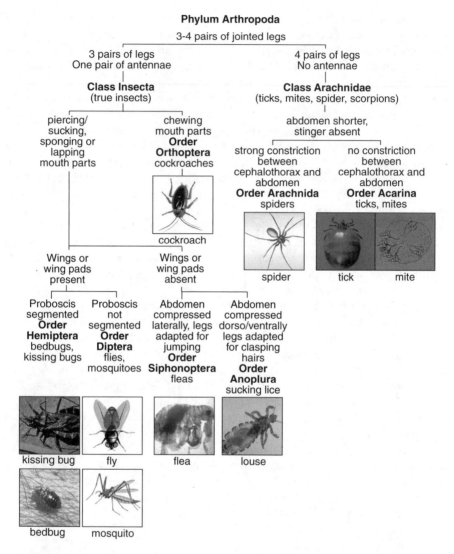

FIGURE 6.1 Pictorial key to selected orders of arthropods of public health significance. (Adapted from U.S. Dept. of Health, Education and Welfare (HEW), *Pictorial Keys to Some Arthropods of Pubic Health Importance*, Washington, D.C., U.S. Department of Public Health Services, 3, 1964.)

sis occurs in insects such as roaches, body lice, and grasshoppers that go through three developmental stages: egg, nymph, and adult. Mosquitoes, flies, and butterflies go through complete metamorphosis and pass through four developmental stages including egg, larvae, pupa, and adult stages. As insects grow and develop, they shed their protective exoskeleton while a new one develops underneath in a process called *molting*. The exoskeleton is a hard outer skin that protects internal organs and serves as an attachment site for muscles to provide movement. The

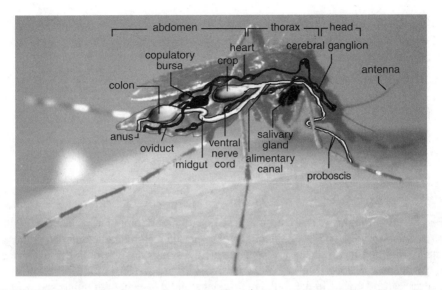

FIGURE 6.2 Internal and external structures of typical insect. (Adapted from Koren, H. and Bisesi, M., *Handbook of Environmental Health and Safety, Principle and Practices,* 3rd ed., Vol. 1, CRC Press/Lewis Publishers, Boca Raton, FL, chap. 4.)

hardened outer body wall of the exoskeleton may be covered with hairs, scales, or spines, and is normally divided into segments joined by flexible intersegmental structures. In insects, these segments consist of a head, thorax, and abdomen, as seen in Figure 6.2. Members of the order Arachnida have only one or two main segments, as in ticks, mites, or spiders. The head usually contains mouth parts, antennae, large compound eyes, and/or simple eyes. The numbers of eyes and antennae help distinguish between species. Mouth parts vary and are adapted to forms of feeding including piercing–sucking (mosquitoes and fleas); chewing and grinding (roaches); and lapping/sponging (houseflies, blowflies). Houseflies and similar sponging insects regurgitate fluids onto solid foods, liquefy them, and suck them up. They can carry disease organisms on body parts or in the digestive tract to food or open wounds and so serve as mechanical vectors of disease. Mosquitoes and fleas can pierce the skin of animals and humans to reach blood. In many instances, the microbial disease agents go through a developmental stage in the vector involving sexual reproduction (i.e., malaria). In these instances, the vector is known as a *biological vector*. In the case of malarial transmission, the infective stage, or sporozoites, migrate to the salivary glands, where they are inoculated into the bloodstream of the next human victim as the mosquito takes a blood meal (see Chapter 7, Figure 7.10). The thorax of insects bears three pairs of jointed legs and may also contain two sets of wings, which are often reinforced with structures called *veins*. The abdomen, or third body part, contains external openings to the respiratory system known as *spiracles*, and also contain external reproductive organs. Internally, the abdomen is composed mostly of reproductive and digestive organs (Figure 6.2).

FIGURE 6.3 Example of typical kissing bug and bedbug.

BEDBUGS AND KISSING BUGS

The insects belonging to this category have three pairs of jointed legs, sucking mouth parts, wings or wing pads, a segmented proboscis, and belong to the order Hemiptera. The kissing bug has well-defined wings, whereas wings are absent in the dorsoventrally flattened bedbug (Figure 6.3). True to their name, bedbugs hide in bedsprings, mattresses, and cracks in the wall, coming out at night to feed on humans and warm-blooded animals. People sensitive to the bites develop large and very itchy welts. No disease transmission has been associated with bedbugs, although here has been a significant increase in the numbers of reported infestations and irritations.

The kissing bug is the arthropod vector for American trypanosomiasis, or Chagas' disease, caused by the flagellated protozoan parasite *Trypanosoma cruzi*. The disease occurs predominantly in Mexico and South and Central America, where hosts include rodents, armadillos, and opossums. The kissing bug lives in the crevices and cracks of stone or mud huts with thatched roofs. This reduviid bug appears at night while people are sleeping, and bites them near the lips (hence the name "kissing bug"). The trypanosome is deposited on the skin surface in the bugs' feces. The irritated skin is scratched by the person, who inadvertently rubs the organism and feces into the open bite wound. The disease affects more then 50 percent of the population in some rural South American communities. Once inoculated, the trypanosomes are carried by the blood to tissues throughout the body, where they can multiply and produce a severe inflammation. Fatalities occur in about 10 percent of those infected. These are usually children, and the heart appears to be the primary target organ disrupted in these fatalities. Damage to the central nervous system with paralysis of certain body functions is also common.[9]

FLIES

Flies belong to the same order of insects (Diptera) as mosquitoes. Adult flies have one pair of functional wings and three distinct body parts including the head, thorax, and abdomen (Figure 6.4). The middle thoracic section is larger than the others because it supports the highly developed wing muscles. Flies may have sponging mouth parts as does the housefly, whereas other fly species may have rasping or sucking mouth parts. All have compound eyes and a single pair of antennae. There are thousands of species of flies. They are significant vectors of human disease and

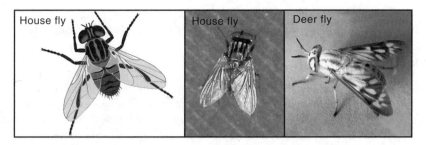

FIGURE 6.4 Typical domestic fly (*Musca domestica*) and the deerfly (*Chrysops discalis*), which is a biting fly that transmits deerfly fever, a form of tularemia.

also cause discomfort. Blackflies, deerflies, sandflies, horseflies, and stable flies can all inflict painful bites or bleeding punctures. Humans have died from anaphylactic shock produced by the bulk of foreign protein injected into the skin from the simultaneous bites of hundreds of flies. Some species may lay eggs or larvae in wounds, resulting in invasion of the living tissue by fly larvae, producing a condition known as *myiasis*, in which maggots grow in the tissue or in open cavities. The common housefly, *Musca domestica*, is an important vector of infectious disease organisms and many foodborne illnesses. Houseflies can travel several miles in a single day, feeding on human wastes, and transporting disease organisms from their bodies and feces to human food and beverages. Diseases that may be mechanically transmitted by flies include cholera, bacterial and amoebic dysentery, typhoid and paratyphoid fevers, salmonellosis, hookworm, pinworm, and whipworm. Biting flies have been responsible for the transmission of a number of diseases including sandfly fever, onchocerciasis (blinding filariasis), African sleeping sickness, deerfly fever (a form of tularemia), and loaisis (African eyeworm disease).

Flies breed rapidly, with the female laying several batches of 75 to 150 eggs onto garbage, manure, or any moist organic material within 1 to 2 weeks of reaching the adult stage. Flies undergo complete metamorphosis, with the eggs hatching in 24 h to a larval stage that lasts 4 to 7 d. The larvae crawl into the soil or under debris, where they enter the pupa stage for 3 d to several weeks, depending on the temperature. The adult stage emerges from the pupa stage ready to start breeding. Several generations can be produced during the warm seasons.

The control of flies begins with the elimination or covering of breeding materials such as garbage, feces, or dead animals. These materials need to be buried, tightly covered, or incinerated to prevent access by adult flies. Tight-fitting screens of 16 mesh size or smaller are effective physical barriers. In heavily trafficked areas, air shields that blow down and away from the entrance have been successfully used to limit access by flies to a room or building. Biological controls have involved the radiation of male screwworm flies in Florida and on the island of Curacao. The screwworms mate once, and if their mating takes place with a sterile male, no progeny develops. This process requires sterilization of large numbers of males by radiation or chemicals and their release into the environment. This is most effective when the area is somewhat isolated, making reinvasion from other areas less likely. Chemical attractants can be used to draw flies onto sticky "fly strips," or DDVP resin strips that are useful in killing flies in areas of 1000 ft³ per strip. As with any of the insecticides, precautions must be taken around foods and sensitive persons.[8]

Mosquitoes

Mosquitoes have one pair of wings with scales, belong to the order Diptera (two wings), and the females are characterized by piercing-sucking mouth parts with an elongated proboscis. Mosquitoes undergo complete metamorphosis. Some of the female species may lay several batches of 50 to 200 eggs, whereas others lay single eggs. Depending on the species, the eggs may be deposited in still water such as swamps, wet depressions, or in the collected water of old tires or containers. The eggs hatch in about 3 d to a larval stage that may molt 3 to 4 times over a 7 to 10 d period. The larvae enter a pupa stage, and in several more days become adult mosquitoes (Figure 6.5). The adult mosquitoes may travel from 100 yd to several

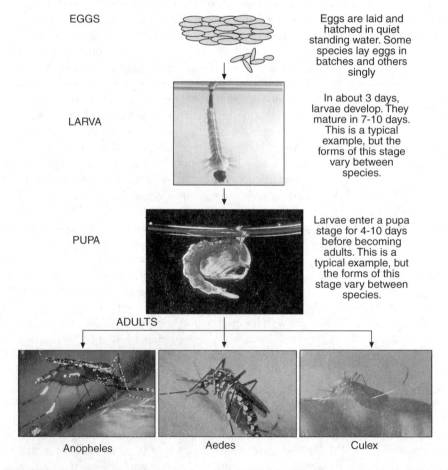

FIGURE 6.5 Developmental stages of a typical mosquito. (Adapted from Koren, H. and Bisesi, M., *Handbook of Environmental Health and Safety, Principle and Practices,* Vol. 1, 3rd ed., CRC Press/Lewis Publishers, Boca Raton, FL, chap. 4.)

miles, depending on the species, to feed on preferred hosts that may include humans, cattle, horses, or other domestic animals, or a variety of wildlife. Once a host is found, the female mosquito takes a blood meal during which time it injects saliva into the skin. The saliva prevents clotting and also causes the itching that is experienced by most people. The time required for digestion of the blood meal, laying more eggs, and then seeking another blood meal may be as little as 2 d. Some disease agents require 10 to 14 d of development in the mosquito before they become infectious to humans or animals. A mosquito may therefore breed multiple times before passing on the disease. There are more than 2600 known species of mosquitoes. Many of the species of public health interest belong to the genera *Anopheles*, *Culex*, and Aedes. *Aedes aegypti* is an important vector for yellow fever. The human virus circulates in the blood only 3 to 4 d during the first stages of illness, and when the Aedes mosquito takes a blood meal during this time, the virus enters the mosquito. The disease can then be transmitted in 10 to 14 d. *Aedes albopictus* is also known as the Asian Tiger mosquito and is thought to have been inadvertently transported from Japan. It was found breeding in the pooled water of tires being transported over Texas highways in 1985.[8,10] This mosquito bites severely, survives cold winters, and is capable of transmitting the agents of dengue fever and dengue hemorrhagic fever, the encephalitides group of viruses, and yellow fever. Although it has not yet become an important agent of disease in the U.S., *Aedes albopictus* has the potential to become a significant public health problem, just as in Asia.

Species of the female anophelene mosquito are important vectors of malaria. Malaria is biologically transmitted from human to human in a process that begins when male and female gametocytes of the malarial parasite *Plasmodium* sp. are taken in with a blood meal by the female anophelene mosquito. The parasite goes through a sexual stage in the developmental cycle within the mosquito that results in the formation of slender sporozoites in a few weeks (see Chapter 7, Figure 7.10). These sporozoites migrate to the salivary glands of the mosquito and enter the host blood during the mosquito's next blood meal. Here, the parasites pass through an asexual cycle characterized in some forms of the disease by the synchronous growth of the parasites in the red blood cell. The periodic destruction of the red blood cells with release of parasites into the circulating blood often causes characteristically periodic episodes, including headaches in the prodromal stage followed by chills and then fever. The chills and fever often occur in regularly timed waves of 36 to 72 h, depending on the malarial species involved and results from inflammatory responses of the body to the explosive release of foreign bodies into the bloodstream.

Culex sp., or house mosquitoes, tend to bite fiercely and continuously on humans, birds, horses, and other animals. They can breed in almost any standing water, especially when contaminated with organic matter such as waste food or sewage. *Culex* sp. is a night feeder and rests by day. They have the capacity to travel several miles, though they usually stay within a mile of the breeding area. They have been implicated primarily in the spread of viral encephalitis.

Control of mosquitoes can occur at the larval and adult stages. Larval stages may be controlled by chemicals and biological and mechanical means. The most effective and environmentally friendly control involves the elimination of standing or stagnant water. Old tires, tin cans, plastic bottles, and land depressions where water can stagnate need to be buried or filled. Irrigation ditches should be smoothed and straightened to permit water to flow easily and drain quickly. In bodies of water too large to fill in, top-feeding minnows have been used to control the larvae. The addition of chemicals such as oil or pesticides to surface water has been effective in the past, but environmental concerns and regulations may restrict most applications. The adult mosquitoes can be temporarily or seasonally controlled by area-spraying with compounds such as carbaryl or malathion. Most community spraying programs now occur in conjunction with a public health threat such as the potential spread of mosquito-borne viral encephalitis. Within the home, screens are an effective control when in good repair and having at least a 16 to 20 mesh. In many lesser-developed countries (LDCs) and in areas where screens are unavailable, mosquito netting is a useful deterrent. In LDCs a contact poison such as DDT has been used on the netting to increase their effectiveness Repellents such as DEET (diethyl toluamide) and Indalone offer some short-term protection against mosquito bites but will vary in effectiveness depending on the mosquito density, the species, and many other factors.

FLEAS

Fleas are true insects with three pairs of legs and no wings. They are small and in the range of 1 to 8 mm (0.04 to 0.3 in.) with a siphon or tube as a mouth part that is used to feed on the blood of warm-blooded animals (Figure 6.6). There are several species of fleas that have the capacity to bite humans and include the dog flea and cat flea (*Ctenocephalides* sp.), the human flea (*Pulex irritans*), and the rat flea (*Xenopsylla cheopis*). The female flea takes a blood meal from its host, mates, and then lays eggs. The eggs drop to the nearest surface such as carpet, concrete floors, or furniture, where they undergo complete metamorphosis to an adult stage within 2 to 3 weeks. At the adult stage, the fleas look for a blood meal and feed on animals or humans, multiplying in large numbers and spreading rapidly through living quarters, livestock pens, and yards. Their ability to take blood meals and move swiftly among humans makes them important vectors of disease, including plague (*Yersinia pestis*), murine typhus, tularemia, and even salmonellosis. The rat flea carries the plague bacillus in its intestinal tract, spreading it from rat to rat and from rat to human. Because of the dark blue-black hemorrhages under the skin seen in bubonic plague, the disease was called the *Black Death* when it ravaged Europe in

FIGURE 6.6 Oriental rat flea (*Xenopsylla cheopis*). The dark area in the abdominal area is where the plague bacillus *Yersinia pestis* has blocked the intestinal tract.

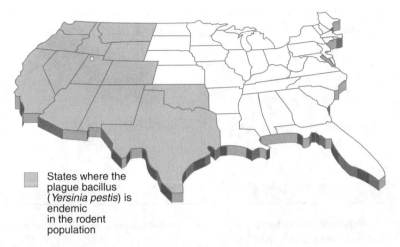

States where the plague bacillus (*Yersinia pestis*) is endemic in the rodent population

FIGURE 6.7 States in which plague is endemic in the wildlife rodent population.

the 14th century and killed more than 25% of the population. India continues to see periodic episodes of plague with nearly 10 million people dying from it in the early 1900s, and many have died in more recent outbreaks in 1994 in the Indian city of Surat.[9] The disease is endemic in the wild rodent population in the far west and southwestern U.S., where it is found in prairie dogs, chipmunks, and squirrels. The disease has continued to spread eastward since its introduction into the wildlife rodent population early in this century, and now has been identified in 13 western states with more than 360 cases of plague identified from these areas over the last 50 years (Figure 6.7).[9,10]

Flea control usually involves treating infested domestic animals and the domiciles in which they live with a pesticidal dust containing either carbaryl or methoxychlor. The mildly toxic pesticides have a prolonged residual time. Powder should be rubbed thoroughly into the animal fur and especially around the head and neck areas.

LICE

Lice that feed on humans or animals are called *sucking lice*. They can bite severely and inject saliva into the bite, resulting in irritation and severe itching on the skin. Lice have no wings, a dorsoventrally depressed abdomen (flat), and legs adapted for clasping hairs (Figure 6.8). Lice pass through an incomplete metamorphosis to adults. The egg, or nit, attaches to the head, pubic hair, or skin, depending on the species. There are three kinds of human lice, including the head louse, *Pediculis humanis capitis;* the body louse, *Pediculis humanis corporis*; and the crab louse, *Pthirius pubis*. The eggs hatch in the presence of body heat and the resulting nymph molts thrice before becoming an adult within 1 to 4 weeks, depending on how long the nymphs stay in contact with the body. The lice are easily and rapidly transferred through person-to-person contact, bedding, brushes, and combs.

Louseborne typhus is caused by *Rickettsia prowazekii* and carried by the human body louse. The organisms grow in the louse intestinal tract and are excreted by it.

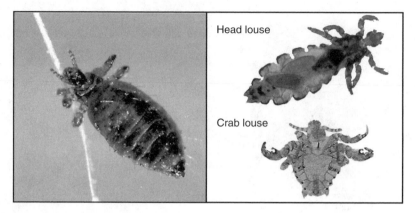

FIGURE 6.8 Typical louse shown clinging to hair shaft: a head louse (*Pediculus humanis capitis*) and crab louse (*Pthirius pubis*).

When the person scratches the itching bites, the feces and bacteria are rubbed into the wound, causing an infection. Inflammation from the infection causes local blockage of small blood vessels and subcutaneous hemorrhaging. Untreated diseases can result in high mortality. The disease is normally seen in very crowded and unsanitary conditions. In addition to spreading epidemic typhus, lice also cause trench fever (*Rickettsia quintana*), and relapsing fever (spirochetes belonging to the genus *Borrelia*).

ROACHES

Cockroaches are true insects having three pairs of legs, one pair of antennae, and chewing mouth parts (Figure 6.9). Roaches are oval and flattened dorsoventrally, with the head usually not apparent or hidden. There are several common species of roaches including the American roach, *Periplaneta americana*; the German roach, *Blattella germanica*; and the Oriental roach, *Blatta orientales*. The largest of these widely distributed roaches is the American variety, which reaches lengths of 2 in. The adults are dark brown to reddish brown in color. Roaches go through incomplete

FIGURE 6.9 Typical American cockroach (*Periplaneta americana*).

metamorphosis involving egg, nymph, and adult stages. The eggs are usually encased in a capsule, carried for a few days to a month depending on the species, and then attached to some object out of sight. The nymphs hatch in 2 to 3 months and pass through many molts over several months before reaching the adult stage. Roaches prefer the dark and are usually concealed during the day or when bright lights are shining on an area. The roaches will conceal themselves behind cabinets, in drawers, behind moldings or furniture, and any place where there is a crack or crevice. The appearance of roaches in lighted areas usually means the area is heavily populated. Roaches feed on most organic material including fecal material, decayed food remnants, glue, and any accessible food. During the process of feeding, they regurgitate. Disease-producing microbes may be carried on the external body parts or in the intestines of the roach and mechanically transferred to food. No specific disease outbreaks have been attributed to roaches but they may serve as vectors and create a level of discomfort for most people.

Roaches are best controlled by making certain no potential food sources exist. Waste foods should be immediately cleaned, and leftovers tightly covered and/or refrigerated. All areas, including the dark reaches of cabinets and closets should be periodically cleaned to remove potential food stuffs from surfaces. Items from food stores including bags or cartons should be discarded as they may harbor roaches. Borax powder has proven effective in controlling roaches by dusting in areas where roaches are likely to hide. Organophosphate poisons such as malathion have also proved effective for roach control.

TICKS AND MITES

Ticks and mites have four pairs of legs, no antennae or wings, and belong to the class Arachnida. A shorter abdomen and the absence of a constriction between the cephalothorax and abdomen place organisms such as ticks and mites in the order Acarina.

MITES

Mites do not have a clearly differentiated head, thorax, and abdomen (Figure 6.6 to 6.10). They are extremely small and may not be easily seen without the aid of magnification. Mites pass through several stages of development that include eggs, larvae, nymphs, and adult within 2 to 3 weeks, which is about their average length of life. Because mites may proliferate in areas where human body cells are shed and where body oils accumulate (mattresses, couches, and beds), they can infest and irritate humans as well as transmit diseases. The mite *Sarcoptes scabiei* burrows

FIGURE 6.10 Typical scabies mite (*Sarcoptes scabiei*).

under human skin to lay its eggs. This burrowing causes slightly elevated, serpentine lines, or a variety of inflammatory skin lesions resulting from secondary infections as a result of scratching. Intense itching, rashes, and redness often develop. The mites are transmitted through close personal contact with an infected person. Dogs and cats are often infested and develop a condition called *mange*. The mites frequently infest birds and rodents, which become a major source for infestations of homes and domestic animals. A type of mite called a *chigger* or *redbug* is the six-legged larvae of mites. Chiggers attach to the skin and inject a salivary secretion that dissolves the surrounding tissue and is then sucked up the tubular stylostome of the chigger. Scrub typhus is a rickettsial disease passed to rodents and humans by the chigger, causing death in up to one half of those infected. The mite-infesting mice has been associated with the transmission of rickettsial pox caused by *Rickettsia akari*. Certain forms of hemorrhagic fever and encephalitis have also been associated with transmission by mites.

Preventing mite infestations begins with eliminating rodent and bird populations in the house or building, keeping everything clean and vacuumed, and allowing plenty of sunlight into the living spaces. When mite infestations have occurred, a pyrethrum bomb should be used, the area cleaned and vacuumed, and malathion applied around windows, doors, and baseboards for residual action.

TICKS

Similar to mites, ticks belong to the class Arachnida. Consequently, they have four pairs of legs, no antennae, and a fused head, thorax, and abdomen (Figure 6.11). Ticks are particularly adapted to feed on blood by (1) the use of barbed feeding organs that pierce the skin to reach the blood and also anchor the tick; (2) a flexible leathery body that easily distends when filled with blood; and (3) uniquely adapted pharyngeal muscles for sucking blood. There are two categories, including soft (family Argasidae) and hard (family Ixodidae) ticks. Ticks undergo a complete metamorphosis often involving a fairly complex cycle, such as in the deer tick, *Ixodes scapularis*, which is the vector of Lyme disease caused by the spirochete

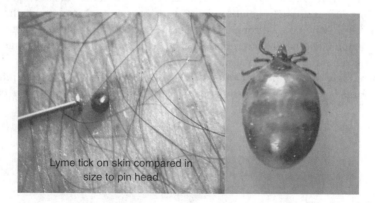

Lyme tick on skin compared in size to pin head

FIGURE 6.11 An engorged tick associated with Lyme disease (*Ixodes scapularis*).

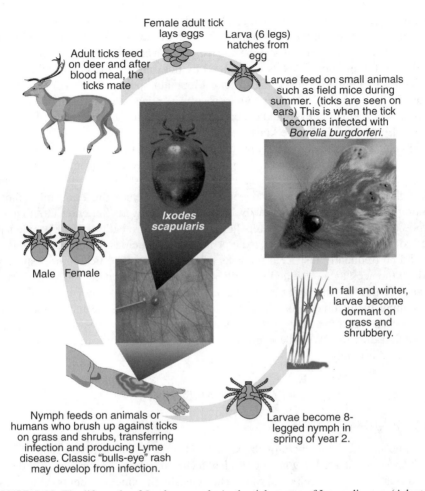

FIGURE 6.12 The life cycle of *Ixodes scapularis*, the tick vector of Lyme disease. (Adapted from Tortora, G.J., Funke, B.R., and Case, C.L., *Microbiology: An Introduction*, 5th ed., Benjamin Cummings, 1995, Chap. 23.)

Borrelia burgdorferi (Figure 6.12). The adult female Ixodes tick feeds once on a deer and then mates. It lays several thousand eggs in a batch. The eggs hatch to larvae in a few weeks to several months. The larvae fall to the ground, where they come into contact with rodents and other small animals. The larvae (now possibly infected with *Borrelia* sp.) would be deposited on grass and leaves, where they become dormant over the winter. During spring of the second year, the larvae develop into eight-legged nymphs that appear on the tips of grass and brushes at the forest edge, lawns, and gardens, where they can cling to humans or animals that brush by them. The ticks then spread the disease to humans on whom they feed.

Ticks can transmit diseases mechanically as disease organisms may be carried on tick mouth parts or other body parts and easily infect other persons who are bitten by the tick. Many other organisms pass through developmental stages in the tick by a process known as *transstudial transmission*. In this process organisms are passed from the tick to the egg and back through the adult tick developing from that egg. Thus, the disease agent may be carried over several generations. Among some of the more important diseases transmitted by ticks are (1) the rickettsial diseases, including tularemia, Q fever, and Rocky Mountain spotted fever; (2) the disease caused by *Borrelia* sp., including Lyme disease and relapsing fever; (3) viral diseases such as Colorado tick fever; and (4) bacillus diseases such as *Pasteurella tularensis*. Ticks also produce a disease known as *tick paralysis*.

Keeping ticks off exposed skin is the best control method. Before hiking, biking, or working in high grass or brush, protect yourself by tucking pants into socks, tightly buttoning or taping long-sleeved shirts at the wrist, and wearing clothing that fits tightly to the neck. It is also good practice to wear light, uniformly colored clothing so that ticks can be easily spotted and removed. Spraying clothing and exposed skin with DEET insect repellent offers some protection. Some people have treated clothing with benzyl benzoate or indalone pesticides. Family dogs and cats should be frequently inspected for ticks using a fine-toothed tick comb, especially around the neck area. Tick and flea collars are helpful, but do not guarantee against some infestation. When ticks are found on humans or animals, they can be removed by pulling them directly out from the body together with the mouth parts. This process can sometimes be facilitated by applying nail polish or Vaseline to the tick. Malathion dust or DDVP have proved effective in controlling ticks when applied over several months.

RODENTS AND PESTS

RODENT CHARACTERISTICS

Rats and mice are members of the order Rodentia, which include mammals with teeth and jaws adapted to gnawing. Rodent teeth can grow 5 to 6 in. a year and must be constantly used to keep them short. The constant gnawing of rodents is the source of much destruction. Other members of Rodentia include beaver, squirrels, porcupines, chinchillas, and lemmings. The two major rat species in the U.S. and many parts of the world include the roof rat, *Rattus rattus*, and the Norway rat, *Rattus norvegicus*. The house mouse, *Mus musculus*, and deer mouse, *Peromyscus maniculatis*, are also pests in the U.S. The two rats mentioned earlier and the house mouse are commensals in that they live in or near humans, eat our food, and share diseases without providing any benefits. Because of their close association with humans, they are classified as murine rodents belonging to the subclass Murinae. The roof rat prefers to live in barns, attics, warehouses, and other human structures. *Rattus rattus* is highly adapted to balancing along narrow beams and squeezing through cracks and crevices or tiny holes that it can create by gnawing. The roof rat is slender, has a pointed snout, large prominent ears, and a tail longer than its body. It weighs between 8 and 12 oz and reaches lengths of 8 to 17 in. (including the tail; Figure 6.13). The Norway rat tends to dig burrows under foundations, along banks of rivers, in fields, under solid waste, and under lumber piles. They live in sewers and between floors and walls. These are muscular rats with thick bodies, a blunt snout, with a

FIGURE 6.13 The roof rat (*Rattus rattus*).

FIGURE 6.14 The Norway rat (*Rattus norvegicus*).

tail shorter than its body. They reach lengths of 12 to 18 in. (including the tail), and weigh up to 16 oz (Figure 6.14). These features are adapted for burrowing. Rats have several abilities that help them survive and adapt around humans. Rats can (1) pass through a quarter-size hole, which is about the size of its head; (2) climb straight up a brick wall provided there are tiny holds for their toes; (3) survive falls up to five stories high; (4) jump over obstacles 3 ft high; (5) cross over and balance on wires; (6) and swim more than a half mile in open water while periodically diving to locate possible escape openings.[1,8]

The house mouse (Figure 6.15) thrives most everywhere worldwide that people live. It weighs from 12 to 34 oz with a length of 6 to 7 in. (including a tail longer than its body). Its eyes and ears are large and prominent, and the nose is pointed (Figure 6.15a). They have a short gestation period of about 29 d, producing up to 8 litters of 5 to 6 offspring per litter. They tend to build small nests of shredded paper, fabric, or grass anywhere, including in stored goods, cabinets, furniture, and drawers, while outside they live in weeds, rubbish, and tall grass.

The deer mouse, *Peromyscus maniculatis*, is a vector of the hantavirus (see Chapter 7). It is a thin, agile creature weighing up to 1.25 oz and reaching lengths of up to 9 in. with a tail longer than its body. Its eyes and ears are large and prominent, and the nose is pointed (Figure 6.15). It has white feet and a white underside with a brownish back. It tends to live apart from humans in the forest and at the forest

FIGURE 6.15 House mouse, *Mus musculus*, and deer mouse, *Peromyscus maniculatis*.

edge or in meadows. They produce 2 to 4 litters per year with 3 to 5 offspring per litter. The deer mouse lives mostly on nuts, acorns, fruits, and insects.

IMPORTANCE AS PESTS AND VECTORS OF DISEASE

Rats and mice are considered to be pests by most societies because they (1) harbor a variety of organisms that can produce diseases in humans; (2) produce crop damage throughout the world; (3) eat and contaminate food produce and cause structural damage; (4) inflict harm directly by biting humans; and (5) cause extreme discomfort in some people by their visible presence.

Carriers of Disease

Rodents harbor and transmit a variety of disease organisms, including bubonic plague, *Yersinia pestis*; rickettsial pox, *Rickettsia akari*; murine typus, *Rickettsia typhi*; rat bite fever, *Streptobacillus moniliformis*; Weil's disease, *Leptospira incetro-haemorrhagiae*; salmonellosis, *Salmonella enteriditis* and other species; and trichinosis, *Trichinella spiralis*. Hantavirus is an emerging disease (Chapter 7) with severe consequences, mortality in humans exceeding 50%. It is normally transmitted when virus particles in the dried urine and feces of the deer mouse are inhaled.

Crop Damage

Rodents produce crop damage throughout the world. In some areas of the world, rats destroy as much as one third of the entire harvest.[11] In the Philippines there have been a number of rat population explosions such as those occurring in 1952 and 1984, with rat populations reaching 200 to 2000 rats per hectare and destroying 90% of rice crops and 20 to 50% of corn.

Food Contamination

Rodents eat and contaminate food and produce structural damage. Rats contaminate 10 times more food and materials than they use, damaging from $1 to $10 worth of food for each of the estimated tens of millions of rats in this country alone. Rats and mice also gnaw their way through walls and doors and eat the coating off wires, causing electrical shorts and fires. Rodents are thought to cause $1 billion of damage every year in the U.S.[8]

Rat Bites

Rats bite an estimated 45,000 people each year in the U.S., producing disfiguring injuries, infections, and some fatalities. In some reported cases, the rat refused to release its human victim and had to be beaten unconscious.[8] As there are no requirements for reporting of rat bites, the actual number of bites may be much higher. The number of rat bites may likely increase as rats thrive in areas of poverty and urban decay, where garbage and trash, poor sanitation, and inadequate rat control measures abound along with increasing population density.

Discomfort

Rats and mice cause extreme discomfort in some people. Persons who have experienced rat bites or who have had a direct confrontation with rats may be traumatized psychologically by the encounter, with lasting emotional consequences.[8] Others are simply terrified by the visible presence of rats or mice. I still recall my otherwise normal mother standing on a kitchen chair screaming as a mouse scurried along the kitchen baseboard.

RODENT CONTROL

Establishing Presence

Murine rodents tend to live close to humans and have a very limited range of 100 to 150 ft for rats and 10 to 30 ft for the house mouse. Deer mice range over 1 to 4 acres. Rodents tend to be most active at night when there is plentiful food. If rats and mice are seen during the day, that is usually evidence of many more in the vicinity. Evidence of the rodents is often established by (1) seeing the rodent or its droppings; (2) seeing nesting materials in drawers, cabinets, near foundations, or other places around or in the structure; (4) seeing holes gnawed into walls, doors, or other structures; or (5) seeing grease trails or runways, tracts, swing, or rub marks along the wall edges (Figure 6.16). Rodents prefer to travel familiar and established routes along room edges and if a route is used frequently, body oils will be left along those edges.

Rodent Proofing

Rodents live in proximity to humans for easy access to food. Any effective control program must begin with the removal or securing of food and denial of harborage. Whether a private home, commercial establishment, or an entire community, every effort must be made to store food and food waste in areas or containers inaccessible to rodents. Refuse should be stored in 30 gal metal containers with tight-fitting

FIGURE 6.16 Rub marks or grease trails left by rats frequently traveling along the edge of a structure. (Source U.S. Department of Health, Education and Welfare (HEW), *Control of Domestic Rats and Mice*, Public Health Service, Communicable Disease Center, 22, 1964.)

lids that are placed on platforms about 12 in. off the ground. The areas should be frequently cleaned and inspected for signs of rodent activity. Dried or canned foods should be stored in clean, dry areas on shelves above the floor or on palettes placed 24 in. from the wall so that inspection of all sides is made easy and cleaning is facilitated. Brush, weeds, and grass should be cut outward from the building to 100 ft if possible. Any rubbish, debris, brush piles, or old lumber should be removed. The structure should be examined for any holes with diameter greater than 1/4 in. The building can be made resistant to rodent entry by installing (1) metal doors or metal edges around doors, (2) metal screens of 19 gauge or smaller, (3) metal cones around pipes or wires entering the building, (4) sheet metal plates where pipes or wires enter the building, and (5) a curtain wall made of sheet metal or concrete around the foundation (Figure 6.17). An L-shaped curtain wall of concrete at least 2 ft deep may be poured along the foundation walls to prevent burrowing. Alternatively, sheet metal can also be used that extends 12 in. above the ground and 6 in. below.

Poisoning or Trapping of Rodents

Poisoning or trapping of rodents is often necessary to complete the task of control. There are numerous varieties of traps available for the homeowner, including some that are disposable plastic shelters where the bait may be inserted at the inside rear, the trap set by pulling a tab at the outside rear, and pushed into place where the rodents frequent. When sprung, the mouse and its container may be discarded without contacting the dead rodent. Traps for rats are less successful because the rats drag them or chew through them or themselves to escape (Figure 6.18). No rodent should be handled with unprotected skin.

Rat and mouse poisons available to consumers usually are composed of powerful anticoagulants such as warfarin, fumarin, chlorophacinone, PNP, and diphacinone. The anticoagulants cause internal hemorrhaging, resulting in the death of the rodent over a period of several days. Consequently, the rats or mice seldom associate the baited food with their illness and the colony continues to feed. The baits may be

Conical metal guard to prevent rodents from reaching structure by wires.

Screening should be galvanized hardwire of 19 gauge (4X4.5 in.).

Keep garbage in tightly-sealed metal container off of ground on pallet.

Openings around pipes covered with sheet metal or concrete mortar.

Metal flashing around wooden, earthen, or adobe dwelling to a height of 12 in. above ground and 6 in. below ground. Concrete curtain can also be used.

Opening below door must be less than 1/2 inch.

Holes greater than 1/4 inch in diameter should be covered with metal sheeting, or plugged with concrete.

Channeling and cuffing of door edges with metal guards.

FIGURE 6.17 Methods of rodent proofing. (Source from U.S. Department of Health, Education and Welfare (HEW), *Control of Domestic Rats and Mice*, Public Health Service, Communicable Disease Center, 22, 1964.)

FIGURE 6.18 Evidence of rat gnawing itself free from rattrap. Rodents have teeth and jaws adapted to gnawing. (From Training Materials from HEW, *Control of Domestic Rates and Mice*, Public Health Service, Communicable Disease Center, 22, 1964. With permission.)

purchased as ready-to-distribute. Further discussion of rodent poisons is given in this chapter under the Pesticides section.

Once the areas of concern have undergone rodent proofing, the removal of food and denial of harborage, and poisoning or trapping has occurred, then continued vigilance is necessary. Cleanliness must be maintained and food made unavailable to the rodents. There should be periodic checks for signs of rodent activity.

PESTICIDES

HISTORY

Pests compete with humans for food, serve as vectors of disease, destroy crops or depress their market value, cause structural damage to buildings and homes, and attack people directly, causing annoyance, injury, or even death. Prior to the 1940s, there was very little most people could do but suffer the consequences of pest invasions and cohabitation. In the middle to late 1800s copper arsenate compounds (Paris Green) and lead arsenate (Bordeaux mixture) were introduced as fungicides and pesticides. These substances remain indefinitely in the environment and are toxic to insects and humans while creating potential harm to the environment through contamination of soils and water.[10] Compounds made from copper, chromium, and zinc were used to control insects while hydrogen cyanide was used to control red scale on plants. All of these compounds are highly toxic, most are long-lived in the environment, and resistant species developed against many of them. All arsenicals were banned following the introduction of better alternatives.[8] The advent of the chemical revolution in pesticides probably began in 1939 with the discovery of the usefulness of dichlorodiphenyltrichloroethane (DDT) by Paul Muller of the Geigy Corporation. It was found to be highly toxic to most insects but not to humans. The use of DDT became critical to the effort of the Allied troops in World War II, who were combating malaria, yellow fever, and a host of mosquito-borne, louseborne, and other vectorborne diseases. Within the next 15 years, more than 25 new pesticides were synthesized, including many chlorinated organics (chlordane, heptachlor, dieldrin, endrin, and iodrin) and the toxic organophosphate, parathion. These compounds were credited with saving millions of lives by suppressing insect vectors of malaria, filariasis, dengue fever, and typhus.[13]

These pesticides were widely used in the U.S. since their introduction after World War II until the 1970s, when it became evident that many of the chlorinated organics produce significant effects on ecology and human health. Consequently, the use of most organochlorine insecticides were canceled for use in the U.S. although DDT and a few other of the organochlorines introduced about the same time remain in use in other parts of the world such as LDCs, where vectorborne diseases remain a major threat and the need for low-cost insecticides is great.[11] Nevertheless, the promise of a world free of vectorborne disease and crops free of insect and fungi infestation has vanished in the avalanche of unforeseen events over the last 40 years. These pesticide-related problems include (1) vector resistance, (2) adverse health and ecological effects, and (3) global proliferation.

INSECT RESISTANCE

The previous four decades witnessed a dramatic shift in the balance of power between people and pests and raised our expectations that the battle against insects, rats, and fungi could be won. After all, there was a revolution in chemical warfare against these organisms that promised low toxicity to humans and high toxicity to insects. These chemicals were more effective and less toxic than arsenicals and heavy metals, while being effective against a broad array of insects. They were revolutionizing agriculture and freeing humanity from vectorborne diseases. However, almost immediately after their introduction, DDT and other organochlorine pesticides surfaced as potential villains, harming ecology and human health. Just as important was the realization that many important pests were developing resistance. U.S. production of chemical pesticides rose from 464,000 pounds in 1951 to levels approaching 1.5 billion pounds today, without any decline in crops lost to pests.[15,16] From 1970 to 1980, the number of arthropods resistant to insecticides nearly doubled from 224 to 428, and the numbers of resistant species of rodents, bacteria, fungi, and weeds are increasing as well.[15] There are presently more than 500 species of resistant arthropods. Farmers use ten times the amount of pesticides, and crop losses are almost double those when DDT was first introduced.[10] Furthermore, many weed species are becoming resistant to herbicides, and it is becoming an international problem as many of the genetically engineered plants associated with the green revolution may be easily overgrown by weed grasses. Such overgrowth could reduce crop yields by one third or more.[17]

The types of resistance most often encountered in arthropods are physiological or behavioral. Physiological resistance may occur from (1) enzymatic detoxification of pesticides into a less harmful form, (2) reduced permeability of exoskeleton to pesticides, and (3) storage or excretion of pesticides without harm. Some insects survive pesticide exposure through such behavior as landing less frequently or changing landing areas, and avoiding of baits such as fly-paper.[8] Some insects such as mosquitoes and flies have developed cross-resistance and are now resistant to both organochlorines and organophosphates.

A related issue with resistant species is the destruction of competitor organisms. Predators and competitors restrain pest populations. The application of pesticides upsets this ecological relationship by killing off predatory and competitive species, which often permit the original target pests to resurge dramatically to even greater numbers when the pesticides are diminished, This process is known as *target pest resurgence*. Very often the target pest is controlled by the pesticide, but once-harmless species grow numerous and become pests themselves to produce a secondary pest outbreak. Examples of this include eruption of the tobacco budworm on crops in Texas, the cottony cushion scale in citrus trees in California, or the growth of the tobacco budworm on crops in Texas and Mexico.[10,14]

THE HEALTH AND ECOLOGICAL EFFECT OF PESTICIDES

Exposures

Humans are exposed to pesticides through (1) contaminated drinking water; (2) eating foods contaminated with pesticides; (3) pesticide use in the home, garden, or lawn; (4) exposure on transcontinental flights; (5) use in agriculture; (6) involvement in the production of pesticides; and (7) in other occupations.[18] Nearly 75% of pesticides are used in agriculture, accounting for 66% of the $7 billion spent annually on pesticides.[19] There are 4 to 5 million Americans working in agriculture, with more than half of these being migrant or seasonal workers, and many of these are children under the age of 16.[18] Poisonings from pesticides have been estimated to affect between 500,000 to 2.9 million people worldwide annually,[20] causing fatalities for 1% of the cases, or more than 10,000 deaths per year worldwide.[21] There may be 300,000 cases of pesticide-related illness in the U.S., although less than 2% are reported, and most of these are neurotoxic pesticides.[18] There are reports that only 25 deaths occur in the U.S. each year from exposure to agricultural chemicals.[22] The major neurotoxic pesticides are organophosphate insecticides, which make up nearly 40% of all pesticide use in the U.S., and is the most commonly reported source of worker illness. Many organochlorine pesticides used prior to the 1970s are now banned in the U.S.[18] Most acute or one-time exposures cause such symptoms as headaches, dizziness, muscular weakness, and fatigue. Poisoning from organophosphates may lead to diarrhea, vomiting, severe abdominal pain, muscular twitching, breathing difficulty, sweating, and in severe cases convulsions, coma, and death.[18] Four to nine percent of acutely poisoned persons experience delayed or persistent neurological or psychiatric effects including agitation, insomnia, confusion, and depression.[18] The consequences of chronic or low-level pesticide exposure are uncertain and unresolved. The concerns for the present tend to be directed at the possibilities of cancer and reproductive problems associated with organochlorine pesticides. There is concern that pesticides may be depressing the immune response in the general population, especially in countries continuing to use pesticides banned in the U.S. Studies in the former Soviet Union found dose-dependent effects of pesticide exposure on immune function and the development of chronic and infectious diseases, and a Canadian study demonstrated effects of pesticide-laden breast milk on babies' immunologic deficiencies.[23] The major groups of people at risk from the adverse risks of pesticides are children and persons in LDCs.

Children at Risk

Children are exposed to pesticides through (1) prenatal maternal exposure; (2) food and water sources; (3) presence in agricultural fields with parents, or working the fields themselves; (4) contact with residues on parents' clothing; or (5) exposure to common household pesticides while in the home. Based on data from the American Association of Poison Control Centers, 74,000 children a year are involved in poisonings in the home from household pesticides.[24] Nearly half of all households with children under 5 had at least one pesticide stored in an unlocked cabinet less than 4 ft off the ground, with bathrooms and kitchens the most likely areas for improper storage. Common pesticides included roach sprays, bath disinfectants, insect sprays, flea and tick shampoos and dips, chlorine bleach, insect repellents, weed killers, and pool chemicals. Children may be protected from exposures to household pesticides through to the steps outlined in Table 6.1. Children are at greater risk from the toxic effects of these chemicals because they (1) absorb more pesticides per pound of body

TABLE 6.1
Recommendations for Preventing Accidental Poisoning

PREVENTING ACCIDENTAL POISONINGS IN CHILDREN

1. Store out of reach in a locked cabinet or shed, use child-proof safety latches

2. Read label and follow instructions for use including precautions and restrictions

3. Before applying pesticides, remove children, toys, and pets until pesticide dries, or according to recommendations on label

4. If use is interrupted (as by phone call), reclose the package and place out of reach of children

5. Never transfer pesticides to other containers

6. Teach children that pesticides are poisonous — something not to touch

7. Keep the number of the Poison Control Center near telephone

8. Alert caregivers to potential hazards of pesticides

Source: Adapted from U.S. EPA, Office of Pesticide Protection (OPP), *Pesticides and Child Safety: Tips to Protect Children from Poisonings Around the Home, Prevention, Pesticides, and Toxic Substances*, EPA#735-F-93-050R, update Nov. 19, 1997, http://www.epa.gov/oppfead/childsaf.htrn.

weight, (2) tend to receive higher doses than adults, (3) have immature detoxification systems compared to adults, and (4) may be more susceptible to neurotoxic effects because the central nervous system is still developing in children.[18] Food may be another potential source of pesticide exposure in children. The Natural Resources Defense Council (NRDC) released a report in February 1989 that attempted to analyze the extent of children's exposure to pesticides and the hazards based on information from the EPA, the Food and Drug Administration (FDA), and the Department of Agriculture (DOA). The report concluded that 17% of preschoolers, or 3 million children in that age group, are being exposed to hazardous levels of pesticides in fruits and vegetables exceeding federal safe-level standards.[25] The concern about cancer-causing pesticides in our food supply and the risk to children continues to be of public health interest,[26] despite the fact that the FDA found that 99% of 19,962 food samples contained no illegal residues.[27] Prompted by concerns of pesticides in foods, President Clinton signed the Food Quality Protection Act in August 1996. The new law requires (1) special consideration of infants and children when setting standards, (2) consideration of endocrine disruptors, (3) consideration of pesticide's benefits, and (3) expanded consumer right-to-know efforts.[28]

Exposures in LDCs

Pesticides are widely misused in LDCs, and it has become a significant problem. It is estimated that less than 25% of the world's production of pesticides is used in these countries, but more than 50% of acute poisonings of pesticide applicators occur in them; overall, 73 to 99% of worldwide deaths due to acute pesticide poisoning occur in LDCs.[18,29] The World Health Organization (WHO) has estimated that every minute some person in an LDC is poisoned by pesticides.[30] The reasons for this are numerous, but may include (1) failure to use protective clothing (sometimes workers are clad only in bathing suits) when applying pesticides; (2) workers may be in the field during pesticide spraying; (3) pesticides are being used in LDCs that are banned or severely restricted in the developed countries; (4) pesticide containers are often recycled to store food, milk, or cooking oil; and (5) pesticide products are labeled in languages the native populations do not understand, many of whom are also illiterate.[18] In an effort to control part of this problem, the United Nations Food and Agriculture Organization (FAO) adopted in 1985, and amended in 1987, a code that included a provision for prior informed consent (PIC). Under PIC, any pesticide restricted or banned in one country cannot be exported to another without its consent, obtained after it has been informed of the dangers. In August 1999 more than 60 countries agreed to implement this as a binding document and under the Rotterdam Convention.[31]

Although DDT has not been used in the U.S. since it was banned in 1972, it is still in wide use in LDCs. In 1999, it was being used in at least 27 countries. Its main purpose in these countries, however, is not to protect crops, but to prevent the spread of illness and death from malaria. Malaria kills 2.7 million people each year, most of whom are children living in LDCs. Normally, environmental protection and efforts to protect the health of populations go hand in hand. In the case of DDT use in LDCs, however, environmentalists and many international public health advocates are at odds with each other.

The United Nations has been attempting to draft a treaty banning the use of DDT (as well as a number of other pesticides) worldwide for several years. A number of prominent medical researchers, however, are fighting to ensure that the treaty will allow for DDT to be sprayed in small quantities on the interior walls of homes, where it acts as a repellent to potential malaria-carrying mosquitoes. Advocates of the ban (including the World Wildlife Fund and Physicians for Social Responsibility) argue that DDT has been found in the breast milk of mothers living in homes sprayed in this fashion. Unfortunately, many of the malaria victims living in LDCs cannot afford alternatives such as drugs available to treat malaria — and drug resistance is increasing in any case. Other pesticides that are less harmful to the environment are available, such as pyrethroids, but there are two significant problems with their use. First, spraying a house with pyrethroids is three to four times more expensive than DDT, and few of the LDCs can afford the cost. Second, a reliance on only pyrethroids may ensure an increase in resistance to the pesticide among mosquitoes. Environmental groups such as the World Wildlife Fund argue, however, that more creative solutions must be pursued, as the short-term benefits of using DDT are not worth the long-term consequences on the environment.[32]

The Counterarguments

The arguments have been advanced that pesticides are designed to be toxic, and that they are made specifically to kill insects, fungi, and other pests. Pesticide in our foods and environment are claimed to cause a variety of toxic effects including reproductive damage, birth defects, interference with the immune and endocrine systems, nervous system damage, and cancer.[23,33] The counterview is that:

1. Trace exposure to pesticides show no evidence of increasing the risk of cancer, birth defects, or other human ailments.[22]
2. No documented cases of deaths or illness from pesticide use in the U.S. has occurred when used in standard and acceptable ways.[22]
3. Many of the banned organochlorine pesticides have shown carcinogenicity only in mice.[22]
4. Pesticides help prevent grains and seeds from being contaminated with fungi that produce carcinogenic aflatoxins, while also providing an abundance of low-cost, attractive fruits and vegetables, which are humanity's best defense against cancer and heart disease.[33,34]
5. Naturally occurring carcinogens may cause an estimated 38,000 of the nation's 500,000 annual cancer deaths, whereas pesticidal residues are thought to contribute to less than 40 cancer deaths per year.[33]

Ecological Concerns

Much more than 1.1 billion pounds of pesticides are used in the U.S. each year, with more than 5 times that amount used globally. The greatest potential for the unwelcome effects on health and ecology come from the contamination of surface and groundwater. Water is one of the major pathways by which pesticides are transported long distances to other parts of the environment, moving from streams, to rivers, to lakes and oceans.[16] Nearly 50% of the nation's drinking water is provided by surface supplies, and surface water also provides vital aquatic ecosystems for fish, birds, plants, and other wildlife. Pesticides enter the waterways from streams through agricultural runoff, from wastewater discharges, groundwater inflow, atmospheric deposition, and spills. Pesticide residues, especially of the persistent organochlorines, are now detected in the tissues of animals in virtually every location on earth, from the Antarctic to the Everglades, and from small New England streams to the deepest ocean trenches. Such is the concern that a program was designed, known as the U.S. Geological Survey's National Water Quality Assessment (NAWQA) Program, to provide information on the extent and consequences of such pollution. The program is designed to provide consistent and reliable information on water resources in 60 important river basins and aquifers across the nation, and was fully implemented in 1997. Early findings on pesticides in the atmosphere show "the pesticides were ubiquitous and generally detected wherever they were sought."[16] Further, a review of studies indicated that most water treatment plants are ineffective in removing the pesticides, and that drinking water derived from some surface water sources contain pesticides above federal guidelines for at least part of the year.[16] Additionally, both the EPA and the National Academy of Sciences and Engineering (NAS/NAE) have developed or recommended water-quality criteria for a total of 43 pesticides in order to protect aquatic life from short- and long-term exposure. One or more of all of these pesticides have exceeded aquatic life criteria in the river basins tested. There are few such criteria for herbicides, and none for fungicides.[16] The significance of pesticides in surface waters has not been assessed because little is known about most pesticide toxicities at low-level exposures, nor the possible cumulative effects of multiple pesticide exposures.[16]

The consequences of surface water contamination are usually evident in their effects on wildlife. At low concentrations, the persistent, bioaccumulative, organohalogen pesticides have disrupted the endocrine systems of birds, fish, turtles, and mammals, causing decreased fertility, metabolic and behavioral abnormalities, feminization of male species, and masculinization of female species. The effects have been manifested in the offspring because of exposures at critical times during embryonic development.[35]

Acute exposures to larger concentrations of pesticides can produce immediate and dramatic effects, including massive fish kills in streams where pesticide runoff or spills have occurred, deaths of geese browsing on pesticide-treated golf courses, or a variety of more than 2 million birds dying from carbamate pesticides contaminating water and plants.[36] Further, the destruction of certain insects or plants may have negative effects when they serve as a food source or habitat for other organisms. This is especially evident in birds that feed on insects that have been killed by pesticides, or use particular plants for nesting that have been eliminated by herbicides.

Groundwater can also be contaminated. Extracting water by pumping it from the depths beneath the ground is no guarantee of chemical or biological purity. Thousands of private wells and those serving entire communities have been contaminated with pesticides and organic solvents. The EPA conducted the National Pesticide Survey in all 50 states between 1988 and 1990. The preliminary results revealed that 10% of the nation's public well supplies had detectable levels of at least 1 pesticide. Federal guidelines were exceeded in 1% of the community wells and 0.8% of the private wells.[8] In 1992, the EPA reported that more than 30% of private (rural) wells exhibited unacceptable levels of pesticide contamination resulting mostly from agricultural use. The complex and usually slow movement of groundwater plumes, combined with the highly variable application of pesticide amounts and types, makes it difficult to assess the full extent of the problem. The increasing use of pesticides in agriculture combined with garden and lawn use may combine with the impending already contaminated plumes to cause more extensive contamination in the future. The Food Quality Protection Act of 1996 established a new safety standard for pesticide residue limits in food, and the EPA is required to periodically reassess these 9721 limits (tolerances). By August 3, 1999, the EPA, as required, had completed 3290 of these reassessments, two thirds of which fell into the category of pesticides most threatening to humans. These reassessments included lower tolerance levels and restricted use for some of the pesticides that represent the greatest risk to humans, including two of the most toxic organophosphates, methyl parathion and azinphos methyl. Still, certain environmental advocacy groups have felt that the EPA's new regulations fall short of its mandate to protect public safety.[37]

The Global Proliferation of Pesticides

More than 300 tons of DDT were shipped to Peru in 1992, 20 years after it was banned in the U.S. The U.S. shipped nearly 12,000 tons of other restricted or banned pesticides in the same year. Since then, the export of pesticides banned for use in the U.S. has escalated by nearly 50%. According to studies performed by the Foundation for Advancement in Science and Education (FASE), about 250,000 tons of pesticides are exported from the U.S. annually, only a quarter of which are reported.[38] Nearly 5000 tons of the exported pesticides were never approved for use in this country, and more than 11,000 tons were severely restricted for use. Many of the LDCs still continue to import harmful pesticides because they are inexpensive and effective. In Kenya, small farms have increased output significantly in the last few years, making it the third-largest source of income from foreign sources. The consequence has been increasing reliance on pesticides banned or restricted in the U.S.[31] The continued use of DDT in Mexico is evidenced in increased levels of related compounds (DDE) in breast milk and tissues, indicating to the authors of the research article that DDT use in Mexico is a public health problem.[39]

The banned insecticides being exported to LDCs are thought to be evaporating or volatilizing to the atmosphere in the warmer climates and "hopping" in stages about the globe to cooler climates such as the Arctic through the global distillation effect.[40] Many of the chemicals banned for decades in the U.S. are being emitted from the hotter climates in LDCs, carried to colder climates in the atmosphere, and condensing, showing up in the bark of shrub trees near the Arctic Circle. The more volatile the chemical, the faster it travels these distances to eventually condense in colder climates. It is thought to be the best explanation for how large quantities of volatile pesticides and other chemicals find their way into the flesh of polar bears, and the breast milk of Inuit natives in the Arctic.[6]

TYPES OF PESTICIDES

INSECTICIDES

Insecticides generally produce their toxic effect on insects by poisoning the nervous system. Well-designed insecticides are harmless to humans, highly toxic to insects, easily dispensed and inexpensive, nonhazardous, and rapidly degradable to nontoxic forms. They often (1) act as contact poisons by penetrating the foot pads or body wall, (2) enter the insect breathing pores as a fumigant, (3) act as a stomach poison after ingestion, or (4) desiccate (dry out) the body wall, causing it to crack or break.[6] Insecticides are also capable of exerting their toxic effects on humans, and the major difference between its use as an insecticide or human poison is the dose. There are several types of insecticides, which include organochlorines, organophosphates, carbamates, and a variety of botanical and biological insecticides.

Organochlorines

The chlorinated hydrocarbons attack the central nervous system, sending insects into hyperactivity, tremors, convulsions, and death. The organochlorines may be categorized into three subgroups: dichlorodiphenyltrichloroethanes (DDT), hexachlorocyclohexanes, and chlorinated cyclodienes.

FIGURE 6.19 Chemical structure of p,p-DDT.

FIGURE 6.20 Chemical structure of lindane.

Dichlorophenylethanes

This subgroup of pesticides consists of DDT (Figure 6.19), methoxychlor, and dicofol. DDT is more acutely toxic and more slowly cleared from the body than methoxychlor or dicofol, and is also more environmentally persistent. Consequently, methoxychlor, being less toxic and persistent, has been used in many household insect sprays. DDT appears to disrupt sodium and potassium permeability in the nerve axon.[14] The storage of DDT in adipose (fat) tissue, biomagnification in the food chain, and disruption of calcium metabolism in fish-eating birds, causing thinning egg shells and declines in bird populations, led to the banning of DDT for use in the U.S. in the 1970s. A closely related pesticide, DDD, is also banned for use in the U.S.

Hexachlorocyclohexanes

There are many isomers for benzene hexachloride (HCB), with the gamma-isomer known as lindane being the most toxic. Lindane (Figure 6.20) is used primarily for treating seeds in the U.S. and is effective in the treatment of ectoparasites such as fleas, lice, and ticks. As such, it is available in a powder, shampoo, lotion, or cream and is often is used in the treatment of scabies, as well as the treatment of fleas on household pets. Part of the toxicity of lindane is due to its ability to inhibit γ-aminobutyric acid (GABA)–dependent chloride flux into the neuron.[14]

Chlorinated Cyclodienes

Chlordane (Figure 6.21), aldrin, dieldrin, endosulfan, and heptachlor belong to this class of compounds, which are characteristically more toxic than DDT and its related

FIGURE 6.21 Chemical structure of chlordane.

FIGURE 6.22 Chemical structure of kepone.

compounds. Because of their toxicity, rapid absorption through the skin, and animal carcinogenicity, the EPA suspended the use of most chlorinated cyclodienes in the U.S. in the mid-1970s. These substances are effective as stomach poisons, fumigants, and contact poisons, and have been used effectively against roaches, silverfish, ants, and termites.[14,36]

Chlorinated Cyclodiene

Kepone (Figure 6.22) and mirex cause liver tumors in mice, and both pesticides were banned for use in the U.S. although they were never widely used there. Kepone has caused severe neurological damage among production workers in Virginia and contamination of the nearby waters including the Chesapeake Bay.[41]

Organophosphates

Organophosphates are chemical derivatives of phosphoric acid developed in Germany during World War II as a poison for the control of aphids. Many of the organophosphates are short-lived in the environment because they are easily biodegraded and are effective against pests resistant to the organochlorine pesticides.[36] These compounds vary dramatically in their toxicity. Parathion is the pesticide most likely to be involved in human fatalities, although it is used extensively in agriculture (Figure 6.23). A single drop in the eye may be sufficient to kill a person, and so only licensed pesticide operators should use the more toxic organophosphate pesticides. In an ongoing search to find organophosphates less toxic to humans, more than 100 compounds have been developed.[42] These compounds vary in their toxicity, with the most toxic being phosdrin, parathion, dichlorvos, and TEPP. Moderately toxic organophosphates include DDVP, dimethoate, baytex, and diazinon. The least toxic of the organophosphates include ronnel, gardona, abate, and malathion.[36] Toxicity of these compounds is based on cholinesterase inhibition at the neuromuscular junction of nerve fibers. Acetylcholine is a neurotransmitter compound that

FIGURE 6.23 Chemical structure of parathion.

accomplishes the transmission of a nerve impulse from the nerve ending to smooth muscle, the heart, and endocrine glands (Figure 6.24). The motor neuron stimulates muscle tissue by releasing the neurotransmitter acetylcholine from synaptic vesicles in the synaptic end bulbs at neuromuscular junctions occurring between the motor neuron and muscle fibers. A nerve impulse causes acetylcholine to be released that diffuses across the synaptic cleft and combines with receptor sites on the muscle fiber. This causes changes in the flow of sodium and potassium ions across the sarcolemma, resulting in a muscle action potential that travels down the sarcolemma, causing muscle contraction. Under normal conditions, the acetylcholine is immediately hydrolyzed by the enzyme acetylcholinesterase. Organophosphates inhibit acetylcholinesterase, causing acetylcholine to accumulate at the neuromuscular junction, producing a variety of symptoms that include involuntary twitching, muscle weakness, paralysis of respiratory muscles, constriction of pupils, involuntary urination, vomiting, abdominal cramps, increased salivation and lacrimation, bronchoconstriction, increased sweating, and increased peristalsis. Symptoms due to accumulation of acetylcholine in the central nervous system may include anxiety, restlessness, headache, insomnia, nightmares, slurred speech, tremors, and coma. Fortunately, therapy is available for diagnosed cases of organophosphate poisoning, which includes the administration of atropine plus 2-PAM. Parathion weakly inhibits cholinesterase but is quickly converted to paraoxon, which is several thousand times more toxic, in tissues. Malathion is much less toxic than parathion because humans produce a carboxyesterase enzyme that renders malathion nontoxic.[14] The highly volatile organophosphate DDVP (also known as vapona or dichlorvos) is often used as a spray or in pest strips. Diazinon is an effective insecticide for use in vegetables and fruits, and is useful against roaches. Some of the more toxic organophosphates have been banned by EPA, whereas others such as parathion are rarely used because of their toxicity. However, organophosphates continue to cause more deaths and poisonings of humans than any other category of pesticides.[14,36]

Carbamates

Carbamates are derivatives of carbamic acid and are contact poisons that inhibit cholinesterase in a manner similar to organophosphates. They vary widely in toxicity depending on the specific carbamate pesticides, as do the organophosphates; but the treatment is not similar. Examples of carbamates include (1) carbaryl, which is mildly toxic to mammals, is used to control fleas on pets, and is also effective in killing the rat flea; and (2) propoxur, or Baygon, which is both a contact and stomach poison that is effective against ticks, flies, sandflies, roaches, and mosquitoes.

Botanical and Biological Insecticides and Other Alternatives

The chrysanthemum plant produces natural, highly insecticidal alkaloids known as *pyrethrins*. They are nontoxic to mammals. Because it is difficult to get enough chrysanthemums to supply the materials needed, a number of pyrethrum compounds have been synthesized, including permethrin, deltamethrin, and cypermethrin. Pyrethroids exhibited a toxicity to insects similar to the mechanism of DDT.

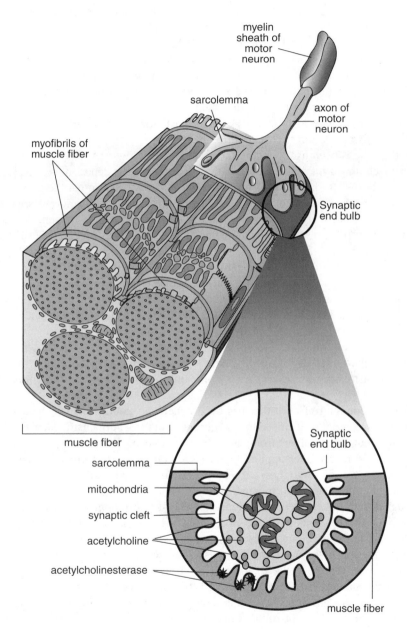

FIGURE 6.24 Acetylcholine, cholinesterase, and the neuromuscular junction.

Another alkaloid naturally derived from the roots of plants grown in South America or South Eastasia is known as rotenone. This pesticide paralyzes insects by closing down chemical respiration in the nerve cell but is fairly harmless to humans. For this reason, rotenone is widely used in home-use insect sprays and other insecticide formulations for the household

Methoprene is a growth regulator for insects that causes the mosquitoes to die in the pupa stage by halting normal metamorphosis. It was first developed in the 1970s to control mosquito larvae, but several versions have been subsequently produced to control some beetles, flies, and fleas.[8,14]

Bacillus thurengensis produces a protein ("Bt" toxin) that is selectively toxic to insects by disrupting ion transport in the gut after they ingest it on the plants they are eating. It has been used with great success against a wide variety of leaf-eating insects, including gypsy moth larvae, although some insects appear to be developing resistance to the Bt toxin. It is also a bioengineered pesticide in that the gene for its production has been inserted into a plant genome that then manufactures the protein Bt toxin that selectively kills insects feeding on the plants. The Bt gene has been inserted into corn, tomatoes, and potatoes and targets insects that feed on these plants.

A variation on bioengineered crops is the development of soybeans that are resistant to a powerful herbicide manufactured by Monsanto called Roundup. Consequently, the crops can be sprayed with herbicide to kill weeds without effecting soybeans. Critics are concerned about the herbicidal resistance being spread to unwanted weeds. They are also concerned that pests will develop resistance to the Bt toxin engineered into the plants.[43] The critics may be justified as similarly bioengineered cotton crops were severely attacked by cotton bollworms that appeared unperturbed by the "Bt" toxin bioengineered into the cotton. This is especially distressing as several pests commonly migrate between the cotton and corn crops.[44]

The introduction of corn that has been genetically modified to include the Bt gene, however, reduced the need to spray crops with the Bt organism, but also raised concern about potential environmental effects. Use of Bt crops reduces use of pesticides and insecticides applied to corn and cotton crops to prevent destruction of the crops by the European corn borer. As mentioned earlier, the use of this technology could lead to an increase in the population of Bt-resistant corn borers. The high dose/refuge strategy is designed to delay the development of resistant pests. By surrounding Bt crops with non-Bt hybrids, the very small number of pests that are resistant to the toxin are likely to mate with non-Bt resistant pests that thrive in the adjacent, non-Bt-treated corn. The offspring they produce should have only low or moderate resistance to the Bt toxin. This strategy, however, depends on certain assumptions about the mating patterns of Bt-resistant pests.[45]

Some concern has arisen over this seemingly harmless pesticide because of its close genetic relation to the anthrax bacterium, *B. anthracis*. Some scientists argue that the genetic similarity points to the potential of the Bt bacteria to mutate into strains that are more dangerous to humans.[46]

Naturally occurring disease agents such as fungal spores (*Metarhizium flavoviride*), pathogenic viruses, and protozoan parasites have been formulated as bait or sprayed on crops. In the case of protozoans and fungi, they grow inside the target pest and dissolve or digest internal organs. A fungal pathogen, *Lagenidium giganteum*, produces a motile biflagellate zoospore that infects mosquito larvae in their native water state, and has been recently shown to decrease populations of selected mosquitoes by 77% in trial assessments.[47]

A novel approach to control of vectorborne diseases is being developed by a team of researchers at the Colorado State University in Fort Collins. Instead of developing methods to kill the mosquito vector for dengue fever, they are using bioengineering techniques to prevent the dengue virus from replicating in the mosquito. The researchers attached a small section of the dengue virus RNA to a harmless virus called Sindbis, and inserted it into a mosquito, where it is replicated. The dengue RNA component halted any replication attempts by complete dengue viruses in the mosquito, thereby establishing a form of intracellular immunity. If the researchers can figure out how to get the mosquitoes to pass the disease resistance onto offspring, this may represent a worthwhile approach to the control of the disease.[48]

A significant alternative to the indiscriminate and runaway use of pesticides is the use of integrated pest management (IPM). IPM is a process that combines the best current knowledge about host-plant resistance, farming practices, and biological control, with some limited chemical controls as required to effectively control weeds and insect pests in an environmentally sound way. It is a process requiring flexibility, strategy, and knowledge. The agriculturist needs to know which pests to treat, when to treat, and how long to treat in order to keep crop losses from becoming economically unacceptable. Organic farming, which employs several different strategies to control pests, is increasing in popularity.[49] Whereas conventional farmers have been and are using genetically modified crops to control pests, organic farmers are using naturally occurring pesticides, predators, crop rotation, cover crops, and other non-petrochemical means of pest control.[50] Conventional farmers are also incorporating some of these methods, using IPM. In part, the conversion to natural methods is due to numerous pests that have evolved resistance to pesticides. Increases in consumer awareness of the adverse effects of pesticides is also increasing market demand for agricultural products grown with reduced pesticide use.

Herbicides

Most herbicides are thought to have low toxicity to humans, although their role as possible carcinogens has been an ongoing controversy.[51] There have been some reports of Hodgkin's lymphomas and other soft tissue sarcomas associated with occupational exposure to phenoxyherbicides.[14] There are sufficient differences in plant species that herbicides can be selective by killing weeds that are broad-leafed dicotyledenous plants while leaving members of the desirable grass family untouched. Atrazine, alachlor, 2,4-D, and 2,4,5-T are examples of such selective herbicides (Figure 6.25). These are chlorophenoxy agents that destroy plants by stimulating abnormal growth and interfering with the transport of nutrients. Here again, there is extraordinary controversy over the potential teratologic effects of these herbicides, spurred by the use of Agent Orange to defoliate areas of jungle in Vietnam.[51] Agent Orange contained 2,4-D, 2,4,5-T, and an unwanted but always present contaminant 2,3,7,8-tetrachlorodibenzo-p-dioxin (TCDD), also known as dioxin. Dioxin is an environmentally persistent substance that is highly toxic to

FIGURE 6.25 Chemical structure of 2,4,5-T herbicide.

FIGURE 6.26 Chemical structure of warfarin.

humans and other mammals and is discussed in more detail in Chapter 5. Other selective herbicides include paraquat, diquat, and atrazine, used on such crops as sugarcane, pineapple, maize, cotton, and potatoes.

Rodenticides

Although rodent control has been discussed earlier in the chapter, discussion of the rodenticide warfarin is offered here. Warfarin (Figure 6.26) is among the most widely used rodenticides because it is effective in small quantities, reasonably nontoxic to humans, and does not promote bait shyness because it is a slow-acting poison. Warfarin is a coumadin compound, and therefore, a vitamin K antagonist that inhibits prothrombin synthesis. Repeated dosing over several days causes an overall reduction in the blood-level prothrombin, which is necessary for coagulation. The consequence of such dosing is internal hemorrhaging, blood in the sputum and stools, and externally visible hematomas. The animals ultimately bleed to death. Some populations of rats have developed a resistance to warfarin. Consequently, more powerful versions known as superwarfarin, with a potency up to 200 times that of warfarin, have been developed that are more effective against these resistant strains.[53] Such compounds include brodifacoum, difethialone, and flupropadine. As might be expected, these compounds have much higher toxicity to humans than warfarin.[54] Other multiple-dose poisons effective in controlling rats include fumarin, PMP, and diphacinone.

REFERENCES

1. Vandermark, P., Oh, Rats, *Boston Globe Magazine,* 24, November 5, 1995.
2. Blank, J., Bombay's Rat Patrol Takes on the Plague, *U.S. News and World Report,* 14, October 14, 1994.
3. Mrotta, A., Rat Meat, Poor Man's Delicacy in Mexico, Reuter's News Service, July 14, 1998, http://www.infoseek.com/Content?am=a2873LBY209reulbl 9980713&qt--rats&sv=IS&Ik=noframes&col=NX&kt=A&ak=newsl486.

4. Gubler, J.G. and Clark, G.G., Dispatch, Dengue hemorrhagic fever: the emergence of a global health problem, *EID,* 1, 2, April–June 1995.
5. World Health Organization, Control of Tropical Diseases (CTD): Malarial Control, Geneva, Switzerland, WHO Office of Information, 1995.
6. Koren, H., *Illustrated Dictionary of Environmental Health and Occupational Safety,* CRC Press/Lewis Publishers, Boca Raton, FL, 407 pp.
7. U.S. Department of Health, Education and Welfare (HEW), *Pictorial Keys to Some Arthropods of Public Health Importance,* Washington, D.C., U.S. Department of HEW Public Health Services, 3, 1964.
8. Koren, H. and Bisesi, M., *Handbook of Environmental Health and Safety, Principle and Practices,* Vol. 1, 3rd ed., CRC Press/Lewis Publishers, Boca Raton, FL, chap. 4.
9. Tortora, G.J., Funke, B.R. and Case, C.L., *Microbiology: An Introduction,* 5th ed., Benjamin Cummings, New York, 1995, chap. 23.
10. Nadakavukaren, A., *Our Global Environment: A Health Perspective,* 4th ed., Waveland Press, Prospect Heights, IL, 1995, chap. 8.
11. Morgan, M., Gordon, L., and Walker, B., *Environmental Health,* 2nd ed., Morton Publishing, Englewood, CO, 1997, chap. 8.
12. U.S. Department of Health, Education and Welfare (HEW), *Control of Domestic Rats and Mice,* Public Health Service, Communicable Disease Center, 22, 1964.
13. Joy, R.M., Chlorinated hydrocarbon insecticides, in *Pesticides and Neurological Diseases,* Ecobichon, D.J. and Joy, R.M., Eds., CRC Press, Boca Raton, FL, 1982, pp. 91–150.
14. Saunders, D.S. and Harper, C., Pesticides, in *Principles and Methods of Toxicology,* 3rd ed., Hayes, A.W., Ed., Raven Press, New York, 1996, chap. 11.
15. Dover, M., Getting Off the Pesticide Treadmill, *Technology Review,* November–December, 1985, p. 53.
16. U.S. Geologic Survey (USGS), Pesticides in Surface Waters, U.S. Geological Survey Fact Sheet, FS-039-97, May 30, 1997, http://wwwrvares.er.usgs.gov/nawqa/NAWQA.ofr94.70.html.
17. Ganesse, L.P. and Puffer, L.A., Herbicide-resistant weeds may threaten wheat production in India, *Resources,* No. I I I (Spring) Resources for the Future, 1993.
18. U.S. Congress, Office of Technology Assessment, *Neurotoxicity: Identifying and Controlling Poisons of the Nervous System,* OTA-BA-426, Washington, D.C., April 1990, chap. 10.
19. U.S. Environmental Protection Agency (EPA), Office of Pesticide Programs, Pesticide Industry Sales and Usage: 1987 Market Estimates, U.S. EPA, Washington, D.C., 1988.
20. Jeyaratnam, J., Health problems of pesticide usage in the third world, *Br. J. Ind. Med.,* 42, 505, 1985.
21. World Health Organization (WHO), Pesticide deaths, what's the toll?, *Ecoforum,* 6, 10, 1981.
22. Whelan, E.M., *Toxic Terror: The Truth Behind the Cancer Scares,* 2nd ed., Prometheus Books, Buffalo, New York, 1993, chap. 4.
23. Repetto, R. and Baligo, S.S., Pesticides and the immune system: the public health risks, *Cent. Eur. J. Public Health,* 4, 263, 1996.
24. U.S. EPA, Office of Pesticide Protection (OPP), Pesticides and Child Safety: Tips to Protect Children from Poisonings Around the Home, Prevention, Pesticides, and Toxic Substances, EPA 735-F-93-050R, update November 19, 1997, http://www.epa.gov/oppfead/childsaf.htrn.
25. Siontral Resource Defense Council (NRDC), *Intolerable Risk: Pesticides in Our Children's Food,* Washington, D.C., 1989.

26. Vegej, J., Cancer causing pesticides abound in supermarkets' bounty — washing, peeling, buying organic can help, *Health News Review*, 14, June 12, 1996.

27. Food and drug administration pesticide program, residues in food, 1990, *J. Assoc. Off. Anal. Chem.*, 74, 1991.

28. Hanson, D., Regulation after Delaney — New Health-Based Standard Replaces Controversial Clause Dealing with Cancer-Causing Pesticides in Foods, *C&EN*, September 23, 1996, p. 38.

29. Conservation Foundation, Opportunities to Assist Developing Countries in the Proper Use of Agricultural and Industrial Chemicals, Washington, D.C., Vol. 1, 1988.

30. U.S. Department of State, *Proc. U.S. Strategy Conf. Pest. Manage.*, Washington, D.C., June 1979.

31. Global Pesticide Campaigner News Note: Prior Informed Consent Update, Vol. 09, No. 2, August 1999.

32. Stolberg, S., DDT, Target of Global Ban, Finds Defenders in Experts on Malaria, *The New York Times*, Saturday, August 28, 1999.

33. Ames, B. and Wiles, R., Pesticide Pros and Cons: Are Residues Really Bad for Your Health?, *Vegetarian Times*, November 1995, *p.*68.

34. Avery, D., How pesticides help prevent cancer, *Consumer Research*, June 1995, p. 10.

35. Colburn, T. and Clement, C., Eds., *Advances in Modern Environmental Toxicology, Vol. XXI. — Chemically Induced Alterations in Sexual and Functional Development The Wildlife/human Connection,* Princeton Scientific Publishing, NJ, 1992.

36. Koren, H. and Bisesi, M., *Handbook of Environmental Health and Safety, Principle and Practices,* Vol. 1, 3rd ed., CRC Press/Lewis Publishers, Boca Raton, FL, chap. 6.

37. Wald, M., Citing Children, EPA is Banning Common Pesticide, *New York Times*, 1999.

38. Raloff, J., The Pesticide Shuffle, *Science News,* 149, March 16, 1996, p. 174.

39. Lopez-Carillo, L., Tores-Arreola, L., Torres-Sanches, L., Espinosa-Torres, F., Jiménez, C., Cebrián, M., Waliszewski, S., and Saldare, O., Is DDT use a public health problem in Mexico? *Environ. Health Perspect.,* 104, 6, 584, June 1996.

40. Naj, A.K., Insecticides found to be traveling throughout the world, *Wall Street Journal, Science and Health,* 1, September 29, 1995.

41. Reich, M.R. and Spong, J.K., Kepone: a chemical disaster in Hopewell, Virginia, *Int. J. Health Ser.,* 13, 227, 1983.

42. Meerdink, G.L., Organophosphorous and carbarnate insecticide poisoning in large animals, *Vet. Clin. North Am. Food Anim. Pract.,* 5, 374, 1989.

43. Tortora, G.J. and Anagnostakos, N.P., *Principles of Anatomy and Physiology,* 6th ed., Harper and Row Publishers, New York, 1990, chap. 10.

44. Greene, R., Hope, Concern, Sprout Over Biotech Crops, *Los Angeles Times, Science, Medicine and the Environment,* October 20, 1996.

45. Beardsley, T., Picking on Cotton, *Scientific American, News & Analysis,* 43, October 1996.

46. Mackensie, D., Friend or Foe?, *New Scientist,* October 9, 1999, pp. 22–23.

47. Renner, R., Will Bt-based pest resistance management plans work?, *Environmental Science and Technology,* 33, 410, October 1, 1999.

48. Cuda, P., Hornbey, J.A., Cotterill, B., and Cattell, M., Evaluation of *Lagenidium giganteum* for biocontrol of *Mansonia* mosquitoes in Florida (Diptera: Culicidae), *Biological Control,* 8, 124, 1997.

49. Adler, T., Keeping Mosquitoes Healthy for Human's Sake, *Science News,* 149, May 11, 1996, p. 295.

50. Bourne, J., The Organic Revolution, *Audobon,* 101, 2, March 1999, p. 64.

51. Morrison, H.I., Wilkins, K., Semenciw, R., Mao, V., and Wigle, D., Herbicides and cancer, *J. Natl. Cancer Inst.,* 84, 1866, 1992.
52. Timoc, R., Parsonnet, J., and Halperin, D., Paraquat poisoning in southern Mexico: a report of 25 cases, *Arch. Environ. Health,* 48, 78, 1993.
53. Haug, B., Schjadtlversen, K., and Rygh, J., Poisoning with long-acting anticoagulant, *Tidsskr. Nor. Laegeforen.,* 112, 195, 8, 1992.
54. Routh, C.R., Triplett, D.A., Murphy, M.J., Felice, L.J., Sadowski, J.A., and Bovill, E.G., Superwarfarin ingestion and detection, *Am. J. Hematol.,* 36, 50, 1991.

7 Emerging Diseases

Escherichia coli on EMB plate

A student, after reading this chapter, should be able to:

1. Differentiate between the emerging infectious diseases in the U.S. and those occurring worldwide.
2. List and recognize the six major reasons associated with the emergence of infectious diseases.
3. Explain the likely reasons for the emergence of specific infectious diseases.
4. Identify, list, and explain the etiological agents, the epidemiology, and the disease characteristics of the major emerging infectious diseases including: influenza, hantavirus, dengue fever, Ebola, AIDS, cryptosporidiosis, malaria, Lyme disease, tuberculosis (TB), streptococcal infections, and *E. coli* infections.
5. Recognize and explain the practical approaches to limit the emergence of infectious diseases.

INTRODUCTION

Infectious diseases continue to be the foremost cause of death worldwide. The Centers for Disease Control and Prevention (CDC) reported a 58% rise in deaths from infectious diseases since 1980. Infections and parasitic diseases caused more than 17 million deaths in 1995, which is more than a quarter of all human deaths.[1]

Although AIDS accounts for most of the increase, deaths from all other infectious diseases rose to a dramatic 22%.[2] These figures are even more extraordinary because the International Code of Diseases lists many infectious diseases in categories other than infectious. For example, meningitis and cirrhosis are classified as diseases of the nervous system and liver, respectively, and only 17% of deaths attributable to infections are actually included in the codes for parasitic and infectious diseases.[3] About a quarter of the visits to physicians in the U.S. each year are attributable to infectious diseases, and the costs associated with these infections are estimated at more than $120 billion.[4]

Despite more than 100 years of scientific progress and the development of sanitary methods, infectious diseases still cause enormous human suffering, deplete scarce resources, impede social and economic development, and contribute to global instability. The potential for even greater human misery looms as an unrelenting danger. Recent outbreaks underscore the potential for the sudden appearance of infectious diseases in currently unaffected populations in the U.S. and the world.

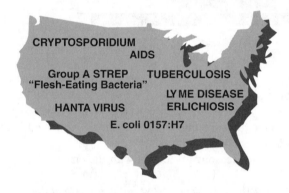

EMERGING DISEASES IN THE U.S.

We are being constantly challenged by the public news media with the terrifying prospect of new and emerging infections. The British tabloids coined the term *flesh-eating bacteria* to describe an invasive necrotizing infection caused by group A streptococci (GAS), and suggested that we were about to experience epidemics from this organism, producing high mortality, amputations, and grotesque disfigurement.[5] New parasitic threats to water supplies have erupted onto the national scenario. One of these parasites, cryptosporidium, has caused more than five major outbreaks in the U.S., with the most severe being in Milwaukee in 1993. This outbreak affected more than 400,000 people.[6] Avian influenza has not yet emerged as a global human disease, but it threatens to do so, and this threat has captured worldwide attention. Avian influenza presents a potential threat because the H5N1 virus recently spread across eastern Asia and other countries as an evolving and highly virulent virus, which is easily transmitted among migrating bird populations with some capacity to afflict mammals, including humans in close circumstances. Most of the human cases are related to exposure to infected poultry, and no sustained human-to-human transmission has yet occurred.[7]

The population of patients with acquired immunodeficiencies is rapidly increasing among certain African-American and heterosexual groups in the U.S. Nearly 225,000 persons over 13 years of age in the U.S. had acquired AIDS in June of 1996, representing a 65% increase over January 1993. The reported AIDS cases by area of residences from July 1995 though June 2005 was 42,034.[8]

AIDS remains one of the leading causes of death among persons of 25 to 44 years of age in the U.S., accounting for 19% of deaths from all causes in this age group.[9] Infection with *Escherichia coli* serotype O157:H7 was first described in 1982 and rapidly emerged as the major cause of bloody diarrhea and acute renal failure. The American public became aware of this disease when 700 people became infected, and four of them died after eating undercooked hamburger from a regional fast food chain in a 1993 outbreak.[10] There have been more than 30 outbreaks in the U.S. since, with many additional cases worldwide.[11] In 1993, a six-year drought followed by heavy rains and flooding contributed to the increase of deer mice in the American Southwest, leading to an outbreak of hantavirus pulmonary syndrome. First appearing in the Navajo reservations, the disease has since spread to 20 additional states and taken the lives of 45 people, nearly half of those infected.[12] Lyme disease is the second most prevalent emerging infectious disease in the U.S.; more than 65,000 cases have been reported to the CDC, since the disease was first described by Steere and colleagues in three Connecticut communities in 1977.[13] Eastern equine encephalitis (EEE) is a viral disease transmitted from birds to humans by the bite of a mosquito such as *Culex* or *Aedes*. Reporting of these diseases began in 1964. Although an average of 5 cases as reported per year in the U.S., there is a 30% mortality rate, and half of those that survive will experience mild to severe permanent neurologic damage.[14] West Nile virus (WNV) is a flavivirus occurring commonly in the Middle East, West Asia, and Africa. It is thought to have been first recognized in New York State in overwintering *Culex* mosquitoes during 1999 before the widespread transmission in 2000.[15] The virus can infect humans, birds, mosquitoes, horses, and some other mammals. It is transmitted to humans primarily by mosquitoes and tends to be seasonal. The disease may be mild or very severe, resulting in death. West Nile encephalitis and West Nile meningitis are forms of severe diseases that affect the nervous system. Encephalitis refers to an inflammation of the brain; meningitis is an inflammation of the membrane around the brain and the spinal cord. There were 33 cases reported to ArboNET by state and local health departments from January 1, 2006, to July 29, 2006. There were 16 cases (48%) of encephalitis (neuroinvasive disease) and 17 cases (52%) of West Nile fever (milder disease).[15] ArboNET is the national, electronic surveillance system established by CDC to assist states in tracking West Nile virus and other mosquito-borne viruses.

EMERGING DISEASES WORLDWIDE

These and other infectious diseases have potential worldwide significance. As an example, the possibility of dengue fever reaching the U.S. is very real. According to the World

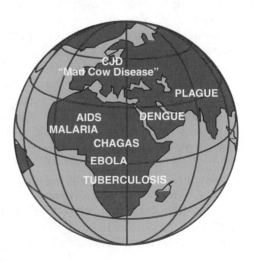

Health Organization (WHO), there are almost 30 "new" recognized infectious diseases including HIV, Ebola, hantavirus, hepatitis C, the *Legionella* bacterium and the *Cyclospora* parasite.[1] Until 1995, dengue fever had been isolated in Costa Rica's coastal mountain regions along the Pacific shore. But rising temperatures permitted the mosquitoes (responsible for dengue fever) to break the coastal mountain barriers and invade the rest of the country. By September, the epidemic had infected 140,000 people and claimed the lives of 4,000. Dengue fever, a debilitating tropical disease transmitted by mosquitoes, is reemerging with such force that it has now reached epidemic proportions in Central America and is bearing down on the U.S.[16] Each year 35 to 60 million people contract dengue fever worldwide.[17]

The Ebola virus grabbed global attention with a flurry of news articles and films, as one the deadliest viruses known, killing nearly 90% of those infected. There is no treatment or vaccine. Ebola and other hemorrhagic fever viruses typically start with sudden symptoms like malaise, fever, and flu-like symptoms, which are often followed by rashes and bleeding, along with kidney and liver failure, as these organs and other body parts dissolve into mush.[18] According to the Pan American Health Organization (PAHO), Chagas is the most important parasitic illness in Latin America, causing over 40,000 deaths each year. The protozoan disease is transmitted by a beetle known as the kissing bug (*Rhodnius prolixus*) that thrives in dark houses, within cracked floors and walls, and comes out at night to feed on blood sucked from the face of the human host. The inoculated trypanosome parasite invades the stomach, heart, and muscles of the human host to produce disabling fatigue, energy loss, and often death.[19] In the 1990s, epidemic cholera reappeared in the Americas after being absent for nearly a century; from 1991 through June of 1994, more than one million cases and nearly 10,000 deaths were reported.[20]

In 1994, an extended period of monsoon season in northern India followed by 3 months of extremely hot weather (100°F) drove rats into the cities, causing pneumonic plague from *Yersinia pestis* in Surat. The subsequent panic claimed many lives and cost India $2 billion.[12]

A cluster of Creutzfeldt-Jakob (CJD) disease cases with a unique neuropathologic picture, among unusually young patients in the U.K., was reported during a widespread epizootic of bovine spongiform encephalopathy (BSE).[21,22] This was notoriously referred to by the press as "mad cow disease," and prompted the destruction of thousands of cows. This appearance of the disease alerted many countries, including the U.S., to update their surveillance for CJD and to look out for similar cases.[23] This increased attention and the fact that CJD is rapid and, invariably fatal, raised public attention toward this emerging disease to a heightened level.[24]

Malaria is the world's largest widespread mosquito-borne illness and is potentially the worst threat to humans in the future. Global warming provides an environment for mosquitoes to multiply, migrate, and increase their metabolism, hence forcing them to feed (bite) more often. Despite the fact that mosquitoes die when temperatures rise above 104°F, so do people and crops. Programs to eradicate malaria, and socioeconomic development, have reduced endemic areas, but malaria has seen an increase in many parts of the world despite continued efforts toward eradication. More than 100 million clinical cases occur annually, and more than 1 million people, mostly children, die each year in tropical Africa.[25] Mainly because of drug-resistant strains, malaria has reemerged as a major international infectious disease, with nearly half a billion people worldwide developing the disease each year.[26] Malaria has also begun to reappear in the U.S., and is now reemerging at a geometric rate in South America.[27]

The WHO reported that efforts to control TB have failed and that drug-resistant strains of *Mycobacterium tuberculosis* are rising. TB is considered one the world's leading infectious diseases, killing over 3 million each year and many more becoming ill.[28]

WHAT IS AN EMERGING INFECTIOUS DISEASE?

New infections are emerging and old ones are expanding their geographical range faster than we can deploy resources to stop the organisms responsible. International travel and commerce are increasing, and technology is rapidly changing the conditions that increase the risk of exposure to infectious agents. The spectrum of infectious disease is quickly changing in conjunction with dramatic societal and environmental changes. Explosive population growth in developing nations, with expanding poverty and urban migration, are contributing to proliferation of both new and old infectious diseases. At the turn of the century, only 5% of the population lived in cities, but by the year 2025, it is estimated that this will increase to 65%. Therefore, more and more microbes will migrate from rural to urban areas.[29] Once expected to be eliminated as a public health problem, infectious diseases remain the leading cause of death and disability-adjusted life years (DALYs). Some of these diseases may be new to the human species. Others may simply have never been seen in North America before. The term *emerging infectious diseases* is now being applied to this phenomenon, and awareness of these emerging diseases has reached international levels. Emerging infectious diseases refers to diseases of infectious origin, whose incidence in humans has either increased within the past two decades or threatens to increase in the near future.[30]

Emerging infectious diseases, often with unknown long-term public health impact, continue to be identified. Table 7.1 lists major diseases or etiologic agents identified within the last 20 years. New agents are regularly added to the list, particularly, with the availability of nucleic acid amplification techniques for detecting and identifying noncultivable microorganisms.[31] In many cases, diseases became better recognized or defined (e.g., Legionnaires' disease, Lyme disease, human ehrlichiosis). Others are completely new. A previously unknown and deadly disease, AIDS, originated in one part of the world and spread globally at a speed that would have been unthinkable in earlier times.[31] Now, 25 years after the first AIDS case emerged, scientists have confirmed that the HIV virus plaguing humans originated in wild chimpanzees, in a corner of Cameroon.[32]

Do you want to learn more about emerging diseases? Point your browser to http://www-unix.oit.umass.edu/~envhl565/weblinks.html, and click on any of the Chapter 7 Web links associated with emerging diseases:

- Emerging Diseases (Michigan)
- FedWorld Information Network Home Page

TABLE 7.1
Major New and Remerging Etiologic Agents of Infectious Diseases and the Reasons For Their Emergence

VIRAL

Viral diseases that have been identified since 1973*

1977 Ebola, Marburg
Origin undetermined (importation of monkeys associated with outbreaks in these primates in Europe and the U.S.)
1980 HTLV Influenza (pandemic)
Pig duck agriculture thought to contribute to reassortment of avian and mammalian influenza viruses
1983 HIV
Transmission by intimate contact as in sexual transmission, contaminated hypodermic needles, transfusions, organ transplants. Contributing conditions that spread the disease include war or civil conflict, urban decay, migration to cities, and travel
1989 Hepatitis C
Transmission in infected blood such as by transfusions, contaminated hypodermic needles, and sexual transmission
1993 Hantaviruses
Increased contact with rodent hosts because of ecological or environmental changes

Viral diseases that have reemerged

Argentine, Bolivian hemorrhagic fever
Agricultural changes that promote growth of rodents
Bovine spongiform encephalopathy (cattle)
Alterations in the rendering of meat products
Dengue, dengue hemorrhagic fever
Travel, transportation, urbanization, and migration
Lassa fever
Conditons such as urbanization that favor rodent host, increasing exposure (usually in homes)
Rift Valley fever
Irrigation dam building, agriculture: possibly change in virulence
Yellow fever
Conditions favoring mosquito vector (in "new" areas)

BACTERIAL

Bacterial diseases that have been identified since 1973*

1977 Legionella disease
Cooling and plumbing systems that allow the organism to grow in biofilms that form on water
1982 Hemolytic uremic syndrome (*Escherichia coli* – O157:H7)
Modern food processing on a large scale permitting contamination through meat storage tanks and unsterile plumbing
1982 Lyme borreliosis (*Borrelia burgdorferi*)
Close contact between homeowners encroaching on forested areas and the mice and deer (a secondary reservoir host) that maintain the tick vector for Borrelia
1983 Helicobacter pylori
Newly recognized as agent involved with gastric ulcers, probably widespread before recognition
1987 Toxic shock syndrome (*Staphylococcus aureus*)
Ultra-absorbency tampons
1992 Cholera (type 0139)
Likely introduced from Asia to South America by ship, with spread made possible by reduced water chlorination; Strain (type O139) from Asia newly spread by travel (similarly to past introductions of classic cholera)

TABLE 7.1 *(Continued)*
Major New and Remerging Etiologic Agents of Infectious Diseases and the Reasons For Their Emergence

Bacterial diseases that have reemerged
Tuberculosis Breakdown in public health measures such as reduction in prevention programs, inadequate sanitation, homelessness, AIDS **Streptococcus, group A (invasive necrotizing)** Unknown, may be increased use of aspirin and related substances
PARASITES
Parasitic diseases that have been identified since 1973*
1976 Cryptosporidium, other waterborne pathogens Contaminated surface water, lack of proper filtration methods
Parasitic diseases that have reemerged
Malaria (in "new" areas) Spread of mosquito vectors, worldwide travel or migration, "Airport" malaria **Schistosomiasis** Agriculture, dam building, deforestation, flood/drought, famine, climate changes

*Compiled by CDC staff. Dates of discovery are assigned on the basis of the year the isolation or identification of etiologic agents was reported.

Source: Adapted from Morse, S., Perspectives: Factors in the emergence of infectious diseases, *EID*, 1, Jan.–Mar., 1995.

- Emerging Infectious Diseases Home Page
- NCID Diseases, Selected Prevention and Program Areas Listing

REASONS FOR THE EMERGENCE OF INFECTIOUS DISEASES

The emergence of so many new diseases and the reemergence of those thought to be controlled, appear difficult to understand or explain. However, rarely do emerging infections surface without reason. There are a number of specific explanations responsible for disease emergence that can be identified in almost all cases.[33] Table 7.1 summarizes the known causes for a number of infections that have recently emerged. Most emerging infections, and even antibiotic-resistant strains of common bacterial pathogens, usually originate in one geographic location and then disseminate to new places. Morse suggested that infectious disease emergence can be viewed operationally as a two-step process: (1) Introduction of the agent into a new host population (whether the pathogen originated in the environment, possibly in another species, or as a variant of an existing human infection), followed by (2) establishment and further dissemination within the new host population.[30,31] The majority of emerging infections are thought to be caused by pathogens formerly existing in the environment and released from obscurity or presented with a selective advantage by altered conditions and given an opportunity to infect new host populations. Occasionally, a variant may evolve and cause a new disease.[30,33] The term *microbial traffic* is given to the process by which infectious agents transfer from animals to humans or spread from isolated groups into new populations.[34,35] There are identifiable activities that increase microbial traffic and promote emergence and epidemics. In a number of instances, including many of the more unusual infections, the agents are zoonotic, that is, crossing from their natural animal hosts into the human population. In separate instances, pathogens existing in geographically isolated populations are able to disseminate further. Very often, the emergence of a disease results from human actions, usually unintended. Changes in climate and other natural causes can also be responsible at times.[36]

Factors responsible for emerging infectious diseases include (1) ecological changes, such as those due to agricultural or economic development or to anomalies in climate; (2) human demographic changes (urbanization) and behavior; (3) travel and commerce; (4) technology and industry (globalization); (5) microbial adaptation and change (resistance); and (6) breakdown of public health measures.[30]

ECOLOGICAL CHANGES

Agriculture

Changes in ecology, such as from agricultural or economic development, are among the most commonly identified reasons in the emergence of infectious diseases. Ecological changes are often present in outbreaks of formerly unrecognized diseases with high case fatality rates. These diseases are often introduced from animal hosts, because people are placed in close contact with a natural reservoir or host where the disease organism is present (often a zoonotic or arthropod-borne infection). People may expand into an area where the animal host thrives, the animal host may expand into human living areas, or the infectious agent may have increased in the animal host population.[30,34,35]

Deforestation is reported to be a major reason for the emergence of Lyme disease in the U.S. and Europe, because it increased the population of deer and the deer tick, the vector of Lyme disease. The movement of people into these areas placed a larger population in close proximity to the vector.[37] In Asia, the field mouse, *Apodemus agrarius* is abundant in rice fields, where people in contact with infected rodents contract Korean hemorrhagic fever usually during the rice harvest. Argentine hemorrhagic fever has increased almost in direct relation to the expansion of maize agriculture in Argentina, which favors the growth of a rodent that carries the disease organism.[38] Current researchers report that pandemic influenza arises when two influenza strains recombine, resulting in a new virus that is infectious for humans. Ducks are a major source of avian influenza. The avian flu virus may enter pigs that are in close proximity, where the gene segments of the viruses may then recombine (or reassort).[39] The agricultural practice of pig and duck farming, as practiced in China, places these animals in contact and creates an ideal setting for the creation of influenza recombinants capable of infecting humans.[40]

Water is also frequently associated with disease emergence. Infections transmitted by mosquitoes or other arthropods include some of the most serious and widespread diseases.[41] Mosquito population growth is often stimulated by expansion of standing water because many of the mosquito vectors breed in water. There are many cases of diseases transmitted by water-breeding vectors, most of which involve dams, water for irrigation, or stored drinking water in cities.

Climate

Virtually all climate scientists believe that an increase in potentially fatal diseases is likely to be the single most important threat that climate changes pose to human health.[42] Extreme weather patterns, as a result of natural fluctuations in the atmosphere or man-made changes (i.e., global warming), have routinely been followed by outbreaks of disease. "Dry spells in wet areas or torrential rains in normally dry spots tends to favor the so-called opportunistic pests — rodents, insects, bacteria, protozoa, and viruses — whereas making life more difficult for the predators that usually control them.[10] "The same synergies that empower microbes also weaken our defenses against them. Heat, increased ultraviolet radiation resulting from ozone depletion, and pollutants like chlorinated hydrocarbons all suppress the disease-battling immune systems, both for humans and for other animals."[12]

The earth's climate is becoming a critical factor in enhancing the spread of infectious diseases. The WHO predicts that the temperature of the planet will rise from 2.7 to 8°F by the year 2030 and profoundly affect human health. One of the predictable results will be the expansion of mosquito-borne disease from the tropics into temperate zones. The expansion may move northward as far as New York and could include malaria, dengue, and yellow fever (this fever has a 50% mortality). In 1990, WHO warned that global warming would enlarge the territories of disease-carrying insects and increase the at-risk population by 620 million by the year 2050.[43]

The Pacific Ocean warming events influence weather patterns and jet streams across North and South America, Western Europe, Asia, and Africa. El Nino events normally begin in the latter part of December and determine weather patterns for a year. The El Nino warming events have historically occurred about twice every 10 years.[44] These sea warming events cause evaporation of water, producing extensive rainfall in some areas and drought in the others. The power of the El Nino events have increased dramatically in recent years producing temperature increases from 2 to 3°C in the ocean, well into the summer months, along with the possibility of extending over much longer periods of time. It is predicted that the El Nino events will be more frequent with greater intensity, onset, and duration, consistent with a continued increase in greenhouse gases.[45] A warmer climate, initiated by increased amounts of carbon in the atmosphere, has caused the El Nino events to last longer and be more dramatic.[46] The sudden eruption of hantavirus infection in the Southwestern U.S. in 1993–1994 that killed 48% of the 94 persons infected has been attributed to the sudden population explosion of deer mice that harbor the organism. The probable events that catalyzed this eruption are heavy rains following 6 years of drought, which caused pine nuts and grasshoppers to flourish, which in turn fed the population eruption of deer mice. This enlarged population was driven from the flooded burrows, increasing the possibility of contact with people.[47]

It has been suggested that certain organisms in marine environments are natural reservoirs for cholera vibrios and that large-scale effects on ocean currents may cause local increases in the reservoir organism with consequent flare-ups of cholera. Pollution and global climate changes, induced by humans, adversely affect the health of marine population and have the potential to harm humans because of the ocean's ability to act as an incubator for diseases like cholera.[42] Changes in coastal ecology are generating "hot systems" in which random mutations of *Vibrio cholera* may be selected, amplified, and then transferred to human hosts through the consumption of contaminated fish and shellfish.[47] Aquatic plants, seaweeds, and free-floating phyto- and zooplankton are known to harbor vibrios.[48] The phytoplankton or algae grow in epidemic proportions, spurred on by warmer sea surfaces and an increase in nitrogen-rich wastewaters from coastal cities.[49] Additionally, the species of algae that are proliferating are less palatable to fish, and the numbers of fish have shown a decline in 14 of the world's 17 major fishing grounds.[47]

HUMAN DEMOGRAPHIC CHANGES (URBANIZATION) AND BEHAVIOR

Changes in human demographics and behavior, or human population movements and upheavals caused by migration or war, are often significant factors in disease emergence. Increased population density in urban areas — migration in hopes of a better, more comfortable lifestyle — has surpassed basic services, including clean water supplies, sanitary conditions such as sewage disposal and adequate housing.

Some of the most profound effects were among the poor of the inner-city slums. It was a frustrating paradox for many in public health that those most susceptible to infection had the worst facilities. Their socioeconomic status led to a heightened risk of disease because of little or no access to clean water, running sewers, and pest control. In one example, until the mid-1950s, large areas around and within Washington, D.C., had no running water and were served by outhouses.[50] These areas not only infected the people who lived there but also acted as a reservoir of disease, which would periodically spread across entire regions of the country. The diseases were incubated by unhygienic practices, as well as brought in by immigrants who could often not afford to live anywhere else.

Depressed economic conditions and the disappearance of viable agricultural land in rural areas in many parts of the world are prompting the mass movement of workers from rural areas to cities. It has been estimated by the United Nations that 65% of the world's population will be living in cities by the year 2025, as this migration continues.[30] Diseases such as HIV are thought to have originated in isolated rural areas and then brought to the densely populated cities, where the disease could spread locally and get carried by infected humans along highways and air routes to cross state and national boundaries. Public health measures in overcrowded cities are often strained or unavailable to large groups of the urban impoverished living in inner-city slums or in shanties on the periphery, thereby increasing the opportunity for emerging infectious diseases.[30,51]

Emerging infections such as HIV, cholera, and dengue have followed this pattern. An example of this is HIV-2 that was isolated from a man living in rural Liberia and thought to have originated from contact with animals, where a virus similar to HIV-2 was isolated from the Sooty Mangabey monkey.[52,53] This monkey is often used for food. Such zoonotic infections may often occur and remain isolated, but the swelling migration of people from rural to urban environments increases the probability of the spread of the disease to densely populated areas and its transportation along highways and airways to reach distant locations (permitting a pandemic), as was the case for HIV-1. Simply transporting the virus with naturally low transmissibility would not be sufficient to produce a worldwide disease.

There were social behaviors that permitted intimate contact with the HIV organism through sexual transmission, intravenous drug use, or the use of technology in the form of blood transfusions and blood products. These factors associated with the very long duration of infectivity have combined to increase the opportunity for transmission, resulting in a pandemic that has affected much of the world and is now spreading through Asia.[30] Human activity influences a number of factors associated with the emergence of disease; the control of such diseases will require the motivation of appropriate individual behavior and constructive action regionally and internationally. Unfortunately, the continued increase of infectious diseases on many fronts points to the need for additional scientific knowledge in human behavior.

TRAVEL AND COMMERCE

The main factor effecting the emergence and spread of new diseases is believed to be transportation.[1] Increased economic growth into national and international boundaries has led to increased travel, contributing to the notion of "diseases without boundaries." The incubation period for infectious diseases can be from a period of hours to weeks. As a result, diseases have the potential to spread from one continent to another in a matter of hours before symptoms occur.

Travel

The spread of HIV through travel has been suggested.[52] Historically, infections introduced into a population in a remote area might, occasionally, be transported to a new place through commerce, travel, or war.[54] The bubonic plague, carried by rats and transmitted to humans by the rat flea, is thought to have been introduced into Europe through the silk trade routes from Asia and believed to have continued with the Crusades. Smallpox was introduced by the Spanish conquistadors into the Aztec nation in the 15th century and decimated more than 80% of the highly susceptible population. The 16th and 17th centuries saw ships transporting slaves from West Africa to the New World infected with yellow fever and carrying the mosquito vector, *Aedes aegypti*, to the new territories. Two centuries later, cholera traveled with its human hosts from its likely origin in the Ganges plain to the Middle East. From there, cholera traveled to Europe and a substantial part of the remaining world.[30] The Asian tiger mosquito, *Aedes albopictus*, entered the U.S., Brazil, and parts of Africa in shipments, in the collected water within used tires shipped from Asia.[55]

The mosquito's introduction is problematic because it bites aggressively, survives in both forest and suburban habitats, and is a vector for many human pathogens. It is a vector for epidemic dengue fever transmission in Asia and is a competent laboratory vector of La Crosse Virus (LACV), yellow fever, and other viruses.[56] Fourteen strains of EEE virus have been isolated from *A. albopictus* in Florida.[57] The mosquito has become established in at least 21 of the contiguous states in U.S. and in Hawaii.[58] Malaria is another disease transmitted by a mosquito vector, and is one of the most repeatedly imported diseases in nonendemic disease areas. Cases of malaria contracted in airports are sometimes identified. Rats have carried hantaviruses virtually worldwide.[59] These infections were originally localized but were carried to exotic and foreign parts of the world. Similar histories are being repeated today with the added problem of greatly increased volumes of traffic moving at enormous speed. Travel may involve short distances or the crossing of international borders. In 1998, an emigrant from Ukraine, in a plane from Paris to New York, infected 13 other passengers with drug-resistant TB.[29] The volume is huge. In the early 1990s, more than 500 million persons crossed international borders yearly on commercial airplane flights.[58] As many as 70 million persons are thought to work either legally or illegally in other countries, and the majority are from developing countries. Many are migrant workers following crops and hence moving seasonally.[60] Large populations move among countries as part of military maneuvers. The consequence of armed conflict and political unrest also displaces millions. In the early 1990s, there were an estimated 20 million refugees and 30 million internally displaced persons (IDP) worldwide.[58] That number has not changed significantly over the years. The difference between refugees and IDPs is that refugees cross international borders, whereas IDPs stay within their own country but may be at greater risk, as they are likely to be persecuted by their own government.[61]

Commerce

Marine organisms may be transported on the hulls of ships and in their ballast water. There were 367 different species identified in the ballast water of ships traveling between Japan and Coos Bay, OR.[62] There is evidence that disease organisms are also transmitted by shipping, such as the introduction of *V. cholerae* into South America.[63] *V. cholerae* was isolated from the ballast, sewage, and bilge of 21% of the cargo ships docked at ports in the Gulf of Mexico. The ships were previously in Brazil, Colombia, and Chile.[64] *V. cholerae* O139 has traveled along waterways in Asia, carried by people and boats generally engaged in commerce.[65] A strain of *V. cholerae* identical to an epidemic strain found in Latin America was detected in Mobile Bay, AL, within oysters and the fish that eat oysters.[61]

Researchers have regularly demonstrated that mosquitoes are present in, and can survive international flights. When random searches were made of airplanes in London, mosquitoes were located in 12 of the 67 airplanes from tropical countries.[60] Arthropods can survive even more extreme environments. In another study, mosquitoes, house flies, and beetles placed in wheel bays of Boeing 747B aircraft survived flights of 6 to 9 h with external temperatures of 42°C.[67] Airplanes have also carried infective mosquitoes that caused human infection outside malaria-endemic areas (e.g., in Europe).

Prerequisites for Disease Development

Travel will permit a potentially pathogenic organism to be introduced into a new geographic area. But, if it is to be established and cause disease, the organism must survive, reproduce, and enter a susceptible host. Fortunately, the majority of such introductions do not result in disease. Microbes that survive exclusively in the human host and are contagious through intimate contact and droplet nuclei can be readily transmitted worldwide. Many diseases such as TB, AIDS, pertussis, measles, diphtheria, and hepatitis are routinely transported around the globe. In those populations where a majority of people have been vaccinated, the diseases find little opportunity to progress. Organisms are much more difficult to transplant when they have complicated life cycles that may involve animal hosts, vectors, or specific environmental requirements. Yellow fever, malaria, and dengue fever require specific mosquito vectors, or they cannot proliferate in a particular geographic area. A snail intermediate host is required for the development of certain life stages in Schistosomiasis, and so the disease organism cannot spread in the absence of this host. Even when all the conditions are present to permit the successful introduction of a parasite in a geographic area, the development of human disease is not an absolute consequence. The hog tapeworm, *Taenia solium,* is often not successfully transmitted in developed nations because there are adequate methods for confining and treating human sewage that prevent the eggs from infecting other people. Briefly, the probability of transmission involves many biological, social, and environmental variables.[58]

TECHNOLOGY AND INDUSTRY (GLOBALIZATION)

High-volume rapid movement characterizes not only travel, but also other industries in modern society. In operations that include food production, the use of products of biological origin or modern production methods increases the chances of accidental contamination and amplify the effects of such contamination.[68] The problem is complicated by globalization, because agents are introduced from far away. Globalization has been defined as the process of denationalization of markets, laws, and politics, in the sense of interlacing peoples and individuals for the sake of the common good.[69] Countries are finding it increasingly difficult to solve the problem of emerging infectious diseases, and public health policy is becoming denationalized to attack this problem. Globalization is influencing public health in three ways[68]:

First, the diseases are moving rapidly around the globe because of technology and economic interdependence, which has increased the international travel and nature of food processing and handling.[58] As an example of technology increasing the risk of emerging diseases, the concentrating effects that occur with blood and tissue products have inadvertently disseminated infections unrecognized at the time, such as HIV and hepatitis B and C. Medical settings are also at the leading edge of exposure to new diseases, and many emerging infections have spread nosocomially in health care settings. Additionally, the common use and misuse of antibiotic treatments have permitted the development of drug-resistant pathogens.[70,71] Food processing is also a contributory factor to emerging diseases. As an example, BSE, which emerged in Britain within the last few years, was probably an interspecies transfer of scrapie from sheep to cattle.[72] This probably occurred when changes in rendering processes resulted in incomplete inactivation of the scrapie agent in sheep by-products fed to cattle.[67]

Second, the funding of public health programs has been reduced because of increased competition in the global market and increased pressures to cut expenditures, thereby negatively affecting public health programs.[73]

Third, public health programs have become international through WHO and health-related nongovernmental organizations. Worldwide health has improved through these international efforts, with the elimination or reduction of many diseases. Unfortunately, these successes have contributed to a population crisis, produced overcrowding, inadequate sanitation, and overstretched public health infrastructures, leading to the expansion of disease.[74]

MICROBIAL ADAPTATION AND CHANGE (RESISTANCE)

Antibiotic Resistance

Different mechanisms of genetic change of microbes are being routinely discovered. Spontaneous mutation is just the beginning. Microbes most often survive in enormous populations and are influenced by a myriad of physical and chemical mutagens. Haploid organisms are able to instantly express their genetic variations. They have numerous repair mechanisms that are subject to genetic control. Some strains that do not repair their DNA are therefore highly mutable; others are relatively more stable. Some organisms are highly flexible in responding to environmental stresses. There are abundant processes that allow bacteria and viruses to readily exchange genetic material. Plasmids now spread throughout the microbial world. They can intersect the boundaries of yeast and bacteria. Lateral transfer is very substantial in the development of microorganisms influencing the expressions of pathogenicity, toxicity, and antibiotic resistance. Microbes have a wide variety of mechanisms. that allow them to exchange genetic information.[75] Microbes are therefore constantly evolving.

Antibiotic-resistant bacteria are emerging from the environment in response to the wide distribution of antimicrobials. This is an evolutionary lesson on microbial adaptation, and a display of the power of natural selection. Selection for antibiotic-resistant bacteria,[76] and drug-resistant parasites have become common, generated by the wide and often unsuitable use of antimicrobial drugs.[77,78] Pathogens can also attain antibiotic resistance genes from other, often nonpathogenic, species in the environment, selected or perhaps even driven by the ongoing pressure of antibiotics.[79] Many viruses demonstrate a high mutation rate and can rapidly evolve to yield new variants.[80] A classic example is influenza.[81] Yearly epidemics result from antigenic drift in a previously circulating influenza strain. An alteration in an antigenic site of a surface protein, usually the hemagglutinin (H) protein, permits the current variant to reinfect formerly infected persons, because the immune system no longer recognizes the altered antigen. Rarely, the evolution of a new variant may result in a new expression of the disease.[34] It is possible that some recently reported characteristics of diseases by group A streptococcus, such as rapidly invasive infection or necrotizing fasciitis, may also fall into this category.[30]

Only during the last 50 of the 3000 years of recorded history have humans been able to cure an individual of a communicable disease. To underline the significance of this apparently mild statement, consider that, in 1925, the number one killer in America was diarrhea and enteritis caused by bacterial infection; 109 of every 100,000 men, women, and children were killed. This is roughly three times the current death rate from all causes combined. In the same year, the top ten causes of death included scarlet fever, pneumococcal pneumonia, whooping cough, diphtheria, and TB. Sixty years later, diarrhea and enteritis in this country were responsible for 1 in 1,000,000 deaths.[82]

There is growing concern that bacterial pathogens are developing a resistance to antibiotics, as a result of patients not completing the prescribed course of treatment or the inappropriate and overprescribing of common antibiotics by physicians. Most of the bacteria are destroyed in the early course of treatment, generally the weakest bacteria. However, the strongest survive and mutate (build resistance) so that the drug becomes ineffective. Thus, bacteria that develop a resistance to a specific antibiotic will not respond to the drug if it is prescribed later.

The use of unsupervised prophylactic tetracycline administration to 100,000 pilgrims, on route to Mecca from Indonesia, is thought to have been significantly responsible for the fact that 50% of cholera strains in that country are now tetracycline resistant. For a country as poor as Indonesia, with an average of $4.52 annual medical expenditure per capita, this is a devastating blow.[83] Tetracycline is one of the easiest and cheapest drugs to produce, which makes it safe and widely available in the Third World. Other drugs often need to be imported and are so expensive as to be out of reach for the general populace.

A different example of resistance developed is in shigella. This disease is the cause of dysentery throughout the developing world. After the development of multidrug-resistant shigella in Africa, it appeared almost immediately in Bulgaria (resistant to four drugs) and on the Navajo and Hopi reservations in the U.S. (resistant to two drugs), and in Vietnam, where 90% of the cases were resistant to three or more antibiotics.[84] Shigella was being carried to the four corners of the earth and breaking out in areas with insufficient water treatment facilities. This demonstrates the disturbing properties of drug-resistant bacteria to travel vast distances.

Whereas it took 15 years, from the introduction of penicillin, for widespread penicillin resistance to be reported to *N. gonorrhoea* in 1962, physicians who began using the flouroquinolone group of antibiotics (ciprofloxacin, ceftriaxone) in the late 1980s began reporting a large number of cases resistant to the drug by 1992.[20] This acceleration in conversion times was not limited to gonorrhea but is universally seen across most common bacterial infective organisms. It suggests something greater than a cause and effect of antibiotic misuse by physician "A" leading to a mutant strain in patient "B;" these drug-resistant strains are displaying an evolutionary trait.

Over the last 50 years, we have not only bred resistant organisms but have also selected organisms that have the ability to develop resistance quickly.

By the last quarter of the 20th century, the threat of bacterial infections had begun to wane from the consciousness of the American mind. People like Jim Henson, the creator of the muppets, who died from a disease such as pneumonia, caused a stir, as they were the exception to the rule that modern medicine could tackle bacterial infections. Part of the popular faith in medicine came from the perception of research as a constant fountain of new compounds. An array of new drugs seemed to appear every few years to be aggressively marketed by the pharmaceutical industry. In the early 1950s, there were three classes of antibiotics to choose from, including the beta-lactams (penicillin), tetracyclines, and aminoglycosides. In the 1990s, there were seven classes, although several are close cousins of one another.

Drug-resistant infections could be treated with other more powerful drugs like vancomycin and rocephin that are considered by hospitals to be "big guns" in the disease war. The appearance of vancomycin-resistant strains in hospitals during the late 1980s has encouraged many epidemiologists and bacteriologists to consider the hospitals themselves as the source of many forms of drug resistance. A recent report by the CDC found that vancomycin resistance measured at 0.3% in 1986 and rose to 7.9% across several facilities in 1994.[84] Intravenous vancomycin is the only treatment for methicillin-resistant *Staphylococcus aureus* (MRSA). MRSA are staphylococci that are resistant to the antibiotic, methicillin, and other commonly used antibiotics, such as penicillin and cephalosporins. There is a unique gene in the staphylococcus that makes them resistant to most currently used antibiotics. Therefore, alternate antibiotics must be used to treat persons infected with MRSA. Vancomycin has been the most effective and reliable drug in these cases, but is used intravenously and is not effective for treatment of MRSA when taken orally.[85] The first infection with MRSA in the U.S. was reported in 1968.[86] Since then, the proportion of *S. aureus* isolates, resistant to methicillin, causing infections in hospitalized patients has risen significantly from 2% in 1974 to about 40% in 1997. The MRSA infections involved patients in hospitals or long-term care facilities, but recent reports of community-acquired MRSA infections raise concern. These community MRSA infections occur in healthy, nonhospitalized persons who have no contact with health care personnel or other colonized patients.[87] Should MRSA become widespread in the community, treatment of these infections will be much more difficult.[85]

Antibiotics in Livestock Feed

Many studies from various sources including the CDC, the FDA, and the Department of Agriculture suggest a strong link between the use of antibiotics in livestock feed and the transmission of drug-resistant bacteria to humans. About 40% of the antibiotics that are manufactured today are given to livestock with 50 million pounds being made each year.[29] Antibiotics have been added to livestock feed for many years to promote growth. Farmers have found that animals are able to absorb more nutrients from the feed and therefore grow more if the bacteria in the gut are killed.

Unfortunately, organisms such as salmonella and campylobacter, which are harmful diseases if contracted by humans, are beginning to appear increasingly more resistant to the antibiotics.[88] Whereas people are not susceptible to all of these organisms, the bacteria are able to transfer their resistance to disease-causing bacteria through genetic material called *plasmids*.

In response to this discovery, the British Ministry of Health in 1989 adopted a white paper on livestock feed and antibiotics that regulated the administration of antibiotics to animals intended for human consumption. Whereas the CDC has considered similar guidelines, they have never been adopted by Congress due to pressure from livestock interests, who fear a competitive burden if they are not able to use prophylactic levels of antibiotics in feed. Whereas these practices do help to prevent some diseases in animals, in the long run, they have been shown to select for resistant organisms that can harm animals as well as humans.

Viruses

Antibiotics have no effect on viruses, and vaccines are often ineffective against bacterial infections. We have traditionally been able to combat such viral pathogens as rabies, polio, measles, and influenza, and have even eradicated smallpox through preventive vaccines. However, there are few antiviral drugs that exist, mainly because drugs that kill viruses also harm or kill its host cell (i.e., healthy cells). In addition, new viruses are continuously emerging, making it difficult for vaccines to be developed in sufficient time.

BREAKDOWN OF PUBLIC HEALTH MEASURES

During the 18th and 19th centuries, advancements in public health vastly improved the overall health of the populace, particularly in urban settings. Improvements included reducing the exposure of drinking water and food to contaminated waste, as well as improvements in the populations' natural mechanisms of resistance though better nutrition.[83] Vector control, chlorination of water, pasteurization of milk, immunization, and proper sewage disposal are classic public health and sanitation measures that have successfully minimized the spread of infectious diseases in humans. The pathogens, however, may exist in smaller numbers within reservoir hosts or the environment and represent a continuing threat for reemergence, if these sanitary methods are not maintained. Well-understood and recognized diseases such as cholera are rapidly increasing, because the once active public health measures have lapsed. This retrogression in sanitary methods, unfortunately, appears frequently now in both developing countries and the inner cities of the industrialized world. The reemergence of infectious diseases may, therefore, signal the disruption of public health measures and should demand vigilance against indifference in the war against infectious diseases.[30] For example, cholera has been fiercely erupting in South America and Africa, for the first time in this century.[89] The swiftly expanding cholera outbreak in South America may have been aided by recent reductions in chlorine levels used to treat water supplies.[90] Cholera and other enteric diseases follow the fecal–oral route of transmission, and so are frequently promoted by the lack of a reliable water supply. Problems such as these are aggravated in developing countries but are not confined to these areas. The U.S. outbreak of waterborne cryptosporidial infection in Milwaukee, WI, in the spring of 1993, with over 400,000 estimated cases, was partially the result of a nonfunctioning water filtration plant.[91] Water purification is a problem found in many other cities in the U.S.[92]

SPECIFIC EMERGING DISEASES

VIRUSES

According to Stephen S. Morse, the viruses with the greatest potential for emergence in the near future include hantaan (hantavirus), dengue, influenza, and HIV.[31] Therefore, these will be discussed in more detail. I have included a discussion of Ebola because it is now globally feared. The number of people who have acquired the disease are exceptionally small compared with all other new and emerging diseases, but the gruesome manifestations of the disease, the high mortality, and the lack of treatment methods has catapulted this organism into the public eye with books, televisions shows, and movies. Therefore, it is discussed here to provide a contrast to the real worldwide killers.

Hantavirus

Background: The saga of hantavirus outbreak in the U.S. began in 1994, when a physician was called to care for a man who collapsed at a funeral in New Mexico and died of acute respiratory failure. Soon after this, the physician found that several others in the Four Corners region (New Mexico, Arizona, Colorado, and Utah) of the Southwest had also died from similarly aggressive respiratory illness dating back to 1993. Initially, the victims were Native Americans located on a huge Navajo reservation. The media called it the "Navajo Flu." However, new cases were reported in the following months from California, Rhode Island, and Florida. The CDC was contacted, as the etiology of the disease could not be determined. CDC personnel descended on the area and began a systematic effort to identify the disease. They collected blood and tissue specimens, victim histories, performed database searches, and tested for many different groups of potential pathogens. They then suspected a hantavirus that is carried by rodents. CDC personnel then trapped and tested rodents from the area, including the deer mouse, *Peromyscus maniculatis*. They found this rodent to be the primary reservoir in New Mexico. The outbreak was brought under control, when people in the area were educated on how to rid their homes of the mice and their droppings.[37] Although there have been isolated outbreaks in New York and Louisiana, associated with different species of rodents, they have been controlled by practicing acceptable rodent control techniques and avoiding enclosed areas where the rodents may thrive.[93]

The disease: The hantaviruses that emerged in the Four Corners region of the U.S. were determined to be the cause of an acute respiratory disease now termed *hantavirus pulmonary syndrome* (HPS). The identification was delayed, because the symptoms varied remarkably from the characteristic hantaviral diseases reported in the Europe and Asia. Therefore, it was an unexpected finding when sera from patients in this area reacted with hantaviral antigens, and genetic identification revealed a novel hantavirus in patients' tissues and in rodents trapped near their homes.[94,95] Initial symptoms of the American version of HPS were flu-like and manifested as fever, chills, headache, muscular aches, and pains. Within a short time, however, the disease turns ugly, and the lungs fill with fluids, causing severe respiratory distress for which there is no specific treatment. In HPS, death occurs from shock and cardiac complications, even with adequate tissue oxygenation.

Epidemiology: Contraction of HPS in the U.S. is uncommon, and fewer than 100 cases have been reported, but the mortality is high and there were 52 deaths among those reported to have contracted the disease. In contrast, people in China have been afflicted with a strain for thousands of years, which annually causes 100,000 illnesses and perhaps as many as 5,000 to 10,000 deaths (5 to 10% mortality).[96] The hantavirus outbreaks in Asia are characterized by kidney failure and have been referred to as *hemorrhagic fever with renal syndrome* (HFRS). In the 1950s, Hantaan virus (HTN) (isolated near the Hantaan River) caused Korean hemorrhagic fever with renal syndrome in thousands of United Nations troops.[97] Still others have been isolated in Scandinavia and called *Puumala* (PUU). About 5 to 10 deaths per year (0.1% mortality) are caused by this strain.[98] Since the discovery of the HPS-causing hantaviruses, intense investigation of the ecology and epidemiology of hantaviruses led to the discovery of many more novel hantaviruses. None of the states where this virus was located wanted it to carry the name of their state or community, so there was bitter controversy over the naming of the virus. Scientists at the CDC finally decided on the term *Sin Nombre* or *No Name* for the virus. Most of the 174 cases of HPS in the U.S. and Canada have been caused by Sin Nombre (SN) virus.[99] Cases of HPS in the southeastern U.S., as well as many in South America, are caused by a newly recognized clade (a group that shares a common ancestor) of viruses that includes Bayou (BAY), Black Creek Canal (BCC), and Andes viruses. These viruses are widespread and present a significant public health issue.[100]

The virus is spread to humans from contact with rodents. The pathway was first revealed in the early 1940s, when a viral etiology for HFRS was suggested by Russian and Japanese investigators who injected persons with filtered urine or serum from patients with naturally acquired disease.[97] These studies yielded the first evidence that the natural reservoir of hantaviruses was a rodent. The Japanese investigators were able to produce this disease in humans by injecting bacteria-free filtrates of tissues from *Apodemus agrarius* or mites that fed on the *Apodemus* mice. In 1978, investigators demonstrated that patient sera reacted with an antigen in lung sections of wild-caught *A. agrarius*, and so confirming a rodent reservoir for the HFRS-causing viruses. They also demonstrated that the virus could be passed from one rodent to another.[101] The successful propagation of HTN virus in cell culture, in 1981, provided the first opportunity to study this pathogen systematically.[102] It is now established that hantaviruses can be carried by at least 16 various rodent species including rats, mice, and voles. Additionally, chickens and cats may also be infected.[103] The viruses tend to be identified with a specific rodent host. In the Four Corners region, the host is *P. maniculatis*, whereas in Florida, the cotton rat (*Sigmodon hispidus*) is thought to be the primary carrier. Hantavirus infection does not appear to incapacitate its rodent reservoir host that tends to produce a significant antibody response against the virion envelope and core proteins, although maintaining chronic, probably lifelong infection.[100]

The emergence of the HPS virus in the U.S. is reported to be associated with environmental factors that promote the multiplication of a natural reservoir (field mouse) and so increases opportunities for human infection.[97] Researchers now report that this outbreak of HPS was probably initiated by climate irregularities associated with El Nino. The warming of the seas off the West Coast resulted in extensive rainfall in many areas including the Four Corners of the Southwest. This dramatically increased the natural foods (pine nuts and grasshoppers) of field mice, and their numbers expanded by a factor of tenfold.[101] The HPS outbreak was due to an increase in the population of field mice that subsequently invaded buildings to produce greater opportunities for indoor contact. The most common route of transmission in humans is by aerosolized mouse droppings containing the viral particles, although there is evidence that bites may also transmit the disease. Investigations have linked virus exposure to such activities as heavy farm work, threshing, sleeping on the ground, and military exercises. Indoor exposure was associated with invasion of homes by field mice during cold weather or to the nesting of rodents in or near dwellings. Persons of lower socioeconomic status appear to be at higher risk, because housing conditions and agricultural activities permit more and immediate contact with mice. However, suburbanization, wilderness camping, and other outdoor recreational activities have spread infection to persons of middle and upper incomes.[97,104]

Dengue Fever

Background: Dengue is considered as one of the most consequential mosquito-borne viral diseases to affect humans. This disease has a global distribution that rivals malaria and has the capacity to infect the more than 2.5 billion people on the planet who at risk from epidemic transmission. There are millions of cases of dengue fever annually, with hundreds and thousands of cases of the more severe dengue hemorrhagic fever (DHF). DHF accounts for the majority of the 5 to 15% mortality that occurs worldwide, mostly in children.[105]

The disease: There are four antigenically distinct viral serotypes that cause dengue and DHF. The serotypes are DEN-1 through DEN-4, and belong to the genus *Flavivirus*.[106] The expansion of dengue/DHF is aided by the fact that there is no cross-protective immunity with any of the viruses, so that it is possible for people to acquire multiple dengue infections during their life-

FIGURE 7.1 Distribution of *Aedes aegypti* (shaded areas) in the Americas in 1970, at the end of the mosquito eradication program, and in 1995. (Adapted from Gubler, D.J. and Clark, G.G., Dispatch, Dengue hemorrhagic fever: the emergence of a global health problem, *EID*, 1, 2, April–June 1995.)

time.[105] Dengue is primarily an urban tropical disease with severe flu-like symptoms that causes high fevers, frontal headache, severe body aches and pains, nausea, and vomiting. The "critical stage" for DHF occurs just at the point where people start to feel better. When the fever eases, patients develop "leaky capillary syndrome" in which the blood vessels leak and untreated patients will go into shock and die. Some cases show no outward symptoms, whereas others show bruising or pinpoint rashes that signal internal bleeding.[107]

Epidemiology: The greatest health menace emerging from dengue/DHF has been in Central and South America, where previous campaigns in the 1950s and 1960s to eliminate the mosquito vector *A. aegypti* to combat yellow fever met with success. However, complacency in mosquito control programs has allowed *A. aegypti* to return with a vengeance, and it is a competent vector for dengue viruses. In the late 1980s, memories of mosquito-borne diseases faded, and other public health programs competed for the monies normally paid for mosquito spraying. A rapid population increase throughout Latin America combined with rapid urbanization, poverty, and insufficient sewage and waste treatment methods resulted in an explosive growth in the mosquito vectors and a 60-fold increase in dengue/DHF as seen in Figure 7.1.[105] Further, air travel has increased the probability of the virus spreading. Based on viral envelope gene sequencing, the DEN-3 virus strain in Latin America appear different than the DEN-3 strains (previously isolated in these areas) and identical to the DEN-3 virus causing major dengue epidemics in India and Sri Lanka.[105] Travel, therefore, appears to have been an important component in the spread of this disease.

Dengue fever and DHF has returned or emerged in many areas of the world (Figure 7.2). A dengue pandemic that began in Southeast Asia after World War II (WWII) has increased dramatically since the 1980s, where it has emerged as the

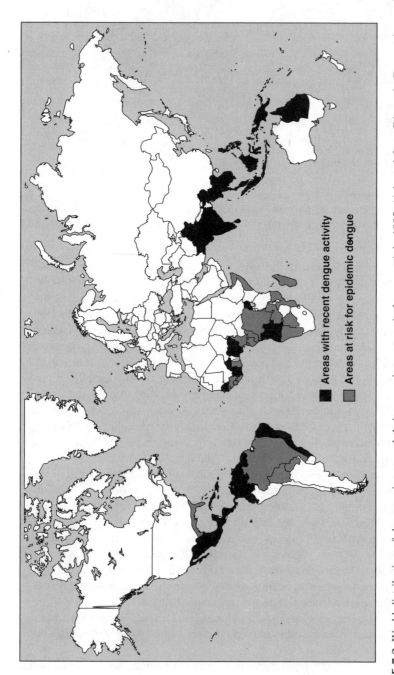

FIGURE 7.2 World distribution of dengue viruses and their mosquito vector, *Aedes aegypti*, in 1995. (Adapted from Dispatch, Dengue hemorrhagic fever: the emergence of a global health problem, *EID*, 1, 2, April–June 1995.)

leading cause of hospitalization and mortality among children. Taiwan and China experienced a series of epidemics after being free of the scourge since the 1950s. When India and Sri Lanka experienced their first major epidemics, DHF once again reemerged in Asia. In 1994, Pakistan reported its first major epidemic with dengue. All four serotypes have been involved in the epidemics in China. In Asia, the Pacific Islands, Africa, and the Americas, the last 15 years have seen increasingly larger epidemics.[105–109] In 1998, 770,000 people in the Americas became infected.[110] In the past 20 years, dengue transmission and the frequency of dengue epidemics has greatly increased in most tropical countries of the American region.[111]

The likelihood of major outbreaks occurring in the U.S. is considered doubtful, but the possibility remains because the competent mosquito vectors are present, including *A. aegypti* and the Asian tiger mosquito, *Aedes albopictus*. These mosquitoes have the capacity to carry and transmit the dengue virus, and such transmission has been detected in 1980 and 1986 in south Texas and in northern Mexico.[112] Dengue fever and DHF have been found within 10 miles of the U.S. border in Reynosa, Mexico, just across the line from McAllen, TX.[107] Additionally, there have been more than 2000 suspected cases of imported dengue since 1977 in the U.S., with more than 20% of these cases confirmed as dengue.[113] The last reported case of dengue fever in the U.S. was in Texas in 1999.[111] The U.S. needs to maintain vigilance in controlling mosquito populations and maintaining the public health infrastructure that prevents uncontrolled urbanization; substandard housing; and inadequate water, sewage, and waste management.

Influenza

Background: There are three types of influenza viruses designated as A, B, and C, along with subtypes of A. Influenza types A and B are responsible for the epidemics of respiratory influenza that occur annually and are frequently connected with heightened rates of hospitalization and death.[114] Type C is quite different from A or B because it either produces very mild symptoms or none at all and, consequently, has not been involved in any epidemics. Therefore, most public health efforts have been focused on controlling types A and B. Influenza is commonly referred to as *the flu* and is caused by these viruses that attack the respiratory system. Presently, there are three strains of influenza viruses that are circulating globally and these include two type A and one type B (Figure 7.3) virus. The viruses contain proteins that can be detected immunologically, and this serves as a useful tool for identification. The two main proteins used for identification are hemagglutinin (H) and neuraminidase (N). Type A viruses are separated into subtypes based on these two proteins and are designated as *A(H1N1)* and *A(H3N2)*. The influenza type A viruses are also capable of undergoing gradual changes, with a succession of mutations or a sudden and significant change in the H or N proteins. A gradual shift in the viral proteins is referred to as *antigenic drift*, whereas abrupt changes in H or N proteins is called an *antigen shift*. Changes in the type B virus appear limited to antigenic drift. Changes in the viral protein through either of these mutation mech-

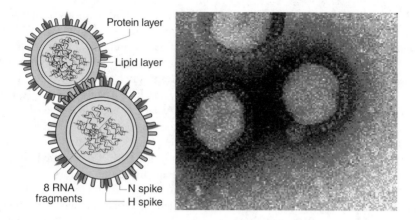

FIGURE 7.3 Photomicrograph of influenza virus. (Source from World Health Organization (WHO), Emerging and Other Communicable Diseases, Influenza A (H5N1) Facts Sheet 188, January 9, 1999.)

anisms permits the virus to evade immune reactions that may have been formed in response to contact with previous but slightly different influenza viruses. The altered viral proteins are no longer recognized by the immune cells and/or antibodies, resulting in a weakened response or no response at all. Influenza A and B viruses are continually expressing antigenic drift, resulting in seasonal flu outbreaks. When the more abrupt antigenic shift occurs, entire populations are left immunologically unprotected, and pandemics may occur.[114]

The disease: The symptoms of influenza can vary from none to fatal. Influenza is normally characterized by a fever (100 to 103°F), respiratory symptoms that include cough, sore throat, stuffy nose, muscle aches and pain, and extreme fatigue. The course of this disease is usually self-limited, and hospitalization is usually not required. Sometimes, the disease produces vomiting and diarrhea in children, and gastrointestinal symptoms are not the routine.[110] Most people recover within 1 to 2 weeks, but pneumonia may develop as a life-threatening complication. There are about 20,000 deaths annually in the U.S. with the majority of serious illness and death occurring in the aged, very young, and debilitated. This is probably strain dependent, however, and recent infections with influenza A(H5N1) indicate severe symptoms in other age groups as well. However, the small number of cases involved with this strain may not reflect unreported and/or milder cases.

Epidemiology: Influenza is an ancient disease, first described by Hippocrates in 412 BC. There have been, to this date, more than 30 pandemics with 3 occurring within the last 80 years. There were epidemics in 1918, 1957, and 1968. The viruses involved were all from animals (pigs or birds). The mortality associated with these pandemics was enormous, with the worst pandemic occurring in 1918–1919 involving influenza A(H1N1) known as the Spanish flu. This disease caused an estimated 500,000 deaths in the U.S. and 20 million deaths worldwide. The 1957–1958 Asian flu, or A(H2N2), resulted in 70,000 deaths in the U.S. followed by the 1968–1969 Hong Kong flu, or A(H3N2), that killed 34,000 people in the U.S. The emergence of the Hong Kong flu in 1968 was relatively mild, in terms of mortality, but it continues to circle the globe and to demonstrate antigenic drifting. Since the virus emerged in 1968, there have been an estimated 400,000 deaths worldwide, with most of the mortality in elderly people. A(H3N2) remains the most destructive of the currently circulating influenza viruses.[114] A new influenza virus, A(H5N1), was isolated from a child in Hong Kong in May 1997, after the child died from Reye's Syndrome (associated with taking aspirin in the presence of the flu virus). This is an avian virus that had a history of killing ducks and chickens. The Avian flu incubated in East Asia for 8 years, where it killed tens of millions of birds and 68 people. Then suddenly in 2005, for reasons unknown, it spread north into Mongolia and Siberia and west into the Ukraine and Turkey.[114] There have been more than 232 human cases worldwide since 2003, with nearly 134 deaths. Ninety three of the cases occurred in Vietnam, followed by Indonesia with 18 cases.[115] Health professionals around the globe are concerned that this continuing spread of (AH5N1) represents a major threat for humans because: (1) it is highly virulent; (2) it has shown limited transmission to mammals and some humans, but could mutate to a highly infectious form in humans; (3) it is being spread around the globe by migratory birds; and (4) we have neither immunologic defenses nor a proven effective vaccine.[116]

Following increased surveillance of A(H5N1) in Hong Kong and Guandong Province in China, there appears to be no risk of this virus being transmitted from raw, chilled, or frozen poultry foods. Investigations are now concentrating on the characteristics of the virus isolated from human cases, whether there are other persons becoming infected, and other natural reservoirs of the virus.

Additionally, efforts are under way to develop high growth reassortment H5 viruses for vaccine production, but no vaccine is being developed specifically for A(H5N1). This is made difficult because the isolated viruses kill the fertilized eggs that are normally used to develop the vaccines. To reduce the risk of pandemic, in the event of the virus becoming transmissible human-to-human, WHO has established a collaborative team that is providing global surveillance, identification of potential pandemic viruses, promotion of high-growth seed virus for vaccine, and the facilitation of vaccine production and distribution.[117,118]

Ebola

Background: Ebola and Marburg viruses belong to a family of viruses called *Filoviridae*. These viruses have the appearance of long threads or filaments (Figure 7.4), and cause severe and often fatal hemorrhagic fevers in humans

FIGURE 7.4 Ebola virus (simulated).

and nonhuman primates. Some strains have produced fatalities of nearly 90% in human outbreaks. Their extreme pathogenicity combined with the lack of effective vaccines or antiviral drugs classify them as biosafety level 4 agents. These viruses are even more mystifying because their natural reservoir remains unfamiliar, and the pathogenicity of the disease is unknown.[119] The Ebola virus was discovered in 1976 and was named for a river in Zaire, Africa, where it was first detected. The widespread public fear of the virus is probably because of the gruesome symptoms and high mortality, which has lead to a flurry of articles, books, and movies. This fear is greatly exaggerated. There have been only 1400 deaths from Ebola in the last 20 years with nearly all of them occurring in Africa. This is in contrast to the millions of cases of HIV that occur in Africa and worldwide.[120]

The disease: The Ebola virus is usually spread by close personal contact with a person who is symptomatic with the disease. This has most often occurred among hospital workers in Africa, where sanitary conditions and infection control were limited, and the reuse of hypodermic needles is common. Ebola fever typically starts immediately within 4 to 16 d after infection, with malaise, fever, and flu-like symptoms that can be followed by rashes, bleeding, and kidney and liver failure. Most patients appear overtly ill, dehydrated, apathetic, and disoriented. The fever associated with Ebola appears to begin with chills, headaches, muscle pain and aches, abdominal pain, sore throat, and chest pain. Generalized bleeding occurs, with massive hemorrhaging of the internal organs, along with bleeding into the gastrointestinal tract, from the skin, and even from injection sites as the clotting ability of the blood is diminished. The Ebola virus produces lesions in the liver, kidney, and spleen with areas of necrosis on the organs and the lymph nodes. The death of the patient usually occurs from shock within 7 to 16 d due to extreme blood loss. The virus is spread through bodily fluids and secretions but not through casual contact.[119]

Epidemiology: Infections from Ebola virus were first reported in 1976 when two outbreaks occurred at the same time, but in different locations, and with different subtypes of the Ebola virus. One occurred in northern Zaire and the other in southern Sudan.[121] The total number of cases in these two outbreaks was 550, with 340 deaths. The case fatality rate from the Zaire subtype Ebola virus was 90% and for the Sudan subtype was 50%. There was also an outbreak in Sudan, with the Sudan subtype in 1979 resulting in 22 deaths among the 34 cases identified for a case fatality rate of 60%, similar to the 1976 outbreaks. The Ebola Zaire strain emerged again on April 10, 1995, in an outbreak in Kikwit, Zaire. A patient admitted for suspected malaria infected the attending medical personnel with the Ebola virus, which spread rapidly, initially, but was brought quickly under control by a coalition of international health organizations including CDC and WHO. The outbreak eventually resulted in 293 cases, with 233 deaths. The U.S. escaped a major epidemic in 1989 when a new Ebola virus infected a colony of cynomolgus macaques that were imported from the Philippines, at Reston, VA, CDC facility.[122] This new virus was named *Reston virus* and was shown to be antigenically and genetically different from the African varieties. The Reston virus was highly pathogenic for the primates but produced no disease in the humans, although some people developed antibodies to the virus.[119] There was a cluster of chimpanzee deaths in the Tai forest on the Ivory Coast (Cote d'Ivoire) of western Africa in November of 1995. A Swiss researcher contracted the disease and subsequently recovered. Researchers have been actively trapping small animals in a 20 mi^2 section of the canopied Tai National Forest, to reveal the natural reservoir for the Ebola virus. Although there are plenty of suspects including bats, red Colobus monkeys, and other small mammals, there has been no success in identifying the virus in any of the animals captured.[120] The reservoir for the Ebola virus is reported to extend throughout the rural and urban areas of Central Africa and might, therefore, be a small terrestrial mammal or a flying animal (bat or bird). No good candidate species has yet been identified, despite extensive studies.[123]

The combination of impoverished populations, inadequate public health services, rapid transmission of the disease, and greater incursions into the African forests increase the possibility of more outbreaks. CDC and WHO have developed sensitive enzyme immunoassays and PCR (polymerase chain reaction) assays for the detection of filoviruses. The availability of rapid and reliable diagnostic tools will permit more effective disease surveillance programs that are necessary to minimize filovirus outbreaks.

AIDS/HIV

Background: AIDS virus is a virus that has been recently discovered to have originated from an endangered species of chimpanzees from Central Africa. It appears that the transmission of the virus from chimps to humans is related to rain forest destruction and the practice of human consumption of chimpanzees.[124] The AIDS virus (Figure 7.5) belongs to a special group of viruses known as retroviruses and is referred to as human immunodeficiency virus (HIV). Its genetic information is not encoded as DNA, but as RNA (ribonucleic acid) and therefore has to be reverse-transcribed into DNA. The tools for this are provided by the host cell itself, except for a little helper protein (reverse transcriptase), which the virus has brought with itself. When the virus enters the body, it prefers to attack certain types of white blood cells (WBCs) that are a component of the cellular immune system. These cells are called *helper T cells* and are an integral component of our immune system. The AIDS virus almost exclusively focuses on these WBCs, as these helper T cells have CD4 molecules on the surface to which the AIDS virus binds. The virus attaches itself to a special protein (CD4) on the surface of the helper T cell, resulting in the fusion of the viral membrane to the host T cell's membrane. The viral genetic information is then able to enter the cell and is transferred to the nucleus. This process may be accomplished in less than 12 h after infection. The viral piece of DNA is then inserted randomly into the host DNA and ready for transcription. At the beginning of AIDS, the viral DNA is being transcribed to form many RNA molecules. The signal that causes this is yet unknown. The accumulating RNA is transported to the cytoplasm of the cell, where it can begin making proteins. The RNA, with the help of the host's resources, begins to make many copies of the different parts of the AIDS virus (the protective shell, the helper, and anchor proteins). Once everything has been copied, thousands of partial virus particles migrate to the surface of the cell

FIGURE 7.5 Human immunodeficiency virus (HIV).

membrane and fuse with it. Finally, a copy of the RNA genetic information is added to the viral particle. Then this section of the cell membrane turns inside out and new viruses leave the cell.[125]

The disease: HIV is transmitted most commonly by sexual contact with an infected partner and can enter the body through the vaginal lining, vulva, penis, rectum, or mouth. As the virus appears in the blood and many body fluids, it can be transmitted by infected blood, even through contaminated needles. The risk of spreading HIV through transfusions has been significantly reduced, since the introduction of heat-treating techniques in 1985. HIV has been transmitted to fetuses during pregnancy and birth. Fortunately, the administration of the drug AZT during pregnancy has dramatically reduced the risk of transmitting the virus to the fetus from its nontreatment level of 20 to 30% infection rates. HIV has also been transmitted to babies through the breast milk of HIV-infected mothers. There is no evidence that casual kissing or contact will result in infection. Simply touching objects such as telephones, towels, and the like that have previously been used by HIV-infected individuals will not spread the disease to uninfected individuals. There is also no known insect vector for HIV.[126]

Many people remain without symptoms for months or years after acquiring the infection. Symptoms do develop in some people within 1 to 2 months after infection. The symptoms may include enlarged lymph nodes, fever, headache, and malaise. The more severe and lasting manifestations of HIV infection may surface from months to years after infection, but during the asymptomatic period, the virus infects and destroys the T4 cells. Gradually, a number of symptoms may emerge that include prolonged enlargement of lymph nodes, energy and weight loss, recurrent sweating and fevers, skin rashes or flaky skin, yeast infections, and pelvic inflammatory disease (PID). A number of victims develop herpes infections with painful sores about the mouth and genitals, or a more painful disease known as shingles. During the early and intermediate periods of the disease, diagnosis depends on the detection of HIV antibodies in the blood. The antibodies may take as many as 6 months to appear in sufficient quantities in the blood to be detected by available tests such as ELISA and Western blot.

The disease will often advance to a stage referred to as AIDS or acquired immunodeficiency syndrome. The U.S. CDC has defined AIDS to include those people who have evidence of HIV and also fewer than 200 T4 cells. The definition also includes more than 25 opportunistic infections that have historically affected people with AIDS because of the immunocompromised state. One of the most common and frequently fatal opportunistic infections is *Pneumocystis carinii* pneumonia. The opportunistic infections produce a myriad of debilitating symptoms from respiratory distress, severe headaches, extreme fatigue, nausea, and vomiting to wasting and coma. The continued decline in T4 cells and associated immunity increases the opportunity for the development of cancer. Many cancers may develop in AIDS patients, with some of the most common being Kaposi's sarcoma or lymphomas. Kaposi's sarcoma is often evidenced by pigmented spots that develop on the skin.[125]

Although treatments continue to enter the war on AIDS including reverse transcriptase inhibitors (RTI) such as AZT, and protease inhibitors, these drugs are likely to produce severe side effects, whereas HIV may become resistant to many of them. A number of drugs are also available to treat many of the opportunistic infections that are likely to arise in immunocompromised situations. No vaccine has been successfully introduced to date, and the best recommendation is to avoid high-risk behaviors that include unprotected sex or the sharing of needles.

Epidemiology: A joint surveillance effort by UNAIDS and WHO estimated in 1997 that over 30 million people were living with HIV infection.[126,127] UNAIDS estimates that in 2006, there are 40.3 million people living with HIV, and 25 million people have died of AIDS.[128] This translates to 1 in every 100 sexually active adults in the ages 15 to 49 worldwide, with about 90% of the HIV-infected people living in the developing nations. This figure also includes over 1 million children under the age of fifteen. HIV continues to spread rapidly at the rate of 16,000 new cases daily. AIDS is responsible for about 20% of all deaths since the epidemic began in the 1970s. More than two-thirds of the total number of people in the world, living with HIV are from sub-Saharan Africa (Figure 7.6). This is contrasted against the 6% decrease in new AIDS cases in the U.S. for the first time ever in 1996, which is continuing to decline. Unfortunately, AIDS cases rose in some sections of American society, including heterosexual African-Americans and Hispanics.[126,128]

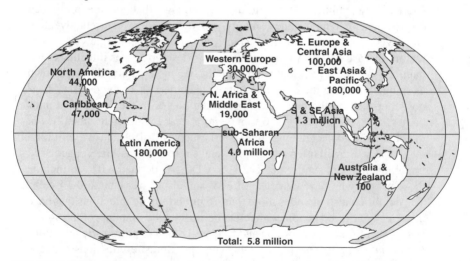

FIGURE 7.6 Distribution of HIV-infected persons globally for 1997. (UNAIDS and WHO, Report on the Global HIV/AIDS Epidemic, December 1997, http://www.unaids.org/unaids/document/epideinio/report97.html.)

West Nile Virus

Background: The WNV was first isolated from a woman in the West Nile
district of Uganda in 1937. In the1950s, the virus became associated
with severe cases of meningitis and encephalitis. During the early 1960s,
WNV was also shown to afflict horses. The WNV virus has been found
in Africa, Europe, the Middle East, west and central Asia, Oceania, and
in North America. WNV made its first appearance in the U.S. in 1999
and appeared as encephalitis in horses and humans. Since then, the virus
has spread by birds and mosquitoes to every state except Hawaii, Alaska,
and Oregon.[129]

The disease: People are usually exposed to WNV through the bite of a
mosquito that has fed on an infected horse or bird. However, a very small
number of cases have been linked to organ transplants, breastfeeding, and
pregnancy from mother to child.[129] About 80% of those people will remain
asymptomatic. About 20% of those exposed will develop mild symptoms
with in 2 to 15 d after exposure, which include fever, headache, tiredness,
body aches, occasionally with a skin rash (on the trunk of the body), and
swollen lymph glands. This is self-limiting and does not usually require
treatment. This is known as West Nile fever and can last from days to
weeks. In about 1 of every 150 people infected, the disease may present
as encephalitis (inflammation of the brain) or meningitis (inflammation of
the membrane around the brain and spinal cord) in humans and horses,
and may also cause mortality in domestic and wild birds. Many state public
health agencies have programs to collect and test dead birds for WNV, as
part of an active surveillance program. Severe symptoms appear in about
1 in every 150 people infected with WNV. The symptoms can include
fever, headache, stiffness of the neck, disorientation, coma, vision loss,
numbness, paralysis, and even death. Those surviving may have lasting
neurological effects.[130]

Epidemiology: There have been a number of outbreaks of WNV encephalitis
in humans, horses, and birds reported since the early 1990s. Outbreaks
among humans have been reported in Algeria, Czech republic, Democratic
Republic of the Congo, Romania, Russia, and more recently in the U.S.
from 1999 to 2003.[129] Neuroinvasive WNV, characterized by encephalitis
and/or meningitis, was reported in 9 states with a total of 16 cases for
2006, as reported to CDC by July 25, 2006.[131] The typically less severe
form of WNV, characterized as West Nile fever without a neuroinvasive
component, appeared in 9 states during the same period in 2006. It is
very likely that far more cases occurred but were underreported because
persons may not have sought medical attention, or their cases were not
authenticated by laboratory tests, or the attending physician did not report
mild cases.[131]

BACTERIA

Escherichia coli

> *Background:* E. coli are Gram-negative, facultative anaerobic, short straight rods that characteristically inhabit the intestines of humans and other animals and belong to the family Enterobacteriaceae. This family of enterics has a number of important pathogens including agents of bacillary dysentery (*Shigella dysenteriae*), gastroenteritis (*Salmonella* and *Shigella*), and the agents of typhoid (*Salmonella typhi*). Members of the enterics cause gastroenteritis, mostly, but have also been implicated in urinary tract infections (UTI), wound infections, pneumonia, septicemia, and meningitis. Strains of *E. coli* have been implicated in many of these disorders. However, most strains of *E. coli* are harmless inhabitants of the intestines of warm-blooded animals and humans. Occasionally, however, a strain develops that is more pathogenic than others. *E. coli* O157:H7 is such a strain. The combination of letters and numbers in the name of this bacterium refers to the specific markers found on the surface of the organism and permits it to be distinguished by serological and other techniques from the hundreds of other strains. Infections with *E. coli* O157:H7 were first described in 1982, and it has emerged along with a few other strains of *E. coli* as the predominant cause of hemorrhagic colitis in humans. The strains of *E. coli* capable of causing hemorrhagic colitis are referred to as enterohemorrhagic *Escherichia coli* (EHEC).[132]

> *The disease:* E. coli O157:H7 is pathogenic for humans and has characteristically produced bloody diarrhea with abdominal cramps; sometimes the infection causes nonbloody diarrhea with very few symptoms. There is usually little or no fever, and the illness resolves in 5 to 10 d. In some persons, particularly children under 5 years of age and the elderly, the infection may progress into a more severe and life-threatening form of the disease known as hemolytic uremic syndrome (HUS). HUS is characterized by the destruction of red blood cells (RBCs) and kidney failure. HUS is a life-threatening condition usually treated in an intensive care unit with blood transfusions and kidney dialysis. Rapid and intensive care can reduce the mortality rate to lower than 5%. About one third of the persons with HUS have abnormal kidney function many years later, and a few may require long-term dialysis. Another 8% of persons with HUS have other lifelong complications such as high blood pressure, seizures, blindness, paralysis and, sometimes, removal of part of the bowel. In the U.S., HUS is the leading cause of acute kidney failure in children, and most cases are caused by serotype O157:H7. This serotype produces a group of cytotoxins collectively known as shigalike toxins (SLTs) that are thought to be factors producing some of its virulence.[133] The toxins are very similar to toxins produced by a pathogenic *Shigella*, another member of the family Enterobacteriaceae.[134]

> *Epidemiology:* In 1993, the American public was startled into an awareness of this disease, following a foodborne outbreak traced to the undercooked

hamburgers eaten at a fast-food chain restaurant.[135] The outbreak involved the infection of 700 persons from 4 different states, with 51 of these persons developing HUS and four people dying from the syndrome. Serotype O157:H7was implicated in this outbreak, but it was first isolated 11 years earlier in 1982. There were more than 30 outbreaks to follow in the next 10 years before the 1993 tragedy.[133] Since the 1993 outbreak, there has been increased reports of O157:H7 infections giving rise to more than 60 out-breaks. Part of this increase may be due to elevated awareness among doctors, microbiologists, and consumers. As there is little evidence that this organism caused infections before the 1970s, it is an example of the emer-gence or evolution of a foodborne pathogen.

The majority of infections with serotype O157:H7 have come from eating undercooked beef products, but many other sources of infection have been identified. Outbreaks have been associated with acidic foods such as fresh-pressed cider in which 27 persons were infected with serotype O157:H7.[136] The apples used to make the unpasteurized cider, without preservatives, were thought to have been contami-nated with cow manure.[137] In another outbreak involving several restaurants and 48 infected persons, the implicated food was mayonnaise that is also acidic to a pH of 3.6.[138] The serotype O157:H7 organism was thought to have been transferred to the mayonnaise by cross contamination with meat products, and this bacteria has been shown capable of surviving up to 55 d in this acidic environment.[138] Serotype O157:H7 infections have also been traced to both drinking water and recreational water. An outbreak in Missouri involving more than 240 people with 4 deaths occurred in 1989 and was associated with contaminated drinking water.[139] Fecal contamination of lake water in Portland, OR, was thought to be responsible for an outbreak of serotype O157:H7 among swimmers, with 21 children being infected. Based on these findings, an infection with serotype O157:H7 is most often associated with eating undercooked beef products. Contamination can also come from drinking contaminated raw milk, or other unprocessed raw beverages that might come into contact with human or bovine feces. Swimming in or drinking water contaminated with sewage is also a risk. Finally, the organism has been found capable of surviving in highly acidic foods that were once thought to be low risk for the transmission of bacterial foodborne infection.

Lyme Disease

Background: Lyme disease is caused by the spirochete *Borrelia burgdorferi*, a Gram-negative, slender, flexible bacteria that is helically coiled (Figure 7.7). The organism is anaerobic and fermentative in its energy metabo-lism, and it is spread to humans by the bite of ticks of the genus *Ixodes* (Figure 7.8). The tick responsible for the spread of Lyme disease in the northeastern and North Central U.S. is known as the deer tick and char-acteristically feeds on the white-tailed deer, the white-footed mouse, other mammals, birds, and even reptiles.[140,141] The black-legged tick is respon-sible for the transmission of the disease along the Pacific coast and in the

FIGURE 7.7 The spirochete, *Borrelia burgdorferi* (simulated).

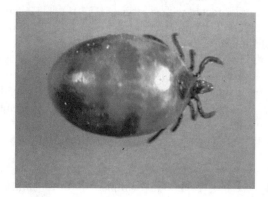

FIGURE 7.8 Female *Ixodes scapularis*. This tick species is capable of transmitting Lyme disease to humans.

southeastern U.S. The *Ixodes* ticks are very small — about the size of the head of a pin in their lymph and larval stages. The adult stages are just slightly larger.

The lymph stage is most likely to be the infective stage of the vector, as it actively feeds and its smaller size allows it to remain undetected for the 24 to 48 h needed to complete a blood meal and transmit the infective spirochete. The tick is usually transmitted from the tips of grass and brushes to persons who brush against them. The ticks will search out any unprotected body parts but often attach to hidden and hairy areas, such as the scalp, groin, or armpits. Here they will insert their mouth parts into the skin and begin a several-day process of feeding on blood, which enlarges their bodies. The life cycle of these ticks is complex and requires 2 years to complete. Typically, the adult ticks feed on large animals such as the deer in the fall and early spring. The females will eventually drop to the ground where they will lay their eggs that hatch into larvae by the summer. The larvae will feed on mice and other small animals and birds throughout the summer and fall. They will become inactive through the winter and will molt to nymphs in the spring. At this stage, they will feed on small rodents, other small mammals, birds, and the unprepared humans who come into contact with them on the tips of grass and brush, usually at the forest edge, but also lawns and gardens. During the normal cycle, the nymphs will develop into adults completing the 2-year cycle.

The disease and epidemiology: Lyme disease was first reported in 1975 near Lyme, CT, following a mysterious outbreak of arthritis. Reports of Lyme disease have escalated since then, placing Lyme disease in the category of an emerging disease. First reported in 1975, the CDC reported 226 cases in 1980, and more than 500 cases in 1983. The number of Lyme disease cases has escalated each year since then, climbing from a total of 12,801 in 1997 to 21,273 in 2003. The states experiencing the highest morbidity from Lyme disease and listed in order of magnitude include New York, Connecticut, Pennsylvania, New Jersey, and Massachusetts.[142]

The disease occurs primarily in the coastal regions of the northeastern U.S., including Connecticut, Rhode Island, New York, Maryland, and Massachusetts. It is also being reported in the western states of California, Nevada, Oregon, and in the many Midwestern states. The early stages of Lyme disease are characterized by headache, fever, chills, swollen lymph glands, muscle and joint pain, and a characteristic skin rash (erythema migrans). This red rash is usually circular and appears 3 to 30 d after the tick bite. The rash often appears on the thigh, trunk, groin, or armpits. This rash (or rashes) will expand to include a large area, and will often show clearing in the center to give a bulls-eye appearance. Diagnosis of Lyme disease is made difficult in the absence of the typical rash, and serologic tests may be required. Untreated, Lyme disease may progress to arthritis with pain and swelling in the large joints, paralysis of the facial muscles, and meningitis characterized by stiffness in the neck, fever, and severe headache. Lyme disease rarely results in death, but chronic Lyme disease can lead to permanent damage to joints or the nervous system.

A newly discovered bacterium carried by the same tick that carries Lyme disease is also emerging as a significant agent of disease. The organism causes human granulocytic ehrlichiosis (HGE). It produces high fever, headache, muscle aches, chills, and sometimes death. Most cases have been identified in Wisconsin and Minnesota, but the disease can be found anywhere there is Lyme disease, and about 10% of patients with Lyme disease may also have HGE. This is important because HGE responds only to tetracycline and doxycycline.[143]

Prevention is best accomplished by avoiding tick-infested areas in the summer. Wearing light-colored clothing that covers arms and legs and is tucked in at the sock and belt line (with tape at the sock line) will help to prevent the ticks from crawling underneath clothing; it also makes them easier to spot against a light background. Spraying with DEET on exposed skin surfaces other than the face will also be helpful. A vaccine has been approved by the FDA called LYMErix,[144] but is no longer available. The FDA recommends that all other preventative measures be taken, whether or not a person has received the vaccine. One out of five people who have had the vaccine will still be at risk for acquiring the disease.

Streptococcus

Background: Streptococci are Gram-positive cocci (spheres) arranged in chains or in pairs and include many species that are pathogenic. There are several groups of streptococci, based on serological reactions to surface antigens, according the original work of Lancefield in 1933. These groups include A–H, J, and K. The GAS are commonly found in the throat and on the skin. They may cause no symptoms but may also cause infections that range from mild to severe and even life-threatening. The major pathogens are included in groups A and B, and their pathogenicity is associated with certain enzymes and surface proteins such as hemolysins, erythrogenic toxins, and M-protein.[145] Hemolysins are enzymes capable of breaking, or lysing, blood cells, and when streptococci are grown on sheep blood agar, the types of hemolysis produced place these organisms into three groups. There are gamma hemolytic strep that produce no hemolysis, alpha hemolytic strep that produce a zone of green discoloration around the colony, and beta hemolytic strep that produce a zone of clear hemolysis around the colony. Beta hemolytic, group A strep are most often associated with strep throat infections. The ability to hemolyze blood cells is associated with powerful hemolysins such as oxygen-sensitive streptolysin O and oxygen-stable streptolysin S enzymes. Streptococci may also produce erythrogenic toxins (A, B, and C) that harm blood vessels beneath the skin. It is thought that the streptococci carry a provirus with the genetic materials capable of coding for the production of these toxins.[145] The streptococci may produce a broad array of enzymes, including neuraminidases, hyaluronidases, streptokinases, ATPases, DNAses, and many others that participate in the destruction and invasion of human tissue.

The disease: The GAS produce a variety of diseases that include strep throat, impetigo, and scarlet fever. Occasionally, the strep invade deeper tissues such as blood, lungs, or even fat and muscles. The more severe of these invasions results in necrotizing fasciitis and/or streptococcal toxic shock syndrome (strep TSS). A popular news media characterized the strep that caused necrotizing fasciitis as flesh-eating bacteria because of their ability to produce a progressing destructive infection of the underlying tissues. Strep TSS is caused by a highly invasive strep that produces shock and injury to internal organs such as the kidneys, liver, and lungs. Necrotizing fasciitis usually begins with a trivial or even unnoticed trauma that appears within 24 h as a lesion with swelling and redness. This is a deep-seated infection of the subcutaneous tissue that progressively destroys the underlying connective tissue and fat, but sometimes sparing the skin and muscle. The early signs and symptoms of necrotizing fasciitis include fever, severe pain and swelling, and erythema (redness) at the site of the wound. The wound deeply colors to blue with blisters that contain a clear yellow fluid. Within 4 to 5 d, the purple areas become gangrenous, and at 7 to 20 d, dead skin separates the margins of infection to reveal marked necrosis of the underlying subcutaneous tissue. In a few cases, this process quickly escalates until several large areas of gangrenous skin are formed, and the patients may become dull and unresponsive because of the toxins in their system. Many of the present GAS cases of necrotizing fasciitis inevitably led to severe systemic illness with high morbidity, even in otherwise healthy patients who receive extensive support in the form of antibiotics, dialysis, and surgical techniques.[145] These recent developments along with the increased mortality from GAS suggest the emergence of a more virulent streptococcus.[146]

Strep TSS is defined as any GAS infection associated with the early onset of shock and organ failure. The initial symptoms of strep TSS is sudden and severe pain in an extremity or may mimic peritonitis, pneumonia, or heart attack. About one fifth of the patients develop flu-like symptoms with fever, chill, nausea, vomiting, and diarrhea. Fever and evidence of soft tissue infection are present in most patients, with a majority of these progressing to necrotizing fasciitis.[145] Patients are often admitted in a state of shock, and despite treatment, this shock continues in combination with renal dysfunction and acute respiratory distress in about 55% of the patients. Mortality rates of 30 to 70% have been reported.[147]

Epidemiology: Beginning in the 1980s, there has been a sudden elevation in the reporting of a highly invasive GAS infection, with or without necrotizing fasciitis, associated with shock and organ failure. The number of these cases appeared to increase from almost none to over 2000 per year.[148] There are about 10,000 to 15,000 cases of invasive GAS in the U.S. annually, resulting in 2000 deaths. The mortality rate for strep TSS is about 60% of the 2000 to 3000 cases reported per year. Annually, about 20% of the 500 to 1500 patients who acquire streptococcal fasciitis have died. These figures should be contrasted to the millions of people who acquire strep infections annually. In fact, GAS is an organism with historic significance that has produced many outbreaks in the past, and the number of people involved were far greater than the several thousand cases of invasive GAS that are presently seen on a yearly basis. The argument has been made that GAS may have become more virulent and hence produced these severe cases. This, however, may not be the only reason for the more intense invasions. Stevens argues that the emergence of a more virulent strain of GAS would have produced many more cases in the form of an epidemic.[146] It has been suggested that people are becoming more susceptible to GAS infections because of the use of nonsteroidal antiinflammatory drugs (NSAIDs) such as ibuprofen. These observations have been anecdotal but reported from more than one medical group.[148] According to Stevens, NSAIDs is a predisposing condition for GAS infections.[149]

Tuberculosis

Background and disease: TB is a chronic infectious disease of the lower respiratory tract caused by *M. tuberculosis*, a slender, acid-fast rod with cell walls containing high lipid levels. The slow growing bacilli are transmitted by aerosols from persons with active disease and are a kind of obligate aerobes and intracellular parasites that survive inside alveolar macrophages in the lungs. After multiplying in the macrophages, they enter the lymph nodes through the lymph channels, where they continue to multiply intracellularly. They can spread further to other organs including the lungs. Symptoms normally begin to develop at this stage from a cell-mediated immunity that walls off the pathogen within multinucleated giant cells surrounded by lymphocytes and macrophages. These structures can be seen on x-ray and are known as tubercles (Figure 7.9). Such tubercles may become scarred or calcified over time, whereas the bacilli remain dormant but alive within. The disease is clinically manifested when there is an immune lapse or reinfection that allows the organisms to multiply. This stimulates the immune response, causing the tubercles to break down and erode through the bronchi wall, spilling mycobacterium-laden liquid caseum into the bronchi. The characteristic cough of active TB derives from this process and is also the reason for the appearance of the organism in the aerosols created from the cough of an infected person.[145]

Epidemiology: In 1999, there were 18,000 cases of TB in the U.S., of which 1% were resistant to antibiotics.[150] Tuberculosis causes more death globally than any other single infectious diseases, with more than 8 million cases being reported each year.[151] The number of TB cases increased by 20% from 1985 through 1992 due, in part, to the onset of the HIV epidemic, an increase of TB cases among foreign-born people, and the appearance of TB strains resistant to multiple drugs.[151] Since 1992, the incidence of TB

FIGURE 7.9 X-radiograph of patient with tuberculosis.

decreased by 40% with a decline every year to a level of 5.6 cases per 100,000 for the lowest recorded incidence in the country.[151] It is thought that this reduction may be due to more effective TB-control programs that lead to prompt identification and treatment of persons with efforts to make certain that therapy is completed.[151] After the introduction of antibiotics, TB was thought to have been under control, but the disease persisted in the poorer reaches of society (inner-city dwellers, the homeless), and it is reemerging dramatically among the foreign-born. Tuberculosis was declared a U.S. public health emergency in 1992, and the CDC is attempting to help control the spread of the world's top infectious disease killer. Tuberculosis kills over 3 million people worldwide each year, and many more are affected.[152] Unfortunately, A WHO report released in March of 1995 reveals that efforts to control tuberculosis globally have been ineffective and that drug-resistant strains of *M. tuberculosis* are on the rise.[153] Even though chemotherapy and effective measures of prevention exist, TB is increasing in the developing world as well as in the industrialized world.[154] WHO estimates 8 million new cases and 3 million deaths per year from TB with 450,000 of those deaths in children under 15 years of age.[155] WHO estimated that we will see 90 million new cases and 30 million deaths with annual rates in infection in developing countries exceeding 2%.[156] South Africa currently has the highest rate of TB in the world and little is being done to curb the outbreak. There are 350 cases of TB per 100,000 population in South Africa with a rising incidence of multidrug-resistant TB and a death rate of 54% of those who develop the disease. In Russia, there were 100,000 cases in 1999, 30% having been resistant to antibiotics.[150] Certain environmental factors appear to accelerate the disease. These factors include crowded hospital rooms shared with TB patients, lack of sanitary conditions, unfiltered air, closed windows at night, and crowded living and working conditions.[157]

Patients often feel better after a few weeks of medicine and stop taking it. The consequence is a reinfection with TB that has a greater probability of being resistant to medication. World TB Day, which commemorates both the struggles and successes in the worldwide fight against tuberculosis was March 24, 2006. Tuberculosis claims the lives of more than 1.7 million people each year and hinders many of the 15.4 million individuals currently afflicted with this disease.[158] The reasons for this continued level of infection may be related to coinfection with HIV, the steady increase in the number of multidrug-resistant cases of TB, and the realization that available pharmaceuticals may be failing.[158] To combat the outbreaks and reemergence of TB, the WHO report titled "Stop TB at the Source" recommends international adoption of a control strategy called directly observed treatment-short-course (DOTS), which is intended to ensure that all patients complete their entire course of treatment. This program requires that patients take their medicine under the eyes of a guardian every day for 6 months and are given regular sputum tests to see if the pills are working. The program has been effective in New York, where TB has been reduced by over 20% to about 3000 new cases per year. Unfortunately, results from studies conducted by the WHO have questioned the effectiveness of DOTS.[158] There is increasingly more support for a new method of treatment called DOTSPlus, which requires taking up to 7 drugs daily for 18 to 24 months.[158] Despite these efforts, there have been a small number of patients resistant to all anti-TB drugs. Starting in 1993, the New York health department has placed patients who resist taking the complete regimen of medicine under "detention until cured."[153] Multidrug-resistant strains of TB will become increasingly less responsive to even the more vigorous treatment programs such as DOTSPlus.[159] Seventeen million new cases and 60 million deaths due to multidrug-resistant TB are expected between 1998 and 2030.

PARASITES

Cryptosporidium

Background: Cryptosporidium is a single-celled microscopic protozoan parasite that belongs to the class Sporozoa. These organisms are not normally motile and asexually reproduce by multiple fission and sexually by gamete fusion. The resistant form of the parasite is called an *oocyst*, which is characterized by an outer protective shell that protects the organism against environmental extremes such as heat, cold, dryness, and chemical insult. There are several species of *Cryptosporidium*, but there is only one pathogenic species, *Cryptosporidium parvum*. This organism is widely distributed in the intestines of animals and humans, and normally causes infections in humans by the fecal oral route. It is estimated that as few as 30, or even 1 oocyst, may cause infection when swallowed.[160] Cryptosporidium is therefore a highly infectious enteric parasite, widely distributed, that is resistant to chlorine and difficult to filter, thereby making it a serious threat to water supplies.

Disease: Persons who swallow cryptosporidium cysts may develop symptoms within 2 to 10 d. The oocysts reach the upper small bowel where they excyst and produce four infectious sporozoites that attach to the surface epithelium of the digestive tract and reproduce, forming more oocysts and sporozoites. Many of the organisms are excreted while others maintain the infection. The symptoms include watery diarrhea, stomach cramps, nausea, and a slight fever. These symptoms are self-limiting and will normally last for 2 weeks or less. The organism may be shed from the intestines for up to 2 months after the initial illness and so spread this disease to others through fecal–oral transmission. Persons who have a weakened immune system such as those with HIV/AIDS, cancer and transplant patients, or those on immunosuppressive drugs, are at increased risk from infection, and may develop serious and life-threatening illness from this organism. There is no pharmaceutical treatment for cryptospordiosis and the more serious illness is managed by fluid replacement.[161]

Epidemiology: Contamination of drinking water by cryptosporidium is a growing concern, especially after the outbreak in Milwaukee in 1993, which affected some 400,000 people. There have been five major outbreaks associated with public water supplies of cryptosporidium gastroenteritis in the U.S. and seven in the U.K. since 1983.[162] There have also been numerous smaller outbreaks. Most of these outbreaks have been drinking and recreational waterborne in nature. Cryptosporidium has also been spread by unpasteurized cider, and by person-to-person in households, day care centers, and hospitals.[163] Cryptosporidium is found in animal droppings and human feces, soil, drinking water and recreational water, food, hands, and surfaces contaminated by such wastes. Many persons have become infected by swimming in chlorinated and filtered pools, because the organism has the capacity to survive chlorination and to pass through most pool filters. Cryptosporidium was first reported as a human pathogen in 1976 and came to prominence in the U.S. in 1982 with the increased infections in the highly immunocompromised population of people living with AIDS.[164] Cryptosporidium is not a new organism but is thought to have emerged because it was newly spread among immunocompromised patients and became symptomatic.[165] In many areas of the world, the organisms have been shown to have infected a large portion of the population. In some areas of Brazil, as many as 90% of the children under five have been infected, and over 50% of the children in certain rural areas of China show serological evidence of infection.[166,167] Similar seroepidemiologic studies suggest that 17 to 32% of persons in Virginia, Texas, and Wisconsin with healthy immune systems showed evidence of cryptosporidium infection by the time they reached young adulthood.[168]

Malaria

Background: Malarial diseases are caused by protozoan parasites belonging to the genus *Plasmodium*. There are four species known to infest humans: *P. falciparum, P. vivax, P. ovale,* and *P. malariae.* These parasites are transmitted from human to human by the bite of a female anopheline mosquito in which the parasite has gone through a complex development taking several days to three weeks depending on the ambient temperature. Because the sexual reproduction of the parasite occurs in the mosquito, it is the definitive host, and humans are the intermediate host and reservoir. The female anophelines become infected when they feed on blood from a person infected that contains male and female gametocytes of the parasite. A complex cycle takes place that involves the union of the gametocytes in the stomach of the mosquito and results in the development of slender, microscopic sporozoites that appear in the salivary glands, and are infective for humans (Figure 7.10). This maturation or sporogenic cycle must be completed within the mosquito before transmission to humans can take place. The night-feeding female anopheline takes a blood meal and injects the malarial sporozoites into the human bloodstream. The sporozoites enter the liver, where they infect liver cells and begin the process of development. Depending on the species of malaria, the stages in the human show some variation. In general, once in the liver cells, the parasites begin the process

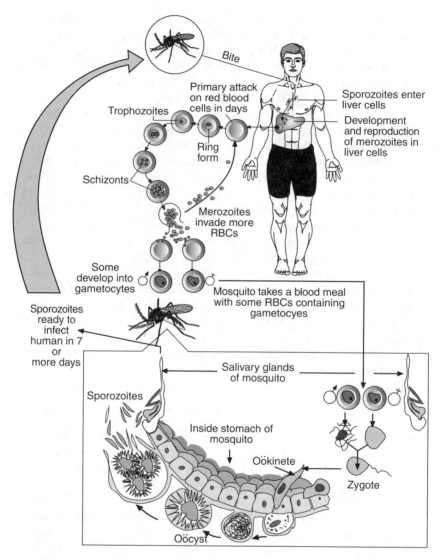

FIGURE 7.10 Life cycle *of Plasmodium* species in female anophelene mosquito.

of schizogeny (splitting stage) and they divide in the liver cells to produce merozoites. These merozoites are infective for red blood cells. The merozoites enter the bloodstream, where they attach to RBCs and invade them. Inside the cell, the parasite form the classical signet ring stage (Figure 7.11) and feed on the cells contents as they grow through the stages of trophozoites and schizonts. Female and male gametocytes are also occasionally produced, and these are the sexual forms of the parasite that continue the cycle of the organism when taken into the anopheline mosquito through a blood meal. While developing within the red blood cell, the parasite converts hemoglobin to hemozoin, thereby depleting the hemoglobin content

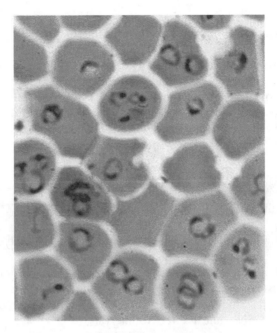

FIGURE 7.11 Ring stage of malarial parasite in red blood cells.

of the blood and giving rise to many of the symptoms of malaria. The parasites proliferate in the RBCs until they can no longer be contained, and the red blood cells are destroyed as they break free into the bloodstream, where they can reinvade more RBCs. This cycle often becomes synchronous, causing regularly timed waves of parasitemia from 36 to 72 h, depending on the particular malarial parasite.[169]

Disease: The symptomatology and pathogenesis of malaria infection are related to the parasite's stage of growth and the host's parasitemia. High parasitemias result in decreased hemoglobin and a lower oxygen-carrying capacity. This quickly leads to fatigue and repository distress. The detritus of lysed RBCs as well as the explosion of foreign bodies in the bloodstream from logarithmic growth of malaria results in massive immune responses, primarily driven by mononuclear phagocyte-released cytokines. Because the explosions of parasitemia occur in regularly timed waves (from as little as 36 h in *P. falciparum* to 72 h in *P. malariae*), the characteristic periodic fever develops. The fever is a response to the massive interferon and interleukin release, which subsides as the body clears the infection and the merozoites invade new cells to begin a new cycle. The systemic damage of malaria is not only from septicemic reactions to cytokines, but also the proliferation of knobs on the red blood cell surface, sometimes called *pigmotic cells*. These projections can result in agglutination of parasites at capillary pathways and ischemia to many organ areas. Untreated infections lead to splenomegaly (enlarged spleen) and, particularly in *P. falciparum*, to cerebral malaria and death. Children are particularly susceptible to cerebral malaria, and the bush doctor cure for urgent cases is an immediate blood transfusion from a close member of the family. This has resulted in a high rate of pediatric AIDS in Africa.

Epidemiology: The WHO estimates that there are 300 to 500 million people worldwide infected with malaria.[170] More than 805,000 deaths occur in sub-Saharan Africa alone on an annual basis. The impact of malaria on the human condition is even more startling when measured in terms of disability-adjusted life years (DALYs). This is an indicator developed to estimate the disease burden so the cost-effective prevention strategies can be developed.[171] On a worldwide basis, malaria has caused the loss of over 35 million DALYs, which is slightly behind tuberculosis and slightly less than half of all cancers globally. The contribution of malaria to global disease and disability is of major public health significance.

The majority of malarial transmission occurs in tropical and subtropical countries (Figure 7.12). The reemergence of malaria as a major threat in these areas is because of a number of reasons. Mosquito control programs have historically been the primary mechanism of attack. However, such programs alone have not been effective in many areas such as sub-Saharan Africa, where the applications of spraying were neither complete nor periodic at effective intervals.

Decreased spraying of the homes with DDT is also blamed for the dramatic reemergence of malaria in South America where there has been a geometric growth in malaria incidence since the 1970s.[172] A major reason for this decreased use is the concern for ecological consequences of DDT spraying. Some authors reason that the use of DDT in homes or on nets used over the sleeping areas at night represents barely a fraction of the DDT used in agriculture. In fact, the amount used to treat 1000 acres of cotton could be used to treat the homes for the entire country of

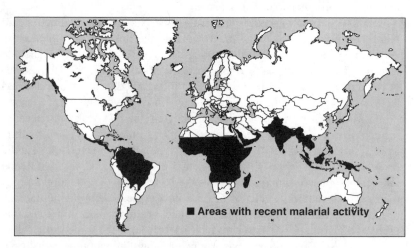

FIGURE 7.12 Global distribution of malaria (Adapted from Zucker, J.R., Changing patterns of autochthonous malaria transmission in the United States: a review of recent outbreaks, *EID*, 2, 1, January–March 1996.)

Guyana. However, most countries have looked to alternative treatments and have found them too expensive. Furthermore, there has been a steady decline in international and U.S. support for malarial control in these countries, and similar reductions in the budgets of the endemic countries themselves.[173] The problem is aggravated by the development of pesticide-resistant mosquitoes in the endemic areas. There is little probability that more money will be allocated for more expensive insecticides.

There are other forces at work to power the reemergence of malaria in the world. These include the appearance of drug-resistant malaria and the predicted expansion of malaria associated with global warming. The malarial parasite has a family of genes called var genes for variability that permits it to change its signature proteins by just enough to stay one step ahead of the host antibodies aimed against it, in a process called *antigenic variation*. These var genes account for nearly 6% of the parasites genome and permit it to mix and match genes to create millions of antigenic proteins. The malarial parasite developed resistance to antimalarial drugs as well. In particular, it developed resistance to the inexpensive and highly effective drug chloroquine, and resistant strains began appearing in the 1960s. Unfortunately, the alternative drugs are too expensive for most people in the malaria-endemic areas, and so the disease continues to kill nearly 2 million people a year.[174]

Global warming spurred by greenhouse gases is an internationally debated concern. One of the potential consequences of such a warming trend may be the increase in the distribution and incidence of malaria.[175] A team of four scientists from the Netherlands created a complex mathematical model that can potentially predict the impact of global climate changes on malarial risk. A simulation experiment using this model showed a widespread increase in the potential for malarial transmission.[176] A global increase of only a few degrees could expand malaria's domain from 42 to over 60% of the planet. Further, the increasing temperatures would increase the metabolism of the mosquito, making it a more voracious biter while speeding up the development of the malarial parasite inside the mosquito.[12]

Americans should not be complacent about this reemergence. Malaria has also begun to reappear in the U.S. There have been at least three outbreaks of locally acquired malaria in densely populated areas of the country that point to the danger of a significant outbreak. Malaria was a once a major disease in the U.S. with over 600,000 cases occurring in 1914, and more than 125,000 cases in 1934 when systematic reporting was in effect.[177] The processes of urbanization, increased drainage, DDT spraying, screening of houses, and introduction of quinine, all contributed to the substantial control of malaria. Malaria was eradicated as a problem disease in the U.S. by the 1950s. Nearly all cases of malaria detected in the U.S. since then have been imported. There have been 76 cases of malaria reported from 1957 through 1994 including the three outbreaks occurring in the densely populated areas of New Jersey (1991), New York (1993), and Texas (1994). In these cases, the weather was more humid than usual, there were larger mosquito populations, and the patients lived in areas with large migrant populations from countries with endemic malaria.[169] The CDC received 1278 malaria case reports occurring among persons in the U.S. and its territories, representing a 4.4% decrease from the 1337 cases reported in the previous year in 2003. U.S. civilians accounted for 767 cases, and there were 306 cases among foreign civilians. In recent years, cases among U.S. civilians have increased and cases among foreign-born civilians have decreased.[178] The likelihood of widespread malarial outbreaks in the U.S. is limited by adequate measures of sanitation, health care, and targeted mosquito control. However, a continued warming trend predicted by global climatic changes could expand the range of malaria by permitting the parasite to complete its cycle in the mosquito. Greater pressures would be exerted on the public health support systems within the U.S.[169]

PRACTICAL APPROACHES TO LIMITING THE EMERGENCE OF INFECTIOUS DISEASES

The emergence of 29 new infectious diseases and reemergence of many others are creating national and international crises. The control of these diseases is made difficult because of the large number of disease-causing organisms, the enormous diversity of geographic areas from which they can emerge, the potential for rapid global dissemination, and the tremendous number of ecological and social factors that influence emergence.[31,35] Primary responsibility for addressing new and reemerging infectious diseases rests squarely with the custodians of public health. Indeed, the fundamental maxim of public health must guide current prevention programs: the health of the individual is best ensured by maintaining or improving the health of the entire community.

D.A. Henderson, the driving force in the successful international smallpox eradication program, proposes the establishment of an international network consisting of disease centers strategically located in tropical areas, especially near metropolitan areas. Each center would have a clinical facility, diagnostic and research laboratory, an epidemiological unit to investigate disease and local response capabilities, and a professional training unit. This would also require an academic and government laboratory necessary for coordinating data collection and analysis, conducting research, providing feedback and support and, when necessary, activate an "international rapid response system."[30] The WHO has created a Division on Emerging Diseases in 1995 to try to deal with a global problem of new and emerging infectious diseases. WHO offices, including the Pan American Health Organization, are developing a regional plan and setting goals that include (1) strengthening regional and national laboratory and surveillance networks; (2) establishing national and regional infrastructures for early warning and rapid response to infectious diseases; (3) promoting and developing applied research for rapid diagnosis, epidemiology, and prevention; and (4) strengthening regional capacity for effective implementation of prevention and control strategies.[179]

The CDC began four Emerging Infections Programs in Oregon, California, Connecticut, and Minnesota to deal with some of the nation's underfunded health threats. Each of these states has different projects that range from foodborne illnesses, and meningitis, to valley fever, ehrlichia, and cryptosporidium. The programs collaborate with federal agencies, local health departments, academic institutions, professional societies, and international organizations. A prevention strategy plan for the U.S. was developed with four major goals: (1) promptly investigate and monitor emerging pathogens, the diseases they cause, and factors of emergence; (2) integrate laboratory science and epidemiology to optimize public health practice; (3) enhance communication of public health information about emerging diseases and ensure prompt implementation prevention strategies; and (4) strengthen local, state, and federal public health infrastructures to support surveillance and implement prevention and control programs.[4]

Ewald reports that control could be made more manageable if it was possible to focus attention on the pathogens most likely to become dangerous. There is disagreement about the ability to predict and prevent the emergence of these new pathogens.[180] Ewald emphasizes that the identification of these pathogens rests upon two major characteristics: (1) an ability to spread easily from human to human through vectors and (2) transmission features that select for high levels of virulence. The transmission features that promote virulence include one or more of the following: (1) being vectorborne with humans as part of the life cycle; (2) durability in the external environment if directly transmitted; (3) being needleborne, or borne by attendants in medical care; and (4) being sexually transmitted and being mutation-prone with an attraction to critical cell types.[158] Knowledge of these evolutionary characteristics should help to distinguish the long-term threat of the emerging disease organism and so adjust the amount of resources invested in keeping with the anticipated threat.

Unfortunately, we cannot accurately predict where, when, or how the next disease will emerge or reemerge and what impact it will have on the individual, society, nation, or the world. Environmental factors such as unusual weather patterns, agricultural practices, deforestation, and other human behavioral factors such as increased population density, compliance, and overprescribing of antibiotics, and mere chance will continue to play a significant role in the emergence or reemergence of disease. Increased vigilance is necessary and the application of more resources is the ultimate answer — contradicting the current erosion of resources to public health being observed nationally and internationally.

REFERENCES

1. Mulvaney, K., The Scourge of Mankind: Infectious Diseases, Once Near\Eradication, Show Remarkable Persistence, *The Environmental Magazine*, 42, 1999, pp. 42–44.
2. Walsham, C., Our Planet is Environmentally Sick, *PlanetENN*, http://www.enn.com/planetenn/I 11 196/feature3.htm, November 11, 1996.
3. Bennett, I.V., Holmberg, S.D., Rogers, M.F., and Solomon, S.L., Infectious and parasitic diseases, in *Closing the Gap: the Burden of Unnecessary Illness*, Antler, R.W. and Dull, H.B., Eds., Oxford University Press, New York, 1987.
4. Centers for Disease Control and Prevention, Addressing emerging infectious disease threats: a prevention strategy for the United States, Atlanta, Georgia, U.S. Department of Health and Human Services, Public Health Service, 1994.
5. Stevens, D.L., Streptococcal toxic-shock syndrome: spectrum of disease, pathogenesis, and new concepts in treatment, *EID*, 1, 3, July–September 1995.
6. Newman, A., EPA Considering Quick Action on *Cryptosporidium, Environmental Science and Technology*, 29, 1, 1995, p. 17a.

7. U.S. Department of Health and Human Services, Pandemicflu.gov, August, 2006, http://www.pandemicflu.gov/general.

8. Centers for Diseases Control, HIV/AIDS Statistics and Surveillance, Table 1. Reported AIDS cases by area of residence, July 1995 though June 2005, http://www.cdc.gov/hiv/topics/surveillance/resources/reports/2005supp_vol11no1/table1.htm, April, 2006.

9. Centers for Disease Control and Prevention, Update: Trends in AIDS Incidence, Deaths and Prevalence, United States, 1996, *MMWR*, 46, 165, 1997.

10. Centers for Disease Control and Prevention, Update: Multi-state Outbreak of *Escherichia coli O157:H7* Infections from Hamburgers-Western United States, 1992-1993, *MMWR*, 42, 258, 1993.

11. Feng, P., Synopsis: *Escherichia coli* serotype O157:H7: novel vehicles of infection and emergence of phenotypic variants, *EID*, 1, 2, April–June, 1995.

12. Linden, E., Global Fever, *Time*, July 8, 1996, p. 56.

13. Steere, A.C., Maliwista, S.E., Snydman, D.R., Shope, R.E., Andirnan, W.A., Ross, M.R., and Steele, F.M., Lyme arthritis: an epidemic of oligoarticular arthritis in children and adults in three Connecticut communities, *Arthritis Rheum.*, 20, 7, 1977.

14. CDC, Division of Vector-Borne Infectious Diseases, Arboroviral Encephalitides, April 2006, http://www.cdc.gov/ncidod/dvbid/arbor/eeefact.htm.

15. CDC, Division of Vector-Borne Infectious Diseases, West Nile Virus, April 2006, http://www.cdc.gov/ncidod/dvbid/westnile/qa/overview.htm.

16. Rohter, L., U.S. is Now Threatened by Epidemic of Dengue, *New York Times International*, September 23, 1995, p. 5.

17. Gubler, D.J., Vector-borne diseases, in *Encyclopedia of the Environment*, Houghton Mifflin, New York, 1994.

18. Altman, L.K., 56 Die in a Month In Zaire Outbreak of Mystery Disease, *New York Times International*, May 10, 1995, p. Al.

19. Mint, C., Insect-Borne Disease: Curing the Carrier, *Science News*, 151, April 12, 1997, p. 223.

20. Satcher, D., Perspectives: emerging infections: getting ahead of the curve, *EID*, 1, 1, January–March 1995.

21. Willq, R.G., Ironside, J.W., Zeidler, M., Cousens, S.N., Estibeiro, K., Alperovitch, A., Poser, S., Pocchiari, M., and Hofman, A., A new variant of Creutzfeldt-Jakob disease in the U.K., *Lancet*, 347, 921, 1996.

22. Centers for Disease Control and Prevention, World Health Organization consultation on public health issues related to bovine spongiforin encephalopathy and the emergence of a new variant of Creutzfeldt-Jakob disease, *MMWR*, 45, 303, 295, 1996.

23. Brown, P. and Cathala, F., Creutzfeldt-Jacob disease in France, in *Slow Transmissible Diseases of the Nervous System*, Hadlow, W.J. and Prusiner, S.B., Eds., Academic Press, New York, 1979, Vol. 1, p. 213.

24. Martens, W.J., Potential Impact of global climate change on malaria risk, *Environ. Health Perspect.*, 103, 5, 458, May 1995.

25. Angier, N., Malaria's genetic game of cloak and dagger, *New York Times*, August 22, 1995, p. C1.

26. Roberts, D.R., Laughlin, P., Hisheih, P., and Legters, L.J., Global strategies, and a malaria control crisis in South America, *EID*, 3, 3, July–September 1997.

27. Nowak, R., WHO Calls for Action against TB, *Science*, 267, March 1995, p. 1763.

28. Drexler, M. and Welch, M.B., Public health-in our battle to stay healthy, we are up against a daunting new array of … clever germs, *Boston Sunday Globe*, Sunday, November 21, 1999.

29. Morse, S., Perspectives: factors in the emergence of infectious diseases, *EID*, 1, 1, January–March, 1995.

30. Satcher, D., Perspectives: emerging infections, getting ahead of the curve, *EID*, 1, 1, January–March, 1995.

31. Morse, S.S., Examining the origins of emerging viruses, in *Emerging Viruses*, Morse, S.S., Ed., Oxford University Press, New York, 1993, p. 10.

32. Associated Press, Scientists trace AIDs origin to wild chimps: Gene tests match virus in primates in Cameroon to first known human case, MSNBC, May 25, 2006, http://msnbc.msn.com/id/12966623/.

33. Morse, S.S., Regulating viral traffic, *Issues Sci. Technol.*, 7, 81, 1990.

34. Morse, S.S., Toward an evolutionary biology of viruses, *The Evolutionary Biology of Viruses*, Morse, S.S., Ed., Raven Press, New York, 1994, p. 1.

35. Morse, S.S., Emerging viruses: defining the rules for viral traffic, *Perspect. Biol. Med.*, 34, 387, 1991.

36. Rogers, D.J. and Packer, M.J., Vector-borne diseases, models, and global change, *Lancet*, 342, 1282, 1993.

37. Barbour, A.G. and Fish, D., The biological and social phenomenon of Lyme disease, *Science*, 260, 1993, p. 1610.

38. Johnson, C.M., Emerging viruses in context: an overview of viral hemorrhagic fevers, in *Emerging Viruses*, Morse, S.S., Ed., Oxford University Press, New York, 1993, p. 46.

39. Webster, R.G., Bean, W.J., Gorman, O.T., Chambers, T.M., and Kawaoka, Y., Evolution and ecology of influenza A viruses, *Microbiol. Rev.*, 56, 152, 1992.

40. Scholtissek, C. and Naylor, E., Fish farming and influenza pandemics, *Nature*, 331, 215, 1988.

41. World Health Organization, Geographical distribution of arthropod-borne diseases and their principal vectors. Geneva: World Health Organization (WHONBC/89.967), 1989, p. 138.

42. Kingsnorth, P., Human health on the line, *The Ecologist*, 29, 92, 1999.

43. Washam, C., Our planet is environmentally sick, *PlanetENN*, November 11, 1996.

44. Galntz, M.H., Katz, R.W., Nicholls, N., Eds., *Teleconnections Linking Worldwide-Climate Anomalies*, Cambridge University Press, New York, 1991.

45. Manabe, S. and Stouffer, R.J., Century-scale effects of increased atmospheric CO_2 on the ocean-atmosphere system, *Nature*, 364, 215, 1993.

46. Kluger, J., By any Measure, Earth is at the Tipping Point, *Time: Special Report on Global Warming*, April 3, 2006, pp. 32–42.

47. Harvell, C.D. et al., Emerging Marine-Diseases-Climate Links and Anthropogenic Factors, *Science*, 285, 1999, p. 1505.

48. Epstein, P.R., Ford, T.E., and Colwell, R.R., Marine ecosystems, *Lancet*, 342,1216, 1993.

49. Anderson, D.M., The Fifth International Conference on Toxic Marine Phytoplankton: A Personal Perspective, *Harmful Algae News*, Int Marine Science (UNESCO), Paris, 62 (Suppl. 6–7), 1993.

50. The Newshour with Jim Leber, An essay in the later half of the show on the troubled history of Washington, D.C., in the 20th century, February 17, 1995.

51. United Nations, World Urbanization Prospects, 1990, United Nations, New York, 1991.

52. Han, J.S., Short, M., Taylor, M.E. et al., Species-specific diversity among simian immunodeficiency viruses from African green monkeys, *J. Virol.*, 65, 2816, 1991.

53. Gan, F., Yne, L., White, A.T. et al., Human infection by genetically diverse SIVSM-related HIV-2 in West Africa, *Nature*, 358, 495, 1992.

54. McNeill, W.H., *Plagues and Peoples*, Anchor Press/Doubleday, New York, 1976.

55. Centers for Disease Control and Prevention, *Aedes albopictus* introduction into continental Africa, *MMWR*, 40, 836, 1991.

56. Moore, C.G., Francy, D.B., Eliason, D.A., and Monath, T.P., *Aedes albopictus* in the United States: rapid spread of a potential disease vector, *J. Am. Mosq. Control Assoc.*, 4, 356, 1991.

57. Mitchell, C.J., Niebylski, M.L., Smith, G.C. et al., Isolation of Eastern Equine Encephalitis Virus from *Aedes albopictus* in Florida, *Science*, 257, 1991, p. 526.

58. Wilson, M., Perspectives: travel and the emergence of infectious disease, *EID*, 1, 2, April–June 1995.

59. LeDuc, J.W., Childs, J.E., and Glass, G.E., The hantaviruses, etiologic agents of hemorrhagic fever with renal syndrome: a possible cause of hypertension and chronic renal disease in the United States, *Annu. Rev. Public Health*, 13, 79, 1992.

60. Siem, H. and Bollini, P., Migration and health in the 1990s, *Int. Migration*, 30, 1992.

61. Interaction: American Council for Voluntary International Action, Assistance to internally displaced persons: Issue Briefs/Facts and recommendations, Humanitarian Policy and Practice Committee, 2002, http://www.interaction.org/refugees/wpdisplaced.html.

62. Carlton, J.T. and Geller, J.B., Ecological Roulette: The Global Transport of Non-Indigenous Marine Organisms, *Science*, 261, 1992, p. 78.

63. World Health Organization, Cholera in the Americas, *Wkly. Epidemiol. Rec.*, 67, 33, 1992.

64. McCarthy, S.A., McPhearson, R.M., and Guarino, A.M., Toxigenic *Vibrio cholerae* and cargo ships entering Gulf of Mexico, *Lancet*, 339, 624, 1992.

65. Rarnamurthy, T., Garg, S., Sharma, R. et al., Emergence of novel strain of *Vibrio cholerae* with epidemic potential in southern and eastern India, *Lancet*, 341, 703, 1992.

66. DePaola, A., Capers, G.M., Moters, M.L. et al., Isolation of Latin American epidemic strain of *Vibrio cholerae 01* from US Gulf Coast, *Lancet*, 339, 624, 1992.

67. Russel, R.C., Survival of insects in the wheel bays of a Boeing 747B aircraft on flights between tropical and temperate airports, Bull. WHO, 65, 659, 1992.

68. Fidler, D., Perspectives: globalization, international law, and emerging infectious diseases, *EID*, 2, 2, April–June 1996.

69. Delbruck, J., Globalization of law, politics, and markets/implications for domestic law; a European perspective, *Ind. J. Glob. Leg. Stud.*, 1, 9, 1992.

70. Centers for Disease Control and Prevention, Addressing Emerging Infectious Disease Threats: A Prevention Strategy for the United States, Atlanta, U.S. Department of Health and Human Services, Public Health Service, 1994.

71. World Health Organization, Communicable Disease Prevention and Control: New, Emerging, and Re-Emerging Infectious Diseases, WHO Doc., A48/15, February 22, 1995.

72. Morse, S.S., Looking for a link, *Nature*, 344, 297, 1992.

73. Wilesmith, J.W., Ryan, J.B.M., and Atkinson, M.J., Bovine spongiform encephalopathy: epidemiological studies on the origin, *Vet. Rec.*, 128, 199, 1992.

74. Berkelman, R.L., Bryan, R.T., Osterholm, M.T., Leduc, J.W., and Hughes, J.M., Infectious Disease Surveillance: A Crumbling Foundation, *Science*, 264, 1992, p. 368.

75. Lederberg, J., Infectious disease as an evolutionary paradigm, *EID*, 3, 4, December 1997.

76. Davies, J., Inactivation of Antibiotics and the Dissemination of Resistance Genes, *Science,* 264, 1994, p. 375.
77. Bloona, B.R. and Murray, C.J.L., Tuberculosis: Commentary on a Re-Emergent Killer, *Science,* 257, 1994, p. 1055.
78. Neu, H.C., The Crisis in Antibiotic Resistance, *Science,* 257, 1994, p. 1064.
79. Davies, J., Inactivation of Antibiotics and the Dissemination of Resistance Genes, *Science,* 264, 1994, p. 375.
80. Domingo, E. and Holland, J.J., Mutation rates and rapid evolution of RNA viruses, in *The Evolutionary Biology of Viruses,* Morse, S.S., Ed., Raven Press, New York, 1994, pp. 161–184.
81. Kilbourne, E.D., The molecular epidemiology of influenza, *J. Infect. Dis.,* 127, 478, 1978.
82. Bureau of the Census, Mortality statistics, 1927, Part 1, Washington D.C., 1927, National Center for Health Statistics: Vital statistics of the United States, 1986, col. 11 — Mortality, part A, Washington, D.C., 1988.
83. Garret, L., *The Coming Plague,* Harper Collins Canada Ltd., New York, 1994, p. 445.
84. Jungkind, D.L., *Antimicrobial Resistance: A Crisis in Health Care. Advances in Experimental Medicine and Biology,* Plenum Press, New York, 1995, p. 33.
85. CDC, National Center for Infectious Diseases, Arctic Investigations Program, Methicillin-resistant *Staphylococcus aureus* (MRSA), July 2006, http://www.cdc.gov/ncidod/aip/research/mrsa.html#what_is_mrsa.
86. Barrett, F.F., McGhee, R.F., and Finland, M., Methicillin-resistant *Staphylococcus aureus* at Boston city hospital, *N. Engl. J. Med.,* 279, 448, 1968.
87. Herold, B.C., Immergluck, L.C., Maranan, M.C. et al., Community-acquired methicillin-resistant *Staphylococcus aureus* in children with no identifiable predisposing risk, *JAMA,* 279, 593–598, 1998.
88. Torassa, U., Antibiotic overuse a growing problem, September 28, 1999, http://www.Sfgate.com/cgi-bin/article.cgi?file=/examiner/hotnews/stories/29/animal.dtl.
89. Glass, R.I., Libel, M., and Brandling-Bennett, A.D., Epidemic Cholera in the Americas, *Science,* 265, 1992, p. 1524.
90. Moore, P.S., Meningococcal meningitis in sub-Saharan Africa: a model for the epidemic process, *Clin. Infect. Dis.,* 14, 515, 1992.
91. MacKenzie, W.R., Hoxie, N.J., Proctor, M.E. et al., A massive outbreak in Milwaukee of *Cryptosporidium* infection transmitted through the water supply, *N. Engl. J. Med.,* 331, 161, 1994.
92. Centers for Disease Control and Prevention, Assessment of inadequately filtered public drinking water Washington, D.C., December 1993, *MMWR,* 43, 661, 1994.
93. Ryan, M., They Track the Deadliest Viruses, *Parade,* April 23, 1995, p. 10.
94. Nichol, S.T., Spiropoulou, C.F., Morzunov, S. et al., Genetic Identification of a Hantavirus Associated with an Outbreak of Acute Respiratory Illness, *Science,* 262, 1995, p. 914.
95. Schmaljohn, A., Li, D., Negley, D.L., Bressler, D.S. et al., Isolation and initial characterization of a newfound hantavirus from California, *Virology,* 206, 963, 1995.
96. Hurt, H., A deadly new virus, *Self Magazine,* 88, April 1993.
97. Gajdusek, D., Virus hemorrhagic fevers, *J. Pediatr.,* 60, 841, 1962.
98. Lee, H.W., Epidemiology and pathogenesis of hemorrhagic fever with renal syndrome, in *The Bunyaviridae,* Elliott, R.M., Ed., Plenum Press, New York, 1996, pp. 253–267.
99. World Health Organization, Haemorrhagic fever with renal syndrome memorandum from a WHO meeting, Bull. WHO, 61, 269, 1983.

100. Schmaljohn, C. and Hjelle, B., Synopses hantaviruses: a global disease problem, *EID*, 3, 2, 95, April–June 1997.

101. Lee, H., Lee, P., and Johnson, K., Isolation of the etiologic agent of Korean hemor Thagic fever, *J. Infect. Dis.*, 137, 298, 1978.

102. French, G., Foulke, R., Brand, O., and Eddy, G., Korean Hemorrhagic Fever: Propagation of the Etiologic Agent in a Cell Line of Human Origin, *Science,* 21, 1981, p. 1046.

103. Epstein, P.R., Emerging diseases and ecosystem instability: new threats to public health, *Am. J. Public Health,* 80, 2, 16, 1995.

104. Wenzel, R.P., A new hantavirus infection in North America, *N. Engl. J. Med.,* 330,1004, 1994.

105. Dispatch, Dengue hemorrhagic fever: the emergence of a global health problem, *EID*, 1, 2, April–June 1995.

106. Gubler, D.J., Dengue, in *Epidemiology of Arthropod-Borne Viral Disease,* Monath, T.P.M., Ed., CRC Press, Boca Raton, FL, 1988, pp. 223–260.

107. Manning, A., Painful, Flu-Like Tropical Illness Can Be Deadly, *USA Today,* September 18, 1995, p. 1.

108. Henig, R.M., The new mosquito menace, *New York Times*, September 13, 1995, p. A21.

109. Gubler, D.J., Dengue haemorrhagic fever: a global update [Editorial], Virus, *Information Exchange News,* 8, 2, 1995.

110. Associated Press, TB, other disease reemerge to threaten millions in the Americas, September 27, 1999, http://my..../Go?template.html.

111. CDC, Division of Vector-Borne Infectious Diseases, Dengue and Dengue Hemorrhagic Fever: Information for Health Care Practitioners, June 30, 2005, http://www.cdc.gov/ncidod/dvbid/dengue/dengue-hcp.htm.

112. Gubler, D.J., Dengue and dengue haemorrhagic fever in the Americas, in WHO, Regional Office for SE Asia, New Delhi, Monograph, SEARO, 22, 1995, p. 9.

113. Rigau-Plkrez, J.G., Gubler, D.J., Vomdarn, A.V., and Clark, G.G., Dengue surveillance-United States, 1986–1992, *MMWR,* 43, 7, 1994.

114. Adler, J., The Fight Against the Flu, *Newsweek*, October 31, 2005, pp. 39–46.

115. World Health Organization (WHO), Cumulative Number of Confirmed Humans Cases of Avian Influenza A/(H5N1) Reported to WHO, July 26, 2006, http://www.who.int/csr/disease/avian_influenza/country/cases_table_2006_07_26/en/index.html.

116. U.S. Dept. of Health and Human Services, Pandemicflu.gov, August 2006, http://www.pandemicflu.gov/general/.

117. World Health Organization (WHO), Emerging and other communicable diseases, Influenza A(H5NI) Fact Sheet 188, January 9, 1999.

118. Centers for Disease Control (CDC), 1998, http://www.edc.gov/ncidod/diseases/flu/fluinfo.htm.

119. Commentary, Re-emergence of Ebola virus in Africa, *EID,* 1, 2, July–September 1995.

120. Troy, K., The Plague that Wasn't, *Newsweek,* December 9, 1996, p. 46.

121. Bowen, E.T.W., Platt, G.S., Lloyd, G. et. al., Viral haemorrhagic fever in southern Sudan and northern Zaire: preliminary studies on the aetiologic agent, *Lancet,* 1, 571, 1977.

122. Jahrling, P.B., Geisbert, T.W., Dalgard, D.W. et al., Preliminary report: isolation of Ebola virus from monkeys imported to USA, *Lancet,* 335, 502, 1977.

123. Allela, L., Bourry, O., Pouillot, R., Délicat, A., Yaba, P., Kumulungui, B. et al., Ebola virus antibody prevalence in dogs and human risk, *EID* [serial on the Internet], March 2005 [date cited], available from http://www.cdc.gov/ncidod/EID/vol11no03/04-0981.htm.

124. Clover, C. and Highfield, R., AIDS Started by Humans Eating Chimps..., *The Daily Telegraph (London)*, Monday, February 1, 1999.
125. Jaeger, P., 1998, http://jaeger.mnith-giessen.de:6423/aids.html.
126. National Institute of Allergy and Infectious Diseases (NIAID), HIV infection and AIDS, US Department of Health and Human Services, Public Health Service, May 1997, http://www.aegis.com/topics/whataidsis.html.
127. UNAIDS and WHO, Report on the Global HIV/AIDS epidemic, December 1997, http://www.unaids.org/unaids/document/epideinio/report97.html.
128. Gerlach, D., AIDS: 25 years later, *Newsweek*, May 15, 2006, p. 41.
129. CDC, Division of Vector-Borne Infectious diseases, West Nile Virus: Background: Virus History and Distribution, April 6, 2004, http://www.cdc.gov/ncidod/dvbid/westnile/background.htm.
130. CDC, Division of Vector-Borne Infectious diseases, West Nile Virus: Questions and Answers, August 16, 2004, http://www.cdc.gov/ncidod/dvbid/westnile/qa/symptoms.htm.
131. CDC, Division of Vector-Borne Infectious diseases, West Nile Virus: Statistics, Surveillance and Control, July 25, 2006, http://www.cdc.gov/ncidod/dvbid/westnile/surv&controlCaseCount06_detailed.htm.
132. Li'eng, P., Synopsis *Escherichia coli* serotype O157:H7: novel vehicles of infection and emergence of phenotypic variants, *EID*, 1, 2, April–June 1995.
133. Griffin, P.M. and Tauxe, P.V., The epidemiology of infections caused by *Escherichia coli* O157:H7, other enterohemorrhagic *E. coli*, and the associated hemolytic uremic syndrome, *Epidemiol. Rev.*, 13, 60, 1991.
134. Karmali, M.A., Infection by Verocytotoxin-producing *Escherichia coli*, *Clin. Microbiol. Rev.*, 2, 15, 1989.
135. Centers for Disease Control and Prevention, Update: multistate outbreak of *Escherichia coli* O157:H7 infections from hamburgers-western United States, 1992–1993, *MMWR*, 42, 258, 1993.
136. Besser, R.E., Lett, S.M., Weber, J.T. et al., An outbreak of diarrhea and hemolytic uremic syndrome from *Escherichia coli* O157:H7 in fresh-pressed apple cider, *JAMA*, 269, 2217,1993.
137. Zhao, T., Doyle, M.P., and Besser, R.E., Fate of enterohemorrhagic *Escherichia coli* O157:H7 in apple cider with and without preservatives, *Appl. Environ. Microbiol.*, 59, 2526, 1993.
138. Weagant, S.D., Bryant, J.L., and Bark, D.H., Survival of *Escherichia coli O157:H7* in mayonnaise and mayonnaise-based sauces at room and refrigerated temperatures, *J. Food. Prot.*, 57, 629, 1994.
139. Swerdlow, D.L., Woodruff, B.A., Brady, R.C. et al., A waterborne outbreak in Missouri of *Escherichia coli* O157:H7 associated with bloody diarrhea and death, *Ann. Intern. Med.*, 117, 812, 1992.
140. Peisman, J., Schreifer, M.E., and Burkot, T.R., Ability of experimentally infected chickens to infect ticks with the Lyme disease sprichete, *Borrelia burgdorferi, Am. J. Trop. Hyg.*, 54, 294, 1996.
141. Levin, M., Levine, J.F., Howard, P., and Apperson, C.S., Reservoir competence of the southeastern five-lined skink *(Eumeces inexpectatus)* and the green anole *(Anolis carolen*sis) for *Borrelia burgdorteri, Am. J. Trop. Med. Hyg.*, 54, 92, 1996.
142. American Lyme Disease Foundation (ALDF), Reported Lyme Disease, 2006, http://www.aldf.com/usmap.shtml.
143. Manning, A., Lyme-causing tick carries worse disease, *USA Today*, September 18, 1995, p. 1.

144. Colburn, D., The clock is TICK-ing, New Lyme vaccine effectively sterilizes the blood-sucking tick, Daily Hampshire Gazette, Tuesday, July 6, 1999.
145. Cano, R.J. and Colonic, J.S., Disease of the respiratory tract, in *Microbiology,* West Publishing Co., MN, 1986, chap. 24.
146. Stevens, D.L., Invasive group A streptococcus infections, *Clin. Infect. Dis.,* 14, 2, 1992.
147. Demers, B., Shno, A.E., Vellend, H. et al., Severe invasive group A streptococcal infections in Ontario, Canada: 1987–1991, *Clin. Infect. Dis.,* 16, 792, 1993.
148. Seachrist, L., The Once and Future Scourge, *Science News,* 148, October 1996, p. 234.
149. Stevens, D.L., Invasive group A streptococcal infections: the past, present and future, *Pediatr. Infect. Dis. J.,* 13, 561, 1994.
150. Hayden, T., Tuberculosis is Making a Comeback, *Newsweek*, November 8, 1999.
151. National Prevention Information Network (NPIN), TB Today, 2006, http://www.cdcnpin.org/scripts/tb/tb.asp.
152. Meyer, T., U.S., Mexico Must Fight Disease That Knows No Boundaries, *Boston Globe,* 1, 15, November 30, 1996.
153. Nowak, R., WHO Calls for Action against TB, *Science,* 267, March 24, 1995, p. 1763.
154. Starke, J.R. and Correa, A.G., Management of mycobacterial infection and disease in children, *Pediatr. Infect. Dis. J.,* 14, 455, 1995.
155. Kochi, A., The global tuberculosis situation and the new control strategy of the World Health Organization, *Tuberc. Lung Dis.,* 72, 1, 1991.
156. Raviglione, M.C., Snider, D.E., and Kochi, A., Global epidemiology of tuberculosis: morbidity and mortality of a worldwide epidemic, *JAMA,* 273, 220, 1995.
157. McNeil, D., South Africa SLow to Abate Spread of Tuberculosis, *New York Times,* October 13, 1996.
158. Sizemore, C.F. and Laughon, B., and Fauci, A.S., World TB Today, March 24, 2006. National Institute of Allergy and Infectious Diseases (NIAID), March 24, 2006, http://www3.niaid.nih.gov/about/directors/news/tb_06.htm.
159. Miller, J., Study says new TB strains need an intensive strategy, *New York Times,* October 28, 1999, p. A6.
160. DuPont, H.L., Chappell, C.L., Sterling, C.R. et al., The infectivity of *Cryptosporidium parvum* in healthy volunteers, *N. Engl. J. Med.,* 332, 855, 1995.
161. Centers for Disease Control and Prevention (CDC), *Cryptosporidium and Water: A Public Health Handbook,* NCID, Mailstop F-22 4770 Buford Highway N.E., Atlanta, GA 30341–3724, 1997, 134 pp.
162. Newman, A., EPA considering quick action on Cryptosporidium, *Environmental Science and Technology,* 29, 1, January 1995, p. 17A.
163. Newman, R.D., Zu, S.-X., Wuhib, T. et al., Household epidemiology of *Cryptosporidium parvum* infection, *Ann. Intern. Med.,* 120, 500, 1994.
164. Nime, F.A., Burek, J.D., Page, D.L. et al., Acute enterocolitis in a human being infected with the protozoan *Cryptosporidium, Gastroenterology,* 70, 592, 1994.
165. Guerrant, R., Synopses cryptosporidiosis: an emerging highly infectious threat, *EID,* 3, 1, January–March, 1997.
166. Zu, S.-X., Li, J.-F., Barrett, L.J. et al., Seroepidemiologic study of cryptosporidiurn infection in children from rural communities of Anhui, China, *Am. J. Trop. Med. Hyg.,* 51, 1, 1994.
167. Ungar, B.L., Gilman, R.H., Lanata, C.F., and Peres-Schael, L., Seroepidemiology of Cryptosporidium infection in two Latin American populations, *J. Infect. Dis.,* 157, 551, 1988.

168. Ungar, B.L.P., Soave, R., Fayer, R., and Nash, T.E., Enzyme immunoassay detection of immunoglobulin M and G antibodies to Cryptosporidium in immunocompetent and immunocompromised persons, *J. Infect. Dis.,* 153, 570, 1986.
169. Zucker, J.R., Changing patterns of autochthonous malaria transmission in the United States: a review of recent outbreaks, *EID,* 2, 1, January–March, 1996.
170. World Health Organization, Control of tropical Diseases (CTD): Malarial Control, Geneva, Switzerland: WHO Office of Information, 1995.
171. Murray, C.J.L., Quantifying the burden of disease; the technical basis for disability-adjusted life years, Bull. *WHO,* 72, 429, 1994.
172. Roberts, D., Laughlin, L., Hsheih, P., and Legters, J., DDT, global strategies and a malaria control crisis in South America, *EID,* 3, 3, July–September, 1997.
173. Pan American Health Organization, Status of Malaria Programs in the Americas, XLIV Report, Washington, D.C., PAHO, 1996.
174. Angier, N., Malaria's Genetic Game of Cloak and Dagger, *New York Times,* August 22, 1995, p. C1.
175. Doll, R., Health and the environment in the 1990s, *Am. J. Public Health,* 82, 933, 1992.
176. Martins, W., Niessen, L., Rotmans, J., Jetten, T., and McMichael, A., Potential impact of global climate change on malaria risk, *Environ. Health Perspect.,* 103, 5, 458, 1995.
177. Pan American Health Organization, Report for Registration of Malaria Eradication from United States of America, Washington, D.C., December 1969.
178. Eliades, M.J., Snehal, S., and Nguyen-Dinh, P. et al., Malaria Surveillance — United States, *MMWR,* June 3, 2005/54(SS02), 25–39, 2003, http://www.cdc.gov/mmwr/preview/mmwrhtml/ss5402a2.htm.
179. WHO, WHO reports on new, re-emerging diseases threatening world health, *The Nation's Health,* Vol. 24, November 1995.
180. Ewald, P.W., Guarding against the most dangerous emerging pathogens: insights from evolutionary biology, *EID,* 2, 4, October–December, 1996.

8 Foodborne Illness

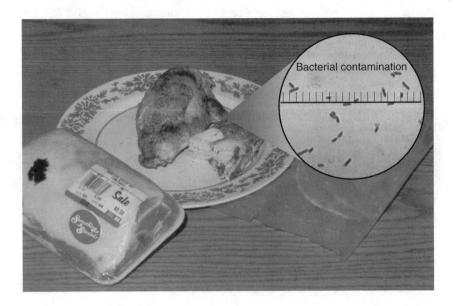

Bacterial contamination

CHAPTER OBJECTIVES

A student, after reading this chapter, will be able to:

1. Recognize, list, and explain the major objectives of food protection programs.
2. List and describe the major categories and subcategories of agents causing foodborne illness.
3. Describe the major foodborne pathogens, including parasitic, viral, and bacterial organisms.
4. Explain the mechanisms by which these pathogens cause foodborne illness, and describe how the life cycles of these organisms are important in this transmission of disease.
5. List and describe the major disease symptoms in humans for these foodborne pathogens.
6. Describe and explain the HACCP system for protection against foodborne disease.
7. Discuss recent regulatory efforts in the area of food protection.

INTRODUCTION

Martin Molina is a health promoter in the small village of El Achiotal located on the Pacific coast of El Salvador. Like most evenings, all lights were off by 8 p.m. and most people had settled quietly in their hammocks, a custom originating from the civil war that had just ended. One night Martin was awoken by a knock at the door; he remained silent and listened, as the knock grew frantic and more rapid. Finally, a familiar voice called his name; it was Mrs. Cruz, an elderly mother of six. Martin was relieved, as in his childhood he had heard stories of those who had answered the door in the night: they were usually recruited immediately into the military or disappeared at the hand of the infamous death squads.

An upset Mrs. Cruz explained to Martin that her son had been vomiting with profuse diarrhea for several hours, a scenario that was all too familiar to Martin. Martin grabbed his small bag, which he constantly kept packed for emergencies and went to see little Santos Cruz, the fifth patient of the day with these symptoms.

Martin had fled to the U.S. when the civil war broke out, and like many of his countrymen, he had managed to return to El Salvador with enough money to buy some property, build a house, and even finance a high school education. Martin did not know why he had returned to his small village with no potable water, no human waste facilities, and an electric current so weak it could power little more than a light bulb. Martin did enjoy the prestige of being the most educated person in the village, and found working as a health worker extremely rewarding.

When Martin arrived at Santos's side, he was naked, dry-heaving into an old bucket, and complaining of stomach cramps. Martin started to prepare a rehydration solution using a powder concentrate and water from a nearby well. Mrs. Cruz explained to Martin that she had been giving Santos the same solution, which she had received at the community meeting concerning the sudden increase of sickness in the village, but Santos always vomited it up. Martin insisted that Santos try to drink the mix because he was severely dehydrated and was sure to only get worse. As soon as Santos swallowed a cup full, he vomited on Martin's feet, who looked down and realized that he was standing in what appeared to be diarrhea from little Santos.

Mrs. Cruz saw the frustration in Martin's face and began to cry, fearing that Santos, like so many other children of the village, would not recover. Martin looked up at the rising moon, aware his village was experiencing an outbreak of some sort, aware, like so many times before, that the origin of the outbreak would remain unknown, and aware that he was ill equipped to remediate an outbreak of this magnitude. Was it the food, the water he was using to mix the rehydrating solution, or some thing else of which he was unaware? Martin looked back down to his feet, thought about the nice apartment in New York which he had shared with four illegal aliens, and wondered what life would have been like if he had not returned.

—Christopher Landry, Peace Corps Volunteer in El Salvador, 1994–1996

Worldwide Distribution of Foodborne Pathogens

Worldwide, 1.5 billion children under the age of 5 suffer from diarrhea and, tragically, over 2 million die as a consequence. Most of these cases are due to foodborne illness, which puts in perspective the severity of this issue.[1] Every year an estimated 76 million cases of foodborne illness and 5000 associated deaths occur in the U.S.[2,3–7] Foodborne illness is a major health and economic problem in the U.S. and other countries. In the U.S., surveillance systems have been put in place to monitor the incidence and impact of important foodborne pathogens such as *Campylobacter jejuni, Clostridium perfringens, Escherichia coli* O157:H7, *Salmonella* spp., and *Staphylococcus aureus*. Internationally, it is very difficult to accurately assess the prevalence of foodborne disease because of the differences in existing surveillance systems, if they exist at all. On an international level, cholera is the only foodborne disease monitored by international health regulations. When comparing statistics among countries, differences in disease classification often make it difficult to find data on specific diseases. For example, the same symptom may be classified under "amebiasis" or "dysentery." The list of notifiable diseases also varies among countries, and whereas some countries report all identified cases, others report only those associated with outbreaks, leaving the sporadic cases unaccounted for. Lesser-developed countries (LDCs) lack the infrastructure and resources necessary to monitor many of the causative agents of foodborne disease, and it may be in the best interest of some governments to minimize reports on foodborne illness to protect the food export and tourism industries. Unfortunately, it is these LDCs that suffer most from foodborne illness.[2]

In spite of these difficulties, organizations such as the World Health Organization (WHO) have attempted to work around the barriers of estimating the incidence of foodborne disease to provide global estimates. When preparing such estimates, certain principles apply. When severe diseases such as botulism are reported, one can assume that the degree of underreporting is small and reported numbers can be considered close to the actual figures. When symptoms of disease are mild or short in duration, such as those caused by *Staphylococcus aureus*, investigators can expect a higher rate of underreporting, which is factored in during the estimation of the disease. When there are little or no acceptable data available, researchers base their estimates on the incidence of disease in infants, or by extrapolating data from similar regions where such information may be available.[1,3]

Differences in the prevalence of foodborne illness among the various regions of the world are due to the following:

- Climate influences life cycles and growth patterns of foodborne pathogens. A colder climate may hinder the growth of certain pathogens; a warmer climate will promote the spread of some pathogens.
- Populations with greater numbers of elderly persons or infants, who are more susceptible to disease, will have higher rates of illness.

- The nutritional status of a culture plays an integral part in determining the health status of a population. People with healthy diets will most likely possess immune systems capable of fending off potential health threats.
- Cultural aspects such as the knowledge, attitudes, stress (such as illness, loss of a significant other, or poor living conditions), and practices of individuals may be most important in determining the risk factors for disease. For example, African indigenous populations retain the custom of eating raw meat or pork, which puts them at high risk of consuming viable *Trichinella spiralis* (roundworm) or *Taenia* spp. (tapeworm) organisms.[4,5]

REASONS FOR FOOD PROTECTION PROGRAMS

The implementation of programs to minimize foodborne diseases is important because of the problems associated with morbidity, mortality, and economic loss. Foodborne diseases escalate health care costs, as those who are ill access the health system for treatment, and they impose an enormous economic burden on related industries that in turn impacts the communities dependent on these industries.

Morbidity and Mortality due to Foodborne Disease

As mentioned earlier, the morbidity (sickness) and mortality (death) rates from foodborne disease are nearly impossible to estimate due to the inadequacies in surveillance and reporting systems. Even though the U.S. has a relatively safe food supply, the Centers for Disease Control and Prevention (CDC) shows there are still 76 million cases of foodborne illness each year (as compiled using death certificates, hospital surveys, and academic studies).[2] Although in the U.S. the accepted number of deaths due to foodborne illness is 5000 annually,[2,7,8] it is difficult to determine how accurate this figure is; foodborne illnesses are often not reported, because people affected do not believe their illness is caused by food and states do not often have the resources to look into all foodborne illness cases.[7] Of the food-related outbreaks in the U.S., the origin of half remains unknown. The causative agents and modes of transmission (means through which an causative agent is spread) are known in less than 1% of the severe gastroenteritis cases, which leaves the true impact on society in doubt.[5] Regardless of the exact numbers, symptoms from foodborne illness range from nausea and mild diarrhea, to severe dysentery and death. Sickness can be self-limiting, as is illness from cryptosporidium, or life-threatening from complications, as seen with *E. coli* O157:H7. The onset (time lapse between exposure and the expression of symptoms) of illness can be as quick as half an hour, or take years, as heavy metals accumulate in the body until they reach toxic levels.

Economic Consequences of Foodborne Illness

An outbreak of foodborne illness is almost always accompanied by a cost that reflects the magnitude of the outbreak. Medical costs incurred during the treatment of an illness and the loss of wages for those who must stay home from work are examples of expenses encountered by individuals suffering from a foodborne illness. If a food

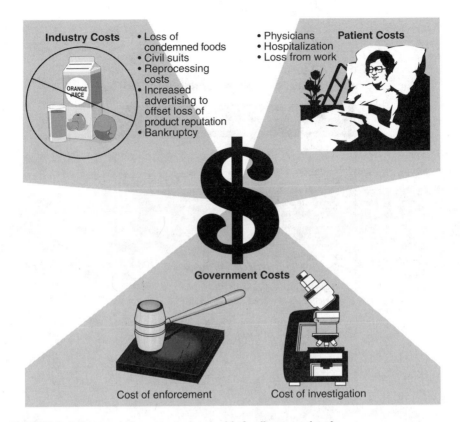

FIGURE 8.1 Economic losses associated with foodborne outbreaks.

product is identified as having been responsible for illness, the industries responsible for producing, processing, or distributing the foodstuff would suffer losses totaling millions, as they would have to halt production, recall the contaminated product for disposal, lose sales, and increase marketing efforts to compensate for losses. Other costs would result from investigations into the source of contamination, increased quality assurance measures to produce safer products, litigation fees from lawsuits, and the loss of trade (Figure 8.1).[6]

CAUSATIVE AGENTS OF FOODBORNE DISEASE

Foodborne illness is defined as any illness due to the consumption of contaminated food. Food contaminants come from a wide variety of sources. Certain plants and animals produce substances that are toxic to humans, chemicals such as arsenic and lead may contaminate food through agricultural or industrial processes, and radionuclides from nuclear fallout are present in food products cultivated at or near testing sites. Parasites, viruses, fungi, and bacteria are all examples of living organisms that have the potential to contaminate a food supply; these agents are termed *foodborne pathogens* (Table 8.1).[5]

TABLE 8.1
Some Causative Agents of Foodborne Disease

SOME CAUSATIVE AGENTS OF FOODBORNE ILLNESS
Chemicals
Antimony
Cadmium
Lead
Mercury
Polychlorinated biphenyls
Packaging materials
Pesticides
Industrial processes
Food Additives
GRAS (Generally Recognized as Safe)
Saccharin
Monosodium glutamate
Nitrates and nitrites
Color additives
Poisonous Plants and Animals
Plant Sources
Alkaloids
Lectins
Saponins
Glucosinolates
Mushroom poisoning
Animal Sources
Paralytic shellfish poisoning
Ciguatera poisoning
Pufferfish poisoning
Foodborne Pathogens
Parasites
Nematodes
Trichinella spiralis, Tapeworms
Protozoans
E. histolytica, G. lamblia, Cryptosporidium
Viruses
Hepatitis, Norwalk-type
Fungi
Aspergillus spp., Penicillium spp., Mucor spp.
Bacteria
Toxin producing
C. botulinum
Staphylococcus aureus
Vibrio cholerae
Infectious
Salmonella, Campylobacter, E. coli

RADIONUCLIDES

Radiation is introduced into the food chain naturally from mineral deposits beneath the earth's surface or from the atmosphere in the form of ultraviolet and cosmic rays. Radiation from these sources is emitted at low levels that plant and animal life can tolerate. Radionuclides, which are deposited in the environment accidentally, or intentionally, as a direct result of human activities are of much greater concern. The levels that result from these activities are tens to hundreds of times higher than those normally encountered by plant and animal life. Exposure to radiation at high levels has the potential to cause irreversible damage to whole organ systems, genetic mutations or deformities in offspring, and if the central nervous system is damaged, symptoms of disorientation, convulsions, and lack of muscle coordination are experienced.

The fallout from the Chernobyl accident of 1986 deposited cesium 137 (^{137}Cs) and iodine 131 (^{131}I) throughout the Northern Hemisphere. The isotopes ^{137}Cs and ^{131}I are absorbed by plants through their roots and foliage, which are in turn used as feed for farm animals. Humans ingest these radionuclides directly from the consumption of fruits and vegetables, or indirectly by consuming meat and diary products.[6] Concern has been recently expressed that the concrete vault meant to contain the Chernobyl radioactivity is crumbling and additional radioactive leakage is imminent. A similar scenario arises from nuclear weapons testing, and although we are aware of the dangers that arise from fallout, nuclear weapons testing continues to be a concern currently. Recently, India and Pakistan conducted a series of nuclear weapons tests in an attempt to achieve the upper hand in an arms race.[6] Although the nuclear testing sites are underground and usually in remote areas where there is little or no agricultural activity, radioactive plumes may travel across national boundaries and fallout could still pose a significant threat to human populations.

CHEMICALS

Ironically, humans are responsible for many chemical contaminants presently found in food. Between 80 and 90% of our exposure to potentially harmful chemicals is from food consumption.[3] Chemicals enter the food from packaging materials, agricultural applications of pesticides and fertilizers, by addition of preservatives or coloring agents to foods, or by the release of industrial chemicals into the environment. Many of these chemicals have become ubiquitous. Table 8.2 lists some of the many chemicals that are found in foods. These chemicals largely originate from human processes.

Packaging Materials

Chemical poisoning from packaging materials can cause illness and death within 30 min. Substances such as plasticizers, stabilizers, and inks can migrate from the packaging to the food, causing illness.[5] Exposure to these substances is usually very low, and the chemicals tend to be nontoxic. In the U.S. commercial packaging materials are tested for chemical migration and can be considered safe. There are a few chemical contaminants of this nature that are worth mentioning.

TABLE 8.2
Some of the Many Chemicals Originating
from Human Processes That May Be Found
in Food

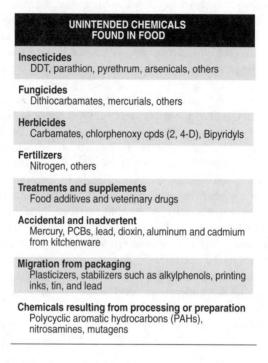

UNINTENDED CHEMICALS FOUND IN FOOD
Insecticides DDT, parathion, pyrethrum, arsenicals, others
Fungicides Dithiocarbamates, mercurials, others
Herbicides Carbamates, chlorphenoxy cpds (2, 4-D), Bipyridyls
Fertilizers Nitrogen, others
Treatments and supplements Food additives and veterinary drugs
Accidental and inadvertent Mercury, PCBs, lead, dioxin, aluminum and cadmium from kitchenware
Migration from packaging Plasticizers, stabilizers such as alkylphenols, printing inks, tin, and lead
Chemicals resulting from processing or preparation Polycyclic aromatic hydrocarbons (PAHs), nitrosamines, mutagens

Antimony

Antimony leaches into foods stored in chipped enamel containers. Acidic foods such as lemon juice cause antimony to leach out into the product. The ingestion of antimony can cause complications of the gastrointestinal, cardiovascular, and hepatic systems.

Cadmium

Cadmium, which is used as a plating material for trays and containers, is also leached into food from containers by acidic conditions. Chronic exposure to cadmium can lead to kidney damage. Cadmium accumulates in both the liver and the kidneys, but in the liver cadmium is tied up by metallothionein, whereas in the kidney, metallothionein is broken down by lysosomes, which allows the unbound cadmium to accumulate.

Lead

Apart from contaminating food through pesticide application, lead is present in the solder of metal containers and cans. Again, acidic foods will cause exposed lead to leach into the food. But in canning there is a coating that prevents the food from coming in contact with the metal. Lead consumption causes neurological complications, kidney failure, and weakens bone integrity. For this reason, lead is being phased out of use in the industry. Other options are being explored.

Industrial Processes

Traces of chemicals from industrial processes are often found in foods. Levels are usually extremely low and, therefore, they go unnoticed. Occasionally, hazardous quantities are dumped into the environment by accident, carelessness, or with malicious intent, causing foodborne illness.

Mercury

Mercury is a by-product of many industrial processes. Chronic exposure to mercury presents as an emotional or psychological disorder. Methyl mercury, however, is an acute toxin that causes tremors, neurological complications, kidney failure, and birth defects. Methyl mercury is the primary form of mercury found on foods.

In Minamata Bay, Japan, industries were dumping waste contaminated with inorganic mercury into the bay. Bacteria in the bay converted the inorganic mercury to methyl mercury, causing over 1200 cases of food poisoning from the consumption of contaminated fish.

Mercury is also a component of fungicides used on animal feed. Animals consume this feed, and mercury enters the bodies of humans when these animals are slaughtered for consumption. Consuming meats and fish containing high levels of mercury can also cause blindness, paralysis, or death.[5]

Polychlorinated Biphenyls

Polychlorinated biphenyls (PCBs) are widely used in industry. They are extremely stable compounds that do not degrade easily, are resistant to heat, and are also highly toxic. In Japan, in 1968, 1000 people suffered food poisoning from PCB contamination in rice oil. During the rice oil process, a heating pipe with a small hole allowed PCBs to contaminate the rice oil. Symptoms included swelling of the eyes, rash, and gastrointestinal illness, which resulted in five deaths. Alternatives to PCBs are being researched, but they are so widespread today that we can expect to see trace amounts for many years to come.

Dioxin

Another chemical that is turning up in foods is dioxin. In early 1999, high levels of dioxin were found in Belgian cattle. The government admitted that it does not inspect cattle that are raised near industrial sites (a common source of dioxin, along with incinerators), nor do they have requirements to measure dioxin levels in beef. Following this discovery, new efforts are being made in Belgium and other industrialized nations to keep dioxin-contaminated beef out of the global market.[10]

Pesticides

Pesticides are introduced into the environment through the air, water, and soil. Although they are used to control insects, fungi, and rodents, they may reduce crop yield. Organochlorine compounds (such as DDT and chlordane, organophosphates (such as parathion and malathion), and inorganic compounds (such as arsenics) have been applied to food in the form of a pesticide. DDT was found to interfere with the reproductive cycle of some birds and was banned in 1972. Since then, the EPA has banned aldrin, dieldrin, heptachlor, and kepone. Yet, traces of these compounds and their metabolites continue to be found in our food. Many Americans are eating imported foods that still contain these potentially dangerous residues. Many of these chemicals banned from use in the U.S. are sold to developing nations, which use them extensively in producing crops for export to the American market.

The characteristics of some pesticides make them especially dangerous. DDT and other chemicals of its class accumulate in the environment. This is why we still find trace amounts of residue in foods produced in the U.S., especially shellfish. In response to this, pesticides that break down more readily were formulated. Because these pesticides spend less time in the environment, they tend to be more toxic. For example, parathion is a threat to farm workers because it can be absorbed through the skin. Carbamate compounds inhibit cholinesterase, which is an enzyme essential for transmitting nerve impulses at the synapses. The EPA has proposed tolerance levels for pesticides applied to food products. Tolerance levels are the maximum permitted amounts of pesticides that can be found on a raw product. Although these efforts are well intentioned, only a small portion of imported products entering the U.S. is tested for pesticides.[10,11] Researchers analyzed data collected by the U.S. Department of Agriculture's Pesticide Data Program (PDP) to compare the relative amounts and toxicities of pesticide residues in different foods. They obtained pesticide residue data on over 27,000 food samples tested by the PDP from 1994 to 1997. The researchers weighted the amounts of residues present to account for differences in the toxicity of individual pesticide chemicals, and computed a toxicity index (TI) for each food. The researchers calculated TI-integrated measures of the frequency of pesticide detection, the levels of residues present, and the relative toxicity of the detected residues, yielding an index of the relative toxicity loading of each food.[12] The researchers suggested that values greater than 100 on the TI scale showed comparatively high pesticide contamination, and those less than 10 indicated foods that are comparatively quite "clean." Fresh peaches (both domestic and imported); frozen and fresh winter squash grown in the U.S.; domestic and imported apples, grapes, spinach and pears; and U.S.-grown green beans were among the foods showing the highest TI values, and 11 of the 12 highest TI scores were for U.S.-grown foods.[12]

FOOD ADDITIVES

Food additives are used to alter taste, color, texture, nutritive value, appearance, and resistance to deterioration. There are thousands of food additives, ranging from simple compounds such as salt to more complex compounds such as monosodium glutamate. Food additives are also considered to be the least hazardous source of foodborne illness, ranking behind pesticides, environmental contaminants, natural toxins, and microbial toxins.

In the early part of the 20th century, increased knowledge of the toxic effects from long-term chemical exposure prompted the passage of the Food and Drug Act of 1906, to which several amendments have been added. In 1958, the Food Additive Amendment to the Food Drug and Cosmetic Act required the manufacturer of any new potential food additive to conduct testing and achieve FDA approval before marketing. Not included in this amendment are "generally recognized as safe" (GRAS) substances, and color additives, which are covered under the Color Additive Amendment of 1950.[10]

Saccharin

Saccharin is a low-calorie sweetener that has been approved by the FDA but still remains very much under scrutiny. In laboratory animals, saccharin promotes bladder cancer after exposure to a bladder carcinogen. In 1977, the FDA proposed a ban on saccharin as a food additive. However, the ban was postponed by Congress pending additional studies to support the results. Saccharin was not removed under the Delaney Clause because various studies have reported conflicting results. A considerably high amount of saccharin is needed to produce the observed effects. It has been speculated that sodium, not saccharin, may be responsible for the bladder cancer.[10]

Monosodium Glutamate

Monosodium glutamate (MSG) is used to enhance flavor in Chinese foods. It is also thought to be the cause of Chinese restaurant syndrome (headaches and possible nausea) and lesions of the retina. An allowable daily intake (ADI) of 120 mg/kg has been established for individuals over 1 year of age, and manufacturers in the U.S. have voluntarily halted the use of MSG in infant foods. The role of MSG in Chinese restaurant syndrome remains unclear, but recent studies indicate that MSG might not be responsible for its development.[19]

Nitrates and Nitrites

Nitrates and nitrites are used to cure ham, bacon, and other meats. They give meats their characteristic taste and color while preventing the growth of *Clostridium botulinum* spores. Nitrate and nitrites can also be found in vegetables, water, saliva, and tobacco. In the body, nitrates can be reduced to nitrites, which in turn oxidize hemoglobin into methemoglobin. If methemoglobin accumulates in the blood, oxygen cannot be transported and anoxia will result.

In food, nitrites react with amines to form nitrosamines. This is of concern because nitrosamines have caused cancer of the liver, kidney, bladder, stomach, and pancreas of laboratory animals. Because nitrosamines are not added to foods, they are not excluded by the Delaney Clause. The FDA has lowered the amount of nitrites permitted to be added to meats, and has suggested the addition of ascorbate or sodium erythorbate to prevent the formation of nitrosamines.

GRAS (Generally Recognized As Safe)

GRAS substances are chemicals that had a history of safe use before the 1958 Food Additive Amendment passed. There are approximately 700 GRAS substances. Currently, the FDA is reviewing the safety of these substances and reclassifying them if necessary. One other piece of legislation worth mentioning is the Delaney Clause, which is included in the Food Additive and Color Additive Amendments of the Food Drug and Cosmetic Act. The Delaney Clause prohibits the use of any food additive found to induce cancer in laboratory animals under appropriate conditions. This clause has been challenged with much resistance from manufacturers because of the controversy surrounding the extrapolation of results from laboratory animals to humans.[10]

Color Additives

Colors are added to foods such as candy, cakes, cereal, ice cream, and soft drinks to give them a color that is more appealing to the consumer. Natural colors such as grape skin extract, caramel, and fruit juices did not have to undergo certification after the Color Additive Amendment of 1960 passed. All synthetic colorings, however, required certification if there were any questions as to their safety, even if the substance had been previously certified. There are seven synthetic colorings that are certified for unlimited use according to good manufacturing practices (GMPs). They are FD&C Blue No. 1 and 7, Red No. 3 and 40, Yellow No. 5 and 6, and Green No. 3. Other colors such as Green No. 1 and Red No. 2 were removed from use because of concerns over their carcinogenicity and other effects from chronic exposure.[10]

POISONOUS PLANTS AND ANIMALS

Although there are numerous species of plants and animals that are poisonous, the majority are not part of our diet. By the process of trial and error, humans have identified plants that are either harmful to humans or possess little nutritional value and excluded them from our diet. Some plants and animals known to be harmful may still part of our diet. Plants such as castor bean, alfalfa, radish, and some mushrooms produce substances that could be harmful if consumed. Currently, there are 1200 species of marine organisms that produce toxins. Fortunately, only a few are important in foodborne illness. Toxic plankton are consumed by shellfish and other marine animals, which in turn are harvested for human consumption. Likewise, a species of the puffer fish, which is important in China and Japan, also produces a toxin. In some instances, the toxic substance is inactivated by properly cooking the food (puffer fish, lectins). However, many plants and animals are served raw, leaving the consumer defenseless against any toxins that may be present in the food.

Plant Sources

Alkaloids

Alkaloids are part of a plant's natural defense. These nitrogen-containing organic compounds ward off predators such as insects and other herbivores that might consume a portion of the plant. Three major groups of alkaloids important to humans are the following:

1. The pyrrolizidine group of alkaloids are found in herbs such as coltsfoot, comfrey, and petasite. These are used in home remedies. Humans consume these herbs in the form of teas and other herbal concoctions, which are a common source of these alkaloids. Pyrrolizidine alkaloids cause extensive damage to the liver and lungs, and are associated with a high mortality rate among children.
2. Solanum alkaloids cause diarrhea, headaches, vomiting, and neurological problems in humans who have consumed green, sprouted, stressed, or storage-abused potatoes. Solanum can withstand normal preparatory procedures such as baking, boiling, or microwaving. It was once thought that solanum played a role in anencephaly and spina bifida (ASB), but research has refuted any such hypothesis.
3. Xanthine alkaloids are found in coffees, teas, and many other beverages in the form of caffeine. The caffeine found in ten cups of coffee (0.5 to 1.0 g) is enough to produce adverse effects in adults, and child mortality is possible after the ingestion of 5 g of caffeine.

Lectins

Lectins are plant proteins that agglutinate red blood cells. More than 800 species of edible plants contain lectins, and over 600 of those species belong to the Leguminosae family. Ricin, found in the castor bean, is a highly toxic lectin. Effects in humans vary from stunted growth to death, according to the source and dose of the lectin. Lectins are destroyed by cooking in wet heat.

Saponins

Saponins are glycosides that hemolyze red blood cells. When alfalfa, which has a high concentration of saponins, is fed to laboratory animals, they suffer from depressed growth rates and anemia. Saponins are of significant concern in the U.S. As we are experiencing a dietary shift to healthier foods such as alfalfa- and soy-based products, we can also expect an increase in incidence of saponin intoxication.

Glucosinolates

Glucosinolates are found in broccoli, cabbage, radish, horseradish, and mustard greens and are thought to cause hypothyroidism (enlargement of the thyroid gland); hence, they are also termed *goitrogens*.

Mushroom Poisoning

There are approximately 100 species of mushrooms thought to be toxic and 12 species known to contain lethal toxins. Mushroom toxicity can be classified into two types of syndromes based on the latency time of the symptoms. Rapid-onset syndrome is rarely severe and produces symptoms such as gastrointestinal illness, paresthesia, and/or CNS syndrome within 3 h of consumption. Delayed-onset syndrome presents symptoms as headaches, diarrhea, hepatic incompetence, muscle pains, and chills 6 h after consumption. Symptoms of mushroom poisoning vary with mushroom species. Mushroom-induced headache is common with European species, whereas profuse diarrhea and hepatic complications are more common in North America.

Animal Sources

Paralytic Shellfish Poisoning

Paralytic shellfish poisoning results when humans consume shellfish that have fed on toxic plankton. Shellfish become toxic to humans when they feed on dinoflagellates such as *Gonyaulax catenella* in numbers greater than 200/ml of water. When theses toxic organisms are present in numbers greater than 20,000/ml, they are visible to the naked eye. This phenomena has been termed *red tide* for its crimson appearance. Saxitoxin, which is just one of the various toxins involved in paralytic shellfish poisoning, can be lethal after consuming only 4 mg. Symptoms include a tingling or burning sensation of the lips and gums, ataxia, and paralysis of the diaphragm. If treatment is sought early enough, death can be avoided.

Ciguatera Poisoning

Species of barracuda, eels, jacks, groupers, red snapper, and sea bass specific to the Caribbean and South Pacific have been implicated in ciguatera poisoning. The ciguatera toxin, like the saxitoxin, is found in a dinoflagellate. The species of marine animals mentioned previously do not feed directly on the dinoflagellate *Gamblerdiscus toxicus*, which contains the ciguatera toxin, but the small feed fish they thrive on consume the plankton as a primary component of their diet. Signs of ciguatera poisoning are a tingling or numbing sensation of the lips, tongue, and throat, which gives way to nausea, vomiting, and diarrhea. In extreme cases, muscle and joint pains develop, making it difficult for the patient to walk.

Puffer Fish Poisoning

The puffer fish, also known as the *blowfish*, is a common delicacy in China and Japan. Although not all puffers are poisonous, those belonging to the genus *Fugu* (fugu fish) contain the lethal toxin tetrodotoxin, and are also a most sought-after delicacy. When the fish is prepared in acidic or alkaline conditions, the toxin is inactivated. If the fish is prepared incorrectly, victims experience symptoms similar to intoxication from saxitoxin, and death from respiratory paralysis can occur from the consumption of only 1 to 2 mg of the toxin.

FOODBORNE PATHOGENS

Until recently, when one thought of a foodborne outbreak, salmonella was assumed to be the causative agent unless investigation proved otherwise. Since 1975, 25 new

TABLE 8.3
Foodborne Pathogens Included in the CAST Report and Fungi

FOODBORNE PATHOGENS	
Viruses Hepatitis A Norwalk Norwalk-like	**Parasites** Anisakid nematodes Cryptosporidium parvum Diphyllobothrium latum Entamoeba histolytica Giardia lamblia Taenia saginata Taenia solium Toxoplasma gondii Trichinella spiralis
Bacteria Aeromonas hydrophila Bacilus cereus Brucella abortus Campylobacter jejuni Clostridium botulinum Clostridium perfringens Coxiella burnetti Escherichia coli Listeria monocytogenes Mycobacterium bovis Salmonella spp. Shigella spp. Staphylococcus aureus Vibrio cholerae Vibrio parahemolyticus Vibrio vulnificus Yersinia enterocolitica	**Toxins** Ciguatera toxin Diarrheic shellfish poisons Domoic acid Histamine Scromboid Paralytic shellfish poisons Tetrodotoxin **Fungi** Aspergillus spp. Penicillium spp. Mucor spp. Rhizopus spp. Candida spp. Cryptococcus neoformans

Source: Adapted from Motarjemi, Y. and Kaferstein, F.K., Global estimation of foodborne disease, World Health Org. Stat. Q., 50, 8, 1997.

foodborne pathogens have been identified. More than 40 potential foodborne pathogens have been listed by CAST, and a number of these are discussed in the following sections along with fungi, an area not considered by CAST (Table 8.3).[3]

Coupled with these new foodborne pathogens are changes in dietary habits and food processing techniques, which are partly responsible for the surfacing of new pathogens and an increase in the number of illnesses due to traditional pathogens. Some of these changes are as follows[11–18]:

- A decrease in lactic-acid-producing bacteria due to processing of milk increases the chances of survival of salmonella or campylobacter, as their competitors have been eliminated.
- When water polluted with raw sewage or manure containing harmful numbers of microbes comes in contact with food, a variety of illnesses is caused.
- Due to the abuse of antibiotics and sulfur-containing drugs in farm animals, some bacteria such as salmonella have acquired resistance to antibiotics.

- In an attempt to eat healthier, people are consuming foods that are lower in fat, but also more susceptible to microbial growth.
- The increase in demand for packaged food with a longer shelf life prevents aerobic growth, but promotes growth of anaerobic pathogens like clostridia, which can be deadly.
- The growing popularity of ready-to-eat or take-out foods that require only minimal heating before consumption involves the following danger: the temperature of the food is raised to the ideal growth range of many bacteria, without the pathogen being destroyed.

The presence of these pathogens in foods, or in the body, can result in illness due to infection or intoxication. For example, infections such as those caused by *Giardia lamblia* arise when the organism colonizes on the intestine wall, reducing the absorption surface and therefore decreasing the intake of nutrients. Infections caused by *Salmonella typhi* or hepatitis A have invasive properties and can spread beyond the gastrointestinal tract, entering the bloodstream or liver, and causing typhoid fever and liver disease, respectively. Intoxication results when an organism releases a toxin while growing on a food, or while inhabiting the gastrointestinal tract of humans. Illness from toxigenic pathogens has a rapid onset and is usually more severe than that from an infection. In many cases, food products that contain a toxin have no altered appearance, taste, or smell, making it especially dangerous. In the following sections, we will examine parasitic, viral, fungal, and bacterial pathogens and the illnesses they cause.[12]

PARASITIC INFECTIONS

Parasites are organisms that require another animal (host) to complete their life cycle. Organisms such as *Entamoeba histolytica* or *Cryptosporidium parvum* require only one host. They are shed in the feces, and infection occurs when food or water contaminated with these feces is consumed. *Trichinella spiralis*, on the other hand, causes infection in those who eat meat infested with the cyst of the organism. In the case of trichinella, the organisms requires two hosts: the intermediate host (pig), which houses the cyst, and the definitive host (human), in which the organism develops into an adult, which is shed in the feces.

The Nematodes

Trichinella spiralis

Species of *Trichinella spiralis* are found in warm-blooded animals such as bear, hogs, walrus, wild boar, and rats. This organism causes trichinosis when meat from warm-blooded animals containing viable larvae are consumed by humans. In the 1940s, there were an average of 400 cases of trichinosis per year, and this number has decreased markedly to just 32 cases in recent years as a result of improved cooking habits, greater access to home freezers, and legislation that regulates the quality of feed for commercial hogs. Trichinella organisms can be destroyed by cooking meat until it turns a gray color, or by freezing. In Idaho in 1995, an outbreak

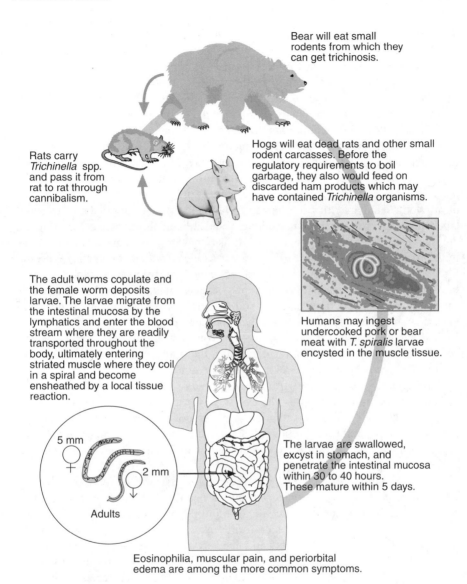

Bear will eat small rodents from which they can get trichinosis.

Rats carry *Trichinella* spp. and pass it from rat to rat through cannibalism.

Hogs will eat dead rats and other small rodent carcasses. Before the regulatory requirements to boil garbage, they also would feed on discarded ham products which may have contained *Trichinella* organisms.

The adult worms copulate and the female worm deposits larvae. The larvae migrate from the intestinal mucosa by the lymphatics and enter the blood stream where they are readily transported throughout the body, ultimately entering striated muscle where they coil in a spiral and become ensheathed by a local tissue reaction.

Humans may ingest undercooked pork or bear meat with *T. spiralis* larvae encysted in the muscle tissue.

5 mm

2 mm

Adults

The larvae are swallowed, excyst in stomach, and penetrate the intestinal mucosa within 30 to 40 hours. These mature within 5 days.

Eosinophilia, muscular pain, and periorbital edema are among the more common symptoms.

FIGURE 8.2 Life cycle of *Trichinella spiralis*.

occurred from the ingestion of cougar jerky. The outbreak was contained to ten people who displayed symptoms of eosinophilia, fever, rash, and overall weakness.[17]

Tapeworms

The beef tapeworm, *Taenia saginata,* and the pork tapeworm, *Taenia solium*, infest humans through the ingestion of undercooked beef or pork that contains viable larvae. The pig is an intermediate host for *T. solium*, and the pig becomes infected when it consumes food contaminated with human fecal material that contains the tapeworm eggs (Figure 8.3). The eggs enter the pig stomach and excyst, after which they penetrate through the mucosa into the lymphatics and bloodstream, where they are then are carried to the striated muscles, the brain, and possibly other organs. They become embedded in the muscle of the animal, where they develop into cysterici (encysted larvae), which are thin-walled, fluid-filled bladders containing an inverted scolex or head of the tapeworm (Figure 8.3). If undercooked pork containing viable larvae are ingested by humans, the larvae excyst in the intestinal tract. The scolex of *Taenia solium* contains four suckers and two rows of hooks that protrude from the cystericus. It attaches itself to the intestinal wall, where it develops into an adult with the formation of proglottids. The progottids mature to develop 7 to 13 lateral uterine branches that produce tens of thousands of embryonated eggs. The most distal (and most mature) proglottids detach from the chain of proglottids and are either discharged from the intestine or migrate out. The adult tapeworm can grow up to 7 m in length and inhabit the intestine for 30 years, during which time it sheds gravid proglottids (tapeworm segments containing eggs) in the feces of the primary host (human). If careless hygiene is practiced, person-to-person transmission is possible via direct contact or by food contaminated with fecal matter. There is also a cycle that may occur in humans whereby the *T. solium* eggs may be ingested from external sources, from self-contamination with fecal matter, or from reverse peristalsis, when the contents of the intestines are back-flushed into the stomach from the process of vomiting. In this case, the eggs excyst and the larvae migrate to the human muscles, the brain, and possibly other organs such as the eye, where they develop into cysticerci.

The cycle of the beef tapeworm, *Taenia saginata*, follows a similar route with the exception that reinfection by the egg stage of the tapeworm does not occur in humans (Figure 8.4). The consumption of undercooked or raw beef containing the cysticerci leads to the development of an adult beef tapeworm in the human intestinal tract. The tapeworm can reach lengths of 25 m, but is usually less than half that size.

Although tapeworm infections are rare in the U.S., they are more prominent in areas where beef or pork is consumed raw. In healthy individuals symptoms often go unnoticed. In malnourished individuals multiple infections can occur, leading to blockage of the intestine. If the tapeworm leaves the intestine and infects organs such as the heart or brain, effects can be severe regardless of the health status of the individual. Tapeworms can be prevented by appropriately disposing of contaminated feces, cooking meat to an internal temperature of 57°C, or freezing it at −10°C for 5 d.[18,19]

The Protozoans[18–20]

Entamoeba histolytica

Amebiasis, which is caused by *Entamoeba histolytica*, affects about 10% of the world's population. Outbreaks occur where sanitation is poor, where risky sexual habits are practiced, and in institutional facilities. In such settings, fecal material comes in contact with water and food products, increasing the likelihood of the organism being ingested. While in the cyst stage of its life cycle, the organisms can survive in the external environment, where it is transmitted by water or the fecal–oral

The hog accidentally consumes the microscopic egg deposited to the ground. Larvae excyst in the intestines and migrate in the blood throughout the body where they encyst in the muscles, brain and other tissues, and organs. This encysted form is called Cysticercus cellulosae.

Cysticercus cellulosae This stage is a fluid-filled bladder with an inverted scolex.

The ingestion of eggs from one's own stools, or that of another infected person, or from vomiting may allow the eggs or oncospheres to penetrate the intestinal tract and enter the brain, the eyes, the muscles, and other organs developing into cysticerci just as in the hog.

Embryonated eggs (oncospheres) contain six-hooked embryo.

Gravid proglottid

Each mature proglottid contains male and female sex organs that produce thousands of eggs. Improperly disposed human waste allows eggs to be distributed to ground or water.

Humans ingest undercooked pork containing the *Cysticercus cellulosae* larvae. The larvae excyst in the intestine and the scolex attaches to the intestinal wall and grows to produce a chain of proglottids.

Hooked rostellum

Suckers

Scolex

FIGURE 8.3 Life cycle of *Taenia solium*.

Cattle accidentally consume the microscopic eggs deposited to the ground. Larvae excyst in the intestines and migrate in the blood throughout the body where they encyst in the muscles, brain and other tissues and organs. This encysted form is called Cysticercus bovis.

Cysticercus bovis This stage is a fluid-filled bladder with an inverted scolex.

Embryonated eggs (oncospheres) contain six-hooked embryo.

There are 15-20 lateral uterine branches of the uterus in this species.

Gravid proglottid

Each mature proglottid contains male and female sex organs that produce thousands of eggs. Improperly disposed human waste allows eggs to be distributed to ground or water.

Humans ingest undercooked beef containing the *Cysticercus bovis* larvae. The larvae excyst in the intestine and the scolex attaches to the intestinal wall and grows to produce a chain of proglottids.

Adult T. saginata may attain a length of 25 m, but usually 12 -13 m.

Suckers

Scolex

FIGURE 8.4 Life cycle of *Taenia saginata*.

route (Figure 8.5). Once ingested, the cysts are broken down by normal digestive action. Motile trophozoites are then released. Trophozoites multiply and invade the large intestine, causing symptoms that vary from mild to bloody diarrhea with a dozen explosive liquid stools a day. This latter case is termed amoebic dysentery. The incubation time of the disease varies from 3 d to a month. In the case of chronic infections, the patient can be asymptomatic while continuously shedding viable cysts, or symptomatic, suffering from bouts of diarrhea and fever until treatment is sought.

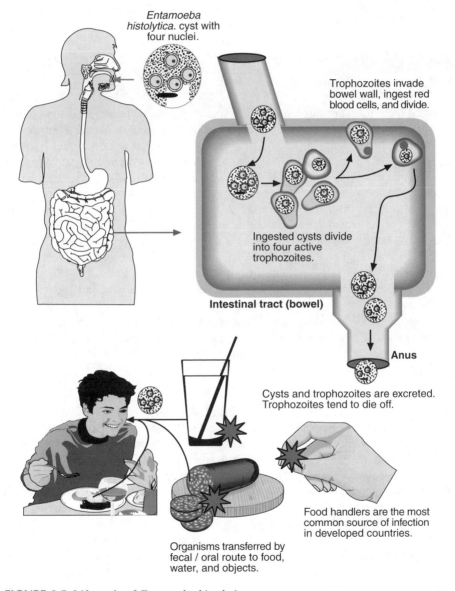

FIGURE 8.5 Life cycle of *Entamoeba histolytica.*

FIGURE 8.6 Trophozoite and cyst stages of *Giardia lamblia*.

Giardia lamblia

Giardia lamblia is a protozoan flagellate found in areas with poor sanitation and in unfiltered surface water supplies (Figure 8.6). Giardia is considered to be the primary agent for diarrheal outbreaks associated with contaminated water supplies having a prevalence of 1.5 to 20%.[16] In the U.S., giardiasis is most common among those who travel to endemic areas, in homosexuals, and in child day care settings where constant diaper changing facilitates the spread from one child to another. The organism is also found in surface water used for drinking and recreational purposes. The cyst stage of the organism resists chlorination and can therefore pass through municipal supplies that do not also have adequate filtration systems. The cysts reach the surface water supplies through the fecal deposits of beaver and muskrats. When ingested cysts reach the small intestine, trophozoites are released. The trophozoites attach themselves to the intestinal wall by adhesive disks, causing inflammation and irritation of the bowel. This inflammation prompts the influx of fluids, causing a loose stool. As a result, food is malabsorbed. The patient suffers from dehydration. Symptoms usually manifest within 1 week of eating the contaminated food and consist of nausea, explosive diarrhea (up to 10 movements per day), and fatigue. Without treatment, symptoms linger and reinfection is possible.[17]

Cryptosporidium

The first reported case of cryptosporidiosis, which is caused by the parasite cryptosporidium, occurred in 1976. Cases were prevalent primarily among immunocompromised individuals whose immune systems were unable to protect the body from infection. In 1993, the city of Milwaukee experienced the most comprehensive outbreak to date, which involved 400,000 inhabitants of the city who drank from the contaminated public water supply. Primarily a waterborne pathogen, cryptosporidium is transmitted via water contaminated with feces of human and agricultural origins. In 1994, the first documented case of foodborne cryptosporidiosis stemmed from fresh-pressed apple cider. Foodborne transmission of cryptosporidium occurs via the fecal–oral route, usually from careless food handlers shedding the hardy oocysts (see "life cycle," Figure 8.7) of the organism. In healthy individuals, symptoms appear as mild diarrhea, nausea, cramps, and a low-grade fever. Immunocompromised patients such as infants, the elderly, or those with AIDS, experience high-

Sporozoite

Trophozoite

Meront

Cryptosporidium parvum

Intestines

Auto-infection

Leaves anus to
enter water, food,
or appears on
objects where they
can be ingested.

Thick-walled
oocyst

Thin-walled
oocyst

Merozoites

Zygote

Sexual stage
includes
microgametes and
macrogametes

FIGURE 8.7 Life cycle of *Cryptosporidium parvum.*

volume diarrhea, weight loss, and severe abdominal cramps. Cryptosporidium is of
particular concern because the oocysts remain viable in the chlorinated water of
swimming pools, and can survive the water treatment processes through which our
drinking water passes. We can expect to see more outbreaks of cryptosporidiosis
until researchers and regulators devise better means of destroying the organism in
our water supplies.[21]

Viruses[22–26]

Viruses are microscopic particles that usually contain a single strand of RNA.
Although viral particles can exist outside a host cell, as during transmission, they
require a host cell for replication to occur. After invading and replicating within a
cell, death or damage to the host cell is common, but this is not always the case.
The two most prominent foodborne viruses of present day are Hepatitis A and
Norwalk-like virus.

Hepatitis A

Hepatitis A is transmitted via the fecal–oral route, and causes liver infection occasionally accompanied by jaundice. On average, symptoms occur 28 d after infection, but infected individuals shed the virus before symptoms manifest, producing an exceptionally risky situation as the infected person might handle food without additional precaution. Contamination can occur by infected food workers handling foodstuffs, or from food products that have come in contact with water polluted with fecal matter, as is the case with shellfish harvested from sewage-infested waters. *Hepatitis A* can also be transmitted from patient contact or intravenous drug use. In each of the last three decades there have been sharp peaks in the number of hepatitis A cases in the U.S. There was a 5% increase in the number of reported cases of hepatitis A from 1992 to 1993. We will have to wait for further data, but it is possible that we are experiencing the beginnings of another upward trend.[22–24] There are new techniques, however, to determine the sources of outbreaks. Genetic analysis can be used by getting samples of *Hepatitis A* from patients in a suspected outbreak area and comparing them to determine if the strains came from similar or different sources. This technology was used most recently to determine the source of hepatitis A outbreaks in Michigan, Maine, and other states in which frozen strawberries used in the schools were responsible for the outbreaks.[25]

Norwalk-Like Virus

Norwalk-like viruses are small circular organisms thought to contain RNA. Little is known about these viruses, which do not multiply in laboratory conditions. In 1982, Norwalk-like viruses were the leading cause of reported foodborne illness in the U.S., responsible for 5000 cases from 2 different outbreaks. Both outbreaks occurred in Minnesota, and the source of contamination in each outbreak was found to be infected food handlers; the food merely served as a vehicle for the spread of the virus. Since 1982, Norwalk-like viruses have not caused any outbreaks of comparable proportions, but they are recognized as a major cause of foodborne illness in Minnesota, New York, and the U.K.

Food products such as creams, cream fillings, and salads are efficient vehicles for viruses because they do not undergo any extensive heating before being served, because of which the virus is not destroyed. In New York, shellfish are the major vehicle of transmission. Norwalk-like viruses display an attack rate near 50% following the consumption of contaminated shellfish or other foods. The onset of illness occurs between 18 and 36 h after ingesting the virus, and typical symptoms include diarrhea and nausea, lasting 24 to 36 h. Although little is known about the Norwalk-like viruses, it is believed that foods that serve as a vehicle for hepatitis A can also transmit the Norwalk-like type of virus.[24]

FUNGI[26–28]

Fungi, such as molds and yeasts, are single and multicelled plantlike organisms that grow on cereals, breads, fruits, vegetables, and cheeses. For years humans have used fungi for their beneficial properties. Antibiotics such as penicillin and vitamins such

A Case History: Almost a Tragedy

Colleen Tobin was a normal schoolgirl growing up in Northampton, MA. When Colleen was 13, she started developing symptoms of stomach aches, diarrhea, fatigue, and loss of weight. She had stopped growing, and her weight was decreasing at an alarming rate. Naturally, her parents had consulted several specialists in the state, but the initial examinations turned up no answers or leads. Doctors had not considered a parasitic infection for two primary reasons: (1) she had not traveled to any country where unsanitary conditions might facilitate the spread of parasitic disease, and (2) she had not been on any backpacking excursions where she might have consumed untreated water. Unable to find any explanations for Colleen's illness, doctors suspected she was deliberately starving herself and suggested that she be admitted to a psychiatric hospital. Appalled by this suggestion, Colleen's parents continued to seek out new doctors, as she had now lost 12 lb and had a body temperature of 92°F. The gastrointestinal specialists from Springfield, MA, ran several tests, all of which were negative. Again, little Colleen was blamed and accused of taking laxatives. In her defense, Colleen mentioned that she did not know what a laxative was and, furthermore, there were no stores nearby where she could purchase laxatives. Finally, Colleen was tested for parasites and the results retuned negative, a phenomenon Colleen's parents learned was not unusual, even among infected individuals.

Nearly 1 year after the onset of her initial symptoms, Colleen spent 10 d at Massachusetts General Hospital. She now weighed only 90 lb, 30 lb less than when she first began exhibiting symptoms. She had a feeding tube down her throat, her fingernails and mouth were turning blue, and her hair fell out in clumps. After having been seen by the hospital psychiatrist again, it was recommended that she be admitted to a psychiatric hospital.

Finally, after more than 2 years of illness, Colleen's family had received word of a similar case in Northampton in which the illness had gone undiagnosed until they sought the help of a well-known parasitologist in New York City. Having exhausted every other resource, the Tobins went to New York City in the hope of gaining some insight into Colleen's condition.

Using a special rectal smear that the parasitologist had developed himself, 3 parasites were identified in Colleen's specimen within 2 h. *Giardia lamblia*, *Entamoeba histolytica*, and *Enterobius vermicularis* parasites had stripped the lining of Colleen's intestine. So, although she was eating, her body was unable to absorb the necessary nutrients. Colleen was put on a regimen of 11 different antibiotics, which initially turned her skin green and yellow. Twice weekly she had to receive an intravenous injection of vitamins and take pills to enhance the absorption in her intestines. Since her treatment, Colleen has regained the weight she lost and begun leading a normal life. The origin of the illness remains unknown.[21]

FIGURE 8.8 Morphology of molds and yeasts.

as vitamin C are derived from molds. Yeasts are used in making wine and bread They also serve as a protein source in foods. However, certain fungi are harmful to humans. The potato blight in Ireland during the 1800s was caused by the fungus *Phytothora infestans*. Certain molds such as *Penicillium* and *Aspergillus* are toxigenic and can produce mycotoxins such as aflatoxin. Mycotoxins are mold metabolites produced on food, which cause illness or death when ingested by man or animals. Other molds such as *Mucor* and *Rhizopus* species are common spoilage organisms that can also be pathogenic to humans. Fungi can grow in conditions that most bacteria could not survive. Molds can tolerate a pH as low as 2.0, and can thrive in low-moisture environments. Most molds are aerobes, and are therefore found on the surfaces of substrates where oxygen is present. In contrast to this, yeasts are facultative anaerobes, which may be found within a substrate, require more moisture than molds, and prefer slightly acidic environments.

The body of a mold, called a *colony* or *thallus* (see Figure 8.8), gives rise to long filaments known as *hyphae*. Hyphae that are divided by septae into cell-like compartments containing nuclei are called *septate hyphae*. If septa are not present, the hyphae are said to be *coenocytic*, with continuous cells and nuclei throughout the appendage. As hyphae grow, they form a tangled mesh called a *mycelium*, which can be divided into two parts, the vegetative mycelium, which is responsible for the uptake of nutrients, and the aerial mycelium, where the reproductive spores can be found. The spores on the mycelium give color and texture to the molds, producing various shades of green, blue, brown, and even black, with a cottony, velvety, or

Penicillium   Aspergillus   Mucor

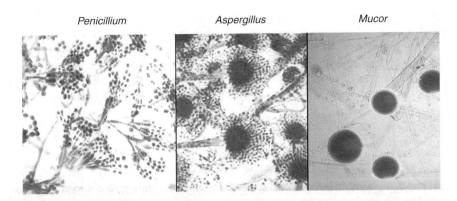

FIGURE 8.9 Microscopic views of slide cultures of *Penicillium, Aspergillus*, and *Mucor*.

waxy texture. When the molds are grown using a technique called *slide culture*, the spores may be viewed microscopically in an undamaged state in relation to the hyphae. These structures are very useful in identifying the molds. Examples of three common molds, *Aspergillus, Penicillium*, and *Mucor*, are shown in Figure 8.9. Yeasts, on the other hand, are not made up of unicellular filaments, but usually come in the form of a sphere. Yeasts reproduce by mitosis, forming a colony of unicellular organisms that normally forms a glistening, smooth, pasty colony on agar. Yeast growth on foods commonly has a white powderlike appearance.[28]

Aspergillus spp.[26,28]

In England during the 1960s, turkey X disease killed thousands of turkeys, ducklings, and baby chicks, which resulted in a huge economic loss to the area. An investigation revealed that the peanut meal feed was contaminated with *Aspergillus flavus* and a toxin that appeared to be produced by the mold. Since this incident, much research has been conducted on molds and the mycotoxins they produce. Aflatoxin, produced by strains of *Aspergillus flavus*, causes hemorrhaging, anemia, ataxia, hematosis, cirrhosis of the liver, and is a very potent carcinogen. There are four primary aflatoxins, B1, B2, G1, and G2, which are found in peanuts, corn, and cotton seed. In the U.S., aflatoxin is of concern during the storage of corn, and in milk from cows that have consumed the aflatoxin. Techniques such as treating milk with ultraviolet light or soaking corn in bisulfite during processing, are ways in which the food industries eliminate the toxin.

Another toxin produced by *Aspergillus* species is the ochratoxin, produced by *Aspergillus ochraceus*, found in peanuts, Brazil nuts, specific grains, red and black pepper, and citrus fruit. The ochratoxin causes swelling of the liver and necrosis of the kidney tubules, leading to blockage. The ochratoxin has been isolated in low-quality corn in the U.S.

Penicillium spp.[27]

Rubratoxin, patulin, and yellow rice toxins are produced by members of the genus *Penicillium. Penicillium rubrum*, which produces the rubratoxin, can be found on plants such as peanuts, corn, legumes, and sunflower seeds. Patulin, which is produced primarily by *Penicillium expansum*, is found in rotting fruit owing to the growth of organisms. Yellow rice toxins are produced by various species of *Penicillium* growing on rice during storage; the toxin causes the rice to turn yellow, and hence the name *yellow rice toxins*. Citrinin and citreoviridin are examples of such toxins, which produce symptoms such as vomiting, difficulty breathing, low blood pressure, and respiratory arrest.[27]

Mucor and Rhizopus spp.

Mucormycosis is the disease caused by fungi in the order Mucorales. Species of the genera *Mucor* and *Rhizopus* are the common species responsible for this disease. Found in the soil, *Mucor* and *Rhizopus* species are common spoilage organisms of bread and fruit. Pathogenic strains such as *Rhizopus oryzae* are opportunistic organisms, infecting diabetics, leukemics, and those with suppressed immune systems. The onset of the disease is rapid and includes the invasion of blood vessels, causing embolisms and tissue necrosis. The disease can spread to whole organ systems, causing meningitis and death within 2 to 10 d.[26,28]

Yeasts

Yeast are sometimes difficult to distinguish from molds. Yeasts are used in industry for the manufacture of beer, ale, bread, and glycerin. Yeasts also produce infections and disease as a result of food contamination. True yeasts, such as *Saccharomyces cerevisiae*, which is used in making beer, produce ascospores after sexual mating. In the laboratory, many yeasts do not display this sexual stage and have been termed *false yeasts*, or yeastlike fungi. *Candida albicans* and *Cryptococcus neoformans* are two examples of yeastlike fungi. They are the most common causes of these infections. Small numbers of *Candida albicans* may be found in healthy individuals, but infection can occur in infants when taking broad-spectrum antibiotics, during pregnancy, or when the body is burdened with a malignant disease. Infections vary considerably in severity. Infection can occur between fingers and toes, causing a rash, in the mouth, vagina, and lungs, or there could be systemic involvement of multiple organs. *Cryptococcus neoformans* is normally a ubiquitous saprophytic organism that is recovered in large numbers from the filth and debris associated with pigeon roosts. The organisms gain entry into the host through the lungs, normally producing a rapidly resolved pulmonary infection. In a few cases, it produces a serious and often-fatal central nervous system infection, in which the organisms appear in India ink preparations of the cerebrospinal fluid (Figure 8.8).[26,28]

BACTERIA

Bacteria are the single-celled organisms responsible for more than 80% of foodborne illness.[5] Bacteria are much larger than viruses, and can be viewed with a conventional microscope. Bacteria are categorized according to their physical and chemical properties. Two broad groups of bacteria classification are Gram-positive and Gram-negative, which are based on the amount of peptidoglycan (layer of sugar derivatives

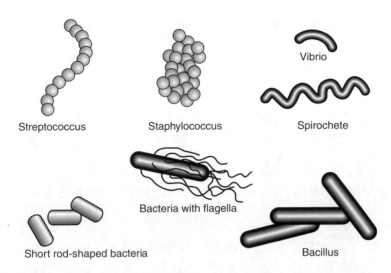

Streptococcus   Staphylococcus   Spirochete

Vibrio

Bacteria with flagella

Short rod-shaped bacteria   Bacillus

FIGURE 8.10 Examples of various bacterial forms.

which gives strength to the cell wall) in the cell wall of the organism. The cell wall of Gram-negative bacteria such as salmonella and *Escherichia coli* contains only 10% peptidoglycan. In Gram-positives such as staphylococci and clostridia organisms, the cell wall contains up to 90% peptidoglycan, which is why they retain the violet crystal stain. After staining, examination under a microscope will reveal the different shapes of bacteria in the form of coccus, rods, spirilium, spirochete, and appendaged bacteria (Figure 8.10). Rod- and coccus-shaped bacteria are most commonly associated with foodborne illness. Another important characteristic useful in identifying bacteria is the ability to grow in the presence or absence of oxygen. Aerobic bacteria require oxygen for growth, whereas the growth of anaerobes such as clostridia is inhibited by the presence of oxygen (Figure 8.11). Other organisms such as campylobacter are microaerophiles, which require low concentrations of oxygen for growth. If the anaerobe *Clostridium botulinum* is suspected, the investigator might search for endospores, which are structures produced during the life cycle of certain bacteria (Figure 8.12). Endospores can withstand extreme heat, harsh chemicals, and are very difficult to destroy. The dehydrated spore consists of bacterial DNA, a little RNA, some enzymes, and several layers around the spore, including a thick protein spore coat that confers resistance to harsh chemicals. The spore cytoplasm also contains dipicolinic acid, which forms a complex with calcium. The dipicolinic acid is a major factor in the heat resistance of spores, and is essential for later resumption of metabolic functions. Apart from these characteristics, several other factors such as motility, lactose fermentation, or growth on a specific medium are properties that aid in the identification of bacteria.

Salmonella spp.[29–32]

Salmonellosis is the disease caused by the salmonella organisms, Gram-negative facultative anaerobes. Although there are approximately 40,000 confirmed cases

Obligate aerobe
Growth occurs at surface where there is oxygen

Facultative anaerobe
Growth occurs in areas with and without oxygen

Obligate anaerobe
Growth occurs only where levels of oxygen are very low or absent

Microaerophilic
Growth occurs only in areas of low concentrations of oxygen

FIGURE 8.11 Different types of microbial oxygen requirements.

Bacterial cell

Bacterial chromosome (DNA)

Peptidoglycan layer

Spore coat

Plasma membrane

Terminal endospore

Clostridium spp. showing endospores

FIGURE 8.12 Endospores of *Clostridium* spp.

each year, investigators have estimated that the actual number of cases is between 2 and 4 million.[32] *Salmonella* species are part of the normal intestinal flora of humans, cattle, poultry, pigs, and many other animals. The disease is transmitted via food and water, putting every person at risk. There are three syndromes caused by *Salmonella* species: typhoid fever, enteric fever, and gastroenteritis.

Typhoid fever, caused by *Salmonella typhi,* accounts for less than 2.5% of all salmonellosis in the U.S. Humans are the only reservoir for *S. typhi*, and transmission occurs through fecal contamination of food and water by an infected food worker or sewage contamination. Infection occurs when ingested organisms invade the epithelial cells of the small intestine. After penetrating the intestine wall, they invade the immune system, where they multiply and then are carried in the blood throughout the body, causing systemic infection. Symptoms include septicemia, high fever, constipation, diarrhea, vomiting, spotting on the chest and trunk, and bleeding from the bowels and nose. Not all who harbor the organism show symptoms, even though they may shed the organism and infect others. These people are called carriers, and the most notorious of these is *Typhoid Mary.* The term originated with Mary Mallon, an Irish immigrant who is thought to have infected hundreds of people by transferring the *Salmonella typhi* organisms she carried through her unsanitary cooking habits until she was placed in exile on an island off Manhattan.[30]

Enteric fever is the second type of syndrome, caused by *Salmonella paratyphi.* This illness comprises only 0.5% of the salmonellosis cases in the U.S., and is similar to typhoid fever, but less severe.

The third and most prominent syndrome of salmonellosis is gastroenteritis, caused by some 2300 serotypes of *Salmonella* species. Only 150 of this number are actually known to cause human sickness, but all are assumed to be pathogenic. Gastroenteritis from *Salmonella* species does not include the invasive properties or systemic infection as seen with the typhi and paratyphi species. These organisms colonize in the small intestine, causing intestinal inflammation, resulting in diarrhea, abdominal cramps, chills, fever, and vomiting, which last 1 to 4 d. The severity of the infection varies with the involved *Salmonella* species, the health status of the individual before infection, and the implicated food. Many healthy individuals who become infected do not seek medical treatment, as symptoms are often mild, lasting 1 to 3 d. This is the primary reason why only 40,000 of the estimated 2 to 4 million cases are reported each year.

Salmonellosis stems from a wide variety of sources, including milk, eggs, ice cream, poultry, beef, and salads, infected with such species as *S. typhirium, S. enteriditis,* and *S. montevideo,* just to mention a few. Investigations of poultry products in the U.S. revealed that *Salmonella* species are present in 30% of the chicken available on the market, but the numbers of bacteria are usually low, which is why we do not experience a greater number of outbreaks. Considerable efforts are being put forth to eradicate salmonella in poultry, as it is implicated in one third of the foodborne outbreaks in the U.S.

In 1997 the FDA approved PREEMPT, the first competitive exclusion product for use in poultry. The concept of competitive exclusion is rather simple. When chicks are hatched, they are especially vulnerable to salmonella infection because their natural intestinal flora, which compete with salmonella and other bacteria for nutrients, have not yet been established. By spraying newly hatched chicks with a mixture of live bacteria, researchers were able to eliminate 99.9% of the salmonella in poultry. Although this technique may eliminate salmonella in poultry, there are other means by which an individual may become infected.[31] Salmonella is also spread by infected people who are shedding the organisms in their feces. Unsanitary habits such as not properly washing after using the rest room facilitate the spread through a handshake or other contact. This mode of transmission is known as the person-to-person mode.

Domestic animals such as turtles and iguanas should be considered potential carriers of salmonella, and handled accordingly. Before 1975, domestic turtles accounted for 280,000 cases of salmonellosis annually. These numbers prompted the FDA to ban the interstate shipment of turtles, which resulted in an 18% reduction of salmonellosis in children of ages 1 to 9.[32]

Salmonella species are commonly found in iguanas (*S. marina*), but the species involved rarely causes human disease. Even though iguanas are imported into the U.S. from many foreign countries, there are no quarantines or health inspections upon entering this country.

Treatment of gastroenteritis from salmonella is limited to hydrating fluids and rest. Antibiotic treatment is offered only to severe cases for two reasons. First, nonselective antibiotics will eliminate the natural flora of the intestine, leaving the body vulnerable to other infections, and second, antibiotic treatment will not kill organisms that have acquired resistance to these drugs, leaving only them to flourish and infect. This second reason is the primary argument of advocates lobbying against the use of antibiotics in meat and poultry.

Staphylococcus spp.[9]

Staphylococcal food poisoning, caused by the Gram-positive cocci *Staphylococcus aureus*, is one of the most common foodborne illnesses in the U.S. Staphylococcal organisms may be found in the nasal passages, pimples, throat, and wounds of humans. The organism can grow between 7 and 48°C, and produces an enterotoxin between 20 and 37°C. Sickness is due to the consumption of the enterotoxin that is produced; therefore, staphylococcal food poisoning is considered an intoxication as opposed to an infection. The intoxication is characterized by nausea, vomiting, and diarrhea, occurring within 30 min to 8 h of enterotoxin consumption. Sickness is usually short in duration and self-limiting, but death has occurred. The enterotoxins can survive prolonged boiling, and are considered to be heat-stable. Therefore, once the enterotoxins are produced in food, they have the capacity to produce illness even if the food is reheated and the staphylococcal organisms are killed.

Contamination occurs through the preparation of foods by infected food handlers. Foods such as creams, cream pies, potato salad, and ham have all been implicated in outbreaks of staphylococcal food poisoning. Cream products were of major importance in staphylococcal food poisoning, but prevalence of infection has decreased as a result of proper refrigeration. Currently, ham is the most common vehicle of transmission for the staphylococcal organisms.

Staphylococcal growth is inhibited by the presence of competitors, but it can survive in high salt and sugar concentrations that destroy other organisms. Precooked ham contains salt in concentrations up to 3.5%, which kills off other organisms, leaving a perfect environment for staphylococci to thrive.

In September 1997 in northeast Florida, 18 individuals attending a retirement party suffered from staphylococcal food poisoning. Samples of leftover ham and rice pilaf were found to contain the enterotoxin produced by *Staphylococcus aureus*. Although investigators never found the source of contamination, it was likely because the whole ham was placed immediately in the freezer after cooking, which would have prolonged the cooling time, allowing staphylococcal organisms to multiply.[33]

Although the foods mentioned previously have been implicated in staphylococcal intoxications, virtually any food can serve as a vehicle if handled carelessly by infected food workers, and left out at temperatures ideal for growth.

Clostridium spp.[34–36]

Clostridium perfringens is an anaerobic spore-forming bacteria found in soil and the intestinal tracts of humans and other animals. *C. perfringens* grows at temperatures between 20 and 50°C. In optimum conditions (45°C) with the right media, *C. perfringens* can double its numbers in as little as 8 to 10 min, making it the fastest-growing bacteria known to humans.[34]

For food poisoning to occur, numbers in excess of 10^8 bacteria must be ingested.[34] When the organism senses harsh conditions, such as the lack of nutrients or water, it sporulates and releases the enterotoxin responsible for the intoxication. Two types of spores that are classified by their sensitivity to heat are produced by various strains of *C. perfringens*. Heat resistance is measured by the D value (the amount of time and heat required to kill 90% of the population). Heat-resistant spores have a D90°C value of 15 to 45 min, which implies that 90% of the organism population is eliminated by heating to 90°C for 15 to 45 min. Heat-sensitive spores have a D90°C value of 3 to 5 min. The enterotoxin produced by *C. perfringens* is destroyed by heating to 60°C for 10 min.

C. perfringens food poisoning usually occurs when cooked foods such as beef, turkey, or chicken are held for several hours at room temperature before serving, or when dense foods are refrigerated after cooking. These conditions allow the vegetative cells to grow to dangerous levels before being consumed.

In March 1993, 156 persons suffered from food poisoning at a St. Patrick's day meal in the state of Ohio. Investigation of the outbreak revealed that the delicatessen serving the meal had prepared corned beef sandwiches early in the day and held them at room temperature until they were purchased. As mentioned earlier, holding foods at room temperature for extended periods of time is a dangerous practice that gives many pathogenic organisms the chance to proliferate.[35]

Food poisoning by *C. perfringens* is mild, usually limited to diarrhea, abdominal pain, and vomiting. Symptoms can manifest as soon as 2 h after ingesting the contaminated food, and last about 24 h. Death from *C. perfringens* is rare but does occur among the elderly and those with an underlying condition.[35]

Clostridium botulinum

Clostridium botulinum, like *Clostridium perfringens*, is a spore-forming anaerobic bacteria found in soils throughout the world. Botulism is the illness that results when *C. botulinum* spores germinate and produce a toxin in food. Outbreaks of botulism are rare in the U.S., and although commercial foods have been implicated, home-canned foods are the most common source. *C. botulinum* organisms are not a threat in food unless they can germinate into the toxin-producing vegetative cells. By destroying the spores in foods before canning or storing products, risk of botulism can be eliminated. The spores, however, are resistant to heat (D90°C is 30 min) and chemicals, making them difficult to destroy. The neurotoxins produced by the organism are fortunately heat-labile and can be easily destroyed by boiling for 3 min or by heating to 80°C for 30 min.

C. botulinum organisms can grow in temperatures between 3.3 and 45°C, but this varies with the type of organism. Apart from heat and chemicals, pH and oxygen concentration also affect growth. A pH of 4.6 inhibits the germination of spores. Foods such as tomatoes may naturally have a pH as low as 4.6, and food of a higher pH may have acids added to them during processing to prevent spore germination. The oxygen concentration of the growth medium is a critical factor for *C. botulinum* growth. Although the conditions on the surface of a food product may be aerobic, anaerobic conditions may exist within the food mass. In products that are smoked, such as fish and sausage, anaerobic conditions can develop to the cell, producing a particularly dangerous situation because the organism may be present in lethal numbers without having affected the appearance of the food. Fresh mushrooms were previously packaged in containers sealed with a cellophane wrap. If aerobic bacteria consumed all the oxygen present in the package, they would die off, leaving anaerobic conditions without competitors, an ideal setting for an outbreak of botulism. The cellophane wrap now used is punched with holes to avoid this scenario.

Currently, there are seven types of *C. botulinum* listed, A–G, which are identified by the toxin they produce. Only types A, B, E, and F have been found to cause foodborne illness in humans. Types C and D cause botulism in cattle and other animals, and G has been reported to cause infant botulism not associated with food. Botulism caused by the A toxin is the most common in the U.S., followed by B, E, and F. The A and B toxins have been isolated in fruits, vegetables, fish, condiments, beef, pork, and poultry. The E toxin has been isolated primarily in fish, and the F toxin only in meat products.

Food poisoning in adults occurs within 12 to 36 h of consuming the toxin. At the onset, symptoms such as nausea, vomiting, and diarrhea, are present, and as the condition develops, fatigue, blurred vision, difficulty speaking, and swallowing are experienced. If the condition remains untreated, death can occur from respiratory paralysis. Infant botulism occurs in children less than 1 year old. Ingested honey has been the source of most food-related cases, but 50% of infant botulism occurs in children who were only breastfed, leaving air, dust, and soil as the likely sources of contamination.[36]

Campylobacter

Campylobacter are one of the leading causes of acute gastroenteritis in the U.S., responsible for approximately 2 million cases annually. They are part of the normal flora of the gastrointestinal tract of warm-blooded animals. Cattle and poultry are of primary concern. Contamination occurs primarily during food processing when the intestinal tract is lacerated, allowing feces to contaminate the food product. Infected food handlers and raw milk are also significant sources of infection. *Campylobacter jejuni* is the organism most commonly implicated in campylobacterial infection.

Campylobacter jejuni is a microaerophile (requiring little oxygen), capable of growing in 3 to 15% oxygen. It has a narrow temperature range for growth (between 30 and 47°C), but more importantly, campylobacter can survive for weeks in refrigeration at 4°C. Although no growth may occur during refrigeration, campylobacter are still a threat because infection can occur from as few as a few hundred cells. Symptoms in infected persons are usually mild, including nausea, vomiting, and bloody diarrhea, but in severe infections, Guillain-Barré Syndrome develops, which causes neuromuscular paralysis.[37]

Escherichia coli[38–40]

E. coli organisms can be found among the normal bacterial flora in the intestine of humans and other warm-blooded animals. The O157:H7 strain of *E. coli* has caused several outbreaks of severe sickness and death, which has given the public the impression that all *E. coli* strains are harmful, when in fact the majority of strains are harmless and some strains can be found in the intestinal tracts of healthy, asymptomatic individuals.

E. coli organisms that are important to foodborne illness can be divided into four groups, enteroinvasive, enterotoxigenic, enteropathogenic, and enterohemorrhagic, the latter being the strain the infamous *E. coli* O157:H7 belongs to. Enteroinvasive *E. coli*, as the name implies, invade the epithelial cells of the intestine, resulting in fever, chills, and bloody diarrhea. In contrast, enterotoxigenic *E. coli*, which are responsible for traveler's diarrhea, induce cholera-like symptoms, with cells colonizing the wall of the intestine and producing a toxin that causes watery diarrhea, vomiting, and abdominal cramps, which can result in dehydration. Enteropathogenic *E. coli* are most commonly found among infant nurseries in developing countries. During the 1940s and 1950s, enteropathogenic *E. coli* were a problem in the U.S., but better sanitation has diminished their importance in recent decades.

Enterohemorrhagic *E. coli*, also known as *E. coli* O157:H7, has become one of the more problematic foodborne pathogens during the 1990s. Cattle are the major reservoir for the organism, and human infection is the result of consuming improperly cooked ground beef, raw milk, or unpasteurized apple cider. Symptoms generally include abdominal cramps, watery to bloody diarrhea, vomiting, and possibly fever. *E. coli* O157:H7 is especially dangerous because it can be accompanied by hemolytic uremic syndrome (HUS), which is the primary cause of renal failure in children.[38] In January 1993, Washington state experienced an outbreak of *E. coli* O157:H7 because of improperly cooked hamburgers from a fast-food chain. The outbreak comprised 501 cases involving *E. coli* O157:H7 with 45 cases of HUS, and 3 deaths.[40] Such outbreaks can be avoided by properly cooking meat to an internal temperature of 68°C. Another outbreak of *E. coli* O157:H7 occurred in 1996 in Japan owing to consumption of raw alfalfa sprouts. There were 9000 illnesses and 17 deaths due to the outbreak. This and several smaller outbreaks in the U.S. have prompted the FDA to put out a warning against consumption of raw alfalfa sprouts until new measures are put in place to ensure sprout safety.[39] There are also new techniques being developed to fight off *E. coli* O157:H7 in milk and other products. Cold pasteurization is a process by which an electron beam is directed at the product, inactivating *E. coli* DNA. Thus far, this new technology has not been found to alter the foods in any significant way.[41]

Vibrio cholerae[42–44]

The disease cholera, caused by the organism *Vibrio cholerae*, is most prominent in LDCs, where unsanitary conditions prevail. In the U.S., most cholera cases are associated with international travelers who either become infected while abroad, or who bring contaminated food items back into the U.S., where the infection may occur. In 1992, the number of cholera cases identified in the U.S. was 102, the most reported to the CDC in a single year. Ninety five of the cases were associated with travel, 75 of which occurred during an outbreak on an Argentinean airline.[42]

In Peru in 1991, an outbreak of cholera spread to 322,562 Peruvians. Following this initial outbreak, cholera spread from Peru throughout most of South America, north through Central America, Mexico, even reaching the U.S. Statistics from the Pan American Health Organization indicate that between 1991 and 1992 there were 645,030 cases, with 5,694 deaths related to this epidemic.[43] A member of the genus *Vibrio*, this Gram-negative rod can proliferate in temperatures between 15 and 42°C. in a pH range of 6 to 10, and in a solution containing up to 6% sodium chloride. Hence, it can be found in coastal waters contaminated with fecal matter. Crabs, shrimp, mussels, and claws can harbor the organism for long periods of time, and have all been implicated as sources of infection.[42]

Vibrio cholerae does not invade the intestinal lining. Instead it colonizes on the lining and produces the toxin choleragen, which causes massive influx of fluids into the intestine. Symptoms precipitate 2 to 3 d after consuming the contaminated food and generally appear as abdominal pains, dehydration, and a characteristic diarrhea that has been termed *rice water stool* for its similar appearance to the water left behind after rice has been cooked. Complications associated with cholera include hypoglycemia, severe dehydration, and even renal failure. Cholera can have a case fatality rate of up to 50% in untreated patients, who usually die of complications from extreme loss of fluid from the body (up to 1 l/h in adults).

Although a vaccine exists, it is not practical for the inhabitants of endemic areas because it is effective for only 3 to 6 months. In those endemic areas, an improvement in sanitation practices would be more practical. However, the vaccine is recommended for travelers spending short periods of time in endemic areas. Antibiotics can be given for treatment, although, as seen with salmonella, many strains have acquired resistance to traditional medicines. Rehydration solution can be used to treat mild cases, which usually subside without any other treatment. In severe cases, rehydrating fluids are introduced intravenously to counter the rapid loss of fluid.[44]

FACTORS FREQUENTLY CITED IN FOODBORNE ILLNESS

During the food processing chain, many controls are put in place to prevent food-borne disease. To produce a food that is 100% free of contamination is not feasible owing to the economic costs as well as the fact that such a rigorous sanitation process would alter the food so much that it would no longer appeal to the consumer. Instead, acceptable limits have been set that allow low numbers of certain bacteria to be present in a food. If the food is treated properly, the presence of such low numbers of bacteria should not pose a threat. In many cases, however, food is abused. The result is often a foodbome outbreak.

Factors that contribute to foodborne illness can be broken down into three categories: (1) time and temperature, (2) poor hygiene practices by food handlers, and (3) cross-contamination. Several factors within these categories that are commonly found to be the cause of largely preventable foodborne outbreaks follow[45,46]:

1. Improperly refrigerated food
2. Improperly heated or cooked food
3. Food handlers who practice poor hygiene
4. Lapse of a day or more between preparing and serving food
5. Introduction of raw or contaminated materials into a food that will not undergo further cooking
6. Improper storage of foods at temperatures ideal for bacterial growth
7. Failure to properly heat previously cooked foods to temperatures that will kill bacteria
8. Cross-contamination of ready-to-serve foods with raw foods, contaminated utensils, or machinery, or through the mishandling of foods

Some suggestions for reducing the hazards associated with the contamination of food are shown in Figure 8.13.

HAZARD ANALYSIS CRITICAL CONTROL POINTS[47]

Although there are several measures in place to prevent foodborne disease, pathogens such as *E. coli* and *Salmonella* still continue to pose a serious health threat. In response to this threat, the federal government has mandated the implementation of hazard analysis critical control points (HACCP) strategies in the seafood, poultry, and meat industries.

In contrast to the traditional end-product-testing methods, HACCP intervenes at various stages in the food processing/preparation chain to ensure a safer product. The objective of an HACCP plan is to identify hazard points within a certain industry that may lead to foodborne illness, implement strategies to minimize the identified hazards, and monitor these practices to ensure that all implemented plans are adhered to and are effective. There are seven key principles to the HACCP system: (1)

Refrigerator/freezer
- Freezer should be kept at 0°F and refrigerator at 38-40°F; check temperature periodically with a reliable metal thermometer
- Store raw meats on bottom, and enclose in plastic wrap or baggies
- Large amounts of hot food should be divided into shallow pans for faster cooling, then covered and placed in refrigerator
- Clean refrigerator on a regular basis.
- Freeze meat, poultry, and fish products if you are not going to use them quickly

Time and temperature
- Serve or discard foods within 4 hours from time when it is removed from temperature control
- The danger zone for food is 40°F to 140°F (4.4°C to 60°C)

Foods
- Never mix different meats, poultry, or fish on counter or in storage
- Package meat and poultry in plastic storage bags to prevent cross contamination
- Do not allow high-risk foods to remain at room temperature for more than a couple of hours, refrigerate or cook immediately
- Do not partially cook foods.
- Cook poultry to 180°F meat and ground meat to 160°F, and reheat leftovers to 165°F or until very hot

Counters/sinks
- Disinfect dishcloths frequently each day by placing in disinfectant soap, or heating in a cup of water in a microwave before cleaning countertops, washing dishcloths in a dishwasher is also good practice
- Wash hands thoroughly before food preparation
- Clean countertops with hot, soapy water between placement of high-risk foods on the same surface
- Thaw high-risk foods such as poultry in the refrigerator or run under cold water

Microwave cooking
- Use thermometer or meat probe to insure proper cooking temperature
- Be sure to let the food stand as instructed since further cooking will take place during this time
- Rotate or stir to assure even cooking throughout

Cutting boards
- Avoid preparing food on wooden cutting surfaces where deep grooves permit high numbers of bacteria to grow
- Use plastic or ceramic cutting surfaces that can be easily sanitzed between uses or between the preparation of high-risk foods, fruits, and vegetables

FIGURE 8.13 Recommendations for reducing hazards associated with food contamination.

TABLE 8.4
The Seven Key Principles of HACCP

HAZARD CRITICAL CONTROL POINT (HACCP) SYSTEM

Assess the Hazards
- Identify potentially hazardous foods
- Follow the flow of food to assess hazards at receiving, storing, preparing, cooking, holding, serving, cooling, and reheating
- Estimate risks

Identify Critical Control Points (CCPs)
- Develop procedure and flowcharts showing the flow of food and all of the CCPs

Set Up Procedures and Standards for CCPs
- Standards must be met at each CCP and should be: measurable, based on fact, correct for the recipe, clear directions with specific actions

Monitor CCPs
- Check to see if standards are met; employees should be involved in process; standards must be met

Take Corrective Actions
- If standard not met, correct it
- Have specific steps for correction

Set Up Record Keeping System
- Blank forms near equipment where they are to be used
- Notebooks to write down actions
- Flowcharts and recipes near work areas

Verify That the System Works
- Identify and assess all hazards
- CCPs selected
- Standards set with monitoring and schedules
- Corrective actions in place
- Monitoring being done
- Flaws or omissions corrected
- Monitoring equipment calibrated

assessing the hazard, (2) identifying critical control points, (3) establishing standard procedures and critical limits, (4) monitoring procedures, (5) corrective actions, (6) record keeping, and (7) verifying that the system is working properly. By adhering to these principles, food establishments can reduce the risk of a foodborne illness to a minimum (Table 8.4).

Are you interested in learning more about foodborne disease? Visit http://www-unix.oit.umass.edu/~envhl565/weblinks.html and click on one or more of the following sites under Chapter 8.

General Links to Foodborne Disease
- FDA/CFSAN Bad Bug Book: Introduction to Foodborne Pathogenic Microorganisms and Natural Toxins
- FDA Center for Food Safety & Applied Nutrition
- HACCP Web site
- Food Safety Education
- Foodborne Illness and Death in the United States
- US FDA / Center for Food Safety and Applied Nutrition (CFSAN)

Assessing the Hazard

Adequately assessing the hazards in a food process is essential to an effective HACCP plan. The person or persons conducting the hazard assessment must review the food on-site to determine which are potentially hazardous, employee activities should be supervised for dangerous practices, and the design of the facility should be examined for potential flaws.

Identifying Critical Control Points

Once the hazards have been assessed, critical control points (CCPs) are identified to control the potential hazards. For example, if raw chicken is being handled, salmonella would be an identified hazard, and the cooking process would be the critical control point, because it is there that they can be eliminated. Depending on the hazard, other critical control points could be refrigeration to prevent microbial growth or fluctuation of the pH to inactivate a toxin.

Establishing Standard Procedures and Setting the Critical Limits

Procedures such as washing hands, sanitizing machinery, and food preparation should be established and well defined to a safe product. Critical limits such as time, temperature, pH, and water activity are parameters that ensure product safety. Operating outside the critical limits could result in foodborne illness. In the case of raw chicken, the critical limits during cooking would be heating to an internal temperature of at least 73.9°C. By not meeting these critical limits, salmonella can remain viable in numbers high enough to cause infection or possibly contaminate another food item.

Monitoring Procedures

The periodic monitoring of procedures is necessary to track the operation of the facility, indicate when critical values have not been followed, and provide documentation in the event of an inspection or other situation that may require verification. For instance, while cooking chicken periodic checks on the oven temperature are performed; if inspection reveals that the oven is operating at 74°C, the time, date, and temperature are recorded, and the supervisor might request that the temperature be raised slightly. By monitoring the CCPs, improperly cooking the chicken has been avoided, preventing a possible outbreak of salmonellosis.

Corrective Actions

Corrective action must be taken when a CCP has operated outside of the critical limits. In some cases it may be necessary to discard a product, whereas other situations may simply require bringing the CCP back within the critical limits. In either case, the deviation from the critical limit should be documented and any corrective action should be approved by the proper regulatory bodies before implementing HACCP.

Record Keeping

If a facility is required to have an HACCP plan, a written copy of the plan must be kept on-site. In the event government inspectors visit a facility, documentation regarding CCPs, critical limits, and deviations from those limits should be available.

Verification That the HACCP System Is Working Correctly

The purpose of verification is to check how effective the plan is in reducing the identified hazards, and possibly reveal any hazards that might not have been previously identified. Verification may include reviewing current documentation for patterns in errors or corrective actions, which may reveal false reporting, ineffective CCPs, or other flaws in the HACCP plan. Detection methods such as those discussed previously confirm the presence or absence of identified microbes, and should reveal if the HACCP plan has resulted in decreased numbers of pathogens.[46]

U.S. Regulatory Efforts with Regard to Food Protection

On December 18, 1997, the FDA required that all seafood processors, domestic and those importing to the U.S., carry out a hazard analysis of their products and processes. In the event hazards are recognized, the facility is required to have a written HACCP plan to address the problem. The ruling also identified eight sanitation concerns that must be monitored and controlled. A formal program of sanitation standard operating procedures (SOPs) is not required by the FDA; however, the FDA does require that records on sanitation SOPs be filed.

The U.S. imports more than 50% of the seafood that Americans eat, and there are an estimated 6000 firms in 172 different countries that are involved in the seafood industry. Any importers of seafood will have to verify that their product was manufactured in accordance with the HACCP ruling. HACCP plans will be reviewed annually to keep abreast of any changes in manufacturing practices and to add more controls if additional hazards are identified. As can be expected, this ruling was met with resistance from the seafood industry. Many importers will have to find new producers who can comply with the ruling, and domestic producers may have to change their current practices to comply with the new requirements. The meat and poultry industries are currently experiencing similar rulings.

On January 27, 1997, the USDA required meat and poultry slaughterers and processing facilities to have sanitation SOPs in place, and that they also conduct generic *E. coli* testing. As of January 26, 1998, large facilities employing 500 or more workers must also adhere to reduction performance standards for *Salmonella* organisms to ensure the HACCP plan is reducing the risk of foodborne illness. On January 25, 1999, facilities harboring 10 to 499 workers became subject to this regulation, and by January 25, 2000, over 6200 meat and poultry plants were operating under this ruling. Pending the success of this legislation, we may expect similar rulings on fruit and vegetable industries.[47]

Surveillance Efforts

Outside of the food industry, considerable efforts are also being put forth to control and better understand foodborne pathogens. The Foodborne Diseases Active Surveillance Network (FoodNet), which is part of the CDC emerging infections program, is the latest surveillance strategy aimed at detecting foodborne illness that would normally go unnoticed under the traditional system of passive surveillance.[48]

Previously, foodborne disease was monitored by a passive approach of reporting cases that had been confirmed in a clinical laboratory. This method only accounted for a fraction of the actual cases; hence, the actual importance of foodborne disease was underestimated. The FoodNet is an active approach of tracking disease, in which health officials contact laboratories with requests for data on new cases of foodborne illness. The health official in turn reports the information to the CDC via a computer network. Surveillance sites have been established in several states where surveys are conducted among clinical laboratories, physicians, and the general population. The purpose of these surveys is to more accurately assess the incidence of foodborne illness, and to monitor the procedures of physicians and laboratory personnel that might influence the number of pathogens which are isolated and reported.

Case control studies are another component of the FoodNet, which provides information on the food items that are most commonly associated with these diseases. Isolates from these studies are sent to the CDC, where they are phage-typed and tested for antibiotic resistance. This information alerts health officials to the presence of new pathogens that have not been previously isolated, or to strains that have become resistant to antibiotics used to reduce their numbers in animals or to combat infections in the human body. The FoodNet has been up and running since January 1, 1996, and since then it has identified outbreaks of campylobacter in California, salmonella in Oregon, and two outbreaks of *E. coli* O157:H7 in Connecticut.[48]

Another agency recently set up in the U.K. is the Food Standards Agency. This agency will be used for creating policies regarding food safety and reporting of food poisoning. There is some controversy over this agency, however, because it is looking into issues that are normally covered by the Ministry of Agriculture, Fisheries, and Food and local governments. Many in England felt that these agencies did not do a good job with the recent outbreaks of both *E.coli* and salmonellosis and pushed for a new agency that would actually investigate the cases and make new regulations, something the other agencies have only advised on.[49]

Along with sources like FoodNet and agencies like the Food Standards Agency, there are temperature indicators that consumers can buy to test their foods. These indicators are adhesive strips that signal when foods have changed temperature and are therefore dangerous. The strips can also be used in restaurants and grocery stores to determine freshness.[50]

REFERENCES

1. Kaferstein, F.K., Food safety: a commonly underestimated public health issue, *World Health Org. Stat. Q.,* 50, 3, 1997.
2. Centers for Disease Control (CDC), Foodborne and Diarrheal Diseases Branch (FDDB), June 13, 2006, http://www.cdc.gov/foodborne/.
3. Motarjemi, Y. and Kaferstein, F.K., Global estimation of foodborne disease, *World Health Org. Stat. Q.,* 50, 8, 1997.
4. Buzby, J.C. and Roberts, T., Economic costs and trade impacts of microbial foodborne illness, *World Health Org. Stat. Q.,* 50, 57, 1997.
5. Theis, M., Case problem: balancing the risk of foodborne illness with patient rights and quality of life, *J. Am. Diet. Assoc.*, 99, 9, 1999.
6. Koren, H. and Bisesi, M., *Handbook of Environmental Health and Safety,* Lewis Publishers, Boca Raton, FL, 1996, chap. 2.
7. Stuart, D., Study Puts U.S. Food Poisoning at 76 Million Yearly, *New York Times,* September 17, 1999, p. 1.
8. Anonomyous, Food Poisoning: The Real Extent of Food-Borne Disease, *Consultant,* 39, 3, 1999.
9. Cliver, D.O., Transmission of disease via foods, *Foodborne Diseases,* Academic Press, San Diego, CA, 1990, chap. 1.
10. MCGeary, J., Nukes … They're back, *Time,* May 25, 1998, p. 34.
11. Ennever, F.K., *Metals, Principles and Methods of Toxicology,* Raven Press, Philadelphia, 1994, chap. 12.
12. Taylor, S.L., Chemical intoxications, *Foodborne Diseases,* Academic Press, San Diego, CA, 1990, chap. 10.
13. Castle, S., Dioxin in Meat is Europe's New Food Scare, *The Independent* (London), Wednesday, September 22, 1999, p. 2.
14. Reddy, C.S. and Hayes, A.W., *Food-Borne Intoxicants, Principles and Methods of Toxicology,* Raven Press, 1994, chap. 9.
15. Hunter, B., Overlooked Threats of Foodbourne Illness, *Consumer's Research Magazine,* North Bergen, New Jersey, 78, October 1, 1995.
16. Massachusetts Department of Public Health, Foodborne illness investigation and control manual, 1997.
17. CDC, Outbreak of Trichinellosis Associated with Eating Cougar Jerky-Idaho, *MMWR,* 45, 10, 205, 1995.
18. Cliver, D.O., Parasites, *Foodborne Diseases,* Academic Press, San Diego, CA, 1990, chap. 21.
19. Markel, E., Voge, M., and John, D., *Medical Parasitology,* 7th ed., W.B. Saunders Company, Philadelphia, PA, 1992, chap. 3.
20. Eckman, M. and Priff, N., *Diseases,* 2nd ed., Springhouse Corp., Springhouse, PA, 1997.
21. Cameron, J., Parasites, a "select epidemic," *Hampshire Daily Gazette,* September 12, 23, 1995.
22. Juranek, D.D., Cryptosporidiosis: sources of infection and guidelines for prevention, *Clin. Infect. Dis.,* 21, Suppl., 1, S57, 1996.

23. Hutin, Y.J.F. et al., A multistate foodborne outbreak of Hepatitis A, *N. Engl. J. Med.*, 340, 8, 1999.

24. Cliver, D.O., Viruses, *Foodborne Diseases,* Academic Press, San Diego, CA, 1990, chap. 20.

25. CDC, 1998, http://www.cdc.gov/ncidod/diseases/hepatitis/h96trend.htm.

26. Tortora, G.J., Funke, B.R., and Case, C.L., *Microbiology, An Introduction,* 5th ed., The Benjamin/Cummings Publishing Company, Redwood City, California, 1995, chap. 12.

27. Marth, E.H., Mycotoxins, *Foodborne Diseases,* Academic Press, San Diego, CA, 1990, chap. 8.

28. Moore, G.S. and Jaciow, D.M., *Mycology for the Clinical Laboratory,* Reston Publishing, Reston, VA, 1979.

29. CDC, Salmonella Serotype Montevideo Infections Associated with chicks — Idaho, Washington, and Oregon, Spring 1995 and 1996, *MMWR,* 46, 11, 237, 1997.

30. Huckstep, R.L., *Typhoid Fever and other Salmonella Infections,* E. and S. Livingstone Ltd., London, 1962, p. 227.

31. Murphy, J., FDA Approves New Anti-Salmonella Spray Developed by ARS, *Food Chemical News,* March 23, 1998, p. 26.

32. Doyle, M.P. and Cliver, D.O., *Salmonella, Foodborne Diseases,* Academic Press, San Diego, CA, 1990, chap. 11.

33. CDC, Outbreak of Staphylococcal Food Poisoning Associated with Precooked Ham-Florida, *MMWR,* 46, 50, 1190, 1997.

34. Johnson, E., *Clostridium perfringens* Food Poisoning, *Foodborne Diseases,* Academic Press, San Diego, CA, 1990, chap. 16.

35. CDC, *Clostridium perfringens* Gastroenteritis Associated with Comed Beef Served at a St. Patrick's Day Meals-Ohio and Virginia, *MMWR,* 43, 8, 137,1993.

36. Sugiyama, H., *Clostridium botulinum, Foodborne Diseases,* Academic Press, San Diego, CA, 1990, chap. 6.

37. CDC, Outbreak of *Campylobacter Enteritis* Associated with Cross-Contamination of Food Oklahoma, 1996, Morbidity and Mortality Weekly Report, 47, 7, 129, 1998.

38. Doyle, M.P. and Cliver, D.O., *Escherichia coli, Foodborne Diseases,* Academic Press, San Diego, CA, 1990, chap. 13.

39. Kurtzweil, P., Questions Keep Sprouting About Sprouts, *FDA Consumer,* 33, 1999, p. 18.

40. Bell, B.P., Goldoft, M., Griffin, P.M., Davis, A., Gordon, D.C., Tarr, P.I., Bartheson, C.A., Lewis, J.H., Barrett, T.J., and Wells, J.G., A multi-state outbreak of *Escherichia coli* O157:H7, *J. Am. Med. Assoc.,* 272, 1349, 1994.

41. Earth Vision Reports, New Pasturization Technology Makes Foods Safer, Missouri, 1999, http://www.earthvision.net/ColdFusion/News_Page1.cfm?NewsID=8790.

42. CDC, Cholera Associated with International Travel, *MMWR,* 41, 36, 664, 1992.

43. CDC, Update: Cholera-Western Hemisphere, *MMWR,* 41, 36, 667, 1992.

44. Doyle, M.P. and Cliver, D.O., *Vibrio, Foodborne Diseases,* Academic Press, San Diego, CA, 1990, chap. 17.

45. The National Restaurant Association, *Applied Foodservice Sanitation,* 4th ed., Educational Foundation of the Restaurant Association, 1992.

46. The National Restaurant Association, *Serving Safe Food,* The Educational Foundation of the National Restaurant Association, Kendall/Hunt publishing company, Dubuque, IA, 1995.

47. Reisman, M., Regulated HACCP Moves In, *Food Quality,* 5, 1, 1998, p. 21.

48. CDC, CDC's Emerging Infectious Program, CDC/USDA/FDA Foodborne Diseases Active Surveillance Network (FoodNet), November 4, 1997, http://www.cdc.gov. ncidod/foodnet/foodnet.htm.
49. Concentrated Drive to End Food Poisoning, *BBC Online*, January 28, 1999, http://new2this.bbc.co.uk/hi/end…Fsafety/newsid%5f263000/263873.stm.
50. Carter, J., New Weapon Aids Food Safety, *Sunday Republican (Springfield)*, Sunday, February 14, 1999, p. 11.

9 Water and Wastewater

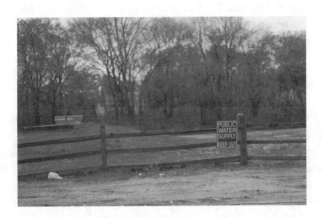

CHAPTER OBJECTIVES

A student, after reading this chapter, should be able to:

1. List and describe the different stages of the hydrological cycle and the relative amounts of freshwater on the planet.
2. List the three main consumers of water, noting each consumer's major use of water.
3. Discuss water scarcity, pointing out the areas of the world at risk of water shortage and potential associated conflict. Describe some methods of water management.
4. List and describe several sources of freshwater, and describe the types of wells used to pump groundwater.
5. Describe groundwater formation, including a discussion of contamination, recharge, water mining, fossil water, and some problems associated with overuse.
6. List the various sources and types of pollution threatening water supplies, noting the difference between point and nonpoint sources.
7. Briefly outline the Clean Water Act and the Safe Drinking Water Act, explaining the purpose of each.
8. Describe the process and purpose of wastewater treatment. List the components of a septic system, and the components of a typical wastewater treatment plant. Define and describe BOD and sag curve.

INTRODUCTION

It is impossible to imagine a world without water. We use water for drinking and cooking; we shower, wash dishes, clean clothes, flush the toilet, water our lawns, and wash our cars with clean freshwater. Water is cheap, accessible, plentiful, and relatively safe to drink. If tap water is not safe for drinking, it is usually simple enough to buy clean water at a store. We are rarely faced with any day-to-day difficulty in obtaining clean water. However, not everyone in the world has this luxury. In many areas of the world, freshwater is scarce, and tremendous time and resources are required to obtain freshwater. All too often, that water is contaminated with waterborne disease agents owing to lack of proper purification or water treatment systems. This situation results in thousands of cases of waterborne illness every year in countries with inadequate clean water.

Despite what many people think, waterborne disease and water contamination are not limited to developing nations. More and more frequently in the U.S., the media report about contaminated water and resulting illness. Water is not always as pure or safe as we think. In countries throughout the world, lakes, rivers, and oceans are used as dumping grounds for all kinds of waste. In the U.S., laws have been passed to protect our water. But these laws might not be enough to stem the tide of pollution discharged from factories, running off fields, and dumped illegally. An awareness of the problem and a strong sense of responsibility are essential for proper management and preservation of a safe, plentiful water supply. However, different cultures and different countries treat water according to their customs and traditions. Not everyone shares the same level of concern about water supply and water pollution. These differences in beliefs about the importance of water quality, water scarcity, and water use can cause conflict and hinder a joint effort to protect the world's water supplies.

THE PROPERTIES OF WATER

Water is a unique compound. It is composed of one hydrogen molecule and two oxygen molecules and has a bent configuration. The oxygen carries a partial negative charge, and the two hydrogen atoms carry a partial positive charge (Figure 9.1). Because of this shape and charge configuration, water exerts a major influence on the earth's environments, dissolving polar molecules easily. Owing to water's high boiling point and a high melting point, large amounts of energy are required to vaporize water or to melt ice. Clouds, rain, and snowmelt in winter all occur from the absorption of heat from the sun, transforming water into different states. This heat absorption leads to the hydrological cycle.[1]

HYDROLOGICAL CYCLE

Consider all the different forms of water. Water exists in the atmosphere as clouds, rain, snow, sleet, and fog; on the ground as rivers, lakes, and oceans; and underneath the earths' surface, as groundwater. These different states of water comprise the hydrological cycle. The hydrological cycle is a process involving the sun, the

FIGURE 9.1 Configuration of water. (Adapted from Turk, J. and Mirk, A., *Environmental Science,* 4th ed., Saunders College Publishing, Philadelphia, PA, 1988, chap. 17.)

atmosphere, the earth, and water (Figure 9.2).[2] The hydrological cycle consists of evaporation, condensation, transportation, transpiration, precipitation, and runoff. When surface water absorbs heat from the sun, it evaporates or vaporizes into the atmosphere. Water vapor or clouds transport water through the atmosphere. Plants also add water to the atmosphere, in a process called *transpiration* or *evapotranspiration*. When water condenses to rain, sleet, or snow, it falls to the earth as precipitation. The precipitation that is not absorbed by the soil becomes runoff, flowing over the land.[3]

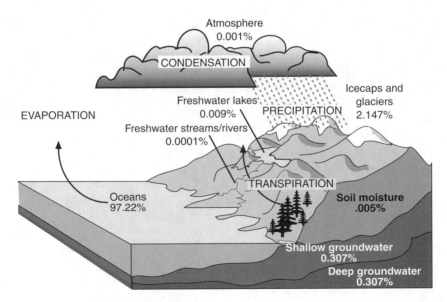

FIGURE 9.2 The hydrological cycle and percentages of water in various earth compartments. (Adapted from Grigg, N.S., *Water Resources Management,* McGraw-Hill, New York, 1996, chap. 1.)

WATER RESOURCES

A view of the ocean would convince even a desert dweller that vast amounts of water cover large regions of this planet. Water covers 71% of the earth's surface. The approximate amount of water on the earth is 1.3 billion cubic kilometers. Ninety-seven percent of this volume is saltwater contained within the oceans and seas. However, humans and animals require freshwater for consumption, which makes up only 3% of the total amount of water. Much of this 3% of freshwater exists as icebergs or deep groundwater. Less than 1% of the earth's water exists as accessible freshwater that is available for human use.[4] Thus, despite the appearance of vast amounts of available water, freshwater sources are limited. In addition, most of the easily accessible sources of freshwater are currently in use. To obtain new sources would require significant expense and effort.[5] Increasing world demand for water, coupled with the threat of water pollution, is creating a growing crisis of water supply. As the demand for water continues to grow, new sources or new water management strategies will be required to meet people's needs.

WATER AND HEALTH

Water comprises approximately 67% of the human body, and is a major component of the blood and lymph flowing through our bodies. Humans must drink water to survive: without water, a human will live only 3 d. Water also plays a major role in hygiene, as simple hand washing can prevent the spread of disease. Access to clean water is critical to human well-being and survival.

Many people in the world lack access to pure freshwater, and their health often suffers as a result. In fact, over 1.7 billion people in the world lack access to clean drinking water.[6] In an effort to provide all people with clean water and sanitation facilities, the United Nations declared the 1980s the International Drinking Water Supply and Sanitation Decade. Unfortunately, much of the world's population remains in need of pure freshwater.[2] In some developing countries, the poor living on the urban outskirts must purchase water from vendors at rates 40 times the municipal rate.[6] Some people must use polluted water for drinking and bathing, as it is the only water source available. The Ganges River in India demonstrates this problem. Revered as a goddess by Hindus, the Ganges serves as a source of water for drinking and bathing, religious rituals, cremation, and sewage outflow. All of this human activity occurring in and along the river leads to rampant waterborne disease in this region, claiming the lives of 2 million Indian children a year. Yet people still drink, cook, swim, and immerse themselves in this holy, but polluted river.[7] Although modern technology may help clean up polluted waters and improve public health, it is not always welcome. A water supply project in Quiescapa, Bolivia, was initially stalled because of villager resistance to abandoning unsanitary traditions, such as allowing animals free access to the water source. Teaching the villagers about water purity helped elicit local cooperation toward completion of the project.[8]

Drought, military conflicts, inadequate government support, and insufficient funds all contribute to the problem of supplying clean water to the world's population. Rural areas are especially hard hit. Worldwide, estimates of investments in water and sanitation projects find that 80% is allocated to urban needs and only 20% to rural facilities. Unfortunately, in developing countries, water facilities in rural areas are often in disrepair. The lack of simple spare parts can render a water supply such as a hand pump useless.[9]

WATER SHORTAGE AND SCARCITY

Water on this planet is not evenly distributed. Some areas of the world have plentiful water supplies; others have little or none. Nearly a billion people in 50 countries live with severe water shortage every day of their lives.[10] A 1995 World Bank report indicated that 40% of the world's population lives in countries facing water shortages. These shortages threaten the public health by causing dehydration, use of polluted water, and lack of proper sanitation facilities.[11] One out of every five people on this planet lacks access to clean water supply. People living in regions of water scarcity often must devote much time and energy to obtaining water, sometimes walking several miles to reach a source of drinking water. These water shortages occur in areas with high rates of population growth, such as Central Asia and sub-Saharan Africa. As increased population means an increasing demand for water, the problems will only escalate. Parts of China, India, and South America are also experiencing water shortages.[12]

Globally, the demand for water has been increasing at about 2.3% annually, with a doubling of demand occurring every 21 years. Such a demand for water cannot be met by our current water supplies. In addition to the pressure of increased demand for water, water pollution also threatens the available water supply. A startling 95% of sewage throughout the world and an increasing amount of industrial waste are discharged untreated into the world's rivers.[12] At some point in the future, worldwide water use will be limited by physical, economic, and environmental limitations.[5]

Water Rights and Conflicts

With clean freshwater in demand, conflicts over water rights and resources are inevitable. Many major water sources cross national boundaries, sometimes causing disputes between countries. Most of the undeveloped sources of freshwater traverse two or more countries. These international waters are actual and potential sources of contention.[13] Asia and the Middle East suffer from water-related disagreements. Such turmoil is not new: In 1950, a quarrel over water occurred between Pakistan and India, bordering on war.[14] Other more recent disputes have involved other countries. Bangladesh and India have locked horns over use of the Ganges and other shared rivers.[15] Turkey's water projects have met opposition from Syria and Iraq. Egypt opposed Ethiopia's plans to use additional water from the Blue Nile for irrigation.[16] The Jordan river basin has increased tension between Israel and Palestine. In the U.S., opposing factions also bristle over water rights. NYC's attempts to control the upstate New York watersheds that supply the city's water angered upstate residents.[17] Water disputes have also occurred in the western U.S., where water is scarce and demand high. The Colorado River serves many cities and agricultural areas, all of which want their fair share of the river.[18] In such cases, choices must sometimes be made regarding water allocation. Clearly, water scarcity and unclear water rights can cause or exacerbate national and international conflict.

These examples of water conflict highlight the importance of water as a pressing global issue. Interestingly, water scarcity has not yet garnered the level of attention global warming and ozone depletion have received, even though it poses a more immediate health threat to billions of people. The countries most affected by water scarcity are developing countries, where a lack of freshwater could seriously limit or hinder growth. The amount and level of water conflict will grow with the level of water scarcity. Proper water management and international cooperation could halt some of the disagreements over water allocation, and even prolong the use of scarce water resources.[19]

Water Consumption and Management

Water is an integral, necessary part of our daily lives. We use water for food production, power generation, dishwashing, and even recreation. Although most people agree that water is a valuable resource, it does not appear to be properly valued. Water abuse and overuse demonstrate that it is not. Many factors impact the amount of water a country uses and the way it is used, including the economy, available technology, level of industry, agriculture, culture, and climate. For example, as standards of living increase in developing countries throughout the world, water use will increase, as more people use flush toilets, dishwashers, and washing machines.[5] Effective water management would improve water efficiency, decrease consumption, and help preserve remaining resources

Water management encompasses a number of techniques to preserve water resources, such as adequate water pricing, community education, and leak prevention. Part of the problem with excess consumption and abuse of water through pollution has to do with inadequate water pricing. The consumer's price of water rarely reflects the actual cost of storage, treatment, and delivery. In the U.S., water is generally inexpensive, which contributes to an attitude of excess. A higher water price that includes the true cost would stimulate conservation, as consumers would bear the cost of overconsumption. If water has no or little value, then it will be wasted without a second thought. As the value of water increases, so does the attention to its effective use. For example, Germany has the highest price for water in all of Western Europe and North America. It also has the lowest overall average water use per person. On the other hand, the U.S., with the least expensive water has the highest per capita use.[20] One drawback to accurate water pricing is the unfair burden it places on the poor. Urban poor often pay more for water than nonurban consumers. Increasing water prices to truly reflect its costs while ensuring that basic water needs are met would stimulate water conservation without detriment to anyone.[21] Education is another key component of a successful water management plan. Informing consumers about the impact of their water use will engender a greater sense of responsibility in the user and hopefully aid in conservation efforts. In addition, large quantities of water are lost through leakage. Stopping this loss would save tremendous amounts of water. All these techniques encourage sustainable water use. Sustainable water use implies meeting current needs without diminishing the resource for future generations and at no cost to the environment.[21] As the world's water demands increase, water management might be an increasingly valuable tool to combat scarcity and pollution.

WATER USE

OVERVIEW

Water use changes over time as countries and economies develop, standards of living rise, populations grow, and technology expands the realms of water use. Water use even varies with time of day and time of year, with the greatest uses occurring during the evening and in summer, respectively. The three major water consumers in the world are agriculture, industry, and household or individual use. In the U.S., the average daily per capita water use from public supplies is approximately 180 gal/d. This includes domestic, industrial, commercial, and public use.[22]

AGRICULTURE

Agriculture consumes the largest portion of the freshwater supply, with over two thirds of the world's water demand used for irrigation. Sixty percent of this water is lost to evaporation or runoff. Such an inefficient system cannot continue as water scarcity escalates in scope. As water resources become scarce, agriculture will need to develop a more efficient system of water usage, using less water to grow more food.[23] In addition, agricultural water use tends to be heavily subsidized, both in the U.S. and throughout the world. Water subsidies provide little incentive to economize on water use and may in fact promote wasteful practices. Droughts can wreak havoc on agriculture, because they rely so heavily on water withdrawals. In 1998, Texas and Oklahoma experienced severe droughts that devastated the agricultural sector through crop loss, costing the state and farmers billions of dollars.[24] Globally, agriculture's share of world water use has been declining.[5]

INDUSTRY

The industrial sector also uses large quantities of water for numerous purposes, including manufacturing, cooling and condensation by power plants, and waste disposal. Approximate industrial water use in the U.S. is over 200 billion gallons per day.[25,26] Worldwide, industrial demand for water is expected to increase approximately 3 times the current rates in the next 20 years.

DOMESTIC

U.S. domestic water use includes cooking, showering, flushing toilets, washing dishes and laundry, and watering the lawn. Water is also used extensively for recreation. In the U.S., household water use is extensive. Technology allows for large volumes of water to be used in dishwashers, washing machines, and flush toilets. Per capita domestic use in the U.S. ranges between 75 and 135 gal per capita

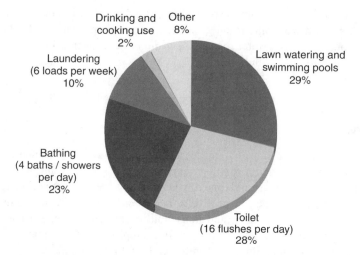

FIGURE 9.3 Water usage in the U.S. based on a four-person family. (Adapted from Grigg, N.S., *Water Resources Management,* McGraw-Hill, New York, 1996, chap. 1.)

per day (gpcd) (Figure 9.3).[27] The bulk of domestic water use serves for flushing toilets, showering, and watering lawns. These domestic uses do not occur in many developing countries, where water availability is limited. In those countries, water serves to meet basic domestic needs, such as cooking and drinking.

SOURCES OF DRINKING WATER

SURFACE WATER

The amount of surface water on the planet is approximately 230,000 km³, or about 0.017% of the total volume of water.[2] The U.S. has vast amounts of surface water, with 3.5 million miles of rivers and streams and 41 million acres of lakes. This surface water serves as drinking water to a large number of Americans. Sources of surface water develop through rainfall or snowmelt. Precipitation that is not adsorbed into the soil flows along the ground as runoff, entering streams, rivers, and lakes. These waterways are sources from which water may be drawn off for human use. The areas surrounding rivers and lakes from which water flows into the waterways are called *watersheds*. Surface water is accessed through intakes, concrete structures on or below the surface of the water that divert or draw water from the source. Because of its exposed state, surface water is prone to contamination from a number of sources, including diffuse pollution such as agricultural runoff. FAO estimates that each year about 450 km³ of wastewater is discharged into rivers, streams, and lakes.[28] Surface water sources and watersheds require protection and management to limit or prevent contamination. Watershed fencing, limited recreational access, and public education can help protect surface water against pollution.[29]

GROUNDWATER

Not all drinking water sources exist above ground. In fact, much of the available freshwater exists as groundwater, below the surface of the earth. Groundwater volume worldwide is estimated to be 8.5 million cubic kilometers, or 0.62% of the total water volume.[2] Groundwater sources supply drinking water to 50% of the people living in the U.S. and to 90% of people living in U.S. rural areas. Groundwater develops in the following manner: When rain falls on the earth, some of the water percolates downward through the spaces in the soil, pulled by gravity. At a certain point, the water reaches an impermeable layer of rock. At this layer, the water stops moving (Figure 9.4). As additional water flows downward, it fills all of the spaces between the rocks and sand that make up the soil above the impervious layer. This area of saturated soil is the zone of saturation. The top of the zone of saturation is the water table. The water table can rise and fall depending on the amount of rainfall and outflow of water from the zone of saturation into a river or a lake. Groundwater sources able to economically deliver water are referred to as *aquifers*. Some groundwater sources have a layer of impermeable rock above and below the zone of saturation. This type of water formation, called an *Artesian aquifer*, is under greater pressure than atmospheric pressure because of the weight of the water.[1]

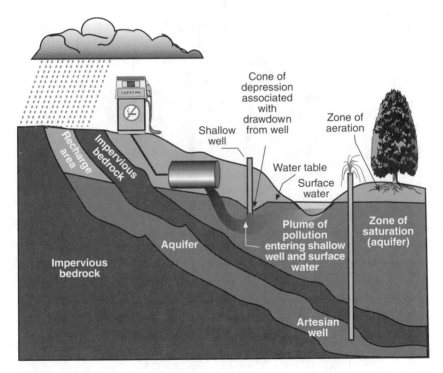

FIGURE 9.4 Groundwater configuration. (Adapted from Koren, H. and Bisesi, M., *Handbook of Environmental Health and Safety*, 3rd ed., Lewis Publishers, New York, 1996, p. 273.)

Wells

Wells provide drinking water to millions of people throughout the world. In some areas of the U.S., no access to a public water supply exists, so wells serve as the central water supply. Wells may be private, serving a single user, or municipal, serving many people. A well site requires careful consideration of several factors: the source of the well water, its purity, the location of the well, and well water protection, both above and below the ground. Several types of wells exist, ranging from the crude to the sophisticated. Some wells are dug (Figure 9.5) or bored; others are driven or drilled (Figure 9.6). Wells vary in their susceptibility to contamination. Open hand-dug wells are frequently contaminated by animal feces, exposing the user to potential waterborne disease. Such wells tend to be shallow and less than 25 ft in depth. The shallow nature of these wells makes them prone to drought and contamination. Covering a dug well decreases the likelihood of contamination. However, if the groundwater itself is contaminated, the well water should be treated before it is consumed. More than 90% of residential wells are drilled. These wells are much deeper, reaching depths of up to 350 ft. They resist contamination and

FIGURE 9.5 Dug well with two-pipe jet pump installation. (Adapted from USEPA, *Manual of Individual Water Supply Systems*, Washington, D.C., 1973, p. 34.)

FIGURE 9.6 Drilled well and submersible pump. (Adapted from USEPA, *Manual of Individual Water Supply Systems*, Washington, D.C., 1973, p. 34.)

yield 2 to 5 gal/min more than 98% of the time. Septic tanks and farm runoff pose a significant pollution threat to well water.[30–32]

Groundwater Contamination

Although groundwater may appear protected, it is not immune from contamination. As far back as the late 1960s, synthetic organic chemicals were found in groundwater sources. Additional instances of groundwater contamination occurred in the 1970s and 1980s. Despite its seemingly inaccessible location underground, groundwater is vulnerable to pollution in several ways. Rainfall may transport surface pollutants deep into the soil, down to the water table. Polluted surface water may infiltrate a groundwater source. Improperly disposed hazardous waste can seep into groundwater. A staggering number of potential pollution sources exist in the U.S., including (1) over 23 million septic systems; (2) between 5 and 6 million underground storage tanks; (3) millions of tons of pesticides and fertilizers; and (4) municipal landfills and abandoned hazardous waste sites. Agricultural runoff, accidental spills, stormwater, and wastewater are all potential sources of groundwater contamination. Sources of groundwater pollution may be divided into planned and unplanned activities, involving surface and subsurface pollution from point and nonpoint sources (Table 9.1). A point source refers to pollutants entering the environment from a specific point such as a pipe or a specific source such as a factory or treatment

TABLE 9.1
Sources of Groundwater Pollution

SOURCES OF GROUNDWATER POLLUTION			
PLANNED ACTIVITIES			
Waste Disposal Landfills NP Open dumps NP Surface impoundments P Waste tailings P Land application of sewage sludge P	**Storage or transport** Material stockpiles P	**Agricultural** Pesticides NP Fertilizers NP Irrigation practices NP Animal feeding P	**Other** Road de-icing NP
Waste Disposal Injection wells P Septic tanks P Cesspools P Other wells P Sewage pipelines NP	**Storage or transport** Undergound storage tanks P Pipelines NP		
UNPLANNED ACTIVITIES			
Urban runoff NP Acid deposition NP Accidental spills P Natural leachates NP			
Saltwater intrusion NP	Groundwater / surface water interaction NP	NP = nonpoint source P = point source	

(Left-margin labels on the table, top to bottom: SURFACE, SUBSURFACE, SURFACE, SUBSURFACE)

plant. Nonpoint source refers to pollutants entering the environment from a broad area and may include scattered sources. The types of pollutants found in groundwater include nitrates, inorganic ions, synthetic organic chemicals, and pathogens. Many of these contaminants pose significant health risks.[35–37]

Recharge and Water Mining

Recharge is the replacement of groundwater by natural processes. Recharge of aquifers can take hundreds and possibly thousands of years. The rate of groundwater use in the U.S. presently exceeds the rate of recharge of these aquifers, which may lead to water shortages in the future. Overuse of some water systems in the U.S. has caused severe water stress. *Water stress* is an actual term defined as the ratio of water withdrawal to water availability. Water stress falls into four categories: low, moderate, medium-high, and high. It occurs when more than 10% of the available freshwater is being used.[27] Currently, the Ogallala Aquifer, which underlies eight states in the western and midwestern U.S., and water systems in Florida are stressed from the large quantities of water being pumped out of these underground sources

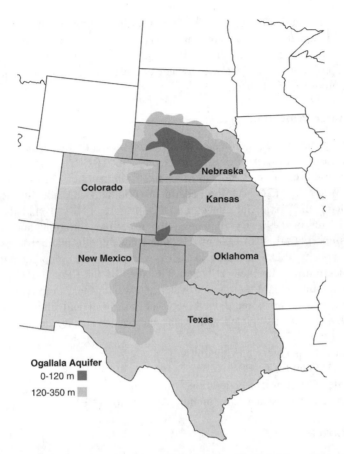

FIGURE 9.7 Geographic location of the Ogallala Aquifer. (Adapted from Turk, J. and Mirk, A., *Environmental Science,* 4th ed., Saunders College Publishing, Philadelphia, PA, 1988, chap. 17.)

(Figure 9.7). In addition, certain sources of groundwater, like coal or oil, are considered nonrenewable resources, because of the length of time they took to develop. The overuse of these "fossil" waters is called *water mining*, because the resource is being permanently depleted. An example of fossil water is the Ogallala Aquifer. Water from this aquifer is used for irrigation by farmers in Kansas, Nebraska, Oklahoma, and Texas. The large quantities of water used in irrigation greatly exceed the rate of replenishment. Unfortunately, rainfall in this region is low compared to other parts of the country: only 25 to 100 cm/yr, compared to the eastern U.S., which receives 100 to 150 cm/yr.[1,36]

Subsidence and Salination

Aquifer depletion can result in subsidence and saltwater intrusion. The first, subsidence, involves a settling of the soil as the water is pumped out. The rocks and dirt settle more tightly together, and the volume of the soil decreases. This decreased volume of earth cannot hold as much water as it did previously. The second problem with aquifer depletion is saltwater intrusion, which occurs in most coastal areas. Freshwater is pumped out of the aquifer, and the zone of saturation decreases at both the upper and lower levels. Saltwater can seep into the aquifer at the lower level, polluting the freshwater.[1]

Groundwater Protection

In the U.S., groundwater is protected directly and indirectly by various laws, including the Resource Conservation and Recovery Act (RCRA), the Comprehensive Environmental Response, Compensation and Liability Act (CERCLA), the Safe Drinking Water Act, (SDWA), the Federal Insecticide, Fungicide, and Rodenticide Act (FIFRA), the Toxic Substances and Control Act (TSCA), and the Pollution Prevention Act of 1990. By regulating the use, transportation, and disposal of thousands upon thousands of chemicals, these laws are an attempt to preserve the quality and purity of U.S. groundwater. However, a national comprehensive groundwater protection plan has not yet been established by the EPA.[37]

Do you want to learn more about water and wastewater? Point your browser to http://www-unix.oit.umass.edu/~envhl565/weblinks.html and click on any of the Chapter 9 Web links associated with water and wastewater including:

- Envirofacts Querying Capability
- Water Quality: USEPA
- Water Environment Federation
- New England Water Environment Association (NEWEA)

OTHER SOURCES

Desalination

Desalinization is a technique of removing the salt from ocean water to provide drinking water. First, filters remove sediments, suspended solids, and salt from the feed water. Applying pressure, this water is then pushed through a semipermeable reverse osmosis filter, resulting in water that is 99% free of salts, minerals, and other ions. The product water then passes through a charcoal filter, removing unwanted tastes and odors. In the last step, the filtration process is completed by an ultraviolet sterilizer, destroying 99.8% of all microorganisms, including viruses and bacteria.[34] Although this process currently carries a high price tag, it does offer a means to access plentiful ocean water. This practice is currently in use in some areas of water scarcity, such as the Middle East.

Bottled Water

Even though Americans have easy access to tap water, as a country, we consume large quantities of bottled water. In 1994, Americans consumed 2.43 billion gallons of bottled water (11.3 gal per person), at a cost of $2.1 billion. That number has escalated beyond 4.6 billion gallons (17 gal per person).[38] Many people believe that bottled water is safer and cleaner than tap water. However, bottled water is sometimes no better and can be worse than regular tap water. Tests by government agencies have discovered arsenic, bacteria, and concentrated chemicals in bottled water. In addition, many of the claims regarding bottled water, such as *pure* and *natural* are meaningless, as these terms are not regulated by any government agency. Although many people visualize crystal clear streams filling their bottles of water, the source could be anywhere, including a kitchen faucet. Despite these drawbacks, bottled water can serve as an excellent substitute for tap water that may be unsafe for drinking. The more responsible bottlers take additional precautions in purifying their water products. Specific wording on the product reveals information about the level of purity. A water labeled *Artesian* is well water from a confined underground body of water. Such water is less likely to be contaminated by chemicals from the surface. Bottled water that has been "purified" has been treated by one of several means of purification, such as distillation or disinfection with ozone or filtration through carbon or by reverse osmosis. Both Artesian and purified water are lead free.[39]

Dams

Dams have long been used as a means of controlling and storing water for human purposes. Reservoirs created by dams allow access to large quantities of water that would otherwise have flowed downstream. In addition to helping meet water demand, dams provide a method to manage river basins. Despite these positives, dams are not popular, as they may disrupt the natural ecosystem. Despite the immediate benefits of increased power production and better navigation, the Egyptian-built Aswan Dam was one such unpopular national project. The Aswan Dam caused the dislocation of 60,000 indigenous people, the flooding of archeological sites, a decline in overall fish populations, and the destruction of many natural bank ecosystems.[40] By the same token, large-scale dams such as the Three Gorges Dam in China have generated tremendous international environmental controversy and protest. Nevertheless, dams continue to provide water access to areas around the world.[41]

Water Reuse

The possibility of reusing water has become a reality in some countries as freshwater resources become depleted. Reuse typically involves the use of treated wastewater for nonconsumptive purposes, such as agricultural irrigation, landscaping, and industrial uses. Owing to the high purity of treated effluents, water reuse does not appear to pose a health threat when used for nonpotable purposes. An excellent means of water conservation, water reuse will continue to expand into areas of the world with increasingly limited water supplies.[42]

Reusing wastewater for human consumption is a more controversial, perhaps less appealing option. However, it is a reality in some U.S. cities, such as Los Angeles and El Paso, TX. In these cities, treated wastewater is injected directly into aquifers, in a process called *artificial groundwater recharge*. This technology has attracted some criticism, as the safety of this practice is still in question.[43] Interestingly, some areas of the world already use wastewater for drinking and bathing, albeit indirectly.

Large cities at the downstream end of major rivers, such as London and New Orleans, draw drinking water from the river. By the time the river reaches these cities, the water has been used several times over for various purposes upstream, including many polluting uses.[44]

WATER POLLUTION

OVERVIEW

Water in its natural state is not pure. Water naturally picks up dirt, minerals, and particles as it flows over the ground. Much of the pollution that threatens our water supply today is anthropogenic, in that it is generated by humans, and not by a natural process. Industry, agriculture, and overpopulation have all contributed to pollution of the world's water supply. In this country, the media reports stories about beaches closed because of dangerously high bacterial counts or about wells contaminated with carcinogenic chemicals. Many different sources of contamination and pollution exist, from microscopic organisms to organic chemicals. In the U.S., laws and regulations exist to try to maintain pure water supplies. Worldwide, water protection and water purity vary from country to country. Some strategies to protect water supplies include protection of areas near sources of drinking water, limitations on pollutant discharges into our waterways, and the processes of chlorination and filtration.[45] However, these strategies are not always effective. Sometimes, a disinfection method, such as chlorination, might not kill all the microorganisms, which might lead to waterborne illness. Other times, a contaminant might go undetected in a public well for years, exposing people to increased risk of cancer from their drinking water. A highly publicized example of well contamination occurred in Woburn, MA, where town wells were discovered to contain trichloroethylene, a suspected carcinogen. Many suspected this contamination was responsible for a cluster of childhood leukemias in the vicinity of the alleged polluting industries.

The potential for water pollution varies in type and scope throughout the U.S. and the world. A country's level of industry, agriculture, population, socioeconomic conditions, water delivery systems, and regulation all impact the degree of water contamination. Historically, water pollution in this country did not garner significant national attention until the 1960s and 1970s. Early legislation in the 1940s, 1950s, and 1960s allocated and expanded federal funding for municipal wastewater treatment plants. But it was not until the passage of the Federal Water Pollution Control Act in 1972, that the U.S. began a serious effort to mend its damaged and abused waterways. The Federal Water Pollution Control Act, renamed the Clean Water Act in 1977 (CWA), established national standards for the nation's waterways and set limitations on allowed pollutant discharges. The intent of the act was to clean up the nation's waterways, including a return of navigable waters to a "fishable and swimmable" condition by July 1983 and a halt of pollutant discharges into waterways by 1985.[45] Subsequent amendments in 1977 and 1987 strengthened this act, tightening regulation on pollutants. Despite this well-intentioned legislation, many rivers and lakes remain dangerously polluted.[46] The EPA estimates that the public and private cost for water pollution treatment is $64 million per year.[47]

WATER QUALITY

Water quality encompasses various characteristics of water, from taste and color to temperature and purity. Different levels of water quality are found regionally and throughout the world, based on available technology and economic resources. Water quality can vary, depending on its intended use: high quality is needed for drinking water, whereas lower quality is sometimes acceptable for irrigation purposes, as in wastewater reuse. Water use may also affect water quality, degrading its purity through pollution or a pH change. With pollution threatening the U.S. water supply, drinking water must be evaluated against certain standards to determine its safety and acceptability. Several different types of pollutants may contaminate a water supply. These can be categorized as physical, chemical, biological, and radioactive contaminants. Each different category poses different types and levels of hazards to the consumer when they are present in the water supply. Consequently, in the U.S., public water supplies employ testing to ensure its purity and safety to the consumer.

Routine biological testing involves the culturing of samples of 100 ml of tap water for the presence of indicator bacteria known as *coliforms*. The coliforms inhabit the intestines of warm-blooded animals including wildlife, birds, and humans. Their presence in the water supply indicates the possibility of fecal contamination and the potential for the presence of pathogens. The coliform test is used because the pathogens are difficult to grow. The absence of the indicator organisms does not ensure that the water is free of pathogens. Chlorination may kill the coliforms while permitting the resistant cysts of parasites such as *Cryptosporidium* or *Giardia lamblia* to survive. However, the coliform test has value as an indicator of potential contamination. The standard for water supplies is less than 1 coliform per 100 ml water. Supplies exceeding this level require additional testing. If the supply continues to exceed the standard, then additional chlorine is usually added. In the event that the coliform remains high even after increasing the chlorine level, the town may issue a boil order requiring consumers to boil their water before consumption.

Tests for physical contaminants include odor, taste, color, turbidity, and total dissolved solids, referring to bicarbonates, sulfates, calcium, and other minerals and matter. These aesthetic water-quality criteria impact the consumer's acceptance of the water. Foamability and excessive heat are also considered pollutants. Turbidity occurs when the water contains suspended clay or silt and may indicate other pollutants. Color in the water results from organic matter, algae, metals, or colored industrial waste. Taste and odor signify organic material in the water and may also indicate the presence of pathogens or other pollutants. Foamability exists when detergents are present in the water supply. Although not a health hazard, foam may indicate the presence of contamination.

Chemical contamination, both organic or inorganic, is tested using a spectrum of methods. Presently, more than 83 chemicals are the basis for testing in municipal water supplies, which are defined under the SDWA as having 15 service connections used by year-round residents or regularly serving 25 or more people. The EPA has established standard or maximum contaminant levels (MCLs) for these chemicals. The SDWA was reauthorized in 1986, also requiring filtration for most public supplies having surface water sources in order to remove pathogens such as cryptosporidia and *Giardia lamblia*. Most of the public water supplies serving more than 10,000 people may require filtration under these requirements, and most of these systems have insufficient funding to implement these federal mandates. Under-funding may also be the reason why a majority of small water supplies have not been adequately checked, and may be operated by persons lacking training or qualifications, resulting in a large proportion of waterborne diseases occurring in these communities. More than 60% of the nation's 60,000 community water supplies that serve less than 500 people are at increased risk of chemical and microbial contamination.[48] The contaminants of concern in drinking water include microbes, volatile organic chemicals, synthetic organic chemicals (including pesticides), inorganic chemicals (lead, mercury, and arsenic), and radionuclides.[22]

RIGHT TO KNOW

Consumers concerned about their tap water now have easier access to water-quality information, thanks to the finalization by the EPA of a new law. This law requires public water systems to inform their consumers about the quality of their drinking water, including any contaminants and related health hazards. Water suppliers that serve the same people year-round must prepare annual water-quality reports (con-sumer confidence reports or drinking-water-quality reports) for their customers. As 200 million Americans receive their water from water utilities affected by this law, its impact will be significant. Utilities are required to send out written reports detailing the quality of the water. As these utilities already report this information to state departments of environmental protection, no additional water testing is necessary. This legislation empowers consumers through access to vital public health information.[49] EPA also has a local drinking water information Web page that offers links to many Internet-accessible water-quality reports (http://www.epa.gov/safe water/dwinfo/index.html). Additionally, the public can use the site to find specific information about its local drinking water supply, including information about the state's drinking water program and source water protection program.

TYPES OF POLLUTION

Inorganic Compounds

Inorganic compounds of many types may be present in water systems. Of particular concern are lead, cadmium, mercury, arsenic, and copper. These substances can cause serious acute and chronic health problems. Lead in particular has been linked with brain damage and nervous system disorders. Mercury has also been implicated in central nervous system problems.[50] Cadmium has been known to cause kidney damage.

Synthetic Organic Chemicals

Organic compounds in water systems can present a serious health threat. The word *organic* refers to a substance containing the element carbon. Carbon can bond with four other atoms, which allows it tremendous variety in the types of compounds it forms. The types of synthetic organic chemicals in existence today number in the tens of thousands. Organic compounds can occur naturally in water, in the form of decaying vegetation and microorganisms. These natural organics may cause taste and odor problems in water. Of greater concern are the tremendous numbers of synthetic organic compounds that may pollute waterways. Sources of synthetic organic compounds include industry, agriculture, and even households. Each time a toilet is flushed or sewage sludge and manure is applied to the fields, large quantities of drugs like antibiotics, hormones, and pain killers, etc., are entering the environment and seeping into our waterways.[50] Disinfection with chlorine may also create organics in a water supply. Synthetic organic compounds can be classified as volatile organic compounds (VOCs) or synthetic organic chemicals (SOCs). VOCs, as their name implies, are volatile and vaporize easily. VOCs migrate quickly in the ground, and travel far from their site of introduction. VOCs present a threat to groundwater, where they are less able to vaporize and can accumulate. A dangerous VOC, carbon tetrachloride, has been found in well water. SOCs threaten surface waters, through both accidental and purposeful discharges into waterways. Atrazine, chlordane, lindane, and toluene are several dangerous synthetic organic chemicals in use today.[51,52] Polychlorinated biphenyls or PCBs are other hazardous organic chemicals that have been found in some U.S. water supplies, even though production in the U.S. ceased in 1977.

Radioactive Material

Radioactive materials from natural and synthetic sources can migrate into water systems. The radiation emitted from decaying radionuclides can have dangerous, even fatal, health consequences. The most common radioactive substances in water are radium, uranium, radon, and certain synthetic radionuclides. Whereas naturally occurring radionuclides appear mainly in groundwater, surface waters are more likely to contain artificial radionuclides from atmospheric fallout.[53]

SOURCES OF POLLUTION

Overview

Pollution sources are labeled *point* sources or *nonpoint* sources. A point source of pollution stems from a single discrete location, such as an outflow pipe. This type of source can be monitored, measured, and controlled. Industrial discharges are generally considered point sources, as they enter waterways at a single location. The EPA regulates point source discharges through a permit system. A nonpoint source of pollution has a wide range of entry into a waterway and is difficult to measure and manage. Agricultural and livestock farming activities are considered nonpoint sources of pollution.[54] Nonpoint sources of pollution have now surpassed point sources as the main culprits of water pollution in the U.S.[45]

Underground Injection Wells

Underground injection wells (UIW) may pollute water. An underground injection well is any hole in the ground that is deeper than it is wide, into which liquids are injected. Five types of UIWs exist, Class I to Class V, each used for a different purpose. The most troubling are the Class V wells, which are shallow injection wells. These simply constructed wells may contain various pollutants, including industrial waste, agricultural drainage waste oil, and wastewater. All these may threaten sources of drinking water. The likelihood of contamination depends on the proximity of the well to an aquifer, the quality of the well construction, and the type and amounts of fluid injected.[37]

Industrial Discharges

Unfortunately, waste creation is part and parcel of the process of manufacturing a product. In the past, the liquid portion of this waste was dumped directly into our waterways or into poorly constructed landfills that leaked dangerous substances into the water. As knowledge about the magnitude and the risks of water pollution increased, public concern over the health effects of chemicals in our water systems also increased. Many people worried that chemical pollutants in our water system would cause cancer and other diseases. Today, restrictions are placed on what and how much can be dumped into waterways. The government has enacted laws, such as the SDWA and the CWA, that require monitoring for specific chemical contaminants in water. The National Pollutant Discharge Elimination System (NPDES) regulates the discharge of wastewater into U.S. waterways. An NPDES permit stipulates the allowable levels of contaminants that may be discharged in an effluent. This level is called an *effluent limitation*. The EPA has set standard effluent limitations for specific industries, such as paper mills and chemical manufacturers. The permit also specifies the required tracking of discharged chemicals.[54] Despite the permit system, chemical contamination is still a growing problem. The vast number and volume of chemicals used by industry make it impossible to monitor water systems for each specific chemical. Even with the limited number of chemicals monitored, each year thousands of public water systems violate one or more of the thresholds, owing to the high cost of monitoring. Adding new chemicals to the hit list every few years increases the financial strain.[54-56]

Agriculture

Agriculture demands an average of 70% of the world's fresh surface waters, making it the largest consumer of freshwater. Not surprisingly, much of the pollution entering our waterways comes from agricultural sources. During rainfall, fertilizers and pesticides applied to crops can flow as runoff into surface waters or percolate through the ground to groundwater. In September 1999, North Carolina was flooded as a result of Hurricane Floyd. The floodwaters managed to shut down one third of the state, an area of 18,000 mi^2, affecting 2.1 million people. The floodwaters also caused huge amounts of hog feces and urine to flow into rivers and enter the groundwater there. Drinking water was tainted by such overflow of artificial lagoons, causing fecal coliform bacteria levels to increase to thrice the normal average.[57] Increasing amounts of fertilizer and pesticides used for agricultural purposes also increases the possibility of water pollution. Information on agricultural pollution in developing countries is limited. However, we do know that pesticide use in developing countries continues to increase, suggesting that incidents of water contamination will occur more frequently.[59] Another area of our country that is affected by agricultural runoff is an area in the Gulf of Mexico known to farmers and fishermen as the *dead zone*. At last estimation this area was at about 7728 mi^2. The source of the dead zone lies inland, stemming from the agricultural runoff of farms, which is then carried down the Mississippi River and out into the Gulf of Mexico. The principal cause of this *dead zone* is nitrogen fertilizer and hog-farm surface runoff. When the nitrogen reaches the Gulf, it stimulates growth just as it would on land, causing major algae growth. As the algae die and decay, they remove oxygen from the water, which results in the death of plant and animal species.[60]

Pesticides

Although pesticides have improved the world's food supply, they do pose potential health risks if they enter the water supply. Some pesticides have been linked to an increased risk of cancer. The U.S. uses 1.1 billion lb of pesticides annually. Pesticides have been found in stormwater runoff from both agricultural areas and urban areas.[56] Pesticide contamination of surface waters occurs when rainfall carries the chemicals in runoff into the waterway. Pesticides may also enter surface waters directly from the atmosphere. Pesticides have been found in U.S. surface waters since the commencement of water tests in the 1950s. Surface-water-monitoring studies conducted throughout the U.S. have detected over 100 pesticides or related compounds. The types of pesticides found in surface waters change from region to region, seasonally, and over time. Organochlorines, most of which the U.S. banned in the 1970s, have been seen in decreasing concentrations in the water supply. Herbicide use has increased, as have the concentrations of herbicides in the water supply, which regularly surpass the maximum contaminant level (MCL) standard following applications.[60]

Fertilizer

Worldwide, fertilizers are increasingly found in water sources, both surface and groundwaters. Nitrate, a major component in fertilizer, has been detected in wells in many rural areas of the U.S., where fertilizer use is greatest. In fact, 9% of wells in agricultural regions of the U.S. exceed the 10 mg/l nitrate limit. Nitrates pose a health risk to pregnant women, infants, and young children. Babies are especially at risk from a condition called *methemoglobinemia*, or *blue baby*, syndrome, in which nitrates act to reduce the ability of red blood cells to transport oxygen. This condition can be fatal.[61–63] Fertilizers in water can also cause eutrophication, which is nutrient overload of a water source. Eutrophication promotes algae growth, disrupts the ecosystem, deoxygenates water, and may result in fish kills. Nitrate contamination of groundwater poses a significant problem, as they are difficult to cleanup.[55]

Stormwater

Defined as *stormwater runoff*, *snow melt runoff*, *surface runoff*, and *drainage*, stormwater poses a threat to drinking water sources. It may contain pathogens, nutrients such as nitrates, synthetic organic chemicals, heavy metals, and salts. Groundwater contamination can occur when stormwater transports pollutants through the soil into the groundwater. Impervious surfaces, such as roads and parking lots, increase both the amount of stormwater and the level of contamination. Currently, stormwater discharges from industrial, municipal, and other small facilities are regulated under the Water Quality Act of 1987. Permits are required for industrial and municipal stormwater discharges, to lessen the possibility of water pollution.[64]

Acid Mine Drainage

Water sources may be contaminated by drainage from mines. Acid mine drainage generally comes from active or abandoned underground mines. This acidic drainage enters streams and rivers, acidifying the water and harming the aquatic environment. Heavy metals may also be leached into the water during drainage formation. Sealing off abandoned mines can alleviate this pollution threat. Reducing water flow from the mines also can limit the amount of acid mine drainage. Treatment of drainage is a third option to reduce its impact on waterways.[65]

Waterborne Disease

Historically, waterborne disease has been a significant threat to human health. As many as 3.3 billion people in the 127 countries of the developing world suffer from water-related diseases. There are almost 6 million deaths from water-related diseases each year, and 50% of people on earth lack adequate sanitation.[10] In the developing world, waterborne illness causes 900 million cases of diarrhea, which claims the lives of several million children annually.[66] Each year, an estimated 4 billion episodes of diarrhea globally result in an estimated 2 million deaths, mostly among children. Waterborne bacterial infections may account for as many as half of these episodes and deaths.[67] Before methods of water purification were established in this country, diseases like cholera and polio were common health problems. With the development of chlorine treatments and other purification methods in the early part of the 20th century, the incidence of waterborne disease decreased significantly in the U.S. However, despite extensive water treatment in the U.S., water may still harbor waterborne disease agents. Over the past 10 years, several outbreaks of waterborne disease have occurred. In fact, in the U.S., waterborne illness causes over 900 deaths annually and more than 1 million cases of gastrointestinal illness. Two reasons for the increase in waterborne disease are inadequate purification and water contamination.[68]

Microbial contamination of water occurs when a pure source of water is polluted by raw sewage or animal feces. As this water is thought to be pure, it is used for drinking water, and subsequently makes people ill. This problem occurs when an undetected contamination source enters the water delivery system of a city or town, polluting the water. The usual culprits are wild animal feces, or sometimes runoff from farms. Contaminated water could cause a host of gastrointestinal diseases, and dehydration from severe vomiting or diarrhea could be fatal in children, the elderly, and people with weakened immune systems.[69] The most common microbial contaminants are *Escherichia coli*, *Giardia lamblia*, and *Cryptosporidium parvum*.

Over the past 15 years, the increased incidence of waterborne disease has forced us to examine our water purification methods. Even when water is chlorinated, it might still harbor disease-causing microorganisms. Certain microorganisms, in particular, *Giardia lamblia* and *Cryptosporidium parvum,* are very hardy and can tolerate the concentrations of chlorine used to treat drinking water. Unfortunately, when ingested, these organisms can cause disease and even death. A major U.S. outbreak of cryptosporidiosis occurred in Milwaukee, WI, in 1993, when over 400,000 developed gastrointestinal illness from drinking water contaminated with cryptosporidium. One hundred people died from the illness during this outbreak. As cryptosporidia live in the water supplies of approximately 45 million Americans, it represents a significant public health threat. Removal of this dangerous parasite from drinking water requires expensive filtration systems.[66] Other occurrences of waterborne illness in the U.S. demonstrate that our water is not free of biological contamination.

WATER TREATMENT

Municipal Water Treatment

In the U.S., most of the drinking water comes from 200,000 community water supply systems. These systems, consisting of approximately 140,000 small-scale suppliers and 60,000 municipal supply systems, supply water to 241 million Americans.[68] Most of this water requires treatment to remove contaminants before it can safely be used for drinking water. As mentioned previously, surface water is exposed to many sources of pollution and requires extensive treatment. Owing to its protected nature, groundwater may require less treatment, though increasing reports of groundwater contamination may suggest additional purification.

Water treatment proceeds as a series of steps, each intended to remove additional unwanted material from the water and improve the water quality. The main steps of treatment are sedimentation, coagulation–flocculation, filtration, and disinfection. Other treatment includes softening, fluoridation, taste and odor removal, algae control, and aeration. After this treatment, the water should be free of almost all contaminants. Sedimentation involves the settling out of large or weighty suspended objects from the water. A settling tank holds the water for 24 h as the objects settle out. The next step in the process, sedimentation–flocculation, involves the addition of chemicals such as hydrated aluminum sulfate to facilitate large-particle formation. These particles, or floc, settle out of the water when the water is allowed to sit. Filtration also removes suspended material, but involves passage of the water through beds of sand or other porous material. Whereas filtration is generally used in large water systems, it can also be used for individual water supplies: filtration units may be attached directly to the faucet. Disinfection, discussed following this, is the next crucial step. Other types of water treatment include aeration and water softening. Aeration is a technique used to remove VOCs from a water supply. An upward stream of air intersects with a downward flow of water. As the air rises through the water, it pulls the VOCs into the airstream and out of the water. Water softening involves removing the minerals calcium and magnesium from a water supply to prevent the buildup of mineral deposits on cooking equipment, washing machines, and water tanks.[53]

DISINFECTION

The most critical step in water treatment, disinfection, should destroy all organisms in the water supply. To ensure proper disinfection, organic matter and other material must be removed prior to disinfection. An important characteristic of an effective disinfectant is that it leaves a residual to prevent regrowth of microorganisms. This residual maintains the purity of the water as it travels to the consumer. Chlorine is the major disinfectant used in U.S. water systems today. It provides an inexpensive, relatively effective method of killing most waterborne microorganisms, and protecting water as it travels through a delivery system. However, chlorine does have some drawbacks. When added to water containing certain organic matter, chlorine can combine with these pollutants to form dangerous disinfectant by-products called *trihalomethanes* or (THMs). These compounds have been linked with an incidence of bladder and rectal cancer. Despite the increased risk of cancer from trihalomethanes, chlorinated water is safer than unchlorinated water, because of the threat of waterborne disease. In the many countries throughout the world that lack purified chlorinated water, waterborne disease is a tremendous public health problem. The World Health Organization (WHO) estimates that several million people die annually worldwide from disease caused by waterborne pathogens. Not chlorinating water to avoid the risks from THMs would increase the incidence of waterborne disease.[71,72]

Although chlorine is the most widely used disinfectant, other disinfection methods exist, including ozonation, ultraviolet light exposure, and chloramination. Ozone can disinfect water 3100 times faster than chlorine and leaves no residual by-products. However, because ozone does not provide a long-lasting residual, ozonated water is treated with a small dose of chlorine. Chloramination, which involves mixing chlorine and ammonia in the water to create chloramines, also disinfects the water and does not form THMs. But because chloramination lacks the disinfectant strength of chlorine, the water usually requires a second dose of chemical.[53]

HOME WATER TREATMENT

In some areas of the U.S., the available water might not meet purity or safety standards. In such situations, people have the option of buying home water treatment systems for their individual water supply. The water treatment market contains many options for the water-wary consumer, from the simplest of systems to the complex (and expensive). A relatively inexpensive option, activated carbon filters, can remove mercury, volatile organic compounds, trihalomethanes, and unpleasant tastes and odors. However, if these filters are improperly maintained, they may act as breeding grounds for bacteria. More expensive home treatment units, such as reverse osmosis or distillation units, remove most heavy metals and bacteria.[66]

REGULATIONS

SAFE DRINKING WATER ACT

In the U.S., drinking water received from public drinking water systems is regulated by the Safe Drinking Water Act (SDWA). The SDWA allows the EPA to set maximum contaminant levels (MCLs) for water pollutants to protect the public health. An MCL is determined by calculating the greatest dose of a chemical that generates no negative health effects in animal studies. Chemicals with an MCL also have a maximum contaminant level goal (MCLG), which is the target level of a chemical in the water. Although MCLs are required, MCLGs are not. MCLs have not been established for all the potential chemical contaminants in treated water. Enforcement of the SDWA is left to the individual states, with oversight provided by the Office of Groundwater and Drinking Water, a division of the EPA. State standards for public water systems must be as strict as EPA standards. All public water systems in the U.S. must comply with the National Primary Drinking Water Regulations, which contain the MCLs. These water supply systems must test the water for specific contamination, including various inorganic and organic compounds, THMs (a specific organic), turbidity, coliform bacteria, and natural and synthetic radionuclides.[73]

The SDWA has evolved considerably since its inception in 1974. Presently, every 5 years the EPA administrator selects at least 5 chemicals from a list of contaminants not currently monitored, but which are known to be in the water supply. The administrator must decide whether to regulate for these chemicals. A substance may also be added to the list if it is a CERCLA hazardous substance or a FIFRA pesticide. Although new requirements provide improved water quality, these requirements place a financial strain on a community's water budget. To properly implement the current goals of the SDWA tens of millions of dollars will be required. The EPA estimates that $138.4 billion will be required over the next 20 years to overhaul the nation's 55,000 aging community water systems.[74]

WASTEWATER DISPOSAL AND TREATMENT

SEWAGE

In many developing countries around the world, human waste pollutes the land and the water. People lacking any sanitation facilities simply defecate on the ground, leaving fecal matter to contaminate food and water sources. The developed world generally has ready access to sanitation in the form of septic systems or municipal sewers. In the U.S., wastewater from household water use is generally routed away from the house through pipes to treatment facilities or septic systems. This wastewater contains human wastes, food, detergents, and household chemicals. Sewage or wastewater is 99.9% water. Organic matter and impurities comprise the remaining 0.1%. Upon entering the ground or a waterway, this 0.1% of matter can drastically impact the quality of the groundwater or surface water. This organic material can serve as food for organisms living in the water.

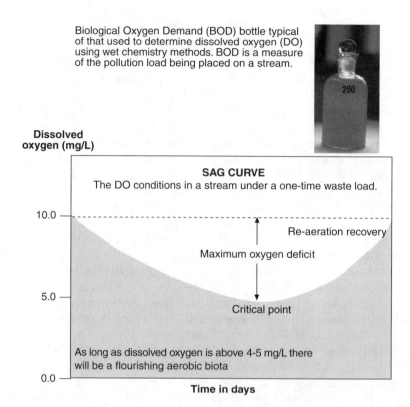

FIGURE 9.8 A typical sag curve resulting from the deposit of high-BOD materials into a receiving body of water.

BIOLOGICAL OXYGEN DEMAND

As microorganisms decompose organic material in surface water, they use oxygen dissolved in the water. When a waterway has plentiful dissolved oxygen (DO) to meet the biological oxygen demand (BOD) of the organic matter, then the water can still support aquatic life. However, if a waterway is overloaded with biodegradable organic pollutants, this decomposition process can deplete the supply of DO. This deoxygenation diminishes the water's capacity to support aquatic life and can result in fish kills and algal blooms. The BOD of waste can be determined through a simple BOD test. Water reaerates through a natural process over time and distance from the waste input. The deoxygenation and reaeration of water can be presented graphically as a sag curve. A sag curve demonstrates the level of dissolved oxygen over time, showing the critical level where aquatic life dies (Figure 9.8).[4]

TYPES OF DISPOSAL

Pit Privies

In many areas of the world, human excreta and wastewater is dumped directly into waterways, for lack of proper sanitation facilities. Such practices contaminate the

Ventilation opening

Floor

Mound draining away from base

Base

Pit is 5-6 feet deep. The waste should be frequently covered with lime. Must be a minimum of 15 m from water source.

FIGURE 9.9 Typical pit privy construction. (Adapted from Chanlett, E.T., *Environmental Protection,* 2, McGraw-Hill, New York 1979, chap. 3.)

water supply and increase the spread of waterborne disease. Rudimentary facilities such as pit privies can protect water supplies from fecal contamination. A pit privy is simply a manually dug hole, with some type of seat placed over it and surrounding walls for privacy Figure 9.9. Used mainly only in developing countries, the pit privy isolates human waste from the water supply. However, if not properly maintained, the privy itself can become a source of contamination.[75]

Septic Systems

Septic systems provide a method of wastewater disposal used by many people in the U.S., especially in rural areas or where no access to municipal sewage treatment exists. A septic system is a two-stage system, made up of the septic tank, usually a precast or reinforced concrete tank, and an absorption field, where the effluent flows into the soil. Sewage flows from the house into the septic tank, where it is held for 24 to 72 h (Figure 9.10). During this time, the heavier materials settle to the bottom, where they form a layer of sludge. Lighter particles migrate to the top of the wastewater, forming a scum layer. The middle layer, consisting of wastewater, flows out of the tank into a series of perforated pipes that lie in the drainage field. The effluent flows out of the pipes and percolates through the soil. The soil in which the effluent flows must allow for adequate percolation. This capacity is crucial for an

FIGURE 9.10 A typical septic system construction. (Adapted from Chanlett, E.T., *Environmental Protection,* 2, McGraw-Hill, New York 1979, chap. 3.)

effective septic system. Soil permeability is determined through a perc test. Filtration through the soil further purifies the effluent as soil organisms break down the organic material into simpler inorganic forms. The minimum septic tank capacity is 750 gal, extending up to 2000 gal for large households. Tanks require periodic cleaning to remove the scum and sludge layer (Figure 9.11). A view of a typical installation of a septic system, including the septic tank and leach field is shown in Figure 9.12.[53]

FIGURE 9.11 A cross-sectional schematic of a septic tank.

FIGURE 9.12 A typical installation of a septic system showing septic tank (A) and leach field (B).

Municipal Sewage Treatment

Sewage treatment provides a method to speed up and control water's natural process of purification, through biooxidation, filtration, and settling. The end product of sewage treatment, called an *effluent*, should have significantly reduced BOD, and poses little or no harm to the waterway or human health. Achieving these goals requires several stages of treatment, including primary, secondary, and tertiary treatments, and sludge disposal (Figure 9.13).

Primary Treatment

Primary treatment is largely a mechanical process, concerned with the removal of solids. First, screens are used to catch large materials as the sewage flows through them. Next, the comminuter, a grinder or a shredder, reduces the size of the solids in the water and removes any coarse objects (Figure 9.14). The sewage flows into a grit chamber, where heavy matter such as stones, sand, and glass particles can settle out. This prevents such materials from entering the pumps and destroying the mechanisms by abrasion. From the grit chamber, the sewage flows to the primary settling tank, called the *primary clarifier*.[73] Here the flow is slowed, and the settleable solids descend to the bottom, where they form sludge that is pumped out for processing and drying. A skimmer removes floating debris, and the liquid portion of the sewage overflows into a receiving channel around the clarifier (Figure 9.15). At this stage, the effluent still contains substantial organic matter causing a high BOD, and numerous microbes. Many municipal systems throughout the U.S. still stop treatment at this primary stage. Chlorine is usually added at this point, and the effluent is pumped to a receiving body of water, where the pollution load from this effluent will be high.

Secondary Treatment

Secondary treatment employs biological oxidation to break down pollutants into more stable inorganic or oxidized organic forms that resist further metabolism, and so do not contribute to the biological oxygen demand when released into a body of water. Trickling filters and activated sludge treatment employ bacteria to breakdown and digest organic material in the sewage. Trickling filters consist of a bed of coarse stones or plastic media, covered with biomass consisting of bacteria, protozoa, and fungi. When wastewater is applied to the stones, dissolved organic matter is adsorbed by the slime and digested by the microorganisms. Activated sludge treatment occurs in an aeration tank, where sewage is mixed with active or bacteria-rich sludge to create *mixed liquor*. This mixture is aerated for several hours as the sludge bacteria digest the organic material in the sewage (Figure 9.16). This mixed liquor is transferred to a settling tank, where the activated sludge settles out, most of which is routed back to the aeration tank. The effluent flows to a sedimentation tank, known as a *secondary clarifier*. An advantage of activated sludge treatment is that it can remove nitrates and phosphorus from sewage, whereas trickling filters do not.[76,77]

Tertiary Treatment

In some cases, additional treatment of the effluent is necessary to further reduce the BOD and suspended solids. A number of tertiary treatments or advanced wastewater

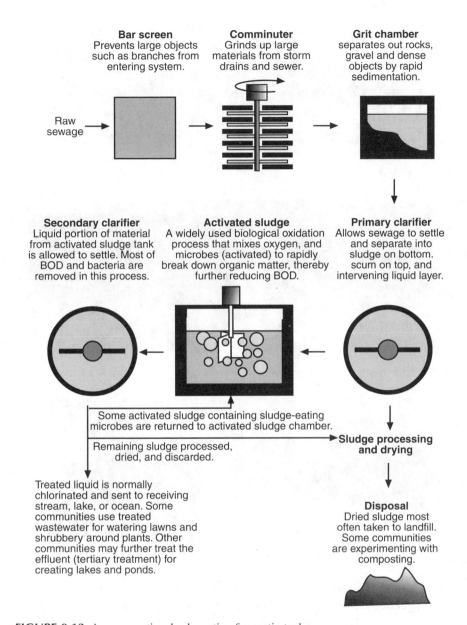

Bar screen
Prevents large objects such as branches from entering system.

Comminuter
Grinds up large materials from storm drains and sewer.

Grit chamber
separates out rocks, gravel and dense objects by rapid sedimentation.

Raw sewage

Secondary clarifier
Liquid portion of material from activated sludge tank is allowed to settle. Most of BOD and bacteria are removed in this process.

Activated sludge
A widely used biological oxidation process that mixes oxygen, and microbes (activated) to rapidly break down organic matter, thereby further reducing BOD.

Primary clarifier
Allows sewage to settle and separate into sludge on bottom. scum on top, and intervening liquid layer.

Some activated sludge containing sludge-eating microbes are returned to activated sludge chamber.

Remaining sludge processed, dried, and discarded.

Sludge processing and drying

Treated liquid is normally chlorinated and sent to receiving stream, lake, or ocean. Some communities use treated wastewater for watering lawns and shrubbery around plants. Other communities may further treat the effluent (tertiary treatment) for creating lakes and ponds.

Disposal
Dried sludge most often taken to landfill. Some communities are experimenting with composting.

FIGURE 9.13 A cross-sectional schematic of a septic tank.

treatment methods accomplish this goal, including air stripping by ammonia and rapid granular filtration.[76] At this stage the effluent may or may not be disinfected before it is released into receiving waters. Controversy exists about disinfecting wastewater with chlorine. Although the chlorine kills disease-causing organisms, it also may kill aquatic life when treated effluent is discharged into a waterway. Disinfection practice varies from state to state.[3]

FIGURE 9.14 Bar screen (left) and comminuter (right) at a municipal wastewater treatment plant.

FIGURE 9.15 A primary clarifier at a municipal treatment plant.

FIGURE 9.16 An activated sludge unit seen full and empty.

Sludge Treatment and Disposal

Sludge refers to the solids and liquids separated out of wastewater during sewage treatment. Wastewater treatment in the U.S. generates approximately 8 million dry tons of sludge annually. Sludge is generally composed of organic matter, pathogens, metals, and chemicals removed from wastewater. Treatment of sludge entails the removal of liquids and digestion of the organic matter. The treatment process includes thickening to increase solid density, dewatering to remove liquid, stabilization to digest or neutralize the sludge, conditioning to promote larger-particle formation, and disinfection. Sludge disinfection is a crucial step, as it destroys pathogens in the sludge to prevent the spread of disease. Digested sludge may be air-dried (Figure 9.17).[74,75]

Studio dried on huge rotating barrel that suctions out moisture, creating blanket of dried sludge

Sludge used in composting effort at the University of Massachusetts

FIGURE 9.17 A sludge-drying apparatus and a sludge-composting effort.

In the past, sludge was either dumped in the ocean or buried in a landfill. Today, sludge disposal poses an increasing problem, as ocean disposal is no longer legal and landfill space is increasingly scarce. One solution to sludge disposal is to use it as fertilizer (Figure 9.17). However, some sludge may contain heavy metals. These should not be used for agricultural purposes. Such sludge may pose a risk to ground-water sources if improperly disposed of in a landfill, as the metals or toxics may leach into the soil and water. Incineration presents another disposal option, although it generates air pollution, requires significant energy input, and creates ash waste.[76,78]

WATER POLLUTION AND HEALTH

The reality of polluted water can be both disruptive and devastating. In Westport, MA, wells contaminated with the cleaning solvents trichloroethylene and chloroben-zene forced homeowners to limit their exposure to their well water through drinking and cooking with bottled water, limiting shower time, and wearing gloves to wash dishes.[79] Increased risks of cancer and uncertainty about the future confront con-sumers of polluted water. Unfortunately, only time can reveal the true dangers of contaminants in our waters, as little is known about the long-term effects of exposure to various water pollutants, such as synthetic organic chemicals. Consumers armed with information about their water sources can protect themselves from exposure to polluted water. Worldwide, water pollution continues to plague many developing countries, inflicting immediate and long-term damage on the public's health.

FUTURE OUTLOOK

To adequately meet water demand in the future, we must recognize the value of this resource today. Increasing population, increased pollution, and competition for water challenge the goal of satisfying everyone's basic needs for water. Conflicting user groups, such as industry and agriculture, will vie for water rights in the face of scarcity. Effective water management, including conservation, accurate pricing, and education can aid efforts to use water efficiently. Throughout the world, the problems of pollution and scarcity will increase over time. The need to protect water from pollution will increase as freshwater supplies dwindle, both in the U.S. and elsewhere. The notion of sustainable water use will continue to gain momentum in the face of a water crisis, encouraging conservation and preservation of our water resources. On a planet covered with water, yet plagued by waterborne disease, drought, and water mismanagement, we must promote conservation, efficiency, and frugality. Every drop counts.

REFERENCES

1. Turk, J. and Mirk, A., *Environmental Science,* 4th ed., Saunders College Publishing, Philadelphia, PA, 1988, chap. 17.
2. Grigg, N.S., *Water Resources Management,* McGraw-Hill, New York, 1996, chap. 1.
3. Nadakavukaren, A., *Our Global Environment,* Waveland Press, Prospect Heights, IL, 1995, chap. 14.
4. Chanlett, E.T., *Environmental Protection,* 2, McGraw-Hill, New York 1979, chap. 3.
5. Biswas, A., Water development and environment, *Water Resources,* Biswas, A., Ed., McGraw-Hill, New York 1997, chap. 1.
6. Johnson, B.K. and Donahue, J.M., *Water Culture and Power,* Island Press, Washington, D.C., 1998, chap. 1.
7. Stille, A., The Ganges' Next Life, *New Yorker,* January 19, 1998, p. 58.
8. Yates, L.D., A water supply development project in Quiescapa, Bolivia, *J. Environ. Health,* 3, 20,1997.
9. World Health Organization, *Implementation of the Global Strategy for Health for All by the Year 2000,* 8th Report on World Health, 1993, chap. 6.
10. Gumpte, P., Water: Not a Drop to Drink, *Newsweek,* March 29, 1999.
11. Anonymous, Thirsty planet, *Environ. Health Perspect.,* 104, 1, 12, January 1996.
12. Hutchings, C., Down to the Last Drop, *Geographical Magazine,* 71, August 1997, p. 24.
13. Pinkayan, S., Asian international waters: From Ganges–Brahmaputra to Mekong, *Asian International Waters,* 4, Biswas, A.K. and Hashimoto, T., Eds., Oxford University Press, 1996, p. 1.
14. Verghese, B.G., Towards an eastern Himalayan rivers concord, *Asian International Waters,* 4, Biswas, A.K. and Hashimoto, T., Eds., Oxford University Press, 1996, p. 29.
15. Nishat, A., Impact of Ganges water dispute on Bangladesh, *Asian International Waters,* 4, Biswas, A.K. and Hashimoto, T., Eds., Oxford University Press, 1996, p. 60.
16. Hinrichson, D., The World's Water Woes, *International Wildlife,* July–August 1996, p. 22.
17. Revkln, A.C., Chasing a Deal on Water with a Few Pitchers of Beer, *New York Times,* November 5, 1995, p. 41.
18. Grigg, N.S., *Water Resources Management,* McGraw-Hill, New York, 1996, p. 493.
19. Biswas, A.K., *Water for Sustainable Development of South and Southeast Asia in the TwentyFirst Century,* 4, Biswas, A.K. and Hashimoto, T., Eds., Oxford University Press, 1996, chap. 2.
20. Hoffbuhr, J.W., Is there enough? *J. Am. Water Works Assoc.,* 6, August 1999.

21. World Health Organization, *Water Consumption and Sustainable Water Resource Management*, Organization for Economic Co-operation and Development 1998, p. 23.

22. Von Huben, H., *Water Sources*, 2nd ed., American Water Works Association, Denver, CO, 1995, chap. 5.

23. Winpenny, J., *Reforming Water Resources Policy*, Food and Agricultural Organization of the United Nations, Rome, 1995.

24. Lyman, P.L., Drought in Texas and Oklahoma Stunting Crops and Economies, *New York Times*, August 12, 1998.

25. Von Huben, H., *Water Sources*, 2nd ed., American Water Works Association, Denver, CO, 1995, p. 98.

26. Easter, K.W., Becker, N., and Tsur, Y., Economic Mechanisms for Managing Water Resources: Pricing, Permits, and Markets, *Water Resources*, Biswas, A.K., Ed., McGraw-Hill, New York, 1997, p. 589.

27. Lindeburg, M.R., *Civil Engineering Reference Manual*, 6, Professional Publications, Belmont CA, 1997, p. 7.

28. Squandering the Planet's Lifeblood, *International Wildlife*, 24, September–October 1999.

29. Von Huben, H., *Water Sources*, 2nd ed., American Water Works Association, Denver, CO, 1995, chap. 3, 7.

30. World Health Organization, *Guidelines for Drinking Water Quality*, 2nd ed., 1997, p. 93.

31. Koren, H. and Bisesi, M., *Handbook of Environmental Health and Safety*, 3rd ed., Lewis Publishers, New York, 1996, p. 273.

32. Koren, H. and Bisesi, M., *Handbook of Environmental Health and Safety*, 3rd ed., Lewis Publishers, New York, 1996, p. 257.

33. Whelan, E.M., *Toxic Terror: The Truth behind the Cancer Scares*, Prometheus Books, Buffalo, New York, 1993, chap. 14.

34. USEPA, *The Quality of our Nation's Water*, Office of Water, www.epa.gov/enviro.

35. USEPA, *Manual of Individual Water Supply Systems*, Washington, D.C., 1973, p. 34.

36. Kahn, J., *Economic Approach to Environmental and Natural Resources*, Harcourt and Brace Co., New York, 1995, chap. 14.

37. USEPA, Is Someone Contaminating Your Drinking Water? Groundwater Protection Council, Oklahoma City, OK, 1997.

38. International Bottled Water Association, U.S. Bottled Water Market, Volume, Growth, Consumption, 1976–1999, 2002, http://www.bottledwater.org/public/volume%20growth%20and%20consumption.htm.

39. Uehling, M.D., How Safe is Your Drinking Water?, *Popular Science*, 10, 1996, p. 63.

40. White, G.F., The Environmental Affects of the High Aswan Dam, 1988.

41. Grigg, N.S., *Water Resources Management*, McGraw-Hill, New York, 1996, chap. 13.

42. Geyer-Allely, E., *Water Consumption and Sustainable Water Resources*, Organization for Economic Co-operation and Development, 1998, 28.

43. Pinholster, G., Drinking Recycled Wastewater, *Environmental Science and Technology*, 29, 1995, p. 174.

44. Asano, T., Wastewater reuse, *Water Resources*, Biswas, A.K., McGraw-Hill, 1997, p. 383.

45. Frederick, K.D., Changing water resources institutions, *Sustaining our Water Resources*, Frederick, K.D., Ed., National Academy Press, Washington, D.C., 1993, p.73.

46. Grigg, N.S., *Water Resources Management*, McGraw-Hill, New York 1996, chap. 14.

47. Palmer, *T., Lifelines: The Case for River Conservation,* Island Press, Washington, D.C., 1994, p. 117.

48. Nadakavukaren, A., *Our Global Environment,* Waveland Press, Prospect Heights, IL, 1995, chap. 15.

49. Pfarrer, S., Water quality mandate praised, *Daily Hampshire Gazette,* Northampton, MA, August 13, 1, 1998.

50. Von Huben, H., *Water Quality,* American Water Works Association, Denver, CO, 1995, chap. 6.

51. Potera, C., Drugged Drinking Water, Forum in *Environ. Health Perspect.,* Vol. 108, No. 10, October 2000.

52. Von Huben, H., *Water Quality,* American Water Works Association, Denver, CO, 1995, chap. 7.

53. Koren, H. and Bisesi, M., *Handbook of Environmental Health and Safety,* 3rd ed., Lewis Publishers, New York, 1996, chap. 3.

54. Ongley, E.D., *Control of Water Pollution from Agriculture,* 55, Food and Agriculture Organization of the United Nations, Rome, 1996, chap. 1.

55. Gallagher, L.M., *Clean Water Act, Environmental Law Handbook,* 14, Sullivan, T.F.P., Government Institutes, Rockville, MD, 1997, p. 114.

56. Tibbets, J., Water pressure, rewriting the safe drinking water act, *Environ. Health Perspect.,* 104, 930, September 1996.

57. Kilborn, P.T., Storm Highlights Flaws in Farm Law in North Carolina; A Threat to Water Supply, *New York Times,* Sunday, October 17, 1999.

58. Ongley, E.D., *Control of Water Pollution from Agriculture,* 55, Food and Agriculture Organization of the United Nations, Rome, 1996, chap. 3.

59. Krafter, C.R., Pesticides in Storm Runoff from Agricultural and Urban Areas in the Tuolumne River Basin in the Vicinity of Modesto, California, U.S. Geological Survey Water Resources Investigation Report 98–4017, 1998.

60. Annin, P., Down in the Dead Zone, *Newsweek,* October 18, 1999, p. 60.

61. Larson, S.J-., Capel, P.D., and Majewski, M.S., *Pesticides in Surface Waters,* 3rd ed., Ann Arbor Press, Chelsea, MI, 97, chap. 1, 3, 4, 7.

62. Mitchell, T.J. and Harding, A.K., Who is drinking nitrate in their well water?, *Environ. Health,* 11, 14,1996.

63. Holton, C., Nitrate elimination, *Environ. Health Perspect.,* 104, 36, 1996.

64. Pitt, R., Clark, S., Parmer, Y., and Field, *R., Groundwater Contamination from Stormwater Infiltration,* Ann Arbor Press, Chelsea, MI, 1996, chap. 1, 2.

65. Kelly, M., *Mining and the Freshwater Environment,* Elsevier Science, Essex, England, 1988, chap. 1, 4.

66. Riley, M.B., Murphy, A.D., and Rosado, M.A.M., A reversal of tides, *Water, Culture, and Power,* Donahue, J.M. and Johnston, B.R., Eds., Island Press, Washington, D.C., 1998, p. 237.

67. Centers for Disease Control, Division of Bacterial and Mycotic Diseases, Preventing bacterial waterborne diseases, June 14, 2006, http://www.cdc.gov/ncidod/dbmd/diseaseinfo/waterbornediseases_t.htm.

68. Wasik, J., Special Report: How Safe is Your Water, *Consumers Digest,* 35, May 1996, p. 63.

69. Thompson, E., Rotting Carcasses Sewage Create Mounting Health in N.C., The Associated Press, September 20, 1999.

70. Joyce, S., Capturing *Cryptosporidium, Environ. Health Perspect.,* 104, 834, August 1996.

71. Whelan, E.M., *Toxic Terror,* Prometheus Books, Buffalo, New York, 1993, chap. 11.

72. Pilkington, N.H., Disinfection: An overview, *Modem Techniques in Wastewater Treatment,* Kolarik, L.O. and Priestly, A.J., Eds., CSIRO Publishing, East Melbourne, Australia, 1995, p. 75.

73. Von Huben, H., *Water Quality,* 2, American Water Works Association, Denver, CO, 1995, chap. 1.

74. Williams, S.E., Safe Drinking Water Act, *Environmental Law Handbook*, 14, Sullivan, T.F.P., Ed., Government Institutes, Rockville, MD, 1997, chap. 6.

75. McGarry, M.C., *Waste Collection in Hot Climates: A Technical and Economic Appraisal in Water Wastes and Health in Hot Climates,* John Wiley & Sons, New York, 1977, p. 243.

76. Barnes, D., Aerobic wastewater treatment systems, *Modern Techniques in Water and Wastewater Treatment,* Kolarik, L.O. and Priestley, A.J., Eds., CSIRO Publishing, East Melbourne, Australia, 1995, p. 92.

77. Casey, T.J., *Unit Treatment Processes in Water and Wastewater Engineering,* John Wiley & Sons, New York, 1997, chap. 16.

78. Cheremisinoff, P., *Sludge Management and Disposal,* PTR Prentice Hall, Englewood Cliffs, NJ, 1994, chap. 1.

79. Humphrey, P., Water Fears Contaminate Westport, *Boston Globe,* February 26, 41, 1995.

10 Air, Noise, and Radiation

CHAPTER OBJECTIVES

A student, after reading this chapter, should be able to do the following:

1. List and explain the reasons why air pollution is considered a national and global threat.
2. Discuss and describe the chemical and physical components of the atmosphere, and explain the mechanisms of dispersion.
3. Describe the regulatory efforts in the U.S. with emphasis on titles of the 1990 CAAA.
4. Discuss the issues behind stratospheric ozone depletion and global warming.
5. List and discuss the nature, sources, and health and welfare effects of the criteria pollutants.
6. List, discuss, and describe the major sources of indoor air pollution, including health effects and methods of control.
7. Define noise pollution and radiation. List the major sources and known health effects of noise and radiation.

INTRODUCTION

The problems associated with air pollution are not recent. The obnoxious stench of burning coal was so annoying to Edward I and Edward II of Great Britain that those caught burning coal were at risk of losing their lives. In fact, a man was hanged during the reign of Edward II for burning coal and fouling the air.[1] Such severe penalties have not been witnessed since then. However, air pollution continues to be a problem, and has worsened with the combined effects of an increasing population and the Industrial Revolution. Generally, air pollution was considered to be an annoyance, resulting in respiratory irritation, deposits of dust and filth on surfaces, and reduced visibility. However, the potentially serious consequences of air pollution were thrust into the public eye in documented episodes that have become "classics" in air pollution literature. These episodes began in the first week of December 1930 in the Meuse River Valley, Belgium. This deep river valley was highly industrialized with manufacturing plants in steel, glass, lime, fertilizers, and sulfuric acid. That week in December, the air became stilled, and the continuously emitted pollutants in the valley were trapped. The residents complained of irritated throats, coughing, and tightness in the chest. More than 60 deaths were attributed to the episode, primarily among the elderly and those with preexisting cardiorespiratory disorders. Nearly 18 years later in Donora, PA, similar conditions prevailed, with steel manufacturing concentrated in a river valley. A stable, slow-moving air mass resulted in the formation of a dense, cool layer of air near the earth with warmer air above in a phenomenon known as an *inversion*. The noxious pollutants of industry were trapped below the interface of the warm air above and the cool air below, and began accumulating on Wednesday, October 28. The first death from this episode occurred on that Saturday, and nearly 20 people died by Sunday evening. Six thousand people, or 40% of the population, developed symptoms associated with the episode, including nausea, vomiting, irritation to the eyes, nose, and throat, headaches, and muscular aches and pains. The severest symptoms, and most deaths were among the elderly and those with preexisting respiratory and cardiovascular conditions. The lungs of those who died showed evidence of capillary dilation, hemorrhaging, purulent bronchitis, and edema.[2,3]

Excess deaths were also reported in London during several air pollution episodes associated with stagnant air. These episodes occurred in 1952, 1956, 1957, and 1962. London's worst episode was in 1952 during which a thick, yellow smog settled over the city for 5 d, resulting in the death of 4000 more people than usual in the absence of air pollution. Again, the deaths were mostly among the elderly and those with preexisting respiratory or cardiac problems. The excess deaths spiked with the air pollution episode and subsided when the stagnant air was eventually swept away by more turbulent conditions.

Episodes of air pollution resulting in measurable levels of excess deaths have also occurred in Los Angeles and New York, primarily in the 1950s and 1960s. The disasters focused public attention on the severity of air pollution. It became clear that exposure to concentrated levels of ambient pollutants could cause severe symptoms and even death. Such awareness prompted greater attention to studying the factors associated with air pollution and then establishing regulations to control and limit the pollutants. Such efforts have made it unlikely that similar air pollution episodes will occur in industrialized countries. However, countries such as China, Mexico, and Eastern Europe are facing substantial air pollution problems as they strive for industrial growth in the absence of vigorous air pollution control policies.

Although progress in controlling air pollution in the developed nations has been made, problems of global concern remain. Government leaders and scientists are working toward a consensus in the midst of controversy that industrial emissions such as carbon dioxide and other greenhouse gases are contributing to global warming with potentially severe consequences. Emissions of nitrogen and sulfur oxides are thought to be major contributors to acid deposition, which is being associated with the disappearance of fish and other biota in the tens of thousands in lakes and streams worldwide, and the decimation of high-altitude forested areas. Chlorofluorocarbons (CFCs), once widely used in air conditioners, in electrical component cleaning, and in the production of foam cups, have been linked to the catalytic destruction of stratospheric ozone in a process known as *ozone depletion.* Such depletion is thought to result in higher levels of damaging ultraviolet radiation reaching the earth. These atmospheric effects of air pollution have mobilized governments internationally to meet and attempt to resolve these issues. Actions that may be taken are likely to have far-reaching effects on economies, standards of living, and the livelihood of hundreds of millions of people. The quality of our lives, the stability of our environment, and the survivability of the planet have been brought forward as arguments in these controversial issues. Is there a global warming problem? Is stratospheric ozone being depleted? Is acid deposition killing forests internationally and destroying aquatic life? These issues are presented in this chapter in a scientific and objective framework for the reader to draw an independent conclusion.

Air pollution not only represents a perceived threat to global ecological damage, but also exerts both acute and chronic threats to human health at exposures within the normal range of pollutants in our cities and towns.[4] These threats include respiratory disorders such as asthma, bronchitis, and emphysema. There are also increased risks of cardiovascular disease, cancer, and respiratory infections. Air pollutants have been implicated in direct injury to living plants, and to materials, including (1) the corrosion of metal, (2) soiling of buildings, and (3) the degradation of paints, leather, paper, textiles, and dyes. These effects result in significant economic losses approaching several billion dollars annually.[5] Based on the preceding overview, the arguments for controlling air pollution are substantial. The student of air pollution requires an understanding of the mechanisms of air pollution; this study must, therefore, begin with a discussion of the atmosphere and methods of pollution dispersion.

THE ATMOSPHERE AND METHODS OF DISPERSION

CHEMICAL CHARACTERISTICS

The earth's atmosphere consists of a mixture of gases, water vapor, and both solid and liquid particles. It is a dynamic system with most components leaving and entering on a continuing basis and with a wide variety of chemicals being created in the atmosphere in the presence of sunlight, catalysts, and moisture. In spite of this continuing exchange of atmospheric ingredients, some of the gases remain relatively constant in their proportions. Nitrogen (N_2) represents a constant 78% of the 500 billion tons of air surrounding the planet, whereas oxygen (O_2) remains steady at 21%, and argon (Ar) at 0.9%. These three gases collectively represent 99.9% of the atmosphere. The remainder of gases that are in constant proportion consist of traces of neon (Ne), helium (He), hydrogen (H_2), xenon (Xe), and krypton

FIGURE 10.1 Atmospheric components.

(Kr) (Figure 10.1). The atmosphere also contains substances that tend to be variable in their concentration and include water vapor, carbon dioxide, carbon monoxide, methane, ozone, ammonia, and nitrogen oxides. Many of these latter compounds enter the atmosphere from both anthropogenic and natural sources, and include substances such as carbon monoxide, ammonia, and nitrogen oxides, which are regulated as pollutants. Additionally many other substances enter the atmosphere from human activities, and those that have the capacity to cause human disease and welfare effects, resulting in disturbances to plant life and physical structures, are collectively known as air pollutants.

Although nitrogen and oxygen, the main components of the atmosphere, have remained relatively constant for millions of years, current scientific thought suggests this was not always the case. The primitive atmosphere is thought to have been rich in carbon dioxide with little free oxygen. Nearly 2 billion years ago, green plants evolved which used carbon dioxide to grow by photosynthesis, producing energy-rich sugars and a waste by-product, now known as oxygen. Organisms adapted to the presence of this waste oxygen and developed into the aerobic species of today, including humans.[6] The earth's atmosphere and climate have achieved a dynamic balance through the removal and recycling of carbon, nitrogen, oxygen and other substances that have sustained plant and animal life for millions of years. Human technology and explosive populations could potentially alter this balance, causing changes in the earth–atmosphere system that jeopardize our sustainability.

FIGURE 10.2 Electromagnetic spectrum and components of solar radiation.

PHYSICAL CHARACTERISTICS

Solar Radiation

Life on earth requires a continuing source of energy. That energy is supplied by the sun, which radiates enormous electromagnetic energy at effective temperature of 6000°F and travels in less than 10 min to the earth at 186,000 mi/sec.[7] More than 99% of the energy from the sun is within the spectral range of 150 to 4000 nm (0.15 to 4.0 μm). Fifty percent of the energy striking the earth is in the infrared spectrum exceeding 0.7 μm (700 nm), whereas slightly more than 40% is in the visible spectrum of 400 to 700 nm (0.4 to 0.7 μm), and about 9.0% is in the energetic and shorter wavelengths of 150 to 400 nm (0.15 to 0.4 μm) (Figure 10.2).[7,8] The solar radiation reaching the earth's atmosphere is absorbed, scattered, and reflected, so that an average of less than 50% reaches the earth's surface. The upper and lower atmosphere absorbs and scatters 15 to 20% of the total incident solar radiation. When skies are clear, carbon dioxide and water vapor strongly absorb infrared radiation, whereas clouds and dust particles in the lower atmosphere scatter and diffuse visible light wave. Clouds cover a major part of the earth at most times, causing significant scattering and reflection of incoming solar radiation. Solar radiation is also reflected back to space by ground surfaces, including oceans and lands. The combined reflective ability of cloud cover and ground surfaces are known as the earth's *albedo*. About 30 to 40% of the solar radiation is reflected by this albedo, with most of it attributed to cloud cover. The amount of solar energy absorbed must ultimately be radiated back to space to maintain a global radiation balance; otherwise, the earth would soon be overheated. This is not normally a uniform process, but, on average, the annual global temperature varies less than 1°C. Solar energy that is absorbed by ground surfaces is radiated back as heat in longer, lower-energy infrared wavelengths. You can observe this effect if you look up a long stretch of blacktop roadway on a hot summer day, when the absorbed energy from the sun heats the air above the pavement, causing air density changes that distort the scenery in the distance. Air density changes cause shifts in optical refraction that appear as "waves" when they distort distant views. The blacktop absorbed the higher-energy wavelengths of the visible range and radiated back longer-wavelength infrared radiation

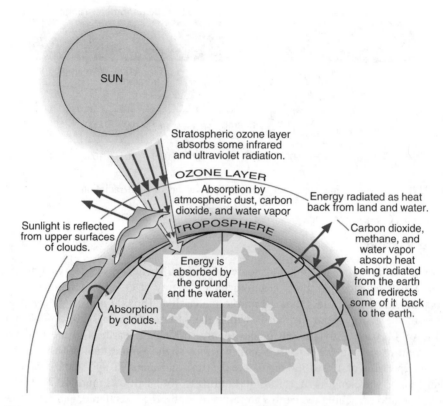

FIGURE 10.3 Mechanisms of reflection and absorption of solar radiation.

in the form of heat. If this reflected heat energy (infrared) is absorbed by infrared-absorbing gases such as carbon dioxide, CFCs, methane, and water vapor, then it traps the warmth and reflects it back to the earth's atmosphere in a process known as the *greenhouse effect* (Figure 10.3). If the greenhouse gases increase in concentration, it is logical that more heat energy will be absorbed, and the average annual global temperature may rise causing a global warming trend. This is the focus of world attention and a matter of discussion later in this chapter.

Vertical Temperature Differences and Atmospheric Regions

Imagine yourself in the basket of a hot air balloon lifting off the ground with a thermometer in your hand. You would normally experience a declining temperature as you gained altitude at a rate of about 6.5°C/km, or a loss of about 65°C over the 0 to 10 km altitude range of this region known as the *troposphere* (Figure 10.4). Of course, at levels of 6 km or higher, you have entered the "killing zone," in which the air has become so thin that supplemental oxygen is required. The thickness of the troposphere is greater over the equator (12 km), and lesser over the poles (8 km). It is the troposphere in which life activities occur and the changes and manifestations in weather and climate are driven by variations in thermal energy, friction,

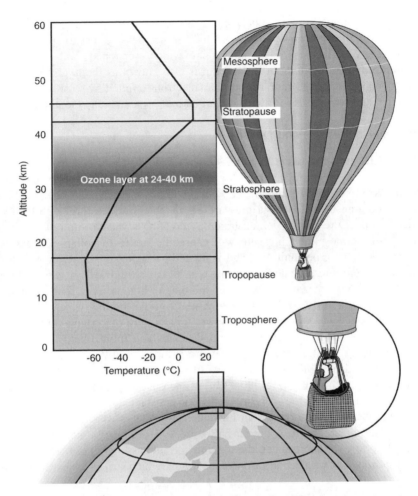

FIGURE 10.4 Atmospheric regions and temperature profile. (Adapted from Chamber, L., *Classification and Extent of Air Pollution*, Stern, A.C., Ed., 3rd ed., Academic Press, New York, 1976, chap. 1.)

and global rotation (Coriolis force). The end of the troposphere is marked by stationary or isothermal temperature as you increase in altitude from 10 km to 20 km in a region known as the *tropopause*. The temperature begins to increase to 0°C beyond this region to an altitude of about 50 km. This area is known as the *stratosphere*, and contains concentrations of ozone approaching 200 parts per billion (ppb) in the densest region at 24 to 40 km (15 to 24 mi). Ozone (O_3) is an isotope of oxygen that absorbs infrared radiation causing the elevated temperature of the stratosphere. Ozone also strongly absorbs ultraviolet radiation in the UV-B region of 290 to 320 nm, thus restricting damaging levels of UV-B from reaching the earth and causing increased skin cancer in humans. The stratospheric ozone concentrations vary geographically and seasonally. Peak ozone levels are found between autumn and spring and are normally most dense over the poles. Levels above the stratosphere include the stratopause, mesosphere, mesopause, and thermosphere. These regions are in the outermost atmosphere where light molecules are lost to space, and the relatively few molecules of N_2 and O. are widely separated, and may become photoionized. The region at 90 to 400 km is also known as the *ionosphere*, where the earth's magnetic field captures prot and electrons emitted by solar storms resulting in the aurora borealis (northern lights) and auroroa australis (southern light).[7]

Atmospheric Pressure and Density

Gravity pulls the atmosphere to earth, so that the densest part of the atmosphere is found closest to the surface. About 99% of the atmospheric mass is below 30 km (18 mi), 90% is below 12 km, and 75% of the atmosphere is below 10 km (Figure 10.5). A column of air exerts a pressure on an object at sea level of 14.7 lb/in.[2] (PSI). This pressure can be demonstrated by creating a vacuum in a plastic bottle. The external atmospheric pressure will cause the bottle to collapse. Pressure is defined as the force per unit area and is related to the density of the air such that there is both low pressure and low density at higher altitudes. This can be demonstrated by boiling water first at sea level and then at high altitude. The temperature remains steady at 100°C as gases are released from the boiling water. Atop a high mountain peak, water would boil at a lower temperature of perhaps 90°C, because the pressure is lower and gaseous vapors can escape more readily under lower pressure (Figure 10.6). Atmospheric pressure is measured by a barometer and normally expressed in millimeters or inches of mercury. The pressure at sea level is equal to 760 mm mercury (Hg), 29.92 in. of mercury, or 1010 mbar. Pressure and atmospheric density in the troposphere is not constant and varies in response to changes in thermal radiation and air movement. Air that is warmed rises in contrast to the cooler air around it. A hot air balloon uses a propane burner to heat the air inside a fabric envelope causing the balloon and its passengers to rise. The energy driving air movement in the atmosphere is the sun, which strikes the earth at varying

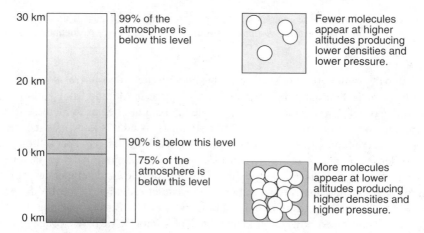

FIGURE 10.5 Atmospheric density as a function of altitude.

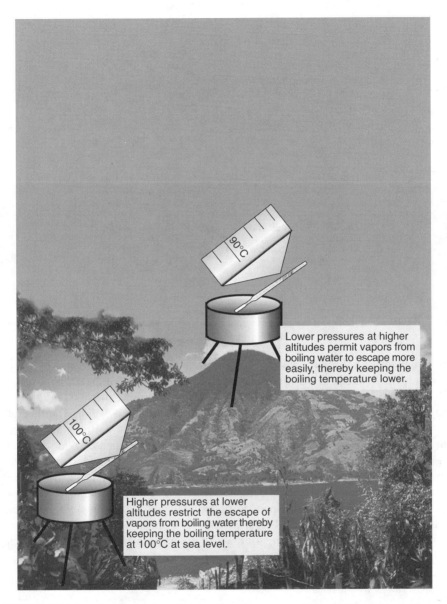

FIGURE 10.6 The demonstration of atmospheric pressure at low and high altitudes.

angles, warming the air unevenly. Warmer air becomes less dense, expands, and rises, thereby creating a column of air that is lower in pressure. Cooler air is denser, settles, and is characterized by higher pressure. Differences in pressure cause air to move in both horizontal and vertical patterns. Air tends to move from regions of high pressure to low pressure. Therefore, as sun heats a body of air and it rises, a region of low pressure is created into which flows cooler air. The greater the temperature differences, the larger the pressure differences will be, and the more

low pressure

Warm air flows into regions
of low pressure and rises.
This pattern is
counterclockwise in the
northern hemisphere.

Cyclonic air motion

High pressure

Cool air descends and
radiates outward.
This is clockwise in the
northern hemisphere.

Anticyclonic air motion

FIGURE 10.7 Cyclonic motion in a low-pressure system and anticyclonic air motion in a high-pressure system in the Northern Hemisphere.

rapid or violent will be the flow of air. If the air flows horizontally, this is known as wind. The direction of the wind is influenced by the rotation of the earth ("Coriolis force") (see Chapter 1), friction, and the differential warming provided by the sun. Such forces cause air to flow into regions of low pressure in a cyclonic motion and then to rise. This cyclonic motion is counterclockwise in the northern hemisphere and clockwise in the Southern Hemisphere (Figure 10.7). When cool air descends in the Northern Hemisphere, it is radiated outward in a clockwise motion known as an *anticyclone* (Figure 10.7). The direction is reversed to a counterclockwise motion in the southern hemisphere. These represent very large-scale air movements and play important roles in the dispersion of air pollutants. Low-pressure migrating cyclones usually are associated with inclement weather, including precipitation, clouds, and windy conditions. High-pressure systems normally signal cooler dry air with sunny conditions.

Atmospheric Inversions

There is an ample supply of naturally occurring and anthropogenic pollutants in the atmosphere. They are normally removed by chemical reactions in the atmosphere, dry or wet deposition, and microbial conversions known collectively as *sinks*. Additionally, such pollutants are normally dispersed by wind and turbulence to very low concentrations that are unlikely to produce adverse health effects. However, sometimes atmospheric conditions become very stable and slow moving, with the effect of causing the pollutants to build in concentration. The concentration of pollutants is markedly influenced by meteorological conditions. The stability of the

FIGURE 10.8 Dispersion characteristics of air having environmental lapse rates that are cooler (good dispersion) or warmer (poor dispersion) than the adiabatic lapse rate.

air directly impacts the mixing and transport of pollutants along with chemical reactions in the atmosphere that form new categories of pollutants. A stable air mass is slow moving with little turbulence and permits the accumulation of pollutants. Certain meteorological conditions promote stability. Under normal conditions in the troposphere, air that is warmed rises as long as the air above and around it is cooler. The warmer the air mass in relation to its surroundings, the more rapidly it will rise. As the air rises, it will lose heat from such mechanisms as radiative cooling at an average of 6.5°C/km, which is termed the *normal lapse rate*. Under theoretical conditions, a parcel of warm, dry air would cool at the rate of 10°C/km, referred to as the *dry adiabatic lapse rate*. The actual change of temperature with altitude is referred to as the *environmental lapse rate*, which is a direct measure of atmospheric stability. This stability can be measured by placing a thermometer at ground level and another several meters in the air to determine the environmental lapse rate (Figure 10.8). If the air cools with altitude at a rate faster than the dry adiabatic lapse rate of 10°C/km, then the vertical motion of the air will be quite rapid, permitting cooler air to sweep in and replace it, thereby creating turbulent conditions with good dispersion of pollutants. Atmospheric turbulence promotes pollution dispersion and results from mechanical obstruction and atmospheric heating. The higher a column of air can rise, the greater the maximum mixing depth (MMD) will be. The MMD varies with the time of day, the season, cloud cover, geography, and the presence of anticyclones. Mixing depths tend to be greater during the daytime, in summer, and in geographical areas where there is little ground cover or water to absorb the heat, such as deserts or the Great Plains. However, if the air temperature does not cool sufficiently with increasing altitude, or even gets warmer at higher elevations, the upward movement of air is prevented, stable conditions predominate, and dispersion is poor (Figure 10.8).[3,9] The warming of air with increasing elevation is a reversal of the normal tropospheric cooling process and is

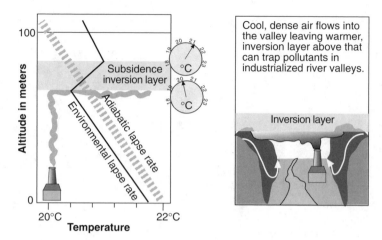

FIGURE 10.9 Environmental lapse rate conditions associated with subsidence inversion, and nocturnal inversion occurring in a river valley.

known as *atmospheric inversion*. It is the most common atmospheric condition associated with stability. The cooler, denser air near the ground is unable to rise through the warm layers above it and is therefore trapped until turbulent conditions can be reestablished. There are two primary types of inversions:

Radiation inversion: This type of inversion normally occurs at night when heat absorbed by the earth during the day radiates out quickly at night, leaving the air nearest the ground cooler than the radiated warmer air above. These inversions are normally not persistent and are easily broken up when the sun rises in the morning to cause convective currents. Such short-lived inversions seldom present a health hazard because there is little time to accumulate pollutants.

Subsidence inversion: Subsidence refers to the descent of air masses. Cool air masses (anticyclones) associated with Hadley cells descend causing air molecules to compress in a layer above the ground. This layer becomes warmed by compression whereas the air at ground level remains unchanged and often cooler than the air above, producing an inversion layer at distances of 500 to 1000 m above the ground (Figure 10.9). Such inversions may occur frequently along coastal areas such as California, which has documented inversions more than 300 d of the year. Such subsidence inversions occur most often during the fall and winter months, and are particularly troublesome because they may persist for days and permit the buildup of potentially dangerous levels of pollution.

Nocturnal inversions also occur when the cool dense air along mountain tops descends into the river valleys during the evening hours and accumulates under a layer of warm air above. The shaded valleys are often covered with mists or clouds, resisting penetration by sunlight until midday when the sun is directly overhead where it can cause convection currents to break up the inversion layer. When such conditions combine with migratory anticyclones, the inversions become very persistent and hazardous. The stagnation of air produced by such migrating anticyclones was responsible for the classic air pollution disasters of Meuse Valley, Belgium, in 1930; Donora, PA, in 1948; and others occurring in England, New York, and Los Angeles.[3,9,10]

Inversions are not alone in meteorological events that influence air pollution. Sunlight and humidity are important in forming chemical reactions in the atmosphere, producing secondary pollutants such as ozone, created from nitrogen oxides in the presence of sunlight. Nitrogen oxides are considered primary pollutants because they are derived from identifiable sources such as smokestacks or automobile exhausts. Rain, sleet, and snow serve as sinks to remove pollutants from air through absorption or adsorption, carrying pollutants to the ground or bodies of water, where they may cause harmful effects. Pollutants emitted into an air mass may not widely disperse, and a migrating air mass may accumulate pollutants as it moves great distances in a process known as *long-range transport*. Much of the acidity from atmospheric deposition falling on the northeastern U.S. has been attributed to industries and power plants in the Midwest burning high-sulfur coal.

THE HISTORY OF AIR POLLUTION CONTROL IN THE U.S.

The control of air pollution was seen as a local control issue until the 1950s, when a federal role began to slowly evolve. It was in 1955 that Congress authorized the Public Health Service in the Department of Health, Education, and Welfare (DHEW) for the purpose of researching air pollution and conducting training programs. However, air quality continued to grow worse and Congress passed the Clean Air Act of 1963, which gave federal authority to (1) develop air-quality criteria for protecting the public health, (2) conduct research on sulfur dioxide pollution, (3) provide grants to establish state air pollution control agencies, and (4) to provide more research, training, and technical assistance with regard to air pollution. Amendments to this act in 1965 established the National Air Pollution Control Agency (NAPCA) and set emission standards for the 1968-model light motor vehicles. In spite of these regulatory initiatives, air pollution continued to worsen, and the 1963 Clean Air Act proved inadequate to deal with the problem. The response by Congress was to pass the Comprehensive Air Quality Act of 1967, which represented the first attempt to develop a regional approach for the control of air pollution through the designation of Air Quality Control Regions (AQCRs). However, states still kept the main responsibility for enforcement and were expected to develop air-quality standards and plans for implementing them. Progress proved to be slow, and as Americans celebrated Earth Day for the first time on April 22, 1970, public attention to environmental concerns reached a high level, stimulating Congress to strengthen air pollution control initiatives. This time, Congress passed the 1970 Clean Air Act Amendments (CAAA), which (1) dissolved NAPCA, (2) transferred air pollution control activities to the U.S. Environmental Protection Agency (EPA), which the president created by executive order, (3) increased federal enforcement authority, (4) established uniform national ambient air quality standards (NAAQS) for six criteria pollutants, (5) required states to develop plans for implementing programs to achieve NAAQS in the states and submit the state implementation plans (SIPs) to the EPA, and (6) gave American citizens the right to sue private and government entities to enforce air pollution requirements. In 1977, additional amendments to the 1970 CAAA that postponed and extended many compliance deadlines for air-quality standards and auto emission standards were made.[11]

However, the 1977 CAAA did authorize the EPA to regulate chemicals suspected of destroying stratospheric ozone, and also provided for the protection of clean air areas by preventing further air pollution through the concept of PSD or prevention of significant deterioration.[11] Despite these efforts and some success under pollution control, three major threats to the nation's environments and to the health of millions of Americans defied these regulatory initiatives. These threats included acid rain, toxic air emissions, and urban air pollution. Additionally, the threats to global environment persisted in the form of global warming by greenhouse gases, and the continued destruction of stratospheric ozone by ozone-depleting chemicals. In June 1989, President George Bush proposed major revisions to the Clean Air Act, which were passed by Congress and signed into law on November 15, 1990. These revisions are known as the Clean Air Act Amendments of 1990. Many of the provisions of this act are still coming on line as there are many progressive and creative new directions included in the amendments.[12] There are several titles to the act that are intended to produce a healthy, productive environment, combined with a sustainable energy policy and economic growth.

TITLES OF THE CLEAN AIR ACT

Title I: Provisions for Attainment and Maintenance of the NAAQS

Building upon the 1970 and 1977 CAAA, the 1990 CAAA attempt to strengthen the provisions protecting the public against seven of the most widespread and common pollutants designated as criteria pollutants (Table 10.1). The criteria pollutants are those for which standards have been set establishing maximum concentrations allowed in the ambient or outdoor air environment and are known as the national ambient air quality standards (NAAQS). The EPA promulgated standards for six classes of pollutants in 1971, and an air-quality standard for lead was promulgated in 1978. These standards are based on air-quality criteria that document the relationship of these pollutants to effects on health and welfare, thus providing a scientific basis for the standards.[11]

Over 100 million Americans live in cities that fail to meet the NAAQS for ozone, and many people continue to be exposed to unacceptable levels of carbon monoxide and particulates. The 1990 CAAA established these as nonattainment areas, which were first described in the 1970 CAAA and refers to those air quality control regions (AQCRs), as designated by the EPA, that exceed the level of one or more of the criteria pollutants during the year. The EPA ranked these areas according to the severity of an area's air pollution problem.[12] A nonattainment status carries serious

TABLE 10.1
National Ambient Air Quality Standards

POLLUTANT	AVERAGING TIME	PRIMARY STANDARD	MAIN SOURCES
Carbon monoxide	8 hours	9.0 ppm	Transportation
Hydrocarbons (corrected for methane)	3 hours (6–9 am)	160 $\mu g/m^3$	Transportation, Industrial processes
Nitrogen dioxide	Annual average	0.05 ppm	Stationary source fossil fuel combustion, Transportation
Sulfur dioxide	Annual average	0.03 ppm	Stationary source fossil fuel combustion
Particulates PM$_{10}$	Annual arithmetic mean	50 $\mu g/m^3$	Multiple sources including stationary source fuel combustion, industrial processes, and transportation; fine particles are associated with photochemicals products of fossil fuel combustion
PM$_{2.5}$*	Annual arithmetic mean	15$\mu g/m^3$	
	24 hr. average	65$\mu g/m^3$	
Ozone	1 hour 8 hour*	0.12 ppm 0.08 ppm*	Secondary pollutant formed in presence of sunlight, NOx, and hydrocarbons; fossil fuel combustion is a major contributor
Lead	3 months	1.5$\mu g/m^3$	Food, dust, older houses with lead paint

•July 1997 amendments to the 1990 CAAA.

implications because new industries or sources that might contribute to pollution would be prohibited and therefore limit economic development. This was economically very confining. The EPA, therefore, developed an offset policy which allowed an industry to build a new facility if it achieved agreements with nearby sources to reduce their emissions so that no net increase in pollution levels occurred. Alternatively, new sources could be developed in a nonattainment region if the pollution emission rates could be reduced maximally by the best available control technology (BACT) as designated by EPA. Under the 1990 CAAA, EPA classified nonattainment areas for ozone, carbon monoxide, and particulates based on the extent to which NAAQS were exceeded. The 1990 CAAA further establishes specific pollution controls and attainment dates for each control. Carbon monoxide and PM 10 are classified as either moderate or serious nonattainment areas, whereas ozone has five nonattainment classifications ranging from marginal (Albany, NY) to extreme (Los Angeles). The more severe the nonattainment problems, the more controls the area needs to apply, and the more time it is allotted to do so.

Title II: Provisions Relating to Mobile Sources

Most automobiles emit up to 80% less pollution than in 1960, but they still account for the largest combined amount of criteria pollutants, including carbon monoxide, hydrocarbons, and nitrogen oxides. The main reason for this is the proliferation of vehicles on the nation's highways. More than one in every five households in the U.S. have three or more cars, and the miles traveled are projected to grow by 2% each year through 2010 as Americans no longer just travel from home to work, but link their trips to shopping, after-school activities, day care, and many other activities. Efforts by government agencies to discourage the use of single-occupant vehicles (SOVs) have been generally discouraging because Americans value their travel independence and continue to enjoy the luxury of inexpensive fuel, affordable cars, and free highways.[13] To combat this problem, the 1990 CAAA reduced the allowable emissions for carbon monoxide, hydrocarbons, and nitrogen oxides from vehicle tailpipes through tighter inspection and maintenance (I/M) programs. The amount of evaporation of gasoline while refueling at the pumps has also been reduced through the use of specially designed recapture nozzles. Programs requiring cleaner reformulated gasoline and oxygenated fuels are also being phased in throughout the country in nonattainment areas for ozone and carbon monoxide, respectively.

Methyl-t-butyl ether (MTBE) is the biggest volume oxygenate used in gasoline and contains at least the 2.7% O_2 by weight required to reduce CO emissions. The additives must be used during the winter months of November through December when levels of CO peak. The amount of MTBE has increased from the 1.8 billion gallons used in 1992 because the program to use reformulated gasoline to combat ozone depletion began in 1995.[14] The law also required California to phase in the use of 150,000 clean-fuel cars (i.e., propane or electric cars) by 1996 and 300,000 by the year 1999. Additionally, 26 of the country's most polluted areas will have to further regulate companies and agencies with centrally fueled fleets of 10 or more vehicles such that vehicle emissions are further reduced. This process began in 1998.

Title III: Air Toxics

Congress recognized that very hazardous chemicals other than the criteria pollutants were being emitted into the nation's air, and directed EPA under the 1970 CAAA to establish a list of such chemicals along with standards and control techniques. Unfortunately, the EPA acted very slowly over 20 years, and listed only eight hazardous air pollutants (HAPs), comprising arsenic, asbestos, benzene, beryllium, mercury, radionuclides, and vinyl chloride and emissions from coke ovens. Congress was aware that more than 2.7 billion pounds of toxic air pollutants were being emitted annually in the U.S. and that exposure to such substances may be causing up to 3000 cancer deaths annually.[12] They obtained the information, which industry was required to supply under the Superfund "Right to Know" rule (SARA, Section 313). This rule was passed in 1986 in response to the deaths of nearly 2000 people in 1984 in Bhopal, India, because of the accidental release of toxic chemicals from a chemical manufacturing facility. This heightened concern prompted Congress to list 189 toxic air pollutants for which emissions must be reduced. The EPA was required to publish a list of source categories that omit certain levels of these pollutants and to issue standards based on the best demonstrated control technology within the regulated industry known as the *maximum achievable control technology (MACT)*. The source categories were to be controlled by the year 2000, although companies voluntarily reducing emissions may obtain a 6-year extension.

Title IV: Acid Deposition Control

Emissions of nitrogen and sulfur oxides are partially converted in the atmosphere to nitric and sulfuric acids, which return to the earth in rain, snow, and fog and on dry particles to cause acidic deposition. Such acidic deposition harms nonbuffered lakes and streams by killing fish and other aquatic organisms, causes forest destruction, corrodes steel, damages concrete and marble structures, reduces visibility, and even contributes to adverse health effects.[12] Despite progress in reducing emissions under the 1970 CAAA, nearly 20 million tons of SO_2 were being emitted annually in 1990, mostly from combustion of fossil fuels by electrical generating facilities. The goal of the 1990 CAAA is to reduce the total SO_2 emissions by half to 10 million tons per year through a phased program requiring power plants to reduce emissions. The 1990 CAAA is the first air regulation to encourage the use of market-based principles, such as emission banking and trading through the use of allowances. It works as follows: Coal-burning electric utilities and other industries are allowed to emit nearly 9 million tons of SO_2, while close to 250,000 theoretical tons are set aside for auction. If a utility thinks it is not able reduce its SO_2 emissions to meet the new federal standard, it could purchase an "allowance" from other utilities whose emissions are lower than that required under federal regulations. An allowance is defined under 1990 CAAA as the right to emit 1 ton of sulfur dioxide. This is a market incentive approach in which allowances, or the right to pollute, could be banked and traded in the market. This concept was originally proposed by members of the Environmental Defense Fund (EDF) as part of the effort to reduce SO_2 emissions 50% by the year 2010. According to EDF representatives, the program has been very successful with emission reductions of 3.4 million tons in 1 year alone at unexpectedly low program costs.[15] However, the market for the sale of allowances has significantly diminished, and auction prices for a 1-ton allowance have dropped from the expected price of $1,000 in 1990 to $250 in 1992, $140 in 1995, and $68 in 1996. Such low prices cause some people to worry that utilities will find it less expensive to buy allowances and so pollute considerably more.[15]

Title V: Permits

The 1990 CAAA also introduced an operating permits program similar to that used in federal legislation designed to protect water, such as the Federal National Pollution Elimination Discharge System (NPDES). This updated the Clean Air Act to make it more consistent with other federal statutes. Under the permit program, air pollution sources regulated by the act must obtain an operating permit through their state environmental program. State programs must be approved by the EPA if they are to administer the permit system. The permit requires that all rules and regulations regarding air pollution control be listed on the permit, making enforcement a much easier task. Additionally, the permit will list the permissible levels of pollutants and the required control measures. The source is required to file periodic reports showing its level of compliance with the permit obligations. They must also pay a fee for the permit, which is intended to cover costs by the state in administering the program.

Title VI: Stratospheric Ozone and Global Climate Protection

Stratospheric ozone is most concentrated in the region of 24 to 40 km, although it is among the least concentrated of naturally occurring atmospheric gases. Ozone is formed when ultraviolet rays split oxygen molecules to release an oxygen atom that combines with an oxygen molecule to form ozone (O_3). Although created by UV light, ozone also absorbs UV energy in the range of 200 to 300 nm, protecting plant and animal life from the harmful effects of intense ultraviolet radiation. In the mid-1970s, concern about the integrity of the ozone layer grew when the scientists Rowland and Molina, from the University of California, reported that CFCs were attacking and destroying the ozone layer. CFCs were widely used as refrigerants, aerosol propellants, and foam-blowing agents in styrofoam containers. The scientists began their work in the 1970s and showed that CFCs could be photolytically destroyed in the atmosphere to release chlorine atoms that could then catalytically destroy O_3 molecules in the stratosphere (Figure 10.10). Paul Crutzen, Mario Molina, and Sherwood Rowland received the Nobel Prize for chemistry in 1995 for their work in establishing that CFCs were destroying the ozone layer.[16] The work of Rowland and Molina erupted from the narrow views of a few esoteric scientists to international attention in 1985 when British scientists discovered a large hole in the ozone over the Antarctic during September to mid-October. This was subsequently confirmed by National Aeronautics and Space Administration (NASA).[17] The Antarctic hole has continued to grow since then, while also lasting longer. In the early 1990s, NASA scientists reported that very low concentrations of stratospheric ozone were appearing over the middle latitudes of 20 to 60°. Although some ozone loss was partially due to volcanic dust and gases hurled into the atmosphere by the eruption of Mount Pinatubo in the Philippines, losses in ozone were being observed over the U.S., Japan, Europe, Russia, and China before and since then. The World Meteorological Organization reports that the ozone hole over Antarctica is approximately as large as the combined area of Canada and the U.S., peaking at 7.7 million square miles and lasting for 50 d.[18] Scientists predict the ozone hole will likely grow larger over the next 15 to 20 years. Despite global efforts to halt the production of

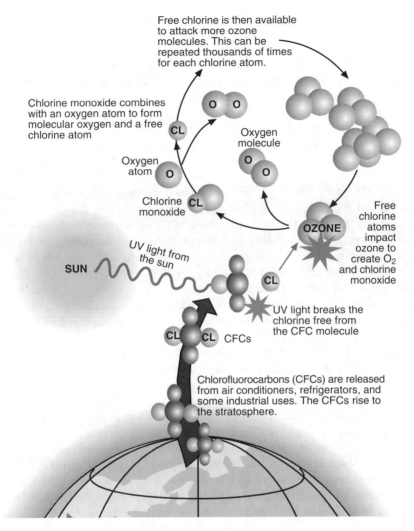

FIGURE 10.10 The catalytic destruction of ozone in the atmosphere by CFCs according to the Rowland–Molina hypothesis.

ozone-depleting CFCs, there remain substantial amounts of CFCs to be released, and CFCs may persist in the stratosphere for many years.[19]

The efforts to control and halt the production of ozone-depleting chemicals have been controversial. However, representatives from 29 nations met in Montreal, Canada, and reached an international agreement (Montreal Protocol) to freeze CFC production at 1988 levels by 1990 and then cut production in half by the year 2000. Mounting evidence that the ozone layer was being depleted resulted in a subsequent conference in London in 1990 and the Montreal Protocol was amended to hasten the phaseout of CFC production by the year 2000. The amendments also called for a phaseout of ozone-depleting halons (used in fire extinguishers) and CCL_4 by the year 2000. Methyl chloroform is to be phased out completely by the year 2005. More than 80 countries signed the 1990 agreement in London. The 1990 CAAA required that Class I ozone-depleting chemicals be phased out on a schedule specified in the 1987 Montreal Protocol. Class II chemicals (HCFCs) were to be phased out by 2030. However, as stratospheric ozone depletion became progressively worse, and apparent over North America and northern Europe, representatives of more than 100 nations met in Denmark and agreed to speed up the phaseout of CFC production, which was to have been completed by January 1, 1996, and that of halon by 1994.[8,11]

The efforts to find substitute chemicals have usually resulted in inferior performance and the requirement to modify existing equipment to make use of the alternatives such as hydrochlorofluorocarbons (HCFCs) or chlorine-free HFC-134a (CF_3CFH_2). These substitutes are not expected to damage the ozone layer, but are still infrared absorbers and contribute to the greenhouse effect. Therefore, the amendments to the Montreal Protocol target many CFC substitutes for phaseout by 2040 through voluntary efforts.[11] However, there was concern that many LDCs would be unable to afford the expense of modifying equipment to accept the substitute refrigerants. The Montreal Protocol allowed for continued production of CFCs in developed nations for use in developing countries to prevent them from building their own CFC-producing factories. American factories may produce 53.500 tons each year for export until 2005. Unfortunately, there is evidence that CFCs have been returning illegally to this country, where a 330-lb cylinder that costs $70 in Europe is sold for $242. This is still less costly than the CFCs recycled in the U.S. Officials believe that as much as 10,000 tons entered the U.S. in 1995 from Russia and other countries in the former Soviet Union.[20,21] Despite these troublesome and controversial events, overall production of CFCs is down markedly and the National Oceanic and Atmospheric Administration (NOAA) has reported a reduction in the rate at which CFCs are accumulating in the atmosphere.[22] A graph is shown featuring the lowest recorded column ozone during September/October of the years 1979 to 1997 as measured by satellite using total ozone mapping spectrometers (TOMS) (Figure 10.11). The minimum value of 88 Dobson units was recorded in 1994.[23] Although there appears to be a slight increase in stratospheric ozone compared to 1994, the recovery of stratospheric ozone is expected to take decades or up to a 100 years.[22]

Title VII: Provisions Relating to Enforcement

Since the start of the clean act movements in the middle of the 20th century, the laws and regulations have changed in response to ongoing research and public policy. In view of the fact that the laws have changed, it is also important to keep the enforcement ability up to date. Included in the 1990 Clean Air Act, there is a broad array of authorities to make the law more readily enforceable, thus bringing it up to date with the other major environmental statutes. Through the Clean Air Act of 1990, the EPA has gained authority to issue administrative penalty orders up to $20,000 and field citations up to $5,000 for lesser infractions. Contained in the Clean Air Act of 1990 is a clause, which actually enhanced civil judicial penalties. Criminal penalties for knowing violations are upgraded from misdemeanors to felonies, and new criminal authorities for knowing and negligent endangerment will be established.

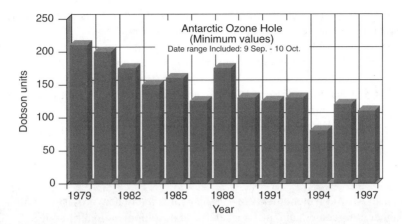

FIGURE 10.11 Antarctic ozone hole minimum values, 1979–1997. (Adapted from EPA, *The Size and Depth of the Ozone Hole*, written by EPA's Stratospheric Protection Division, June 10, 1998, http://www.EPA.gov/ozone/science/hoe/size.html.)

In addition, polluting sources must remain within compliance and annually certify their compliance with the EPA. The EPA has also gained the ability to administer subpoenas for compliance data. EPA will also be authorized to issue compliance orders with compliance schedules of up to 1 year.

Health and Welfare Impacts of Ozone Depletion

If the stratospheric ozone that protects the earth from UV-B radiation (280 to 320 mm) were brought to sea-level pressures, it would be about in. thick. The destruction of this very delicate layer could lead to increased UV-B radiation, which is thought likely to result in the following:

1. Increases in basal and squamous cell skin cancer; such cancers now affect 500,000 people annually along with 25,000 new cases each year of malignant melanoma.
2. Suppression or weakening of the human immune response system.
3. Damage to the cornea and conjunctiva of the eye.
4. Reduction in plant leaf size, total dry weight, and stunting of plant growth.
5. Decreased amounts of phytoplankton and zooplankton; studies showed that phytoplankton production was reduced by 6 to 12% under the ozone hole.

Less phytoplankton can have several negative impacts. Because phytoplankton photosynthesis is reduced, less carbon dioxide will be removed from the atmosphere, further enhancing the greenhouse effect. Also, because phytoplankton is at the base of the food chain, a reduction could threaten the food source of the world's oceans.[24,25] It needs to be mentioned that pilot studies in southern Chile, where people have been periodically exposed to high levels of UV-B, found no convincing evidence to support the reported anecdotal acute adverse health effects due to ozone thinning.[24] The authors conclude that exposure to UV-B at current levels in Punto

Arenas, Chile, was not likely to produce adverse health effects, but cautioned that increased levels may be of concern if they should occur.

The reduction of CFCs is important for the protection of stratospheric ozone. CFCs also contribute to global warming. Greenhouse gases' contribution to global warming is as follows: CO_2 55%, CFCs 7 to 24%, methane 15%, and nitrous oxide 6%.[8] CFCs are strong infrared absorbers and are up to 20,000 times as efficient as carbon dioxide in producing warming effects. Efforts at reducing CFCs are doubly important to protect the stratospheric ozone layer and to reduce the impact on global warming.

Other Titles to the 1990 CAAA

The Clean Air Act Amendments of 1990 contains many new provisions with regard to tougher enforcement and new research and development in air monitoring. There are also unemployment benefits through the Job Training Partnership Act for those workers who may have lost their jobs because industries had to comply with the new provisions of the 1990 CAAA. Efforts under the act also attempt to reduce the problem of pollution-created haze areas at National Parks and other parts of the country.

Revised Ozone and Particulate Standards

On July 17, 1997, the EPA announced new NAAQS for ground-level ozone and particulates. Ground-level ozone is a major component of smog that is photochemically produced as a secondary pollutant of the stratosphere from the interaction of sunlight, nitrogen oxides, and hydrocarbons. Ground-level ozone will be discussed in more detail later in the chapter. The previous ozone standard was last revised in 1979 and set at 0.12 ppm for 1 h. However, based on more than 3000 new studies on the ecological and health effects of ozone published since 1980, it had become clear that the standard did not adequately protect the public health, and the EPA replaced the previous standard with an 8-h standard set at 0.08 ppm.[26]

Similarly, the EPA announced a new standard for particulate matter (PM) under the national ambient air quality standards (NAAQS) based on the review of hundreds of scientific studies. EPA has added a new annual PM 2.5 standard set at 15 $\mu g/m^3$ and a new 24-h PM 2.5 standard set at 65 $\mu g/m^3$. The EPA is retaining the current annual PM 10 standard set at 50 $\mu g/m$.[3,27] Health effects of particulates are discussed in detail in a following section.

These changes to the standard were met with significant opposition, from businesses such as the auto industry, oil corporations, and from some state governors. Areas of the country producing more smog and soot than considered healthy under the proposed standards would face economically strangling restrictions, including controls on truck operations, mandates to reduce pollution-causing auto emissions to zero, and requirements that vehicles be run on fuels other than gasoline, strict controls on power plants and factories, and increased emission checks to autos. The new standards are considerably tougher than the ones developed under the 1970 CAAA and 1990 CAAA which are still not met by many states across the country.[28]

Bush Clear Skies Initiative

The Bush administration developed a plan called the Clear Skies Initiative and submitted it to Congress in February 2003 as a proposal to amend the Clean Air Act, which is the primary federal law governing air quality. According to the EPA, The emissions reductions from Clear Skies would help to alleviate our nation's major air pollution–related health and environmental problems, including fine particles, ozone, mercury, acid rain, nitrogen deposition, and visibility impairment. The goals of Clear Skies include the following:

- Cut sulfur dioxide (SO_2) emissions by 73%, from year 2000 emissions of 11 million tons to a cap of 4.5 million tons in 2010 and to a cap of 3 million tons in 2018.
- Cut emissions of nitrogen oxides (NOx) by 67%, from year 2000 emissions of 5 million tons to a cap of 2.1 million tons in 2008 and to a cap of 1.7 million tons in 2018.
- Cut mercury emissions by 69% — the first-ever national cap on mercury emissions. Emissions would be cut from 1999 emissions of 48 tons to a cap of 26 tons in 2010 and to a cap of 15 tons in 2018.[29]

The president's Clear Skies bill has been stagnant in both the Senate and the House since it was first introduced in late 2002. "Just a few months ago, the Senate rightfully rejected the president's bill. We hope that the House will do the same and protect public health for all Americans."[30] Some argue that if Congress passes the Clear Skies bill, the result will be to weaken and delay health protections already required under the law. The Clear Skies legislation sets new targets for emissions of sulfur dioxide, mercury, and nitrogen oxides from U.S. power plants. But some argue that the targets are weaker than those that would be put in place if the Bush administration simply implemented and enforced the existing law. According to these authors, when compared to current law, the Clear Skies plan would allow three times more toxic mercury emissions, 50% more sulfur emissions, and hundreds of thousands more tons of smog-forming nitrogen oxides. The authors also suggest that the plan would delay cleaning up this pollution by up to 10 years compared to current law and force residents of heavily polluted areas to wait years longer for clean air compared to the existing Clean Air Act.[31]

THE ISSUE OF GLOBAL WARMING

The Hot Air Treaty, Kyoto, Japan

The global warming treaty concluded in December 1997 (Kyoto, Japan), asked Western nations to reduce greenhouse gases to pre-1990 levels by 2010. The U.S., if it were to comply, would need to reduce greenhouse emissions to 7% below the level of the year 1990 or by about 33% or, one third of its otherwise projected use in the year 2010.[32] Western European nations and Japan are assigned similar goals. According to the agreement, reductions would not start taking effect until 2008.

Even if the nations complied, total world production of CO_2 would be reduced by only 1.1 billion tons by the year 2010. The treaty opened for signature on March 16, 1998, and closed on March 15, 1999. The agreement came into force on February 16, 2005, following ratification by Russia on November 18, 2004. As of April 2006, a total of 163 countries have ratified the agreement (representing over 61.6% of emissions from Annex I countries). The U.S. and Australia are exceptions. Other countries, such as India and China, which have ratified the protocol, are not required to reduce carbon emissions under the present agreement.[33] The LDCs have the urgent problem of providing for exponentially increasing populations, and further reducing use of fossil fuels is unlikely. Inexpensive labor and unregulated industry could attract business from the West, thereby shifting jobs, machinery, and CO_2 production to the LDCs with no net overall decrease in greenhouse gases, while greatly disturbing Western economies. The U.S. Congress has already voted down any treaty that does not include the LDCs, and if the accord is not ratified, global emissions of greenhouse gases could become double their preindustrial level by the year 2050.[32] As an example, China, with its heavy reliance on coal, will be the largest single producer of climate-warming carbon dioxide by 2010 to 2030. Despite rising health problems among its citizens because of increased air pollution, there appears to be no slowing in China's coal burning practices as it races to feed its billions of people and enter the world's industrial market place.[34]

Global Warming: The Controversy

The question of global warming has been raised by climatologists since the early 1970s. It has been argued that human activities have upset the balance of atmospheric carbon dioxide through (1) the combustion of fossil fuels, which releases carbon oxides, (2) the clearing of forests, which produces CO_2 and removes a vital consumer of CO_2, and (3) the destruction of phytoplankton by pollution of the oceans. Since the preindustrial age, a steady increase of CO_2 in the atmosphere from 280 ppm to 370 ppm has alarmed specialists and prompted reflection on what to do with the excess CO_2.

Greenhouse gases such as carbon dioxide, methane, ozone, CFCs, and nitrous oxide are strong absorbers of infrared radiation. When the sun's energy reaches the earth, much of it is absorbed by ground cover and water, and reflected back as lower-energy infrared radiation. An increasing blanket of carbon dioxide around the planet absorbs some of this infrared energy, trapping it and causing the earth to warm in a process known as the greenhouse effect (Figure 10.12). This concept of global warming caused by greenhouse gases is undergoing tremendous scientific debate and has come to public attention, starting in 1995 because (1) icebergs the size of small states have broken off the Antarctic ice shelf, (2) the annual average global temperature has risen by about 0.5°C (1°F) since the 19th century,[35] (3) U.S. and French scientists report that satellite measurements show the global sea has risen faster (3 min/year) in recent years than in previous decades,[36] (4) climate scientists reconstructed average annual temperatures back to year 1400 and found that 1990, 1995, and 1997 were the warmest in 600 years,[37] and (5) mountain glaciers around the world are rapidly retreating such as the one at Austria's highest mountain peak,

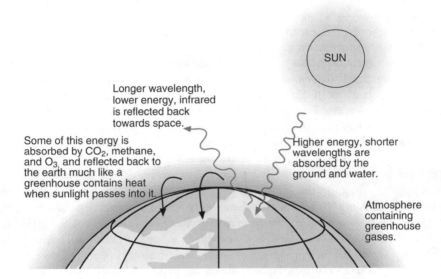

Longer wavelength, lower energy, infrared is reflected back towards space.

Some of this energy is absorbed by CO_2, methane, and O_3, and reflected back to the earth much like a greenhouse contains heat when sunlight passes into it.

SUN

Higher energy, shorter wavelengths are absorbed by the ground and water.

Atmosphere containing greenhouse gases.

FIGURE 10.12 The absorption of reflected infrared energy to produce the greenhouse effect.

FIGURE 10.13 Glaciers at Gross Glockner in Austria have receded rapidly in the last few years. This is a view of the visitors' center overlooking the receding glacier.

Gross Glockner (Figure 10.13), whereas those at Glacier National Park in Montana have shrunk 50% since 1850.

There have been various measures proposed for combating greenhouse gas increases. Alternatives methods include the following: (1) featuring other sources of energy including wind, solar, and nuclear power, (2) promoting increased numbers of trees and woody plants in deforested areas to serve as natural CO_2 sinks, (3) implementing the newest method of carbon sequestration, which takes CO_2 from stationary sources, such as a chemical factory or power plant, storing it for thousands of years either underground or in deep oceans. The ocean contains at least 50 times more carbon than the atmosphere does, making it a more efficient reservoir for carbon dioxide, yet having less effect in altering the chemistry of the water. It will take hundreds of years for the CO_2 to reach the surface of the ocean, rendering it ineffective as a greenhouse gas.[38]

Researchers at the Max Planck Institute for Meteorology in Hamburg, Germany, say that it is unlikely that natural forces (less than 5% chance) caused the temperature rise.[35] Other reputable scientists are not ready to point the finger to anthropogenic sources such as greenhouse gases as the source of global climate change. In fact, while international leaders battle over treaties to resolve a perceived global threat from greenhouse gases, scientists still line up along a mark in the sand that support or reject the role of human-induced global warming. Recent developments favor the theory that greenhouse gases play a role in increasing global temperatures, but uncertainties remain in the studies aimed at showing this, and a definitive answer may not be possible.[35] The scientific community will likely never depend on a single study or observation to accept the concept of greenhouse warming, but will gradually accept the concept when there is a consistent accumulation of evidence.[35] The combined search for such proof encouraged paleoclimatologists at the University of Massachusetts to reconstruct temperatures over the past 600 years using a method of indirect indicators (proxy climate indicators), including ice core samples, coral records, tree rings, and historical temperature measurements. They compared estimated temperatures to factors that effect climate, including the atmosphere's carbon dioxide supply, and discovered CO_2 to be the dominant factor over the past several decades.[37] Many factors influence climate as mentioned in Chapter 1, and global climate is likely influenced by many complex factors much greater in their effect than anthropogenic greenhouse gases. Climate is affected by (1) increases in atmospheric gases that absorb energy, (2) changes in the earth's orbital geometry, (3) changes in ocean temperature, (4) volcanic activity with increased atmospheric and reduced sunlight penetration, and (5) variations in solar radiation.

Are you interested in learning more about air, noise, and radiation? Visit http://www-unix.oit.umass.edu/~envhl565/weblinks.html and click on one or more of the following sites under Chapter 10.

General links to air, noise, and radiation:
- EPA Region 1 — Ground-Level Ozone Information
- Air pollution — U.S. EPA Office of Air and Radiation
- Radiation Protection Division Home Page
- Kyoto Protocol
- EPA — Global Warming Site
- EPA Montreal Protocol Site
- Montreal Protocol
- Bush "Clear Skies" Initiative — National Resource Defense Council (NRDC)
- EPA — Clear Skies Initiative

Although the meaning of the rise in global temperature and effects such as increasing ocean levels continue to be disputed, the spread of certain diseases has been clearly documented. Those diseases have been found at latitudes where they have not been previously identified. The increased incidence of hantavirus, malaria, cholera, and toxic algal blooms suggest a change in ecological balance and predator–prey relationships that can influence the risk of disease in humans. Regardless of whether there is a proven relationship between global warming, climate changes, and the emergence of new patterns of infectious disease, the public health community is obligated to investigate these new trends and determine what additional analysis and intervention are necessary.

FACTORS EFFECTING GLOBAL CLIMATE CHANGE

Orbital Geometry as a Factor Effecting Climate

You may recall the frigid weather of 1994, which plummeted temperatures to all time lows of 20 to 30°F in cities around the U.S. This caused people to wonder if the greenhouse effect was real, and some considered that glaciers may be returning. In the 1840s, the idea that glaciers covered much of the Northern Hemisphere was very controversial, but scientists have shown abundant evidence that thick ice sheets covered much of the North American continent, gouging out lakes, taming mountain ranges, and recreating the landscape. However, recent discoveries are even more surprising. Using core samples from seafloor sediments and ice core samples from Greenland and Antarctica, scientists have shown that (1) carbon dioxide rises in positive, almost perfect, correlation with rising global temperatures and (2) climate changes are normally abrupt and often extreme over the history of the planet.[38] The records show that mean global temperatures fluctuated widely with transitions from warm to cold often measured in decades (Figure 10.14). The interglacial period we are now in has lasted 10,000 years, and this stable period of warmth is very unlike the unstable climate history of the planet. It is probably no coincidence that human population marked its beginning just about 10,000 years ago with the climate stability. Humans were certainly not sufficient in number or technology to affect global climate over the previous 160,000 years, so what events caused such drastic changes? Among the more widely accepted theories is that of Milankovitch, a Yugoslavian astronomer, who theorized that such climate shifts were the result of subtle changes in the earth's orbit around the sun (eccentricity) and slight variations in the earth's tilt (obliquity), which produce climatic cycles (ice ages) of earth at 23,000, 41,200, and 100,000 years. The Milankovitch theories gained support from examination of oxygen isotopes in ocean-floor sediments revealing climate conditions over eons of time. Heavy oxygen isotopes indicate warmer weather, whereas lighter ones predominate in cooler weather.[17,38] The Milankovitch theory does not explain all ice ages, nor the rapid shifts in global temperature evidenced by the Greenland ice core samples.

Changes in Ocean Temperature

Wallace Brodkerad and George Denton from the University of Maine have advanced one of the more interesting theories about climate changes involving ocean currents. They propose that a very salty, dense stream of water flows northward under the

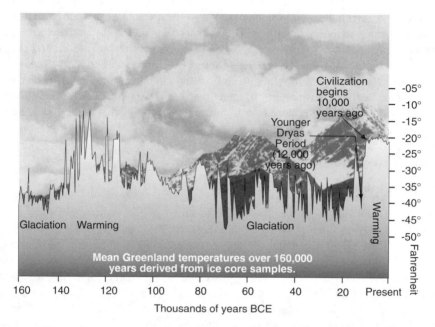

FIGURE 10.14 Record of average annual temperatures in Greenland derived from ice core samples. (Adapted from Lemonick, M.D., The Ice Age Cometh? *Time*, January 31, 1994, p. 79.)

Gulf Stream carrying heat from the tropics. It rises to the surface in the far north as water above is swept away by harsh weather and deposits its heat to warm the arctic, cools, sinks, and returns to the tropic. This continuing process keeps the arctic free of glaciation. Disturbances to this process could lead to a new period of glacier formation. This was thought to occur during the period of glaciation known as the Younger Dryas period, about 12,000 years ago. During this period, huge quantities of freshwater from the North American continent emptied into the North Atlantic for over 100 years.[38]

Some of greatest extremes in weather have been attributed to ocean currents known as *El Nin-o* ("the Little Boy," i.e., the Christ child because it was observed to occur during the holiday season). El Nin-o is a change or shift in ocean temperatures along with atmospheric conditions in the tropical Pacific that changes weather patterns all around the world (Figure 10.15). Scientists believe that high atmospheric pressure in the eastern Pacific sends trade winds westward pushing Pacific water ahead of them. Consequently, the water on the Asian and Australian coasts is about 1 to 5 ft higher than of the Americas. When the pressure drops, the water moves east again forcing downward a cool layer of water underneath called a *thermocline*, while permitting the warmer, unmixed water on top to carry its heat and moisture toward the Americas, and away from Indonesia.[39] Because of this process, Indonesia faced its worse drought in 50 years with fires that consumed more than 1 million acres of forest and threatened hunger for many, while western coasts of the Americas suffered heavy precipitation, powerful storms, and incredible winter blizzards. Torrential rain fell in east Africa, powerful hurricanes swirled in the Pacific, flash-floods killed hundreds in Peru and

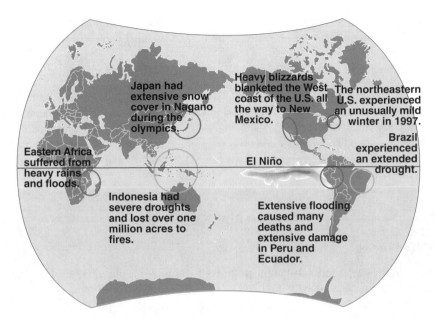

FIGURE 10.15 Global effects of El Nino. (Adapted from Noshi, J.M., The Fury of El Nino, *Time*, February 16, 1998, p. 67.)

Ecuador wiping out entire villages, and typhoons smashed into Japan. The ocean currents associated with El Nin-o produced remarkable global effects, and scientists are concerned that El Nin-o is changing dramatically from its cycle of 2 million years, although somewhat chaotic, to recur far more frequently and persist much longer. Some climatologists argue that the increase in severity and frequency of El Nin-o is a result of global warming.[40] When the waves associated with El Nin-o bounce off the Americas and return, then bounce back off Asia, they lift up the thermocline, causing a cooling effect that ends the El Nin-o effect. This causes shifts in weather that include a cold and wet north, and dry, hot south, which is opposite the effects of El Nin-o. This shift in Pacific ocean temperature is known as *La Nin-a*.

Volcanic Activity

Volcanic eruptions in the modern era may have extreme localized effects on land, and may cause short-term global changes in weather patterns as sunlight is inhibited by a layer of particles thrust into the atmosphere. However, ancient volcanoes are thought to have erupted in Siberia some 250 million years ago and lasted for a million years. This coincided with the Great Dying, occurring at the end of the Permian period, which killed 95% of the life in the oceans and more than 70% of the vertebrates on land. Ocean levels dropped by 300 ft, and acid rain enveloped the earth. Some scientists theorize that this prolonged volcanism blocked out the sun causing extensive glaciation and poisoning the atmosphere with sulfur oxides.

Scientists claim that other factors may have contributed to this mass extinction that developed, the events leading to the extinction of the dinosaurs some 65 million years ago, but accumulating evidence points toward extended volcanic activity as having dramatic effects on the earth's climate and its inhabitants.[41]

Solar Radiation

Scientists have recently discovered that the sun is not constant in its solar output, but varies with changes in the sun's luminosity, which is in turn associated with sunspot activity. Reduced sunspot activity appears to result in a cooling effect whereas increased activity increases warming of the earth. The sunspots show cycles of 11 and 22 years that correlate with nearly half of the global warming evidenced over the last 100 years.[17]

THE CRITERIA POLLUTANTS

INTRODUCTION

The EPA has reported substantial progress in improving the air quality in many states including the heavily polluted areas of Southern California, Hawaii, and regions in Arizona and Nevada. Despite major increases in population and acute travel over the past decade, air pollution has decreased by about one third overall in these regions. The greatest reductions have been recorded for lead (93%), followed by carbon monoxide (35%) and particulate matter (26%). Although most pollutants are being steadily reduced, nitrogen oxides, ground-level ozone, and fine particulates remain a problem in many areas.[42–44]

The total emissions of criteria pollutants fell 32% since 1970 in the face of a 29% increase in the U.S. population, a 121% escalation in vehicle miles traveled, and a gross domestic product increase of 104%.[42] Still, nearly 46 million people live in counties that fail to meet the air-quality standards for one or more of the criteria pollutants (Table 10.2).[42] Fine particulates from power plants, motor vehicles, and photochemical reactions in the atmosphere kill as many as 64,000 Americans each year from lung and heart disease.[45] Similarly, the American Lung Association released a survey of 13 U.S. cities linking exposure to ground-level ozone (smog) to as many 50,000 emergency room visits and 15,000 hospital admissions.[41] The health and welfare problems associated with criteria pollutants have not been completely resolved and many issues remain. A summary of criteria pollutant sources and health and welfare effects is presented in Table 10.3.

PARTICULATE MATTER (PM)

Particulate pollutants include airborne particles in liquid solid form that range in size from visible fly ash greater than 100 μm to particles 0.005 μm in size (Table 10.4). Particles may be produced naturally, such as pollen or sea spray, or by human activities, such as industrial processes, agricultural activities, fossil fuel combustion, and traffic. Particulates include dust, smoke, soot, (carbon), sulfates, nitrates, trace metals, and condensed organic compounds. Particulates produce a number of effects adverse to human interests, including (1) respiratory and cardiac health hazards to

TABLE 10.2
Percent Changes in Ambient Concentration of Air Pollutants from 1970 to 1996

CRITERIA POLLUTANTS	PERCENT CHANGE IN AMBIENT CONCENTRATIONS FROM 1970-1996
Carbon monoxide	-31%
Nitrogen dioxide	+8%
Sulfur dioxide	-39%
Particulates (PM_{10})	-73%
Ozone	-38%
Lead	-98%

Source: Adapted from USEPA, National Air Quality and Emission Trends Report, 1996, USEPA Office of Air Quality Planning & Standards, USEPA document 454/R-97-013 1/13/98, www.USEPA.gov/oar/aqtmd96/trends.html.

humans, (2) the deposit of grime and soot on buildings, (3) reduction in sunlight, thereby causing a regional or global coding effect, and (4) reduced visibility in areas of extensive smoke or particulate pollution.[46]

The characteristics for coarse particles (2.5 to 10 μm) are very different from fine particles (<2.5 μm) and relate to the adverse effects they produce. Coarse particles originate from wind-blown dust coming from deserts, unpaved roads over which vehicles travel, and agricultural fields, pollen, mold spores and plant debris. These particles tend to be mostly minerals consisting of aluminum, silicon, potassium, and calcium, which are chemically basic. Fine particles (<2.5 μm) are mostly emitted from fossil fuel combustion in industry, residences, and motor vehicles. They are also created in the atmosphere from gases of nitrogen and sulfur oxides and volatile organic compounds forming sulfates, nitrates, and aliphatic and aromatic hydrocarbons, including aldehydes, ketones, phenols, esters, terpenes, and phenylacetic acids.[27,47] Coarse particles can build up in the respiratory system, causing respiratory disorders such as asthma in women. Smaller particles are more apt to be associated with (1) increasing lung function in persons who have preexisting conditions such as asthma, (2) increasing premature death among the elderly and those with cardiopulmonary disease, (3) deterioration of respiratory defense mechanisms and adverse changes in lung tissue and structure, and (4) increased respiratory disease and symptoms.[27,43,48] The recent community studies provided support for issuing new standards for particulates to protect the public health and welfare. The new standard of PM 2.5 is set at 15 μg/m³ annual arithmetic mean, and 6.5 μg/m³ for a 24-h average because these fine particles are more closely linked to mortality and morbidity effects than the previous PM 10 standard.

TABLE 10.3

Sources and Health and Welfare Effects of Criteria Pollutants

CRITERIA POLLUTANTS AND SOURCES	HEALTH EFFECTS	WELFARE EFFECTS
Carbon monoxide Incomplete combustion of fossil fuels as in vehicles, kerosene heaters, boilers, and furnaces; cigarette smoking, forest fires, and biological decomposition	Interferes with oxygen transport in blood by binding with hemoglobin; causes headaches, fatigue, cardiovascular disease, and central nervous system disorders	Effects on plants or materials are not evident
Nitrogen dioxide Emitted from the combustion of fossil fuels in vehicles, industrial boilers, and electric generating utilities	Causes increased risk of respiratory infections and aggravates symptoms in persons with asthma and chronic bronchitis	Produces a reddish-brown haze over cities which reduces horizon visibility, causes leaves to yellow, and is a precursor to acid deposition and tropospheric ozone
Sulfur dioxide Fossil fuel combustion especially in coal-burning electric power utilities, metal smelters, oil refineries, and industrial boilers	Causes irritation of the throat and lungs and aggravates symptoms in persons with asthma and chronic bronchitis	Causes corrosion and deterioration of metals, brittleness of paper, paint discoloration, damages textiles and leaves of plants, and is a precursor to acid deposition
Particulates PM_{10} and $PM_{2.5}$ Fossil fuel combustion emissions, industrial processes, photochemical reactions in atmosphere, mechanical abrasion	Aggravates asthma, heart disease, and chronic lung disease; alters lung's natural cleansing mechanisms; smaller particles associated with the most severe symptoms	Causes soiling of materials, grime deposits and reduced visibility; particulates from volcanic eruptions can reduce solar energy and produce temporary cooling effect
Ozone A product of NOx emissions from motor vehicles, power utilities, and industries burning fossil fuels, combined with hydrocarbons and sunlight in the atmosphere	Causes breathing difficulty, irritation to mucous membranes, and increases risk to respiratory infections; acute exposures causes respiratory pain, bronchoconstriction, lung edema, and abnormal lung development	Corrodes rubber, paint; weakens fabrics, rubber, and produces leaf damage and retardation of plant growth
Lead Historically emitted from vehicles burning leaded gasoline; emissions have been reduced by 98% since 1974; most lead exposures in U.S. today are not airborne	Damage to nervous system, blood forming tissues, kidneys; evidence of neurobehavioral disorders inlcuding learning disabilities and antisocial behavior	No known effect on vegetation or materials

TABLE 10.4
Sources and Characteristics of Coarse and Fine Particulates

PARTICULATE SIZE	PARTICULATE SOURCES	CHEMISTRY
Coarse Particles These are coarse particles from 1-100 μm including the course fraction PM_{10} (2.5-10 μm)	Industrial and mechanical processes such as fragmentation of matter and atomization of liquids, agricultural and forestry activities, and dust from unpaved roadways, mold spores, and wood ash	Silicon, aluminum, iron, potassium, and calcium are common components; the coarse particle samples tend to be alkaline
Fine Particles These are fine particles less than 1 μm, but including much of the mass of the $PM_{2.5}$ fraction; the $PM_{2.5}$ fraction consists of fine particles less than 1.0 μm and some coarse particles in the 1-2.5 μm range	Industrial and residential combustion of fossil fuels; secondary particles produced by direct, catalytic, and photochemical oxidation of nitrogen and sulfur compounds, and volatile hydrocarbons to produce sulfates, nitrates, and oxyhydrocarbons	Elemental and organic carbon such as from fuel combustion (soot), sulfates, nitrates, condensed organic compounds, oxyhydrocarbons, and trace metals; the fine particle samples tend to be acidic

OZONE AND THE PHOTOCHEMICAL OXIDANTS

There are two types of ozone, which are known on the popular media as "good" ozone and "bad" ozone. Good ozone is the layer of ozone in the stratosphere previously discussed that absorbs UV-B radiation and protects the earth from excess amounts of these damaging rays. The bad ozone is formed on the troposphere (nose level) by a complex series of reactions, including sunlight, nitrogen dioxide, and volatile hydrocarbons. The reaction occurs as outlined in Figure 10.16.

The photochemical oxidants resulting from such reactions include primarily ozone, nitrogen oxides, and alkyl peroxy radicals (RO_2). The alkyl peroxy radicals are produced by oxidation of hydrocarbons. An oxidant is a substance that readily gives up an oxygen atom, or removes hydrogen from a compound. Photochemical refers to the initiation of these reactions by sunlight. This mixture of photochemical oxidants is often referred to as photochemical smog and is most concentrated in areas with heavy traffic, intense sunlight, and stable air conditions. Such conditions are especially prominent in Southern California and to a lesser degree along the northeast coast, especially during May through September. Ozone is the primary indicator of smog and in concentrated form is a dense (1.6 times as heavy as air), violet blue gas with a sharp metallic odor. It is a powerful oxidant capable of breaking molecular bonds and rapidly degrading many structured materials and plant and human tissues.

FIGURE 10.16 Proposed mechanism for the production and maintenance of tropospheric ozone. (Adapted from Godish, T., *Air Quality*, 3rd ed., CRC/Lewis Publishers, Boca Raton, FL, 1997, chap. 2.)

The ozone standard was revised in 1979 to be 0.12 ppm for 1 h. However, more than 3000 studies since the 1980s show evidence that adverse health effects occur at lower exposures than the previous standard, and that these symptoms are aggravated by longer exposures. Exposures to ozone are associated with increased hospital admissions from asthma, and also accounted for up 20% of summer-time admissions of respiratory distress. Respiratory infections and inflammation are aggravated by ozone as well. Children and adults who play and work outside during summer months are at elevated risk of ozone exposure and may experience coughing, chest pain, and a reduction in lung function. Exposure to ozone over long periods has been shown to cause inflammation of the lungs, impairment of lung defense mechanisms, premature aging of lung tissue through irreversible damage, chronic bronchitis, and emphysema. Consequently, the EPA has set a new 8-h standard at 0.08 ppm based on the decision that this level will protect the public health.[49]

CARBON MONOXIDE

Carbon monoxide is a colorless, odorless, tasteless gas that is produced from the incomplete combustion of fossil fuels, the greatest contribution by far being from motor vehicles. In 1993, motor vehicles accounted for nearly 75% of anthropogenic carbon monoxide emissions.[47] Between 1986 and 1996, carbon monoxide concentrations in ambient air have decreased 37% whereas overall emissions decreased by nearly 20%. These improvements were made while vehicle-miles traveled increased 28% during the same period.[50] Carbon monoxide enters the blood streams through the lungs and combines with hemoglobin of red blood cells to form carboxyhemoglobin. Carbon monoxide has a stronger affinity for hemoglobin than oxygen, therefore as levels of carboxyhemoglobin rise, the adverse effects associated with oxygen deficiency increase. The health threat from chronic low-level exposure is most serious for people who have preexisting cardiovascular disease.[51] Inhalation of slightly elevated levels of carbon monoxide can produce symptoms ranging from dizziness and headaches to visual impairment, reduced work capacity, poor learning ability, lowered manual dexterity, and difficulty in performing complex tasks. Higher levels of CO inhalation can lead to paralysis of motor function and death.[50]

LEAD

Lead is a systemic heavy metal poison, which enters the human body through inhalation of contaminated air and ingestion of lead in food, water, soil, or dust. Lead accumulates in the bones, blood, and soft tissues, with target organs being the central nervous system, the blood-forming tissues, and the gastrointestinal tract. Lead exposures have been associated with neurological impairments, mental retardation, and behavioral disorders. Even at very low doses, children and fetuses suffer from central nervous system damage. The association of lead with behavioral problems and reduced intellectual ability caused lead to be placed on the list of criteria pollutants in 1977 when the Clean Air Act was reauthorized. Because automobiles burning leaded gasoline were the largest contributor to lead exposure, the phaseout of leaded gasoline has been the predominant control strategy. Lead emissions from highways have decreased 99% since 1987, whereas overall ambient lead concentrations decreased 95% since 1977, and 75% since 1987.[42]

SULFUR OR OXIDES

Health and Welfare Effects

Sulfur, phosphorous, carbon, oxygen, hydrogen, and nitrogen are the main components of most living things and consequently are found in fossilized forms, including the fossil fuels such as coal and oil. The combustion of fossil fuels results in the oxidation of sulfur to produce sulfur dioxides which reacts with moisture, oxidizing agents, and reactive hydrocarbons and other substances in the atmosphere to produce sulfates, sulfites, and sulfuric acid. The primary source of emissions is electric utilities. These compounds may produce a variety of adverse health effects. Sulfur oxides are also involved in corroding metals, disintegrating marble, limestone, and dolomite, attacking fabrics, reducing visibility, and impairing the growth of plants and forests. The most prominent health concerns associated with SO_2 include respiratory illness, effects on breathing, a reduction in lung defenses, and aggravation of existing cardiovascular diseases. Persons with asthma, cardiovascular disease, and chronic lung disease are most sensitive to the effects of sulfur dioxide. Children and the elderly are also at increased health risks from the inhalation of sulfur oxides.[50] Sulfur oxides are among the main precursors to acid deposition, with nitrogen oxides being the second greatest contribution.

Acid Deposition

Acid deposition is a term that represents a more inclusive understanding of the acidification process than the term *acid rain*, because acidity may be found in rain, sleet, snow, fog, clouds, and adsorbed-to particles. Consequently, the term acid rain is being replaced by the term acid deposition although the two are presently used synonymously. The problems associated with acid deposition were first recognized about 30 years ago in Sweden when changes in the plant and animal life of freshwater lakes were associated with increased acidity from acidic deposition.[17] The revelation of acidic deposition was not widely accepted by the world scientific community initially, but as measurements of rainfall pH demonstrated greatly expanding areas of increasing acidity, the problem became the subject of domestic and international debates over acid rain controls. The U.S. passed the 1990 CAAA with major provisions for sulfur and nitrogen oxide controls although atmospheric acidity still presents a major environmental problem.

Rain, snow, clouds, and sleet are normally somewhat acidic because of the presence of CO_2 in the atmosphere, producing carbonic acid with pH levels of nearly 5.0 in the absence of pollutants. The pH scale is logarithmic and is based on the negative log of the hydrogen ion concentration. Consequently, a pH of 3.0, as has been found in precipitation in the Ohio Valley, is 100 times the acidity of rain with a pH of 5.0. About 70% of sulfur oxide emissions originates from large electric utilities burning high-sulfur coal, whereas other major contributors to NOx and SOx include industrial boilers, metal smelters, and automobiles.[17] Sulfur acids have contributed about 65% to the acid deposition problem whereas nitrogen oxides contributed 35%. However, as regulatory controls become more effective, the levels of sulfur oxide in the atmosphere have decreased by 39% since 1970 whereas ambient concentrations of NO_2 increased 8% over the same period.[50] Consequently, the emission of NOx from automobiles and electric power generators is taking on greater importance.

Effects of Acid Deposition on Ecology

Acid deposition is a product of fossil fuels with most SOx occurring from electric utilities burning coal or oil. Many of these plants employ the strategy of building tall stacks so that the discharged gases will be carried out of the vicinity. Although locally effective, the gas streams are carried great distances by prevailing winds to be deposited far from the origin in a process known as long-distance transport. Heavily industrialized areas of the Ohio River Valley, upper Midwest and areas of Illinois, Pennsylvania, West Virginia, and Indiana are often considered to be the source of significant acid plumes, which travel eastward to deposit all along the Northeast coast off the U.S. and parts of Canada. The heavily industrialized areas of Great Britain and central Europe and parts of China create similar conditions for polluting distant locations. Sweden and Norway point to central Europe as a major cause of the acidity falling on them. Although long-distance transport contributes significantly to acidic deposition, local sources are also important contributors, and just how much each contributes to the problem has been strongly contested by those allegedly producing the pollutants, and those who receive them.

The New England areas, including upstate New York, the mountainous areas of West Virginia, Pennsylvania, Virginia, Kentucky, Wisconsin, Minnesota, parts of Canada, and the western mountainous areas of North America are sensitive to acidity

FIGURE 10.17 Acid-sensitive regions of North America and Canada. (Adapted from Godish, T., *Air Quality*, 3rd ed., CRC/Lewis Publishers, Boca Raton, FL, 1997, chap. 4.)

because these areas are deficient in magnesium and calcium carbonates. Therefore, the soils and water have low acid neutralizing capacity and are referred to as *acid-sensitive ecosystem* (Figure 10.17).[17]

Aquatic Ecosystems

When lakes and streams receive acid deposition in acid-sensitive areas, they are unable to neutralize or buffer the acidity. Consequently, aquatic systems in the Adirondacks or New York, New England, and other sensitive areas become chronically and progressively acidified or receive sudden massive doses of acidity in spring thaws, resulting in a process known as *shock loading*. As the acidity of a body of water increases, there is a reduction in the diversity of species and a shift in species composition. The smallest of the organisms, unicellular phytoplankton, disappear followed by the benthic invertebrates. These organisms serve as important food sources and nutrient recyclers (decomposers), so that large aquatic creatures are adversely affected by their disappearance. Acidification can lead to mobilization of aluminum and possibly other toxic metals, which, in combination with the increased acidity, leads to reductions in fish populations. At pH levels below 5.5, fish populations decline, with the most sensitive juvenile fish disappearing first. The eggs, larvae, and juvenile fish fail to thrive in a process called *recruitment failure*, often leaving acidified streams and lakes with a diminishing population of older, larger fish. Eventually, entire lakes and streams become devoid of fish, such as in the Adirondacks, mid-Atlantic mountains, New England, Ontario, Quebec, Sweden, and Norway.

Effects on Forest and Plants

The causal role of acid deposition in the decline of forested areas in North America, Germany, and central Europe has been a matter of considerable investigation and debate. Acidic deposition at levels of pH 4.0 to 5.0, which are most common, do not appear to cause widespread adverse effects on forest ecosystems.[17] However, conifer forests such as Red Spruce on mountain tops in New Hampshire, Vermont and in the Appalachians have been more than 80% decimated at the cloud line. Severe damage is also evidenced in central Europe in such places as the Czech Republic and Poland, where 60 to 70% of forests show evidence of damage associated with sulfur and nitrogen disposition. The mechanisms of such destruction are not immediately obvious but may be attributed highly to combinations of ozone and acid clouds (as low as pH 2.2), which (1) directly damage leaves, (2) mobilize toxic metals in soil, such as aluminum, which adversely effect roots, leaching nutrients from soil, and (3) overstimulate plants with excess nitrates, which aggravates deficiencies of other nutrients. These factors may combine to increase forest susceptibility to insect and fungal pathogens.[52]

Current Directions in SOx Control

Ambient concentration of SO_2 decreased 37% since 1987, whereas emissions decreased by more than 14%. In October 1999, EPA conducted a study to determine whether emission regulations impacted the recovery of lakes and streams in North America and northern Europe, which had suffered years of acid deposition. The study found significant declines of 1 to 6% sulfate per year in many lakes and streams. However, reduced emissions do not automatically translate to immediate improvement in streams, lakes, and rivers. For example, some waters in Maine continue to acidify even though sulfuric acid concentrations in rain have decreased. The scientists concluded that several factors contributed to continued lake acidity: (1) variations in climate, (2) increasing levels of nitrogen compounds in precipitation, such as nitric acid, (3) declines in the ability of a watershed to neutralize an acid, and (4) the short length of time for which data have been collected.[53]

Reductions in SO_2 emissions are due mostly to controls implemented under EPA's Acid Rain Program.[50] This program features a number of options, including (1) switching to low-sulfur fuel, (2) using scrubbers that remove SO_2 from the stack emissions, (3) washing coal, which removes up to 50% of the sulfur content, and (4) using advanced combustion technologies such as fluidized-bed combustion to remove sulfur and nitrogen oxides.

Nitrogen Oxides

Nitrogen and oxygen are combined in the high-temperature confines of industrial boilers and automotive engines to produce nitric oxide. This is oxidized in the atmosphere to produce nitrogen dioxide, an orange-yellow to reddish brown gas, which can often be seen from a distance enveloping cities. Nitrogen oxides are an important precursor for ozone and acidic deposition (nitric acid). Nitrates arising from NOx emissions fall from the sky to stimulate algae growth (i.e., red tide), causing toxic conditions for aquatic life in bays and estuaries. Nitrogenous compounds also contribute significantly to the formation of fine particulates. Nitrogen dioxide irritates lung tissues and reduces resistance to respiratory infections, possibly increasing the incidence of acute respiratory disease in children.[50]

HEALTH IMPLICATIONS OF AIR POLLUTANTS

The criteria air pollutants have been implicated in a number of adverse health problems including reduced lung functions, exacerbating problems, such as asthma, increasing risk of chronic bronchitis and emphysema, and aggravating acute respiratory infections. Carbon monoxide, sulfur dioxide, and particulates have also been shown to contribute to cardiovascular disease.[43,45,48,54] It is calculated that 125 million Americans breath unhealthy air, resulting in approximately 70,000 premature deaths in the U.S. alone.[55] Fine particulates from motor vehicles and power plants are reported to kill some 64,000 Americans a year and may be a major contributor to the epidemic of childhood asthma sweeping the country, and developing nations worldwide.[45]

Asthma occurs in persons with hyperactive airways when the bronchial tubes respond to allergens, pollution, irritants, cold air, and even exercise by small muscle contraction, dilation of blood vessels, and mucous secretion. The result is suffocating paroxysms in which the asthma sufferer struggles for each breath through closed airways clogged with mucus (Figure 10.18).[50] Asthma has increased by 42% since the mid-1950s. In 1998 alone, 17.3 million people suffered from severe asthma, 4.8 of which were children.[56] According to the Centers for Disease Control (CDC), asthma rates among children below 17 years of age increased 72% between 1982 and 1994.[57] In reality, there has been a twofold increase in the number of children who have asthma since 1980. This statistic alone has spurred concern among public health researchers.[58] It has become more severe, and research shows that children, other minorities, and urban poor are at greatest risk. Asthma now tends to be concentrated in urban pockets where children live under poor conditions and are frequently exposed to allergens such as mites and cockroaches,[59,60] and to episodes of air pollution involving fine particulates and ozone (Figure 10.19).[61] Air pollutants, particularly ozone and fine particulates, were found to be consistently associated asthma exacerbation in a prospective cohort study of children with asthma.[62] However, this link to specific air pollutants remains controversial. The prevalence of asthma is rising in both industrialized and developing countries, and the two countries with the greatest increase in asthma prevalence are Australia and New Zealand, neither of which has extensive air pollution.[63] Serious questions remain to be

ASTHMA ATTACK
Beginning Signs
• Wheezing
• Coughing
• Difficulty in breathing

Extended Attack
• Sweating
• Pulse becomes rapid
• Skin turns blue
• Arms and legs become chilled or cold

Bronchiole

Muscles contract and
membranes swell around
bronchioles.

Mucus forms in airways.

Narrowed airway makes
breathing difficult.

Alveoli

NORMAL AIRWAY   AIRWAY IN SPASM

FIGURE 10.18 Symptoms and effects of asthma.

answered on this controversy. Have all the responsible gases been identified? Is there an interaction of gases and particulates? What are the mechanisms by which pollutants exacerbate asthma?

Because of the rapid increases in severe asthma cases, alternative plans and programs have been started to aid asthma research. NIEHs has developed several research programs, including the Inner-City Asthma Study, the National Survey of Lead and Allergens in Housing (NSLAH), the Intervention Studies, Dust Mite

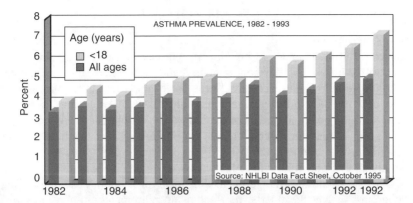

FIGURE 10.19 Trends in the prevalence of asthma from 1982–1996. (Adapted from Friebek, E., *Environ. Health Perspect.*, 104, 1, 22, 1996.)

Intervention Trial, the Subject-Collected Dust Sampling Study, and many more studies.[56]

Main Mechanisms of Air Pollutant Effects on Respiratory System

Air is generally laden with dust, microbes, and gaseous emissions that may cause adverse effects to human health. However, the human respiratory system has a number of defense mechanisms to ward off such attacks. The first line of defense includes nasal hairs, which trap larger particles. A combination of mucus and cilia line the bronchi and bronchioles, which trap and move smaller particles out of the respiratory tree. Irritating dusts and gases may cause the smooth muscles of the bronchioles to constrict, restricting heavier exposure in the short term. However, prolonged or exaggerated constriction is often combined with mucous secretions, leading to difficulty in breathing. Particles that reach the deeper portions of the lung, such as the alveoli, may be confronted by the alveolar macrophages which can ingest, destroy, and discard many microbes. Pollutants may produce their adverse effects by (1) inhibiting and inactivating mucociliary streaming, (2) killing or neutralizing alveolar macrophages, (3) constricting airways, (4) causing vasodilation and excess mucous secretion, or (5) causing changes in alveolar cell-wall structure through abscesses and thickening, which causes scar formation. The formation of scar tissue is characteristic of emphysema in which the walls of alveoli disintegrate and the compartments become unnaturally enlarged. Elasticity of the living tissue diminishes, and the lungs are unable to recoil and expel air, reducing the efficiency of the lungs in enhancing oxygen and carbon dioxide. Victims ultimately may become wheelchair bound, gasping for each breath, and unable even to expel sufficient air to blow out a candle. There is no cure for emphysema.

INDOOR AIR POLLUTION

Imagine suffering through episode after episode of pneumonia within a few short months, or suffering an attack of breath-restricting asthma every time you enter the basement of your home, or developing nausea and violent headaches upon entering your place of employment each day. These are examples of afflictions associated with indoor pollution in private homes and public places.[64] At one time, most people were aware of the potential threats of outdoor air pollution, but today many experts are cautioning against the effects of indoor pollutants. There has been a growing body of scientific literature in the last several years that points to indoor air as being more polluted and potentially more hazardous than outdoor or ambient air.[65] Additionally, people spend an average of 90% of their time indoors, although some at-risk subgroups such as the elderly, very young, and chronically ill may spend nearly all their time indoors.

SOURCES OF INDOOR AIR POLLUTION

Sources of indoor pollution come from several major categories, including combustion sources, such as the following (Figure 10.20):

1. Acid, coal, gas, oil, and kerosene
2. Household cleaning products
3. Furnishings such as cabinetry or furniture made of pressed wood, newly installed carpeting that emits volatile organic carbons (VOC) from adhesives
4. Building materials and paints that off-gas volatile compounds
5. Products for personal care and hobbies
6. Cooling and humidification systems
7. External sources such as radon, tobacco smoke
8. Moisture, which supports the growth of molds and other biological contaminants

Sources such as air fresheners, paneling, cabinets, and rugs may emit pollutants at a fairly continuous rate for some time, whereas other pollutant sources or activities release pollutants intermittently, such as cigarette smoking, use of household cleaners, or pesticide sprays. How long pollutants remain in the air and their relative concentration depends on several factors. Elevated heat and humidity increase the release of some pollutants into the air such as formaldehyde. Concentrations are greatly influenced by the amount of ventilation as well.

Air enters a house in three ways, including the following:

1. Natural ventilation: Air enters through open windows and doors, ridge vents, on the roof, ventilation pipes, or other portals designed for air access.
2. Infiltration: Air is drawn into the house through cracks around doors and windows, spaces in construction joints, or spaces or cracks in the foundation.

POTENTIAL SOURCES OF INDOOR AIR POLLUTANTS

1. Moisture	11. Tobacco smoke
2. Pressed wood furniture	12. Wood stove
3. Contaminated humidifier	13. Wood paneling
4. Moth repellants	14. Asbestos pipe insulation
5. Personal care products	15. Unvented dryer
6. Room air fresheners	16. Radon infiltration
7. Chemical cleaners and	17. Pesticides
disinfectants	18. Auto exhaust
8. Pressed wood cabinets	19. Auto cleaners and additives
9. Unvented gas stove	20. Paints, thinners, and stains
10. Household chemicals	

FIGURE 10.20 Sources of potential indoor air pollutants. (Adapted from Anonymous, *Environ. Health Perspect.*, 104, 21, 138, 1996.)

3. Mechanical ventilation: Air maybe drawn out of the building at home through externally vented fans or into the structure by whole-house fans in air handling systems.

The rate at which air is replaced in the structure by external air is termed the *air exchange rate*. This average rate for an American home is 0.7 to 1.0 air changes per hour. Houses that are tightly sealed without provisions for an exchange may experience as little as 0.2 air changes per hour, whereas drafty houses may approach up to 10 times that amount.[51] More recently constructed office buildings and non-residential structures may have sealed windows that restrict air infiltration. Although ventilation and cooling systems are intended to maintain comfort, air that is recirculated in a building without sufficient outside makeup air may exhibit increased levels of CO_2 and pollutants. Before 1973, at least 15 ft³/min (cubic feet per minute, CFM) per occupant was required for makeup air. However the standard was reduced after 1970 to 5.0 CFM, but this level may be insufficient to provide a safe and healthier environment.

SIGNS OF INDOOR POLLUTION

Signs of indoor air pollution may include physical or health signs or both. Physical symptoms may include (1) heating or cooling equipment that is dirty or moldy, (2) moisture condensation on walls and windows, (3) air that is stuffy or has an unpleasant odor, and (4) signs of water leakage anywhere in the building with growth of molds. Health indicators of indoor air pollution may include immediate or acute effects such as eye irritation, dry throat, headaches, fatigue, sinus congestion, sun irritation, shortness of breath, cough dizziness, nausea, sneezing, and nose irritation. These symptoms are vague and may originate from many sources. Some people recognize that the symptoms occur when they enter a structure and diminish when they leave. In many cases, however, there is not a readily identifiable pattern or illness. Additionally, chronic illnesses such as emphysema, cancer, and heart disease may not appear for years, and any association with exposure to substances in a particular building is usually not recognized. When a number of occupants of a building display acute symptoms without a particular pattern and the varied symptoms cannot be associated with a particular source, the phenomenon is often referred as sick building syndrome (SBS). Further, maladies may not be the cause of indoor air pollution but may be related to psychological stress, inadequate lighting, excessive heat or cold, or even noise.[66]

When well-defined illnesses occur in a building and they can be traced to specific building problems, the illnesses are referred to as building-related illnesses (BRI). The well-defined illnesses normally include asthma, hypersensitivity pneumonitis, humidifier fever, or Legionnaire's disease. The problems associated with SBS or BRI have apparently increased as concerns about energy conservation have escalated resulting in the construction of air-tight buildings with thicker insulation, magnetically sealed doors, sealed windows (office buildings), or triple-glazed windows that all serve to limit the exchange of air with the outside. Office buildings are at risk of exposure to indoor air pollution from (1) poorly designed or improperly used or maintained ventilation systems, (2) the presence of pollution sources such as harsh chemicals, improperly vented photocopiers, or biological contaminants from dirty ventilation systems, and (3) uses of the building for which the building was not designed, such as a paint shop or automotive repair service in the basement of an office building.

COMMON SOURCES OF INDOOR AIR POLLUTION

The most common sources of indoor pollution include environmental tobacco smoke, radon, biologicals, nitrogen dioxide, carbon monoxide, organic gases, formaldehyde, respirable particles, and pesticides.

Environmental Tobacco Smoke (ETS) and Other Combusted Materials

Although smoking has been under vigorous attack in the U.S., and the opportunities for smoking in public buildings has greatly diminished, smoking still contributes to nearly 500,000 deaths each year in the U.S.[67] The smoker is exposed to nearly 4,700 compounds in mainstream smoke through direct inhalation. The sidestream smoke that originates from the burning end of the cigarette and exhaled smoke make up environmental tobacco smoke (ETS) to which nonsmokers may be exposed in a form referred to as *passive* or *second-hand* smoke. Such involuntary exposures have prompted numerous legal actions, further restricting the public places where smoking is allowed. However, people continue to be involuntarily exposed to ETS in the home and many commercial establishments, increasing the risk of respiratory disorders. Second-hand smoke has been linked to the rise of lung cancer risk by an average of 30% in nonsmoking spouses.[64] Other sources of combustion products include gas and kerosene space heaters that are unvented and improperly vented wood stoves, fireplaces, and gas stoves. Pollutants from these sources include carbon monoxide, nitrogen oxides, particulates, and a variety of aromatic hydrocarbons and acid aerosols. Unintended deaths occur every year from the buildup of CO in the home because of improper ventilation of home-space heaters or furnaces. Lesser concentrations may lead to headaches and dizziness, and aggravate preexisting heart disorders. Nitrogen oxide exposure can increase the risk of respiratory infection and emphysema, and may aggravate respiratory disorders such as asthma.

Radon

Radon is a colorless, odorless gas that occurs naturally by the decay of radium-226, a product of uranium decay. It is found in most places but is highest in concentration in uranium and phosphate ores. Areas with lesser concentrations of radon are found in limestone, shale, and granite. Radium-226 is ubiquitous in low concentrations and does not present a health issue until it is trapped and concentrated in areas such as the basement of a building or house built in areas where uranium is present in the soil or rock.

As the uranium naturally decays radioactively, it releases radon gas that further decays into short-lived radon daughters and gamma rays (Figure 10.21) The radon daughters posing the greatest risk are the alpha emitters polonium-218 and polonium-

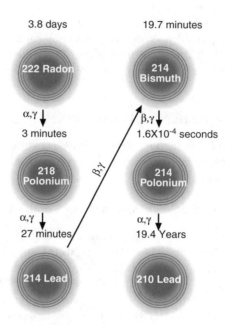

FIGURE 10.21 Radioactive decay of products of radon.

214. The radon progeny quickly attach to particles in the air, which may then be inhaled to lodge in sensitive areas of the upper respiratory tree. Once lodged in human tissue, the radioactive materials increase the risk of lung cancer, causing from 5,000 to 20,000 excess cancer deaths a year in the U.S.[65]

Recent evidence released by the National Research Council points to radon as a significant public health problem causing as many as 21,800 deaths per year with the greatest risk being to people in homes where cigarettes are smoked. Radon is considered to be the second-leading cause of lung cancer exceeded only by cigarette smoking.[68] Radon in soil normally enters structures under slight negative pressure through cracks in concrete walls and floors, sumps, floor drains, dirt floors, joints, and crawl spaces. Radon may also be present in well water or even foundation materials. It is estimated by EPA that as many as 8 million homes in the U.S. have elevated radon levels. Although some regions are at greater risk than others, buildings and households should be checked because even adjacent houses may show remarkably different levels of radon. Most hardware stores have radon detection kits for sale. The kits are either alpha track detectors or chemical canisters. The kit instructions usually suggest that you hang the detector in a closed cellar at 3 to 4 ft above the ground for several days to a month. The kits are returned by mail to the manufacturer for analysis and a report is usually returned within 2 to 3 weeks to the landlord or building manager. The federal action guidelines recommend that corrective actions be taken at 4 pCu/l or above. Radon concentrations are reported as picocuries per liter (pCu/l). A picocurie is the equivalent to one trillionth of a curie representing slightly more than 2.0 radioactive disintegrations per minute. The average concentration of radon and its decay products in ambient air is about 0.25 pCu/l.

Some of the highest levels of radon have been measured in a region called the Reading Prong located in an area covering southeastern Pennsylvania, and parts of Connecticut and New Jersey.[69] The highest level ever recorded in a household (2600 pCu/l) was identified in southeastern Pennsylvania. The northeast has higher mean levels than most other sections of the country, but the presence of radon in a home is unpredictable. Households must be tested to determine if they exceed the recommended action level of 4 pCu/l. Surveys have shown that 1 in every 15 homes in the U.S. exceeds the action level. At this level, the estimated number of lung cancer deaths due to radon exposure would be 13 to 50 per 1000 for an average lifetime exposure. This risk is equivalent to smoking five cigarettes a day.[65,70]

The mitigation of radon may involve actions such as installing subslab or basement ventilation; covering sump holes; sealing cracks and other openings in floors, joints, and walls; ventilating crawl spaces; installing air-to-air heat exchanges; or installing a drain tile suction system around the outer foundation walls. In some cases, more than one method may be used.

BIOLOGICAL CONTAMINANTS

The most common biological contaminants include molds, mildew, viruses, bacteria, dust mites, cockroaches, pollen, animal dander, and cat saliva. The sources for these agents is numerous including people-transmitted viruses and bacteria, household pets, contaminated air handling or ventilation systems, moist surfaces on which organisms can grow, dried fecal matter from cockroaches and other insects, and soil and plant debris. The major threat to the biological contaminant of the home is moisture. Water may enter through a porous basement floor or walls or a leaky roof, or remain in unventilated bathrooms or as a reservoir in dirty humidifiers. In these situations, water stimulates the growth of mold, dust mites, and other infectious or allergenic organisms.[71]

Surveys have shown that 30 to 50% of homes surveyed in the northern U.S. and in Canada have problems with moisture, whereas structures in the South may be at higher risk. Wet surfaces, moist carpeting (often associated with concrete floors), water-damaged materials, and standing water can all support the growth of molds, mildew, bacteria, and insects such as house and dust mites. Dust mites are microscopic insects that feed on shed human skin, thrive in moist environments, and are powerful allergens. Molds are also potent allergens, and they can provide nutrients to bacteria, spread viruses, emit chemicals such as cancer-causing phenols and trichloroethylene, increase susceptibility to a host of diseases by weakening the immune system, and aggravate asthma.

Possible symptoms of illness caused by biological contaminants include runny nose, colds, flu-like symptoms, headaches, unexplained fatigue, and digestive problems. Allergic rhinitis, hypersensitivity pneumonitis, asthma and other serious allergies are common. Some biological contaminants transmit infectious illnesses such as influenza, measles, and chicken pox whereas molds and mildews release disease-producing toxins.[65,71] Homes with water stains or other signs of moisture, leaking plumbing, mold, mildew, or dust that remains after cleaning may have a possible water problem.

Reducing Exposure

Because moisture always migrates to colder surfaces, there are ample steps, such as using fans, warming basements, ventilating attics and bathrooms, and insulating exterior walls, that can check the unwanted movement of water. Gutters may be used to drain water away from buildings. Other methods of controlling biological vectors include washing bedding frequently in hot water, maintaining humidity between 30 and 50%, using a high-efficiency particulate-filtering vacuum cleaner, enclosing pet litter boxes, discarding wet carpeting and other water-damaged materials, and practicing good housekeeping procedures.

Carbon Monoxide, Nitrogen Oxides, and Respirable Particles

These substances are manually emitted from unvented kerosene and gas heaters, wood stoves, fireplaces, environmental tobacco smoke, automobile exhausts in attached garages, and other combustion sources. These have been discussed previously under combustion sources.

Organic Gases and Pesticides

Most households and many businesses have a wide variety of products that can emit potentially hazardous materials, including paints, strippers, disinfectants, cleaners, repellants, automotive products, hobby supplies, volatile office supplies, and pesticides. Such substances often exceed levels found outside the home by a factor of up to 5 times whereas some activities such as paint stripping may produce concentrations of 1000 times background outdoor levels.[65,72] The potential health consequences are numerous based on the extraordinary variety of chemicals available in homes and businesses. These symptoms range from irritation to mucous membranes, multiple organ damage, headaches, nausea, cancer, and many other maladies. Exposures to such substances can be limited by (1) properly disposing of unneeded products, (2) following instructions on a label and using it in a well-ventilated area if so advised, and (3) using nontoxic alternatives where possible.

Formaldehyde (HCHO)

Formaldehyde is found in pressed wood products such as cabinets and furniture made from plywood, particleboard, wall paneling, and fiberboard. It is also found in urea formaldehyde foam insulation (UFFI), some textiles, and environmental tobacco smoke. However, formaldehyde-emitting products are now being manufactured with much less HCHO emission potential, and emissions from pressed wood products are presently 10% of those used 20 years ago.[69] Because older products and structures have off-gassed most of the volatile HCHO, and new products have very little HCHO, the significance of formaldehyde as a public health issue has been markedly reduced. Exposures still can occur, which is aggravated in the presence of a confined or poorly ventilated space, combined with high heat and humidity. Adverse health effects include irritation to the mucous membranes, severe allergic reactions, fatigue, wheezing, and coughing.

NOISE

INTRODUCTION

People spend an average of 90% of their lives indoors and are exposed not only to chemicals and biologics, but also to sounds that enter their homes from heavy traffic, descending or disembarking airliners, trains, and loud radios. Outside, they are confronted with the additional sounds of snowmobiles, Jet Skis, and motorcycles in areas previously thought to be reserves of quiet. Sound itself is not a pollutant, but when it interferes with tasks, when it distracts, annoys, or disturbs, or when it causes loss in hearing or alters physiology in a negative way, then it becomes unwanted sound, or noise. Noise may be considered a pollutant whose primary medium of transmission is air, and so it is considered in this chapter as another form of air pollutant. Organizations such as Noise Network have been created for people who cannot endure sources of noise disturbance any longer. The organizations will help clients relocate from an area of high noise to one of a lower frequency of occurrence. Tragedies have been the result of normally noisy areas. For example, a 78-year-old farmer was driven to the unthinkable after months of enduring noisy neighbors. He reached his breaking point, shooting two of the neighbors, killing one and wounding the other. He was later convicted of manslaughter and is presently serving two consecutive life sentences. This is an uncommon effect of noise pollution; however, it is true and could be a cause for alarm down the noisy road.[73]

THE PHYSICS OF SOUND

Sound is a form of energy that is produced by the vibration of objects, which compress and expand air, water, or solids to produce waves. When the vibration occurs in air, variations are produced in the normal atmospheric pressure, which are carried to the eardrums and cause the membranes to vibrate. These vibrations are passed to the inner ear and ultimately to the brain, which interprets these vibrations of acoustical energy as sound. Because it depends on the movement of molecules causing positive and negative pressure waves, sound cannot travel in a vacuum. As an object vibrates, waves radiate outward in an expanding sphere much as the ripples produced when a stone is dropped in water. Waves are produced, consisting of alternating peaks and troughs, which may vary in their length, peak, and amplitude. These wave characteristics produce changes in frequency (or pitch) and amplitude, which translates to loudness. Frequency is directly related to the speed of vibration. The faster the vibration, the more frequent the waves are per unit time and the higher the pitch (Figure 10.22). Frequency is expressed as cycles per second, and the standard unit of expression is hertz (Hz), which is equivalent to one wave per second passing over a particular point. The normal range of human hearing is in the frequencies of 50 to 20,000 Hz. Sound waves also vary in the height or amplitude and reflects the amount of energy formed in the sound wave (Figure 10.23). Soft sounds have a low amplitude, whereas loudness is characterized by large amplitudes. This amplitude intensity or loudness is measured in decibels (dB). The scale of dB ranges from 0, which is the threshold for hearing among the normal population, to 194, which is the threshold for pure tones.[74,75]

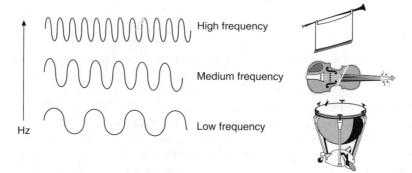

FIGURE 10.22 Frequency differences of sound. Higher frequencies are associated with more cycles per second (hertz), and are perceived as higher pitch.

FIGURE 10.23 Amplitude differences of sound. Higher amplitudes or wave heights are associated with greater energy and perceived as loudness.

The perception of sound (hearing) is not uniform across frequencies (Hz) and amplitudes (dB). Human hearing is most acute in the range of human speech, which is about 400 to 4000 Hz. At frequencies on either side of the range, hearing is less sensitive, and greater amplitudes are required to hear these frequencies. At amplitudes below 0 dB, humans cannot detect sound pressures even though they exist. As sound increases above 70 dB, it becomes increasingly difficult to participate in conversation, whereas sounds above 90 dB may lead to hearing loss, The pain threshold for noise begins at 120 to 130 dB for most people. Because the perception of loudness doubles with each 10-dB increase, this is not a linear relationship, and small differences in dB readings are heard as very large changes.[75,76] The relationship of decibels to common sounds is shown in Figure 10.24.

PHYSIOLOGY OF SOUND AND HEALTH EFFECTS

Sound compression waves are directed by the external auditory canal (meatus), where they strike the tympanic membrane, causing it to vibrate in concert with the

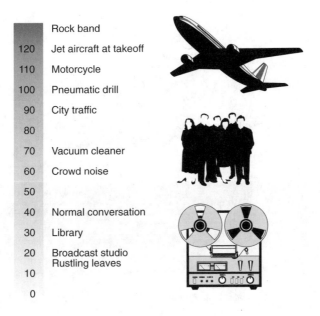

	Rock band
120	Jet aircraft at takeoff
110	Motorcycle
100	Pneumatic drill
90	City traffic
80	
70	Vacuum cleaner
60	Crowd noise
50	
40	Normal conversation
30	Library
20	Broadcast studio
	Rustling leaves
10	
0	

FIGURE 10.24 Sound levels of various activities.

waves. The tympanic membrane is connected to a series of three very small bones in the middle ear known as the malleus, incus, and stapes, which transmit the vibration to the oval window of a small-shaped structure called the *cochlea* (Figure 10.25). The transmitted vibration causes corresponding waves in the liquid (perilymph), which under appropriate conditions cause the basilar membrane to vibrate. This in turn causes hair cells of the spiral organ (organ of Corti) to move against the tectorial membrane. The hair cells convert a mechanical force (pressure) into a nerve impulse (electrical signal), which is ultimately transmitted to the auditory area of the temporal lobe of the cerebral cortex.[76] Excessive sound pressure (loud noises) can destroy the delicate hairs in the spiral organ. This process usually begins with the hairs at the base of the cochlea, most sensitive to high-pitched sounds, and with continued exposure to high-dB sound, the damage extends to other hairs and corresponding lower pitches. Such continued exposure can lead to increased damage of hair cells until 30 to 50% of the hair cells are destroyed, at which time measurable hearing loss becomes evident. Such damage resulting in loss of hair cells produces permanent hearing loss and is known as *permanent threshold shift (PTS)*. Permanent hearing loss can result from prolonged exposures to sound levels of 70 to 85 dB, which are at or above the level of vacuum cleaners or city traffic. Higher sound intensities require much less time to cause damage; being on a landing field, where jet planes are disembarking, without ear protection could cause measurable hearing losses in less than an hour. Noise can also result in a temporary threshold shift (TTS), which is a reversible hearing loss normally lasting from moments to hours, although longer periods of temporary hearing loss are possible. Very loud explosions or gunfire can cause acoustical trauma, which can rupture or damage eardrums, middle ear bones, and even the organ of Corti.

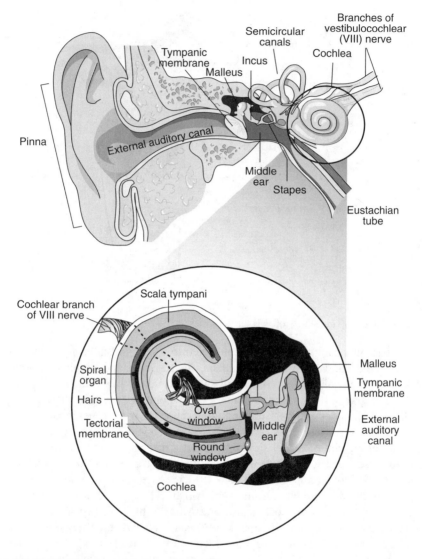

FIGURE 10.25 Structure of the ear and hearing apparatus.

The measurement of hearing loss is usually performed on an instrument called an *audiometer*, whereby a person is placed in a soundproof room and listens on earphones to sounds produced at different pitches and decibels by a trained audiologist. Persons with hearing problems will be unable to detect sounds at normal threshold levels, and the intensity will have to be increased. If normal values are 0 to 25 dB, then hearing the same sounds at 26 to 40 dB would indicate a mild hearing loss, whereas a severe hearing loss would require 71 to 90 dB.

Noise has also been associated with adverse health effects, apart from hearing loss. Noise may be an annoyance. The annoyance level appears to escalate as a function of intensity (loudness) and pitch. Therefore, loud, high-pitch sounds (above 5000 Hz) tend to be the most annoying.[76] Related to this annoyance is the creation of stress and associated physical effects such as increased blood pressure, elevated heart rate, headaches, colitis, and ulcers. Such responses may seem unreasonable, but in the presumably quieter realm of historic humans, sudden loud sounds prompted immediate physiological responses to the presence of danger as a mechanism of survival. Unfortunately, prolonged exposure resulting in continued exaggerated physiological response leads to damaging health consequences.

Noise can lead to interference with speech communication; most of us have been in the presence of a loud band or noisy crowd, when we had difficulty hearing a person yelling at us within a few inches from our ears. Noise ratings measuring its ability to interfere with speech communication have been developed on a scale known as the *preferred octave speech-interference level (PSIL)*. This is useful in determining what changes in PSIL are necessary to make communication easier.[67] Noise levels have also been implicated in decreased learning ability, work performance, and increased safety hazards and sleep disruption.[72,74]

REGULATION OF NOISE

The regulation of sound requires that it be measured according to a standard. The measuring device most frequently used is a sound pressure level meter that converts sound captured by a microphone to electrical signals, which can deflect a needle on an indicating meter. Most instruments contain A, B, and C scales. The A-weighing network is biased to high-frequency noise and best correlates with the annoyance effects and hearing damage in humans. Thus the sound reading on this scale would be expressed in dBA units. This is the scale most frequently used in assessing noise exposure. Exposures are regulated for the occupational environment by the OSHA (Occupational Safety and Health Administration) Department of Labor with a permissible exposure limit of 90 dBA for an 8-h d, 40-h work week. Permissible exposure times decrease half for each 5-dB increase.

Federal agencies that regulate noise include the Federal Highway Administration (FHA), Housing and Urban Development (HUD), Occupational Safety and Health Administration (OSHA), Federal Aviation Agency (FAA), and EPA. Each of these agencies is responsible for the regulation and control of noise to specific operations or segments of society under their direction. OSHA regulations are designed to protect workers from occupational noise-induced hearing losses. The EPA is mandated under the Federal Noise Control Act of 1972 and Quiet Communities Act of 1978 to abate noise in the ambient environment and in communities by (1) identifying levels of environmental noise considered protective of human health, (2) identifying major noise sources requiring regulation, and (3) providing financial assistance to state and local governments to investigate noise problems and develop control measures in the communities. Many programs were initiated in the 1970s, but the elimination of federal funding and associated programs in the 1980s all but eliminated state and local noise control programs throughout the country. Many federal regulations under the NSAD 1972 noise abatement program were never promulgated, and it is unlikely that significant regulatory attention will be paid to issues of noise until there is an increasing public outcry focusing on the increasing noise pollution in our environment.[75]

RADIATION

INTRODUCTION

Marie and Pierre Curie termed the mysterious phenomenon *radioactivity* and isolated two radioactive substances, polonium and radium, from uranium ore sample. These discoveries, along with the discovery of "x-rays," quickly led to the birth of a whole new field of scientific inquiry with far-reaching implications in physics, biology, medicine and, unfortunately, warfare.

IONIZING RADIATION

Early investigations of radioactivity revealed that the observed emissions were of several different kinds, consisting of subatomic particles — protons, neutrons, or electrons — which are released when atoms decay, later becoming known as particulate radiation. The particulate radiation includes alpha and beta particles. Atoms are the basic units of elements and consist of a small dense center called a nucleus surrounded by a cloud of negatively charged electrons (Figure 10.26). The positive charge in the atom is attributed to the prot or atomic number. The proton account for approximately half of the mass of the atom, with the other half composed of neutrons that carry no net electrical charge and are therefore neutral. The total number of prot and neutrons is equal to the mass number of the atom. A particular element may have more than one mass number. Lead may have a mass number of 214 or 210, although both have the same atomic number. The difference in mass number results from different numbers of neutrons. Such atoms are called *isotopes* (Figure 10.26). Many elements have stable and unstable isotopes. Unstable isotopes or radioisotopes (Figure 10.26) spontaneously emit energy in the form of radiation, which was originally detected by their ability to expose photographic film in the dark. When a radioisotope decomposes, it releases energy in the form of electromagnetic radiation (gamma or x-rays), and energy of motion from particles. Naturally occurring gamma rays consist of highly energetic short wavelengths of electromagnetic radiation, a form of energy that also includes ultraviolet light, visible light, infrared waves, and microwaves.

RADIOISOTOPES

Ionizing radiation includes alpha and beta particles, x-rays, and gamma rays. Because the particles or rays involved are sufficiently energetic to dislodge electrons from the atoms or molecules that they encounter, ionizing radiation's ability to destroy chemical bonds gives it special significance. An atom which is missing one or more electrons is referred to as an ion, and energetic radiation capable of doing this is called ionizing radiation. X-rays and gamma rays are included in the category, and are capable of breaking chemical bonds and ionizing molecules. When critical molecules are disrupted, the cell will undergo degenerative changes or even die. Deoxyribonucleic acid (DNA) is a critical molecule in human and animal tissue. Injury

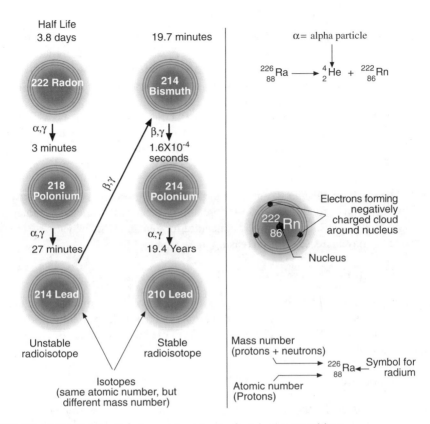

FIGURE 10.26 Radioactive decay of products of nuclear composition.

results from the ionization of water molecules in tissues forming ion pairs and free radicals, which instantly react with the cell components to produce cross-linking of macromolecules, causing changes in the tertiary or three-dimensional structure of the molecule. These changes may disrupt the function of proteins, DNA, and other molecules, causing mild reactions to cellular death. The disruption of the DNA molecule may cause genetic errors to occur that will be evidenced in abnormal growth and maintenance of the affected cells. Death of the cells is often the ultimate outcome.

RADIATION EXPOSURE

Every individual comes into contact with ionizing radiation from three general types of sources: (1) naturally occurring, (2) naturally occurring but enhanced by human actions, and (3) human generated.

Natural Sources

The earth, air, water, and food all contain traces of radioactive materials, which constitute a source of "background" radiation. The following are sources of background radiation:

Cosmic radiation: High-energy particles are composed primarily of proton and electrons that stream toward the earth both from outer space and from the sun following episodes of solar flares. Exposure to cosmic radiation is considerably less at sea levels than at high altitudes (for this reason, residents of Denver receive about twice as much cosmic radiation as do the residents of Los Angeles or Miami).

Radioactive minerals on the earth's crust: Uranium, thorium, potassium, and radium are all found in soils and rocks in many parts of the world. Inhabitants of areas such as the Rocky Mountains are exposed to higher levels than those on the East Coast, where radioactive minerals are in much less abundance.

Radionuclides in the body: A number of radioactive substances enter the body by ingestion of food, milk, or water or by inhalation, which are incorporated into body tissues, in which their concentration may be sustained at a steady rate or may increase with increasing age.

Enhanced Natural Sources

The impact of human activity on levels of exposure would be much lower under other conditions. Examples of enhanced natural sources include uranium mill tailings, phosphate mining, and jet airline travel (from the increased exposure to cosmic radiation at high altitudes).

Human-Generated Sources

These include medical applications, nuclear weapons fallout, nuclear power plant emissions, and consumer products. Medical applications include medical x-rays and radiopharmaceuticals for both diagnostic and therapeutic purposes. Nuclear weapons fallout from the first atomic bomb to the end of atmospheric testing of nuclear weapons by several major powers of the world, primarily the U.S. and the former Soviet Union, resulted in significant amounts of radioactive fallout worldwide. Very small amounts of exposure are received from home smoke detectors, static eliminators, airport security checks, TV receivers, and tobacco products.

HEALTH IMPACTS OF IONIZING RADIATION

During the early years of work with radioactivity was the medical community itself, particularly radiologists, whose enthusiastic use of their new machines exposed their patients and themselves to large amounts of radiation on a daily basis, with a much increased incidence of leukemia. This was due to their enthusiastic use of new machines that exposed them and their patients to large amounts of radiation on a daily basis, with a much increased incidence of leukemia.

Dosage

Because ionizing radiation can neither be seen nor felt, human exposure is measured in terms of the amount of tissue damage it causes. Under the International System of Units, the term gray (Gy) is the unit of absorbed dose, used to quantify the amount of energy form ionizing radiation absorbed per unit mass of material. There is little evidence that present average levels of exposure to ionizing radiation are having a serious adverse effect on the health of the general public. However, higher doses can be injurious to living tissues. The International Congress of Radiation Research has been involved in a long-running argument over what is considered a safe dose of radiation. U.S. radiation experts from the University of California (UC) at Davis have stated that there are safe values of exposure to radiation (i.e., threshold), which the body can tolerate without the development of cancer. Conversely, the International Commission on Radiological Protection stands behind the linear no threshold theory, stating that the indicated safe dose proposed by the UC Davis researchers does not exist. These arguments about the validity of the linear no threshold theories are important because regulators use the theory when setting radiation exposure standards.[77]

Dose Rate

The time span during which a given amount of radiation is delivered is thought to be very significant in determining the extent of tissue damage incurred, perhaps even more important than the total dose.

Radiation-Induced Mutations

Exposure to low levels of ionizing radiation can result in gene mutations. There is a linear relationship between increasing radiation dose and increased number of mutations. Nuclear power plant workers and medical personnel who receive small doses of radiation on a daily basis have been shown to exhibit a marked increase in the frequency of mutations in their chromosomes.

Radiation and Birth Defects

A number of congenital abnormalities, as distinguished from defects caused by mutant parental genes, are known to result from x-ray exposure *in utero*. Sensitivity to radiation is greater during fetal development.

Radiation-Induced Cancer

The potential for health damage as a result of radiation exposure was first recognized in connection with the development of skin cancers among early radiologists, of whom nearly 100 fell victim to this disease within 15 years following the discovery of x-rays. Leukemia is the classic example.

Radiation and Nuclear Power Generation

Most attention is focused on the nuclear power plants, because of their perceived threat. Nuclear power production involves a number of steps, referred to as the *nuclear fuel cycle*. These steps include mining the uranium, crushing and processing or milling the uranium, and converting yellowcake to gaseous uranium hexafluoride, enriching uranium to increase the concentration of the desired isotope from its original concentration.

Although highly controversial, nuclear power is the cleanest fuel known to man. The controversial problems with nuclear power do not deal with the toxic gas or water production as other fuels sources do, but to the production of the waste. The waste for a nuclear reactor, depending on the isotope used for production, has varying half-lives. These half-lives correspond to the time it takes for the isotope to decay to one half of its original limit. Nuclear power is used throughout the U.S. for mainly the production of public electricity. Another major source of nuclear power is through the naval defense departments of the world. International navies harness the nuclear energy for use in submarines and other sea-going vessels. The use of nuclear power allows submarines to stay submerged for longer periods of time, thus allowing a more efficient use of time and federal money.

In 1982, Congress enacted the Nuclear Waste Policy Act which delegated responsibility for high-level radioactive waste management to the federal government and designated the U.S. Department of Energy as the lead agency to coordinate the effort to site, construct, and operate the nation's first permanent repository for such wastes.

ULTRAVIOLET RADIATION

Wavelengths of the electromagnetic spectrum ranging from 40 to 400 nm in length are categorized as *ultraviolet (UV)* light. Certain portions of this range are strongly absorbed by living tissues, particularly by DNA, which constitutes the major target of UV damage. Injury to the hereditary material of cells is the reason for the lethal or mutational effects that excess UV exposure can provoke in living organisms. Light in the ultraviolet range does not penetrate beyond the upper epidermis, but the lack of melanin in people without color places them at increased risk to skin cancers.[74]

Three major types of skin cancer account for over 700,000 new cases of the disease diagnosed in the U.S. each year. Basal cell carcinoma and squamous cell carcinoma represent the vast majority of skin cancers, accounting for nearly one third of all cancers in the U.S. today. Chronic exposure to sunlight (especially the UV-B component of sunlight) is recognized as the cause of over 90% of these two cancers. Malignant melanoma is much less common but far more deadly. Americans born in the 1990s are 12 times at risk of developing melanoma as those born 50 years ago and twice as likely as those born just 10 years ago. Unlike the non-melanoma skin cancers, malignant melanoma appears to develop as a result of occasional severe sunburn, rather than prolonged low-dose exposure to sunlight.

The beneficial effects of ultraviolet light include the production of vitamin D, which prevents the formation of bone deforming rickets, and the germicidal properties of UV, which reduce bacterial and viral infections. This has been demonstrated in operating rooms and in the treatment of bacterial skin diseases including acne and boils.[74]

REFERENCES

1. Chamber, L., *Classification and Extent of Air Pollution*, Stem, A.C., Ed., 3rd ed., Academic Press, New York, 1976, chap. 1.

2. Morgan, M.T., Gordon, L., Walker, B. et al., *Environmental Health*, 2nd ed., Morton Publishing Co., Englewood, CO, 1997.

3. Whelan, E.M., *Toxic Terror*, Prometheus Books, Buffalo, New York, 1993, chap. 10.

4. Godish, T., *Air Quality*, 3rd ed., CRC/Lewis Publishers, Boca Raton, FL, 1997, chap. 5.

5. Liu, B. and Xu, E., Physical and Economic Damage Functions for Air Pollutants by Receptors, EPA/600/6-76-011, 1976.

6. Gribbon, J., Carbon Dioxide, Ammonia-and Life, *New Scientists*, 94, 1305, 1982, p. 413.

7. Godish, T., A*ir Quality*, CRC/Lewis Publishers, Boca Raton, FL, 1997, chap. 1.

8. Nadakavukaren, A., *Our Global Environment: A Health Perspective*, Waveland Press, Prospect Heights, IL, 1995, chap. 11.

9. Godish, T., *Air Quality*, CRC/Lewis Publishers, Boca Raton, FL, 1997, chap. 3.

10. Nadakavukaren, A., *Our Global Environment: A Health Perspective*, Waveland Press, Prospect Heights, IL, 1995, chap. 12.

11. Godish, T., *Air Quality*, CRC/Lewis Publishers, Boca Raton, FL, 1997, chap. 8.

12. EPA, The Clean Air Act Amendments of 1990: Summary Materials, EPA Office of Air and Radiation, Reproduced by Library of Congress, Congressional Research Service, November 15, 1990.

13. Kessler, J. and Schroder, W., Meeting Mobility and Air Quality Goals: Strategies That Work, EPA, Office of Policy Analyses, October 13, 1993.

14. Anderson, E.V., Health Studies Indicate MTBE is Safe Gasoline Additive, *C&EN,* September 20,1993, p. 9.

15. Bradley, J., Buying High, Selling Low, *E/The Environmental Magazine*, July–August 1996, hhp://www.emagazine.com/2curr3.

16. Chandler, D.L., MIT Scientist Shares Nobel for Identifying Ozone Damage, *Boston Globe*, October 12, 101, 1995.

17. Godish, T., *Air Quality*, CRC/Lewis Publishers, Boca Raton, FL, 1997, chap. 4.

18. Associated Press (AP), Ozone Hole Grows to Record Size, CNN Interactive, November 2, 1996.

19. National Oceanic and Atmospheric Administration (NOAA), Stratospheric Ozone: Monitoring and Research in NOAA, last update March 3, 2006, http://www.ozone-layer.noaa.gov/index.htm.

20. Wald, M.L., Smuggling of Polluting Chemicals is Resolved, *New York Times National*, September 17, 1995, p. 1.

21. Begley, S., Holes in the Ozone Treaty, *Newsweek*, September 25, 1995, p. 70.

22 .Stevens, W.K., Scientists Report an Easing in Ozone-Killing Chemicals, *New York Times*, August 26, 1993.

23. EPA, The Size and Depth of the Ozone Hole, written by EPA's Stratospheric Protection Division, June 10, 1998, hhp://www.EPA.gov/ozone/science/hole/size.html.

24. Suitil, K., Holey War, *Discover*, 75, January 1993.

25. Schein, O., Vincencio, C., Munoz, B., Gellat, K.N., Duncan, D.D., Nethercott, J., Honeyman, J., Koren, H.S., and West, S., Ocular and dermatologic health effects of ultraviolet radiation exposure from the ozone hole in Southern Chile, *Am. J. Public Health,* 85, 4, 546, April 1995.

26. EPA, EPA's Revised Ozone Standard, Fact Sheet, EPA, Office of Air and Radiation, Office of Air Quality Planning and Standards, July 17, 1997, http//ttnwww. rtpnc.EPA.gov.

27. EPA, USERA's Revised Particulate Standard, FACT Sheet, EPA Office of Air and Radiation, Office of Air Quality Planning and Standards, July 17, 1997, http://ttnwww.rtpnc.EPA.gov.

28. Gerstenzank, J., Clean-Air Plans Fuels Backstage U.S. Fight, *Los Angeles Times: Science,* Sunday, April 26, 1997 http://www.latimes.com/home/news/science/environ/ x000035587.html.

29. EPA, Clear Skies, last revised, March 2, 2006, http://www.epa.gov/clearskies/.

30. Ledford, A., On Today's House Subcommittee on "Clear Skies," Statement of Angela Ledford, Director of Clear the Air, A National Campaign Against Global Warming, press release, May 26, 2005.

31. Natural Resources Defense Council (NRDC), *Dirty Skies: The Bush Administration's Air Pollution Plan*, last revised September 5, 2003, http://www.nrdc.org/air/pollution/ qbushplan.asp#clearskies

32. Easterbrook, G., Hot Air Treaty, *U.S. News and World Report,* December 22, 1997, p. 46.

33. Wilkepedia, The Free Encyclopedia, Kyoto Protocol: The Kyoto Protocol to the United Nations Framework Convention on Climate Change, last modified June 6, 2006, http://en.wikipedia.org/wiki/Kyoto_Protocol.

34. Tyler, P., China's Inevitable Dilemma: Coal Equals Growth, *New York Times,* November 29, 1995, p. AI.

35. Monastersky, R., Dusting the Climate for Fingerprints: Has Greenhouse Warming Arrived? Will We Ever Know?, *Science News,* 147, June 10, 362,1995.

36. Monastersky, R., Satellite Detects a Global Sea Rise, *Science News*, Vol. 146, December 1994, p. 388.

37. Mann, M.E., Bradley, R.S., and Hughes, M.K, Global scale temperature patterns and climate forcing over the past six centuries, *Nature,* 392, 779, April 23, 1995.

38. Lemonick, M.D., The Ice Age Cometh? *Time,* January 31, 1994, p. 79.

39. Noshi, J.M., The Fury of El Nin-o, *Time*, February 16, 1998, p. 67.

40. Pfarrer, S., Expert fears El Nin-o global warming link, *Daily Hampshire Gazette,* Friday, April 17, 1998, p. 9.

41. Nash, M., When Life Nearly Died, *Time,* September 18, 1995, p. 95.

42. EPA, National Air Quality and Emission Trends Report, 1996, EPA Office of Air Quality Planning and Standards, EPA document 454/R-97-013, January 13, 1998.

43. Suplee, C., Dirty Air Can Shorten Your Life, Study Says, *Washington Post,* March 6, 1995, p. Al.

44. News and Analyses, Air Pollution in the U.S., *Scientific American,* April 27, 1997.

45. Allen, S., More Dying of Dirty Air than in Cars, Study Finds, *Boston Globe,* May 9, 1996, p. 14.

46. Washington (CNN), America's Dirty Little Secret: Smog Still a Health Problem, CNN Interactive, June 21, 1996.

47. Godish, T., *Air Quality,* 3rd ed., CRC/Lewis Publishers, Boca Raton, FL, 1997, chap. 2.

48. Raloff, J., Heart-y Risks from Breathing Fine Dust, *Science News,* 148, 5, July 1, 1995.

49. EPA, EPA's Revised Ozone Standard, FACT Sheet, EPA Office of Air Quality Planning and Standards, July 17, 1997, http://ttnwww.rtpnc.EPA.gov.

50. EPA, National Air Quality and Emission Trends Report, FACT Sheet, EPA, Document Number 454/e — 97–013, EPA Office of Air Quality Planning and Standards, 1, 23, 1998, wysiwyg://l lhttp://www.EPA.gov/out/aqtmd96/trendsfs.htrffl.

51. Anonymous, Carbon monoxide-heart failure link, *Env. Health Perspect.*, 104, 21, 138, 1996.

52. Flynn, J., The following forest, *Amicus J.*, 15, 4, 1, 1994.

53. Shaw, R., Recovery Begins From Acid Rain Damage, Environmental News Network, Monday, October 18, 1999.

54. News and Analyses, Air Pollution in the U.S., *Scientific American*, April 7, 1997, p. 27.

55. Reese, A., Blue Skies, *E/The Environmental Magazine*, November–December 1999.

56. National Institute of Environmental Studies (NIEHS), Asthma Research — NIEHS, Last updated Oct. 13, 2005, http://www.niehs.nih.gov/airborne/home.htm.

57. Cole, C.L., High Rate of Asthma Inspires New Study, *Boston Sunday Globe*, November 24, 1999, p. 1.

58. Shell, E.R., Does Civilization Cause Asthma? *Atlantic Monthly*, 285, 2000, p. 90.

59. American Lung Association, *Life and Breath: Answer and Questions*, item 1245, 15, January 1990.

60. Rosenstreich, D.L., Eggleston, P., Kaltan, M., Baker, D., Slavin, R., Gergen, P., Mitchell, H., McNiff-Mortimer, K., Lynn, H., Ownby, D., and Malveaux, F., The role of cockroach allergy and exposure to cockroach allergen in causing morbidity among inter-city children with asthma, *N. Engl. J. Med.*, 336, 1356, 1997.

61. Friebek, E., The attack of asthma, *Env. Health Perspect.*, 104, 1, 22, 1996.

62. Thurston, G.G., Lippman, M., Scott, M.B. and Fine, J.M., Summertime haze air pollution and children with asthma, *Am. J. Respir. Crit. Care Med.*, 155, 154, 1997.

63. Barner, R.I., Air pollution and asthma, *Postgrad. Med. J.*, 70, 319, 1994.

64. Lyons, B., Something in the air, *Ladies Home J.*, 102, February 1995.

65. EPA, *The Inside Story: A Guide to Indoor Air Quality*, EPA Office of Air and Radiation, and the Consumer Product Safety Commission, U.S. Government Printing Office, LM/4uc/1-88/004, 1988, 32 pp.

66. Menzies, D., Tumbly, R.M., Nunes, R., Hanley, J., and Tamblyn, R.I., Exposure to varying levels of contaminants and symptoms among workers in the office buildings, *Am. J. Public Health*, 86, 1629, 1996.

67. Godish, T., *Air Quality*, CRC/Lewis Publishers, Boca Raton, FL, 1997, chap. 6.

68. Leary, W., Research Ties Radon to as Many as 21,800 Deaths Each Year, *New York Times National*, February 20,1998, p. A13.

69. Godish, T., *Air Quality*, 3rd ed., CRC/Lewis Publishers, Boca Raton, FL, 1997, chap. 11.

70. EPA, *A Citizens Guide to Radon*, U.S. Government Printing Office, EPA-86-CC4, 1986.

71. Hager, M.G., Mites and Molds, *Self Magazine*, 167, May 1997.

72. Holusha, J.A., Study Calls Household Materials Especially Toxic, *New York Times*, March 15, 1995.

73. Geary, J., Mad About the Noise, *Time*, July 27, 1998.

74. Nadakavukaren, A., *Our Global Environment: A Health Perspective*, Waveland Press, Prospect Heights, IL, 1995, chap. 13.

75. Godish, T., *Air Quality*, 3rd ed., CRC/Lewis Publishers, Boca Raton, FL, 1997, chap. 12.

76. Toroza, G.J. and Anagnostakos, N.R., *Principles of Anatomy and Physiology*, 6th ed., Harper and Row Publishers, New York, 1990, chap. 17.

77. Birchard, K., Experts still arguing over radiation doses, *Lancet*, 354, 400, 1999.

11 Solid and Hazardous Waste

CHAPTER OBJECTIVES

A student, after reading this chapter, should be able to:

1. Discuss and explain the consequences of improper solid waste disposal.
2. List and characterize the typical municipal waste stream.
3. Describe and discuss the methods of reducing the solid waste stream through reuse and recycling efforts.
4. Describe the methods of collection and disposal of municipal solid waste (MSW), including the benefits and problems associated with landfills and incinerators.
5. Differentiate the types of hazardous waste, and discuss reasons for proper disposal, giving some case examples.
6. List and describe the various methods of hazardous waste control emphasizing waste reduction, volume or hazard reduction, and long-term storage and disposal options.
7. Discuss the positive and negative aspects of cleanup efforts under "Superfund" and some of the major concerns regarding its future operation.

INTRODUCTION

It was just another day on the Yangtze River of China when Zhoa Bangying, 21 years old, working as a uniformed attendant of a Russian-built hydrofoil boat, threw a large bag of trash out the window and into the river. China is entering an era of economic development complete with fast food, single-use disposable items, and a throwaway culture. Throwing trash out the window into the river, or onto the streets, is so commonplace, that no one is surprised. Garbage and rubbish floats down China's rivers, chokes its canals, flies out of passenger train windows, and creates mountains of trash.[1] China is not alone in this disregard for proper disposal. The constant influx of rural people to the city of Nairobi has doubled its population in the last 10 years and overwhelmed its ability to provide basic services such as rubbish collection. Littering the streets with solid waste is commonplace, and heaps of uncollected stinking, fly and rodent-infested garbage appear everywhere.[2] Once renowned as "the green city in the sun," Nairobi is now riddled with heaps of uncollected garbage, and is called by many "the stinking city in the sun."[3] The failure to properly manage the solid wastes in Nairobi is thought to be contributing to the increased incidence of shigellosis, paratyphoid, and cholera.[4] Such diseases are easily spread as accumulating garbage has clogged drainage systems, thus flooding the cities, polluting water sources, and converting the Nairobi River into an open sewer devoid of higher life-forms.[5]

There are a number of reasons given for the indiscriminate dumping, including poor management of finances and resources, a lax attitude of city employees, and little concern among residents for the cleanliness of their city.[2] The value systems of people moving to Nigeria, or Wuhan, China, or many other areas of the world, change in the face of economic development and industrialization. China once had a culture of thrift; rags and cloth diapers were reused, and national cleanliness was monitored and enforced. Now it is giving way to a hurried, money-driven, throwaway society.[1] This is not unusual for cities. Cities have historically been centers for filth and disease. Cities in England and Western Europe were disease-ridden centers of filth up until the 19th century. It was the norm to cast garbage, rubbish, and body wastes out of the windows, onto the streets below, or left on the floor, about the living space. The consequence of increasing refuse, filth, insects, and rodents resulted in disease outbreaks with terrible consequences, such as the bubonic plague, or "Black Death," that ravaged Europe in the 13th and 14th centuries. The relationship between disease outbreaks and accumulated filth gradually clarified in the eyes of municipal leaders, so that by the early 1900s, efforts were undertaken in Europe and North America to improve sanitation. One of the major efforts was directed at promptly collecting urban wastes and disposing of them at some distance from the population to reduce the probability of disease. This normally meant dumping the collected refuse on the outskirts of towns. Unfortunately, much of the world continues to follow this pattern of refuse disposal, which (1) invites the proliferation of rodents and insects, (2) becomes a source of contamination to groundwater, (3) pollutes ambient air when combusted, (4) facilitates the spread of debris around the dumping site, (5) lowers property values about the site, and (6) encourages the spread of disease from microorganisms and toxic chemicals. It has only been within recent years that the U.S. has adopted waste management policies that recognize the need for safer and more effective methods of collecting, storing, transporting, and disposing of the unusable or unwanted materials generated by society. This author recalls as a child, living in a rural area of Massachusetts and helping to load the domestic trash from our household onto a neighbor's truck. The trash was then discarded at the town dump located in a local wetlands area reached by an unpaved road. There was no soil cover and no barrier to the soil beneath, and so it was called an open dump. Rats, flies, and mosquitoes were rampant. Periodically, the dump was set on fire to reduce the volume and temporarily discourage the proliferation of pests.

A large percentage of MSW was thrown into open dumps up through the early 1970s, with very little recycling effected. Larger communities with diminishing open land for dumping opted to construct incinerators, and there were more than 300 municipal incinerators operating in the U.S. during the 1960s. The introduction of strict air-quality regulations in the 1970s caused many of the incinerators to close, as they represented major sources of air pollution. These same regulations imposed restrictions on the burning of open dumps. During the mid-1970s, incineration capacity declined, open dumps disappeared, and the generation of MSW increased dramatically. During the same period, the recovery of materials, such as from recycling, grew very slowly.[6] The federal government passed regulations in 1976, which forbade open dumping while introducing the concept and use of the sanitary landfill. This law is known as the Resource Conservation and Recovery Act (RCRA) (pronounced "rickra") of 1976 (PL 94–580). There are three distinct programs under RCRA. Subtitle D encouraged states to develop comprehensive plans for the management of solid wastes with emphasis on those of a nonhazardous nature, such as household wastes. Subtitle C was designed to control the improper disposal of hazardous waste through a manifest process known as the cradle-to-grave approach. Subtitle I was designed to minimize the contamination of groundwater from underground storage tanks through leak detection requirements, mitigation and prevention of leaks, and new performance standards for underground storage tanks. Subtitle D was soon found to be inadequate despite regulations, which dictated the conditions for the proper siting, construction, and management of sanitary landfills. Nearly 94% of 17,000 land disposal sites surveyed in the mid-1970s failed to meet minimum requirements.[7] The consequences included substantial odors and debris, contamination of groundwater, proliferation of pests, and other environmental pollution problems. Seeing the need for additional control over the disposal of solid wastes, Congress passed the Hazardous and Solid Waste Amendments Act of 1984 (PL 98–616). This provided strict requirements for the proper siting of landfills to minimize surface water and groundwater pollution. However, in an effort to achieve an improved and sustainable waste management program, the EPA promulgated new RCRA Subtitle D landfill requirements that went into effect in October of 1993. The landfills were to be constructed with double liners, systems for collecting the liquids percolating through the wastes (leachate), groundwater monitoring wells, and methane venting and detection systems. The costs for siting and operating landfills escalated dramatically as the number of landfills diminished in the face of rapidly growing amounts of MSW. Not only has the population of the U.S. increased dramatically since the 1960s, but a person in the U.S. today generates more than 4.3 lb of refuse per day compared to 2.7 lb in 1960. This exceeds 1500 lb per person per year, or more than in any other country.[8] The problem of diminishing refuse disposal capacity has been further exaggerated by the community opposition to landfill siting, which has become known as the "not in my backyard" (NIMBY) syndrome. The cost for a truck to unload its waste at a landfill (tipping fee) has approached $56 per ton in the Northeast U.S., making other disposal alternatives more attractive.[9] Although sanitary landfilling continues to be among the most widely used method of disposal, the EPA's Agenda for Action endorsed the concept of integrated waste management by which MSW is reduced or managed through several different practices that include:

1. Source reduction (including reuse of products and backyard composting of yard trimmings)
2. Recycling of materials (including composting)
3. Waste combustion (preferably with energy recovery) and landfilling[6]

The concepts of MSW and MSW management are discussed in the following sections.

DEFINITION AND CHARACTERIZATION OF MSW

DEFINITION OF MSW

The EPA's Office of Solid Waste uses a materials flow methodology to estimate the amounts of MSW generated. This methodology is based on production data (by weight) for both materials and products in the waste stream and is adjusted for imports and exports. Materials in MSW include paper and paperboard, yard trimmings, food wastes, plastics, glass, metal, and wood wastes.

MSW may also be categorized by the products generated which include durable goods (e.g., appliances), nondurable goods (e.g., newspapers), containers and packaging, food wastes and yard trimmings, and miscellaneous organic wastes. Such wastes generally come from residential, commercial, institutional, and industrial sources. Examples of the types of MSW in each of these categories are listed in Table 11.1.

CHARACTERIZATION OF MSW

Contrary to popular belief, MSW does not include everything that is landfilled in Subtitle D landfills (RCRA Subtitle D), but excludes municipal sludge, industrial nonhazardous waste, construction and demolition waste, agricultural waste, oil and gas waste, and mining waste (Figure 11.1). The term *generation* refers to the weight of materials and products as they enter the waste management system from residential, commercial, institutional, and industrial sources, and before materials are recovered or combusted. The total amount of MSW materials generated (thousands of tons) from 1960 to 2003 are shown in Figure 11.2. Total MSW generation has

TABLE 11.1
Sources of MSW and Example of Products from Each Source

SOURCES	EXAMPLES OF PRODUCTS
Residential, including single and multifamily houses	Nondurable paper items such as magazines, newspapers, and advertising flyers, plastic and glass bottles, aluminum and steel cans, packaging, food wastes, yard wastes
Institutional, including schools, hospitals, prisons, and nursing homes	Food wastes, papers from classrooms and offices, disposable tableware and napkins, paper towels from restrooms, and yard trimmings
Commerical, including retaurants, office buildings, and stores (wholesale and retail)	Food wastes, paper products from offices, restrooms, and serving tables, disposable tableware, corrugated and paperboard products, and yard wastes
Industrial packaging and administrative wastes	Wooden pallets, office paper, corrugated and paperboard, plastic film, and food wastes (from cafeterias)

FIGURE 11.1 Municipal solid waste as a component of RCRA Subtitle D wastes. (Adapted from United States Environmental Protection Agency, *Characterization of municipal solid wastes in the United States: 1997,* Update, prepared for USEPA by Franklin Assoc. Ltd., Prairie Village, KS, Report No. EPA 530-R-98-Oct-May, 1998, chap. 1.)

increased to 236.2 million tons.[10] Paper and paperboard materials represented the largest component at 35.2%, followed by yard trimmings (12.1%), and food wastes (11.7%) (Figure 11.3). The amount of MSW generated in 2003 and categorized as products is shown as Figure 11.4. The major proportion of generated waste is categorized as containers and packaging representing 31.7% or 74.8 million tons. Nondurable goods represented the second largest amount (26.3%) followed by durable goods at 16.7%.[10]

FIGURE 11.2 The total amount of municipal solid waste generated (thousands of tons) from 1960 to 2003. (Adapted from United States Environmental Protection Agency, Solid Waste and Emergency Response (5305W), *Municipal solid waste generation, recycling, and disposal in the United states: Facts and Figures for 2003*, EOA 530-F-05-003, April 2005, www.epa.gov/osw.)

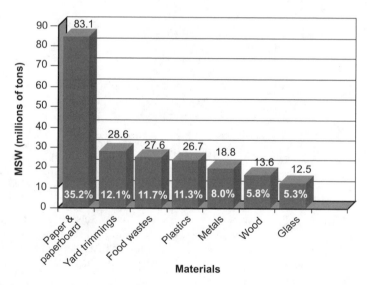

FIGURE 11.3 Major material components of municipal solid waste by percent and weight (millions of tons), 2003. (Adapted from United States Environmental Protection Agency, Solid Waste and Emergency Response (5305W), *Municipal solid waste generation, recycling, and disposal in the United states: Facts and Figures for 2003*, EOA 530-F-05-003, April 2005, www.epa.gov/osw.)

FIGURE 11.4 Products generated in municipal solid waste by weight and percent, 2003. (Adapted from United States Environmental Protection Agency, Solid Waste and Emergency Response (5305W), *Municipal solid waste generation, recycling, and disposal in the United states: Facts and Figures for 2003*, EOA 530-F-05-003, April 2005, www.epa.gov/osw.)

COLLECTION AND DISPOSAL OF SOLID WASTE

COLLECTION OF MSW

Whether collected by private companies under contract, or by municipal employees, the removal of solid waste is the responsibility of government, which must develop and enforce regulations that protect the public health by proper collection and disposal of municipal waste. The collection of MSW consumes $7.5 billion of the $10 billion annual costs for the collection and disposal of solid waste. This effort requires nearly 170,000 employees distributed among public and private sectors.[11] Factors that must be considered in collection include (1) whether the waste will be collected by the municipality or under private contract; (2) the frequency of collection; (3) what materials will be salvaged; (4) the types and location of collection containers or bags; (4) what materials will be forbidden from routine disposal (i.e., tires, automobile batteries, mercury-containing batteries, fluorescent lights, durable white goods); (5) whether wastes should be separated prior to collection as required by certain recycling efforts; and (6) what type of equipment may be required. Collection vehicles often include manually loaded compacting bodies that increase the size of the load that can be carried, and facilitates emptying at the disposal site (Figure 11.5). Communities that offer recycling programs and curbside collection will usually employ specialized trucks that have several compartments or sections necessary for the segregation of collected recyclables (Figure 11.6).

Many rural communities do not offer a collection service, but require the residents to bring their refuse and/or recyclables to a transfer station. A transfer station is a site where solid waste is concentrated before taken to a processing facility or a

FIGURE 11.5 Waste-compacting collection vehicle.

FIGURE 11.6 Compartmentalized truck assigned for pickup of recycled materials.

sanitary landfill (Figure 11.7). The concentration most often involves compaction by placing the waste into a metal channel where a ram compresses the waste into a roll-off collection container (Figure 11.8). The container may then be hailed onto a tractor bed and transferred for processing or landfilling. Some transfer stations also feature recycling bins where separated items such as metal cans, plastic milk bottles (HDPE), paper and cardboard, and glass containers may be collected and periodically removed for further processing and resale (Figure 11.9).

FIGURE 11.7 Transfer station where waste compaction and separation take place.

FIGURE 11.8 Close-up view of compaction ram at a transfer station.

FIGURE 11.9 Recycling bins at a transfer station.

MANAGEMENT OF MSW

Once waste is collected, it must be processed in some way for final disposal. Presently, most states use some combination of waste management that involves one or more components of landfilling, combustion, or recycling. Nationally, more than half (54%) of the total MSW collected is landfilled, about 30.6% is recovered for recycling (or composted), and 14% is combusted with most combustion systems employing energy recovery (Figure 11.10).

LANDFILLS

Design

Landfills have historically been sources of: (1) pollution to streams and groundwater; (2) producers of methane gas, which can leak into basements in potentially explosive amounts; (3) odors; and (4) breeding grounds for insects, rodents, and gulls. Stricter requirements for the siting and construction of landfills under RCRA Subtitle D mandate that all new landfills or ones that are to be expanded must be constructed with double liners consisting of impervious layers, leachate collection systems, monitoring wells to detect potential migration of pollutants from the landfill site, and methane detection systems also designed to detect lateral migration of methane (Figure 11.11).

FIGURE 11.10 How municipal solid waste is managed in the U.S. (Adapted from United States Environmental Protection Agency, Solid Waste and Emergency Response (5305W), *Municipal solid waste generation, recycling, and disposal in the United states: Facts and Figures for 2003*, EOA 530-F-05-003, April 2005, www.epa.gov/osw.)

FIGURE 11.11 Typical construction of landfill and potential leachate routes.

The reason for such extensive design concerns is that a landfill is a depression in the ground (or constructed on the surface to resemble a stadium), which is designed to hold accumulated wastes. Its purpose is to contain wastes and prevent pollutants from reaching surface or groundwater. Landfills have been described as bathtubs that can leak from the bottom or spill over the top.[12] Landfills with two liners are like a bathtub within a bathtub, and are designed to further restrict the migration of pollutants.

Solid wastes that are deposited in the landfill will tend to mix and degrade chemically and biologically to produce a variety of gaseous (methane), liquid (leachate), and solid (metal oxides) products. The amount of moisture, oxygen, temperature, and contents of the fill affect the level of decomposition and the products produced. The major products of concern include leachate and methane. Leachate is usually a dense, dark liquid produced when surface water and groundwater infiltrate through the solid waste and pick up solid matter, chemicals, and waste products of microbial decomposition that serve as a potential serious source of contamination to surface and groundwater. Methane gas is produced when organisms anaerobically decompose organic waste such as food waste and sewage sludge disposed in the landfill. Methane gas can migrate laterally, to enter the cellars of nearby structures, and produce potentially explosive concentrations.[11] Although biodegradation does occur in a limited amount in landfills, the marketing of biodegradable products to the public may be overstated. Archaeologist William Rathje has been excavating and researching landfills for years. His discovery of 25-year-old hotdogs, guacamole, steaks, and even older intact newspapers has Rathje theorizing that very little biodegradation takes place in a landfill. He concludes that most garbage is mummified and takes up space indefinitely, because microbes have insufficient moisture and air to decompose the bulk of landfill ingredients.[13]

The critical elements of a landfill includes (1) a bottom liner or double bottom liner, (2) a leachate collection system, (3) a cover, (4) the natural hydrogeologic setting, and (5) monitoring systems for detecting methane and leachate plumes (Figure 11.11). The national hydrogeologic setting should be away from surface water and aquifer with impervious material below (fractured bedrock should be avoided), while also having a geology simple enough to permit the installation of wells to intercept escaping pollutants. The bottom liners may be composed of one or more layers of clay or a synthetic flexible membrane (or some combination of both) (Figure 11.12). The synthetic liners are often constructed of high-density polyethylene (HDPE). However, other materials may be used for plastic liners (referred to as flexible membrane liners or FMLs). Leachate seeps to the bottom of the sloped landfill and is collected by a system of pipes that are laid along the bottom. The leachate is intended to be periodically pumped and transported to a wastewater treatment plant, where solids are removed and liquids treated. The cover or cap of the landfill is designed to restrict the entry of water from precipitation and usually is constructed of several layers sloped away from the center to promote the shedding of water. The first layer is usually a membrane, covered by a layer of permeable sand or gravel, and then a layer of topsoil where vegetation may root.

FIGURE 11.12 Installation of a bottom plastic liner in a landfill.

Landfills Are Not Secure

Landfills are vulnerable to leakage from several different sources. Landfill liners, even constructed of HDPE, can be expected to leak more than 20 gal per acre per day through tiny holes created during manufacture, and those created when the seams are welded together during installation.[14] Furthermore, such liners are also permeable to a variety of dilute solutions including xylenes, toluene, trichloroethylene (TCE), and methylene chloride, which pass through the solid matrix of 100-mil thick HDPE liners in less than 2 weeks. Such liners also become brittle and will fracture or crack from stress and chemical assault in under 2 years.[15] Even benign chemicals such as margarine, shoe polish, vinegar, and peppermint oil may attack plastic liners.[13]

Leachate collection systems can fail as the polyvinyl chloride pipes fail from chemical assault,[14] are crushed by the weight of the landfill, become clogged from silt or mud, microbial growth, or precipitation of minerals. Once clogged, leachate accumulates, causing increased pressures, which lead to leaks or overflows.[12]

Landfill caps or covers are meant to be impermeable, but are disturbed by (1) burrowing and soil-dwelling mammals, insects, worms, and reptiles; (2) erosion by precipitation, freeze-thaw cycles, and wind; (3) roots of vegetation that may penetrate the cover; (4) uneven settling that causes rips and tears on the membrane or cracks in the clay surface; (5) upward migration of chemicals or piercing objects; and (6) sunlight on exposed surfaces. Caps that are not maintained will permit collection of precipitation causing leachate overflow or penetration through the bottom liner.

The failure of leachate collection systems, bottom liners, and caps ultimately cause the contamination of groundwater, which is likely to remain contaminated for thousands of years.[15] Because groundwater is a diminishing resource, alternative methods of waste management must be employed. The combination of increased costs for landfill construction, the difficulty in siting landfills, and the moratorium on landfill development and expansion by some states have resulted in a significant decline in the number of landfills since a peak of 8000 community landfills in 1988.[10,16] There are less than 2400 MSW landfills today, although the total landfill capacity has not declined considerably. Newly constructed landfills are much larger. The Northeast U.S. has the fewest landfills (208) and the fewest remaining years of landfill capacity (Figure 11.13).[6]

One of the northeastern states, Massachusetts, has adopted a solid waste master plan designed to reduce alliance on municipal landfills and to enter the next century with a waste reduction and recycling-based integrated solid waste management system in place.[16] This plan is consistent with a national moratorium on new landfills and the closure of existing unlined ones, with the anticipated result being significant reduction in leachate (Figure 11.14).[16] As landfills are closed, Massachusetts seeks to increase its efforts in recycling and source reduction, while nonrecovered waste is diverted to waste-to-energy facilities and composting facilities.

Source Reduction

Activities that reduce the amount of the toxicity of wastes prior to entering the waste stream are referred to as *source reduction*. Source reduction activities may include:

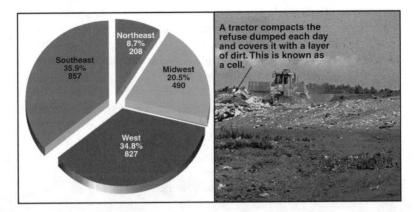

FIGURE 11.13 Landfill capacity by region in the U.S. Also shown is the formation of a cell in a typical landfill operation.

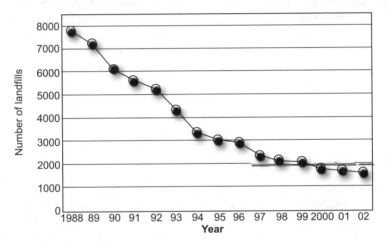

FIGURE 11.14 Number of operating landfills in the U.S. by year. (Adapted from United States Environmental Protection Agency, Solid Waste and Emergency Response (5305W), *Municipal solid waste generation, recycling, and disposal in the United states: Facts and Figures for 2003*, EOA 530-F-05-003, April 2005, www.epa.gov/osw.)

1. Products package reuse — the use of glass bottles has declined in favor of plastic containers, but glass beer or soft-drink bottles represent an excellent example of containers that can be reused for their original purpose. Some chain grocery stores offer reusable cloth shopping bags. Clothing, furniture, and appliances are often informally reused among family and friends, sold at yard sales, or donated to agencies that will redistribute them.

2. Package or product redesign that reduces material or toxicity — containers may be redesigned with lesser materials, such as aluminum beverage cans which are nearly 20% lighter than they were in 1976, or reducing the

thickness of plastic liners used on cereal boxes.[6] The consumer may choose to purchase less toxic materials, such as citrate-based cleaners or nontoxic pesticides. Manufacturers may use less toxic packaging by avoiding the use of highly bleached white corrugated in favor of brown corrugated boxes printed with nontoxic inks.

3. Reducing use by modifying practices — consumers have wide latitude to reduce material use. Many stores now offer the opportunity to purchase foods or other materials in bulk, thereby minimizing packaging. Paper can be used for copying on both sides, while electronic mail (e-mail) reduces the need for hardcopy mail. Many products are now offered in concentrated form, including juices, detergents, and fabric softeners that can be reconstituted in a larger and reusable container.

Trends in source reduction are difficult to estimate because the sharing or passing of goods among family and friends can't be quantified. However, trends in the use of nondurable papers (newspapers, office papers, etc.) and packaging containers (excluding corrugated boxes) have been used as indicators. The generation of packaging on a per person basis from 1960 through 1996 shows an increase in 1970, followed by slight decreases in the early 1980s, and a relatively flat response thereafter (Figure 11.14). The generation of nondurable paper wastes increased through the late 1980s and early 1990s and has decreased only slightly in recent years (Figure 11.14). The advent of computers has not diminished the per capita generation of newspapers, magazines, and other nondurable papers as expected. The generation of nondurable papers continues to be a matter of concern in recycling efforts.

Recovery for Recycling (Including Composting)

The act of removing materials from MSW for productive use is referred to as *recycling* or *resource recovery*. It can be measured by determining the amounts of purchases of postconsumer recovered materials, including the net exports of such materials. The recovery of materials for recycling is one of the most preferred waste management alternatives and continues to be among the most effective as well.[6,10] Citizens perceive recycling as desirable and are willing to participate in programs that are minimally demanding on time and effort. Recycling is perceived to (1) conserve resources by reducing the need for virgin and nonrenewable materials, (2) reduce the amount of pollution by using secondary materials that require less energy to process (using recycling aluminum cuts pollution and energy uses by more than 90% compared to the use of virgin materials), and (3) save energy by using recycled materials, because less is required for processing. The EPA considers the recycling of materials in MSW among the more important waste management strategies, and many states have adopted recycling into their integrated waste management plans. These states have set recycling goals, including the percentage of MSW to be recycled and the dates by which this will be accomplished. The state of Massachusetts now recycles 33% of the MSW, but has set the ambitious goal of 46%.[16]

Recyclable materials must be collected before they can be processed. Collection normally involves curbside collection, drop-off programs, container deposit pro-

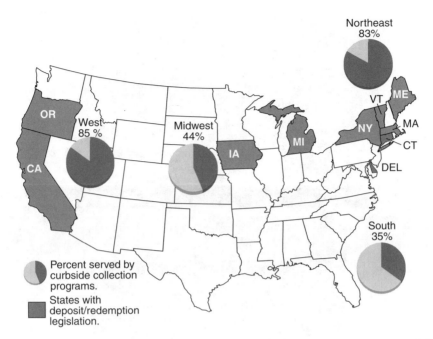

FIGURE 11.15 States with deposit/redemption legislation and regions served by curbside collection programs. (Adapted from United States Environmental Protection Agency, *Characterization of municipal solid wastes in the United States: 1997,* Update, prepared for USEPA by Franklin Assoc. Ltd., Prairie Village, KS, Report No. EPA 530-R-98-Oct-May, 1998, chap. 1.)

grams, and buy-back operations. Curbside collection requires residents to separate their refuse into recyclables and discards, and to place the separated recyclables in collection containers at the curb. Some states have adopted unit-based pricing systems, also known as pay-as-you-throw programs. A typical program would require the resident to purchase plastic trash bags at selected community vendors, and identified with the name of the town on it. The cost may be 0.50 to $1.00 each. Discards would go into the bags, and the resident would have a financial incentive to recycle and avoid filling the bags quickly. Such programs have been shown to be the greatest single stimulus to increasing residential recycling.[17] Nationally, slightly more than half of U.S. population (about 135 million people) have access to one of the 9000 curbside collection programs, although most are not based on pay-as-you-throw programs. The Northeast serves more than 43 million people, or over 80% of its population, with a curbside recyclables program (Figure 11.15).[6] There are in excess of 10,000 locations where citizens may drop off materials for recycling. There are currently nine states that have container deposit systems in place. These are sometimes referred to as "Bottle Bill" states, which require consumers to pay a refundable deposit on soft drink and beer containers. Such programs have documented a marked increase in recycling of these containers with over 35% of all recycled beverage containers occurring in these 9 states.

Once the recyclable materials are collected, they must be further processed and separated to create large amounts of homogeneous secondary materials that are

FIGURE 11.16 Recyclables awaiting processing at a typical MRF.

FIGURE 11.17 Manual separation of recyclables at a typical MRF.

attractive to manufacturers as alternatives to virgin materials. The facilities that perform the function of preparing recyclables for marketing are referred to as *materials recycling facilities* (MRFs) (pronounced murfs). In a typical MRF, recyclables are dumped to a cement floor (Figure 11.16), picked up by a front-end bucket loader, and placed in conveyor belts, which carry the material past magnetic separators to remove steel cans and steel objects, and then past a line of people who remove recyclable materials manually from the stream (Figure 11.17). The materials often require further processing. Glass bottles are separated into clear, brown, and green glass, and ground into small smooth pellets called *cullet*. An example of green glass cullet is shown in Figure 11.18. There are more than 480 MRFs operating in the U.S. processing nearly 30,000 tons per day of material.[18] Some of these MRFs employ more sophisticated technologies to sort recyclables including magnetic pulleys, optical sensors, eddy currents, and air classifiers.

Some MRFs process mixed waste as it would normally go to a landfill complete with food scraps, dirty diapers, and mustard jars. There are only 58 such facilities in the country with most of them being in the West. These dirty MRFs tend to be popular in areas of country where administrators have opted not to involve citizens

FIGURE 11.18 Green glass cullet as a recyclable product after processing.

in recycling efforts. Theoretically, the percentage of material recycled would be very high, because all municipal refuse would be delivered to such facilities.

Trends in Resource Recovery

Materials recovered from MSW have been measured by the purchases of postconsumer recovered materials plus net exports of the materials. Since 1960, there has been a consistent increase in the weight of MSW recovered, and in the percentage of total MSW recovered (Figure 11.19).

Recovery of materials from MSW remained relatively flat from the 1960s through the mid-1980s. However, escalating costs of landfilling, concerns about environmental pollution, and legislative initiatives motivated a renewed interest in

FIGURE 11.19 Recovery of MSW: 1960 to 2003. (Adapted from United States Environmental Protection Agency, Solid Waste and Emergency Response (5305W), *Municipal solid waste generation, recycling, and disposal in the United states: Facts and Figures for 2003*, EOA 530-F-05-003, April 2005, www.epa.gov/osw.)

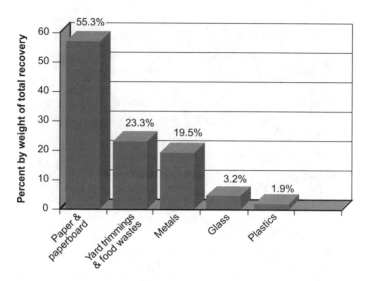

FIGURE 11.20 Materials recovery in 2003 (by weight of total recovery). (Adapted from United States Environmental Protection Agency, Solid Waste and Emergency Response (5305W), *Municipal solid waste generation, recycling, and disposal in the United states: Facts and Figures for 2003*, EOA 530-F-05-003, April 2005, www.epa.gov/osw.)

recycling, which was once strong during the era of World War II. Recovery rates increased significantly from the mid-1980s to the present national level of over 30% (Figure 11.19). Most of the recovery appears to be in paper and paperboard, accounting for 55.3% of the material recovered. This is followed by yard trimmings and food wastes, which are composted (23.3%), and metals at 9.5% (Figure 11.20).[10] Both prices and markets have varied considerably over the past several years making further overall gains in recycling more problematic. Downturns in global economies have reduced the demand for many recycled products. The amount of recovered materials, especially paper and paperboard, has increased dramatically, contributing to a fall in prices. Very often, virgin material capacity, such as virgin pulp will be sufficient to lower the price and so compete with recycled products. Manufacturers usually specify the types and levels of acceptable contamination in such materials as glass, paper, steel, plastics, and aluminum. In many cases, the levels of acceptable contamination are very narrow. The market competition for products also influences recyclability. For instance, the production of glass containers has declined since 1993 by 1500 tons per year causing recycling efforts for glass containers to remain relatively level. Long transport distances for recycled materials also increases cost. This, combined with recycling efforts abroad, diminish export markets. Another factor influencing the recycling effort is the changing capacity to use recovered materials. There are numerous plastic resin types, but most recycling of plastics is directed at unpigmented HDPE (high density polyethylene) bottles for milk, water, and juice, or PET (polyethylene terephthalate) used in soft drink bottles. The recycled PET is used in the manufacture of fibers for carpets and fiberfill for garments, whereas recovered HDPE is used to manufacture new bottles, drainage pipe, trash bags, and plastic lumber. Most other plastic wastes have a limited market because of the technology to recover and reuse them has not been adequately developed. Research is continuing to find new methods for the processing and reuse of mixed plastic waste and selected individual resins. Another factor influencing the recycling market is legislation that fosters recycling. Federal agencies and many state governments mandate a minimum recycled content for targeted products such as newspapers, or for products they purchase such as office paper. The state of Massachusetts has increased its purchase of recycled products from $8.2 million in fiscal year 1994 to nearly $34 million in fiscal year 1997.[16] President Clinton signed an executive order in 1993 requiring all federal agencies to purchase paper containing a minimum of 25% recycled paper as of the year 2000.

Composting

There are approximately 2200 municipal composting programs nationally directed at yard trimmings. Most of these programs are located in the Northeast and Midwest, and handle more than 25,000 tons per day.[10] There are 14 mixed waste composting facilities (eight in the Midwest) that remove large items and metal from the mixed waste and compost the combined paper, food waste, yard trimmings, and wood. *Composting* is a controlled process of degrading organic matter by microorganisms into a humuslike material. The composted material is often low in plant nutrients but is useful for conditioning soil by improving soil porosity and aeration, and increasing water retention. Some communities are experimenting with adding domestic sludge to the mix and cocomposting the sludge and biodegradable refuse together. Unfortunately, public acceptance of this cocomposted material is not universal because of the real or perceived dangers of pathogens and/or toxic agents surviving in compost. The composting process is normally performed by removing inorganic materials such as glass, metal, tires and shredding the organic material. In some sophisticated operations, the entire process from shredding to aeration, curing, and finishing are performed in an in-vessel composting system (Figure 11.21). Most composting use a windrow process in which rows of material are placed next to each other outdoors and periodically mixed to incorporate air into the mix (Figure 11.22). This permits the organisms to aerobically digest the materials, increasing the heat of the compost pile to temperatures sufficient to kill most pathogens and insect eggs. When the materials have been left long enough to cure so that cellulose and lignin are broken down, the material is normally screened to size and either packaged or used in bulk for soil enhancement.

Compost that is consistently high in quality and blended to meet the needs of end users is being successfully marketed. Compost is being used for wetlands mitigation, land reclamation, storm filtrates, soil amendments, mulches, and low-grade fertilizers.[18] The majority of composted material is used in landscaping, agriculture, and for landfill cover. Many of the markets have seen little penetration (Figure 11.23). However, as the quality and consistency of compost improve, there is likely to be higher consumer acceptance when combined with education on the benefits of compost use.

FIGURE 11.21 In-vessel composting system. This system is used to grind mixed food and plant wastes, aerobically digest them, and provide a finished compost product in days to a few weeks.

FIGURE 11.22 Compost piles arranged in rows at different stages of digestion and curing.

Combustion

The treatment of MSW by combustion includes the combustion of mixed MSW, the preparation of refuse-derived fuel (RDF), or a separated component of MSW (such as rubber tires) with or without the recovery of energy.[8] Generally, combustion with the production of energy is called *waste-to-energy* (WTE), while combustion of MSW without energy recovery is called *incineration*. The burning of unsegregated wastes (mass burn) in incinerators was very popular prior to the 1970s, before air-quality regulations made the operational costs prohibitive. However, improved methods in pollution control and monitoring, combined with escalating costs of alternative refuse disposal, have encouraged the growth of the refuse-combustion industry. It is estimated that 17% or 35 million tons per year of MSW are combusted. Most of facilities combusting MSW also incorporate recovery of energy as steam or electricity, or both. The sale of energy helps to defray operational costs. These WTE plants are generally mass burn, although many incorporate recycling activities that

Compost markets

FIGURE 11.23 The market distribution and estimated market penetration for compost. (Adapted from United States Environmental Protection Agency, *Characterization of municipal solid wastes in the United States: 1997,* Update, prepared for USEPA by Franklin Assoc. Ltd., Prairie Village, KS, Report No. EPA 530-R-98-Oct-May, 1998, chap. 4.)

remove noncombustible items, such as metals and glass prior to burning. A state-of-the-art WTE facility integrates with a statewide master plan to combust only waste that cannot be recycled or composted. An example of such a facility is operated by Springfield Resource Recovery, Inc. (SRRI) in Springfield, MA. The facility is designed in stages to minimize emissions. The first stage incinerates MSW by controlled combustion, which minimizes combustion pollution. The second stage recovers energy and converts it to useable steam and electricity, and the third stage minimizes or neutralizes pollutants coming from the waste (Figure 11.24). Independent testing firms use EPA-approved testing methods to evaluate the levels emissions including: dioxins, furans, lead, mercury, beryllium, trace metals (cadmium, chromium, copper, manganese, nickel, tin, zinc), inhalable particles (PM_{10}) and total suspended particulate (TSP) matter. This has been made necessary because of increasing concerns about possible toxic air emissions from these facilities. The EPA approved tighter emissions controls in 1991 for combustion facilities in accordance with the Clean Air Act Amendments. Systems now employ specialized reactors, bag houses, computerized combustion control systems, gas scrubbers, and continuous monitoring systems to control and evaluate emissions. The fly ash captured from the bag house may be mixed with the bottom ash to reduce the probability of it being designated as hazardous waste.

There were 110 WTE facilities in 1996 with a total design capacity of 100,000 tons per day. Most of the facilities are located in the Northeast and southern U.S. Another 5400 tons a day are prepared as refuse-derived fuel (RDF), which is a

Combustion | Energy Recovery | Pollution Control

Tipping floor

Control room

Feed hopper

1700°C   1850°C

Combustion air fan

Retention chamber

Isolation dampers

Boiler

Economizer

Reactor   Bag house   I.D. fan   Stack

Refuse burned in stages   Multiple hearth furnace

Preseparated MSW is deposited to the tipping floor. MSW is swept to the incinerator feed hopper and burned in a multiple hearth furnace in stages for 3 hours until fully combusted. A lean atmosphere and controlled temperatures reduce the formation of CO, NOx, and dioxin emissions.

Hot gases from the incinerator pass through water tubes in the boiler which heat water to create steam. The steam expands through turbines generating electricity.

Economizers extract heat and recycle it for plant uses. Acid gases are neutralized in a dry absorption reactor. Particulates are removed by filters in the bag house. Fly ash and bottom ash are collected and deposited in a monofill. All systems are continuously monitored for CO, CO2, NOx, SO2, and opacity. (Opacity is an indirect measurement of smoke indicated by the degree of obstruction of a beam of light.)

Tipping floor   Control room   Baghouse and stack

FIGURE 11.24 Stages and operations of a modern waste-to-energy facility. This diagram is based on the design of the waste-to-energy facility in Springfield, MA.

pelletized fuel derived from MSW that can be mixed with coal and burned in a standard boiler. The future of constructing additional WTE facilities is in some doubt, as operational costs rise and combustible products are diverted to increasing recycling efforts. Courts have overturned flow-control ordinances passed by communities to guarantee minimum tonnage of combustibles to WTE facilities and incinerators. Therefore, the overall operating capacity for WTE facilities is less than 85% of design capacity.[6]

HAZARDOUS WASTES

BACKGROUND

During the spring of 1997, heavy rains raised the groundwater level and turned the Love Canal region in Niagara Falls, NY, into a muddy swamp mixed with some of the 19,000 tons of toxic wastes buried in the abandoned canal by Hooker Chemical Company. The chemically laced ooze seeped into basements, onto playgrounds, and over lawns creating a storm of protests, accusations of severe health effects, and costs over $37 million to relocate families from the area, and another $30 million to aid families. More than $150 million and 12 years were been spent on cleanup efforts. Love Canal became a major turning point in directing public attention to other chemical wastelands scattered across the country in abandoned warehouses, manufacturing facilities, landfills, and processing plants.[19] In 1982, scientists recommended a government buyout of Times Beach, MO, after waste oil dealer Russell Bliss sprayed the dirt roads in 1971 with dioxin-laced oil to keep the dust down.[20] The events at Love Canal and Times Beach are not unique. The Escambia treating company near a Pensacola, FL, neighborhood closed its doors years ago, but a toxic mess piled in a 60-ft mound by the EPA sat in the neighborhood since 1991, covered only with a polyethylene bag. This is a black, working-class neighborhood, and the citizens felt that their backyards were being contaminated by dioxin and chemical preservatives such as creosote and pentachlorophenol. The EPA announced in 1996 it was planning to spend $18 million to relocate people from 158 houses and 200 apartments making it the largest move since Love Canal and Times Beach, which took place more than a decade earlier. Agency officials acknowledged that they had to consider environmental justice in the decision. The idea of environmental justice developed from studies that indicated minorities were being subjected to environmental pollutants disproportionately. This prompted President Clinton to issue an executive order in 1994 that decreed agencies must address the notion of environmental justice when considering the disposal of hazardous wastes. Additionally, the order required minorities be given a voice in environmental regulation and that the cleanup of the worst toxic waste sites in minority areas be given priority.[21] Despite these efforts, an Escambia grand jury report issued on Tuesday, May 4, 2004, revealed that the EPA, the Florida Department of Environmental Protection, and the Escambia County Utilities Authority failed to do all they could to prevent and remove pollution in Escambia County. The owner of Escambia Treating Co., a wood-preserving facility in central Escambia County, abandoned the site in 1991, leaving it to the federal EPA to clean up. Thirteen years later, the EPA continued to "study" the extent of groundwater contamination as a prerequisite to the cleanup of more than 800 acres of city property.[22]

In the early and mid-1970s, cases of acute lymphocytic leukemia began to surface in the town of Woburn, MA. Although no conclusive link was established between discarded organochlorine chemicals appearing in the drinking water and elevated leukemia levels, the involved companies agreed to pay nearly $70 million in cleanup costs. The story, as written by Jonathan Harr, became a bestseller titled *A Civil Action*, and was made into a film starring John Travolta, released in early 1999. These events point to increased public awareness and concern about hazardous wastes as one of the major environmental challenges. The EPA estimates that an excess of 270 million tons of hazardous waste is generated every year in the U.S., with most of it produced by the petrochemical industry.[23] Discarded or released hazardous substances have contaminated surface water and groundwater, soils, and air. Drinking water has been contaminated, food sources polluted, and wildlife habitats destroyed or compromised.

What Is a Hazardous Waste?

The proliferation of hazardous wastes, industrial sources, and community outcry prompted Congress to pass the RCRA of 1976. This was the first federal regulatory program targeted at controlling present and future hazardous waste. This act was intended to prevent the improper disposal of hazardous waste using a manifest tracking system that attempts to follow the waste from its point of generation to its final treatment and disposal. The process is called the cradle-to-grave system and is meant to protect public health by (1) defining what wastes are hazardous; (2) tracking wastes to the point of disposal; (3) ensuring that treatment, storage, and disposal (TSD) facilities meet minimum national standards; and (4) and making certain TSDs are properly maintained after closure, and that facility owner/operators are financially responsible for hazardous waste releases that may occur at their facility.

Everyone who handles the waste, including the generator, the shipper, and the manager of the TSD facility, must sign a multipart manifest with copies going to the state and federal regulatory agencies and back to the generator after various signatures are applied. The final disposal of wastes has become expensive, and persons wishing to profit by dumping the waste along roadways during rainstorms, down storm sewers, or on privately owned land may realize significant profits by failing to report these activities at the outset. However, the penalties are severe, and include criminal prosecution and hefty fines. The definition of hazardous waste in RCRA includes any discarded material that may pose a substantial threat or potential danger to human health or the environment when improperly handled. The EPA has established that a waste is to be considered hazardous if it meets either of following qualifications.

First, the substance should be listed as one of the 400 wastes or waste streams under Title 40, Part 261, 31–33 of the Code of Federal Regulations (CFR). Such substances include wastes generated by chemical manufacturers, vehicle maintenance shops, the paper industries, metal manufacturers, cleaning agents and cosmetics manufacturers, the construction industry, the printing industry, and many others. These are called *listed wastes*.

Second, wastes are considered hazardous if they possess certain properties or characteristics. These characteristic wastes are subject to regulation if they are toxic, ignitable, corrosive, or reactive according to the specifications outlined in RCRA. A substance is considered toxic if it is potentially poisonous to humans and can cause acute or chronic health problems. Wastes are considered toxic if they produce sufficient amounts under the toxic characteristic leachate procedure (TCLP) described by the EPA. This process requires acidified liquid to be percolated through waste in a laboratory-created landfill condition for 24 hours. Samples of the leachate are tested for the amounts of the target substance released. Substances are ignitable if they have a flash point below 60°C or can cause fires spontaneously by friction, moisture, or chemical change. Certain plastics, oils, solvents, and paints fit this category. Wastes are corrosive if they have a pH below 2.0 or above 12.5, and include such wastes as battery acid, drain cleaners, and other compounds that will dissolve living tissue, metals, and other materials. Reactive wastes may include discarded metallic sodium, or wastes from the manufacture of certain explosives that may react vigorously with air or water to generate toxic gases or even explode.

HAZARDOUS WASTE REGULATIONS

The method of dealing with hazardous wastes in the late 1970s and early 1980s was based on the premise that such wastes are inevitable and acceptable methods of treatment and disposal were the answer to protecting the public health. However, the rapidly increasing generation of hazardous wastes, the discovery of tens of thousands of hazardous waste sites, and growing public concern demanded an alternative to this end-of-pipe approach. Consequently, amendments were made to RCRA in 1984 that significantly expanded its regulatory powers. The amendments are known as the Hazardous and Solid Waste Amendments (HSWA), and they are focused on protecting groundwater by (1) restricting the TSD of hazardous wastes in land management facilities; (2) mandating stricter requirements for landfills accepting hazardous waste; (3) requiring a schedule for determining if the landfilling of untreated hazardous waste should be phased out; (4) creating a new program for the detection, control, and management of hazardous liquids (primarily petroleum) in underground storage tanks; and (5) increasing the numbers of people who fall under RCRA regulations by including small-quantity generators. Small-quantity generators would include those who produce 100 kg (220 lb) per month of hazardous waste, or 1 kg (2.2 lb) per month of extremely hazardous wastes. Previously, only large-quantity generators who produced more than 1000 kg (2205 lb) per month of hazardous waste were subject to RCRA provisions. Congress was concerned about the reliance on land disposal of hazardous wastes and wanted such disposal to be minimized or eliminated; and land disposal, particularly landfills and surface impoundments should be the least favored method of handling hazardous wastes. There was concern that all land disposal facilities, including those that are designed to be secure, will ultimately leak.[24] The same HSWA amendment called for the reduction of hazardous wastes generated at the source, and such activities should have priority. This effort was further strengthened by the passage of the Pollution Prevention Act of 1990 that emphasized a federal policy of preventing the formation of hazardous wastes, followed in priority by the recycling or reuse of hazardous materials, and the last activity in the hierarchy would be treatment and disposal in an environmentally safe way. The EPA strongly committed to this policy by making waste minimization its highest priority, and requiring hazardous waste generators to certify their efforts at waste minimization on the hazardous waste shipping manifests.[25] Similar requirements exist for owners and operators of TSD facilities. The EPA has created The Waste Minimization National Plan, which is a long-term national effort to reduce the quantity and toxicity of hazardous wastes. The goals of the plan are to:

1. Achieve a 50% reduction in the most persistent, bioaccumulative, and toxic (PBT) chemicals in the nation's hazardous waste by the year 2005 as compared to the baseline year of 1991. Bioaccumulative (B) refers to chemicals that tend to concentrate in animal and plant tissues. Many chlorinated organics and heavy metals tend to fit in this category.
2. Prevent the transfer of chemical releases from one medium (air, water, and land) to another, give priority to source reduction (reducing waste prior to its being generated), and environmentally sound recycling over waste treatment and disposal.[26]

Are you interested in learning more about solid and hazardous wastes? Visit http://www-unix.oit.umass.edu/~envhl565/weblinks.html and click on one or more of the following sites for Chapter 11:

- Massachusetts DEP Publications
- Envirofacts Querying Capability
- Scorecard Home
- The Basics of Landfills
- Hazardous Waste — Introduction
- EPA/OSW — RCRA Online
- EPA/OSW — Hazardous Waste

THE MANAGEMENT OF HAZARDOUS WASTES

The three major options for managing hazardous waste are (1) reducing the production of waste by reducing the amount generated or recycling/reusing the hazardous material after its generation; (2) reducing the volume and/or hazard the waste; and (3) long-term storage or disposal (Figure 11.25).[25]

Reduction of Generation of Hazardous Waste

The highest priority item in the management of hazardous waste is to reduce the production of waste at the source. The generation of hazardous waste can be minimized by (1) eliminating or substituting raw materials for less hazardous ones; (2) changing the manufacturing process to reduce or eliminate hazardous waste; and (3) separating or segregating waste at the source to prevent the contamination of

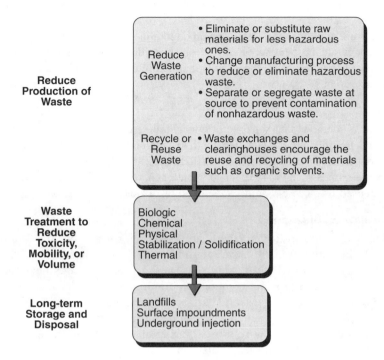

FIGURE 11.25 A desirable hierarchy for the management of hazardous wastes. (Adapted from Stephens, R.D., Hazardous Wastes, in: *Principles and Practice of Environmental Medicine*, Tarcher, A.B., Ed., Plenum Medical Book Co., New York, 1991.)

nonhazardous waste. Recycling and reusing hazardous materials before they enter the waste stream is also strongly supported by the EPA and most state regulatory agencies. Waste clearinghouses and waste exchanges have been organized to facilitate the transfer of discard materials from one industry to another that may use it as raw material in their process. Organic solvents are among the most widely recycled and reused materials.

Technologies for Hazardous Waste Treatment

Biological

Microorganisms are used to break down or degrade organic compounds as they use waste materials for a food source and convert them into water, carbon dioxide, and simple inorganic and organic molecules. This process is known as *bioremediation* and has been used successfully to clean up oil spills and other organic wastes where soil permeability allows good oxygen penetration. Waste degradation may be enhanced by adding oxygen and/or nutrients to stimulate the waste-eating organisms. Treatment may also occur in above-ground facilities or reactors where conditions may be better controlled and higher levels of decontamination can be achieved on wastes fed through the system. Bioremediation is being extensively tested under EPA or state oversight at more than 150 sites throughout the U.S.

Chemical

Treatments are applied that neutralize acid or alkaline wastes, detoxify, dechlorinate, oxidize, coagulate, precipitate, or concentrate hazardous wastes. Chemical treatment is used most often to change the chemical structure of waste and render it nontoxic or less hazardous.[27]

Physical

Nonchemical methods that separate or concentrate components of the waste stream are termed *physical methods*. Such methods include activated carbon to remove organic materials from liquid waste, distillation, filtration, coagulation, centrifugation, reverse osmosis, and ion exchange.[28] Once separated, the hazardous components may be reused or more efficiently destroyed or disposed.

Solidification/Stabilization

Both physical and chemical methods are used to form a solid, inert substance that limits the migration of hazardous materials from a landfill. Compounds amenable to fixation include organic salts, heavy metals such as lead, mercury, and zinc, and many inorganic compounds. Binding agents may include organic polymers, silica, or cement-based materials that are used to encapsulate materials in a water-resistant matrix, which can then be more safely disposed in a landfill. Stabilization can also be achieved by inserting electrodes into the contaminated soil and passing huge quantities of electricity through it, which melts and fuses the soil into a glasslike block. This process is called *in situ* vitrification (ISV) and such stabilized blocks are often left in place because the danger of pollutant migration is largely diminished.

Thermal Treatment

Incineration accounts for less than 2% of the hazardous wastes managed, although it offers the advantages of (1) completely destroying many forms of hazardous waste, (2) reducing the volume of waste, and (3) recapturing energy in the form of steam or electricity. Thermal treatment may include incineration or pyrolysis. Pyrolysis combusts materials in an oxygen-starved atmosphere and is not as widely used as incineration, which is conducted in an oxygen-rich environment. Incineration is normally conducted at temperatures of 425 to 1650°C (800–3000°F) in a turbulent atmosphere of sufficient duration that will insure good mixing and effective destruction of the waste approaching levels of the EPA current standard of 99.99%.[27,28] Incinerators have been used successfully to treat waste paints, plastics, mineral oils, pesticides, solvents, sludge, resins, greases, and waxes. The threat of hazardous emissions, including acids, dioxin, and heavy metals, have led to strict regulations on emissions from incinerator facilities.[29] The materials generated in an incinerator include carbon dioxide, water, sulfur, and nitrogen oxides, acid gases such as hydrogen chloride, and ash. Gas streams must be monitored, and hazardous materials must be removed from the emissions using various scrubbers, filters, and electrostatic precipitators. This has prompted very high construction and maintenance costs, but has become a useful alternative as the options for land disposal continue to decrease.

Hazardous Waste Disposal

Even after extensive treatment, some amount of hazardous wastes will remain in the form of ash, chemically treated residue, physically separated materials, or even in original waste form. Such waste may be economically reusable or recyclable and is usually dispersed to landfills, surface impoundments, or underground injection wells. Nearly 80% of hazardous wastes are disposed to land with 96% of these wastes treated or disposed at the site where they are generated.[23,30] Historically, such wastes were deposited to unlined pits and lagoons (surface impoundments), which presented extensive environmental risk from overflows, leaching, and volatilization.

Landfills

A common method for hazardous waste disposal is the secure chemical landfill as designated in the 1984 HSWA of RCRA. Secure landfills require groundwater monitoring, double liners, a leachate collection system, and financial guarantees for postclosure activities (Figure 11.26). Just as with MSW landfills, current technologies almost guarantee the leakage of hazardous materials from commercial hazardous waste landfills.[27] Increased costs of construction, operation, and liability have curtailed the number of active commercial operations to less than 20 landfills. These landfills are further restricted by legislation to accept mostly pretreated wastes that have been reduced in toxicity and/or treated to prevent leaching. The difficulty in siting for commercial hazardous waste landfills in New England and New Jersey because of public resistance and other factors has resulted in long-haul shipments at very high costs for businesses, schools, and commerce in these areas.

Deep-Well Injection

The process of deep-well injection involves pumping hazardous wastes through wells with double or triple casings into porous rock formations at depths of 1,000 to 10,000 ft below the surface, and well below drinking water aquifers (Figure 11.26). New hazardous waste wells were banned in 1983. Most injection wells are used by petroleum refineries and petrochemical plants located in the Midwest and Texas, who pump nearly 30 million tons of hazardous waste into the ground each year. Despite legislation that prohibits the construction of injection wells within a quarter mile of a drinking water source, there remain concerns about such wells. Detection of leaks and methods of correcting them are severely restricted by the depth. Further, the earth is a dynamic entity with continents drifting, tectonic plates shifting, and local geology under variations in pressure that combine to produce fractures and fissures, which may release injected wastes to combine with groundwater at some future time. For more information see http://www.dep.state.fl.us/water/uic/index.htm.

Surface Impoundments

Surface impoundments are depressions in the ground or areas with dikes around it that contain the liquid wastes pumped into it. They vary in size and shape ranging from several hundred square feet to hundreds of acres (Figure 11.26). Most of these pits, lagoons, or basins were historically unlined, and located on-site at manufacturing facilities. Legislatures recognized the high risk of groundwater contamination and regulations now require such impounds to be equipped with double liners,

FIGURE 11.26 Typical methods for hazardous waste treatment and disposal.

leachate collection systems, and groundwater monitoring wells. However, just as with landfills, such impoundments will eventually leak and attention is focused on reducing hazardous waste through reduction, reuse, and recycling efforts.

CLEANING UP

Before the 1980s, it was common practice for industries and municipalities to haul the wastes to a depression in the ground, dump them, and cover them over. In years previous to that, many local landfills were not even covered with soil. In many cases, drums of toxic wastes were simply stored in piles on-site (Figure 11.27). Many of these sites were abandoned, forgotten, and even built on, as neighborhoods expanded with the pressures of a rapidly increasing population. Then the buried toxic sludge began to appear, as rusted 55-gal drums seeped their contents into the soil to contaminate groundwater, or work their way to the surface where the true drama of poisonous mud and rusted containers could be seen. Driven by the horrors of Love Canal, Chem-Dyne, Swartz's Creek, and dozens of such hazardous waste sites, Congress acted in December of 1980 by passing the Comprehensive Environmental Response, Compensation, and Liability Act (CERCLA). This Act, also known as "Superfund," authorized the federal government to spend $1.6 billion over a 5-year

FIGURE 11.27 Stored (and discarded) drums of hazardous waste.

period for emergency cleanup activities. The Superfund program also provided for response to the release of hazardous substances from leaking underground storage tanks, spills from transportation accidents, and pesticide-contaminated groundwater.[31] The Superfund program also provided for activities to oversee long-term containment of hazardous sites. Despite the term *compensation* in the law, no funds were provided for victims whose health may have suffered from exposure to toxic waste sites. The term Superfund is attributed to the fact that the bill created a trust fund, financed primarily by excise taxes on chemicals and oil, and an environment tax on corporations. The EPA was charged with the responsibility of identifying those hazardous waste sites in most need of cleaning up. Superfund requires that for each site, the EPA identifies potentially responsible parties (PRPs) who are considered to be any entity, however large or small, that has anything to do with the site. The liability rules applied include retroactive liability, joint and several liability, and strict liability. (Figure 11.28) This may force individuals, companies, organizations (entities) to pay for cleanups that have had little or even no part in causing the damage. In practice, state or federal agency personnel will attempt to identify entities with the largest pockets (most financial resources) as PRPs. Such rules have been

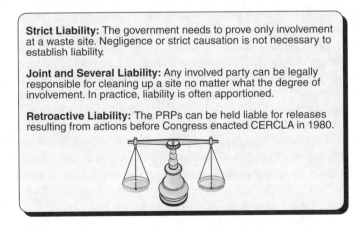

FIGURE 11.28 Definitions of liability under Superfund. (Adapted from U.S. Congress, Congressional Research Service, Superfund Reform: Cleaning Up America's Toxic Waste Sites, *Congressional Digest,* Vol. 68, March 1998.)

held responsible for 36 to 60¢ of every dollar spent in Superfund going to legal and other transactional costs.[32,33]

There are presently nearly 33,000 hazardous waste sites that have come under Superfund authority. More than 1300 of these sites are on the EPA's national priority list (NPL). The NPL was established because sites vary in size, hazardous substance content, containment, and degree of threat to the surrounding populations and environment. The EPA established a hazard-ranking system (HRS) based on the estimated hazard potential of the hazardous waste site. The factors used to make this estimate include the waste characteristics (toxicity, quantity, solubility, and reactivity); the distance to the local population; and the proximity to surface water and groundwater, and drinking water supplies. Other factors include the level of containment such as the condition of the containers (i.e., whether the drums are rusted), and whether there are leachate collection systems, or liners. The NPL is updated annually.

Environmental groups buttressed by a large contingent of concerned citizens charged that RCRA was not being vigorously enforced. These actions caused Congress to respond by passing the Superfund Amendments and Reauthorization Act of 1986 (SARA), which increased the program's funding to $85 billion and provided new and stricter standards. A series of delays, lawsuits, and accusations followed demanding that the Superfund program be repaired. Superfund had spent billions of taxpayer money and private sector money since 1980, to clean 550 sites by 2005.[34] This cost was expected to escalate to $42 billion in the next 70 years to clean up an estimated total of 2300 NPL sites.[33] At the end of 1995, the tax authorization for Superfund expired. The balance of some $6 billion in the Hazardous Substance Superfund Trust Fund was being supported by (1) taxes on domestically produced and imported oil; (2) a tax on feedstock chemicals; (3) a corporate environmental income tax; and (4) general revenues, reimbursements, penalties, and interest on the trust fund.[19] These fees generated 1.5 to $2 billion annually for cleanups, and the trust fund reached a high of $3.6 billion in 1995. Unfortunately, in the same year, the Superfund fee system lapsed, and Congress has failed to reinstate it. The trust fund is diminishing and more of the cleanup burden is falling on taxpayers' shoulders, whereas the pace of cleanups has declined dramatically. The administration has failed to request reinstatement of Superfund's fee system for three years running. In its 2004 budget request, the administration requested $1.39 billion for the Superfund program, expecting taxpayers to shoulder approximately 80% of program costs. Taxpayers paid only 18% of program costs in the last year before the Superfund fee system expired. Industry has avoided paying about $4 million a day, totaling over $10 billion since the Superfund fees expired in 1995. More disturbing is that oil companies are exempted from liability under the Superfund law.[35] In fiscal year 2005, the EPA listed 1303 sites including 59 proposed sites awaiting final agency action, 54 in the general Superfund section and 5 in the federal facilities section.[36] In fiscal year 2005, EPA's Superfund program obligated $524 million to perform construction and postconstruction activities and to conduct and oversee emergency response actions. They conducted or oversaw 665 ongoing construction projects (by EPA, potentially responsible parties, and federal facilities) at 422 sites. Federal facilities accounted for 220 of these ongoing projects. The EPA Superfund program completed construction phase of cleanup at 40 sites across the country for a total of 966 or 62% of the sites on the NPL. Further, Underscoring EPA's commitment to the "polluter pays" principle, the Agency secured private party funding commitments of more than $1.1 billion in fiscal year 2005. Of this amount, potentially responsible parties agreed to conduct more than $857 million in future response work, and to reimburse EPA for $248 million in past costs.[37]

REFERENCES

1. Tyler, P.E., When It Comes to Trash, Chinese Just Say Throw, *New York Times International,* Sunday, December 8, 1996, p. 5.
2. Mwanthi, L.O., Nyabola, E., and Tenambergen E., Solid waste management in Nairobi City: Knowledge and attitudes, *Environ. Health,* 60(5), 23, 1997.
3. Schernding, V.Y. and Yack, D., Environmental health impacts associated with rapid urbanization in South Africa, *Public Health Rev.,* 9(1-4), 295, 1992.
4. Waiyaki, P.G., Cholera: its history in Africa with special reference to Kenya and other East African countries, *East Afr. Med. J.,* 73(1), 40, 1996.
5. Otieno, F.A.O., Quantity and quality of runoff in Nairobi: the wasted resources, *Proceedings of the Sixth International Conference on Rainwater Catchment Systems,* Nairobi, Kenya, 21, 1993.
6. U.S. Environmental Protection Agency (USEPA), *Characterization of Municipal Solid Waste in the United States: 1997 Update,* Prepared for U.S. Environmental Protection Agency, Municipal and Industrial Solid Waste Division, Office of Solid Waste, by Franklin Assoc. Ltd., Prairie Village, KS, Report No., EPA 530-R-98-Oct, May 1998, chap. 3.
7. Nadakavukaren, A., *Our Global Environment,* 4th ed., Waveland Press, Prospect Heights, IL, 1995, chap. 16.
8. Grove, N., Recycling, *National Geographic,* 186, 1, July 1994, p. 92.
9. Solid Waste Digest, 11(12), National edition, Chartwell Information Publishers, Alexandria, VA, 2001.
10. United States Environmental Protection Agency, Solid Waste and Emergency Response (5305W), Municipal solid waste generation, recycling, and disposal in the United States: Facts and Figures for 2003, EOA 530-F-05-003, April 2005, www.epa.gov/osw.
11. Koren, H. and Bisesi, M., *Handbook of Environmental Health and Safety: Principles and Practices,* Vol. 11, 3rd ed., Lewis/CRC Publishers, New York, 1995, chap. 2.
12. Pellerano, M.G. and Parkhurst, B., The Basics of Landfills, Environmental Research Foundation, 1995, http://www.enviroweb.org/enviroissues/landfills/index.html.
13. Anonymous, Garbologist goes where few scientists dare, *Los Angeles Times,* July 24, 1993, p. 1.
14. Bonoparte, R. and Gross, B.A., *Field Behavior of Double-Liner Systems, in Waste Containment Systems: Construction, Regulation, and Performance,* Bonoparte, R., Ed., Geotechnical Special Publication, No. 26, American Society of Civil Engineers, New York, 1990, chap. 2.
15. Lee, G.F. and Jones, A., Municipal Solid Waste Management in Lined, "Dry Tomb" Landjills: A Technologically Flawed Approach for Protection of Groundwater Quality, EL Macero, CA: G. Fred Lee and Associates, March, 1992, available from G. Fred Lee and Associates, 27298 East El Macero, Drive, El Macero, CA, 67 pp.

16. Commonwealth of Massachusetts, Department of Environmental Protection, Executive Summary: 1997 Solid Waste Master Plan Update, Executive Office of Environmental Affairs, Department of Environmental Protection, 1997, http://www.magnet.state.ma.us/dep/bwp/dswm/dswmpubs.htn#swmp.

17. United States Environmental Protection Agency (USEPA), *Characterization of Municipal Solid Waste in the United States: 1997 Update,* Prepared for US Environmental Protection Agency, Municipal and Industrial Solid Waste Division, Office of Solid Waste, by Franklin Assoc. Ltd., Prairie Village, KS., Report No., EPA 530-R-98-Oct.-May, 1998, chap. 4.

18. Yoon Roe-Hoan and Camicia, N.T, Recycling of Municipal Solid Wastes in the United States Center for Advanced Separation Technologies (ryoon@vt.edu), Industrial Waste Recycling R&D Center, 2001.

19. Congressional Research Service, Superfund reform: cleaning up America's Toxic Waste Sites, *Congressional Digest,* March 1998, p. 68.

20. Howlett, D. and Tyson, R., Toxicity of Times Beach: EPA Verifies Mo. Town's Worst Fears, *USA Today,* September 13, 1994, p. 10A.

21. Nassiter, A., Villain is dioxin: relocation is response, *New York Times,* October 20, 1996.

22. *Pensacola News Journal,* Top 10 findings of Escambia Co. Grand Jury, Thursday, May 6, 2004, http://www.fluoridealert.org/pollution/1894.html.

23. Stephens, R.D., Hazardous wastes, in *Principles and Practice of Environmental Medicine,* Tarcher, A.B., Ed., Plenum Medical Book Co., New York, 1991.

24. Council on Environmental Quality (CEQ), Environmental Quality, 15th Annual Report 1984, Washington, D.C., 1986.

25. United States Environmental Protection Agency, Safer Disposal for Solid Waste: The Federal Regulations for Landfills, Report EPA/530-SW-91-092, Washington, D.C., 1993.

26. U.S. Environmental Protection Agency, Waste Minimization National Plan: Reducing Toxics in Our Nation's Waste, EPA Solid Waste and Emergency Response, EPA530-F-97-010, Washington, D.C., 1997.

27. Senkan, S.N. and Stauffer, N.W., What to do with hazardous waste, *Technol. Rev.* 84, 34, 1987.

28. Harris, R.H., English, C.W., and Highland, J.H., Hazardous waste disposal: Emerging technologies and public policies to reduce public health risks. *Am. Rev. Public Health,* 6, 269, 1985.

29. Rose, J., Incineration Regulations Set To Tighten, *Environmental Science and Technology,* 28, 12, 1994, p. 512A.

30. U.S. Congress, Office of Technology Assessment, Technologies and Management Strategies for Hazardous Waste Control, OTA-M-197, Washington, D.C., 1983.

31. U.S. Congress, Office of Technology Assessment (OTA), Assessment Superfund Strategy, OTA-ITE-282, Washington, D.C., 1985.

32. U.S. Congress, Congressional Research Service, Superfund Reform: Cleaning Up America's Toxic Waste Sites, *Congressional Digest,* Vol. 68, March 1998.

33. The Congressional Institute, Inc., Superfund, July 17, 1, 1996, http://www.conginst.org/cidoc-l.nsf/e74d.,167f730.d9718525636coo79ao93?Open Document.

34. USEPA, Superfund Redevelopment Program, At A Glance, Approximately 550 Superfund sites are ready for reuse, or have already been returned to productive uses, update March 2006, http://www.epa.gov/superfund/programs/recycle/index2.htm.

35. Zdeb, S., Polluters Must Pay: Superfund Reauthorization, Green Scissors, http://www.greenscissors.org/other/superfund.htm, June, 2006.

36. Smith, R., Superfund, Cleaning up the Nation's Hazardous Waste Sites, USEPA, Superfund, Washington, D.C., April 18, 2006, http://www.epa.gov/superfund/.
37. USEPA, Superfund, Superfund National Accomplishments Summary Fiscal Year 2005, as of November 22, 2005, last updated on Wednesday, March 1, 2006, http://www.epa.gov/superfund/action/process/numbers05.htm.

12 Assessing Human Risk

CHAPTER OBJECTIVES

A student, after reading this chapter, should be able to:

1. Define risk and discuss the uncertainties associated with environmental risk.
2. Discuss the characteristics of risk.
3. Define risk analysis and describe the tools used to perform risk analysis.
4. Explain the concepts of dose, extrapolation, and acceptable daily intakes (ADIs).
5. List and discuss the process of risk analysis, including hazard identification, dose–response assessment, exposure assessment, and risk characterization.
6. Outline and discuss the major components of risk management and risk communication.

INTRODUCTION

A farmer stands in his field, staring in horrified awe at the stream of black waste flowing onto his land from an adjacent river. This scene occurred on April 29, 1998, when a waste reservoir in Southern Spain burst, releasing toxic sludge into waterways and onto surrounding farmland. The reservoir served as a holding area for zinc, lead, iron, and cadmium wastes from a nearby Canadian-owned zinc mine. The dangerous waste posed both an immediate and long-term threat to the health and livelihoods of locals and to the environment. Crops were ruined and land contaminated by the toxic mess. Locals complained of burning eyes and throats. Ten tons of dead fish and shellfish were carted away from the acidic waters. With shock and dismay, people wondered how this unexpected event occurred. A shift in the earth apparently

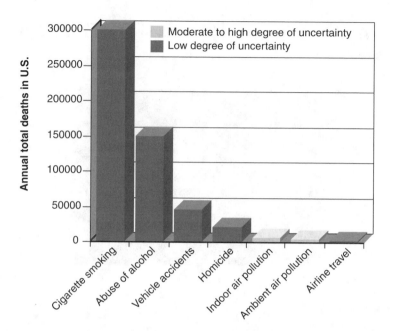

FIGURE 12.1 A comparison of some common risks. (Adapted from Moeller, D.W., *Environmental Health,* revised ed., Harvard University Press, Cambridge, MA, 1997, chap. 16.)

sent the reservoir wall tumbling, releasing millions of cubic feet of metal waste into the surrounding region. This event typifies the multitude of environmental risks facing the world today. These types of risks are uncertain and unexpected, and the consequences can be severe, even deadly.[1]

Risk is an integral part of life, permeating every aspect, from eating, breathing, sleeping, socializing, working, and playing. Risk can be of a financial, personal, social, health, and environmental nature. Some industries, such as insurance, gambling, and finance, are based on risk. The word "risk" encompasses a range of activities, situations, and concepts, from drinking a glass of red wine daily to skydiving and extreme skiing, to chemical exposure. Risk can imply chance consequence, danger, and opportunity.[2] In everyday language, risk is commonly used to describe types of people or situations. For example, the term *risky business* implies shady dealings, a *risk taker* suggests a brave, confident, and perhaps foolhardy, individual. *Risk-averse* describes someone who shies away from risk. *Risk-free* describes an activity involving no uncertainty or negative consequences. A comparison of some of life's risks is shown in Figure 12.1.

Many of the risks in our lives are from voluntary choices, such as our diet, sunbathing, and smoking. People accept certain risks because they enjoy the benefit they receive from the behavior or activity. People risk money in the stock market in the hopes of making a profit; people live in earthquake- and hurricane-prone areas in exchange for the lovely climate often found in those regions. We accept many risks in our lives, because they seem both mundane and remote: We cross streets, consume pharmaceuticals, drive in cars, and fly in airplanes on the assumption that

FIGURE 12.2 Extreme sports are becoming popular because of the excitement associated with high risk. (From Corel sampler, limited edition CD-ROM.)

these activities are safe. However, even these everyday activities carry some level of risk. But because these activities are familiar and routine, they do not appear threatening. However, not everyone is content with life's daily dose of risk. Some people seek out extraordinarily high levels of risk, engaging in skydiving, bungee jumping, rock and ice climbing, and other extreme sports (Figure 12.2). People choose these activities because they enjoy the element of danger from the risk, and the adrenaline "rush." They have defined their own "acceptable" level of risk, which might be too high for most people, who reach their risk threshold more easily.

ENVIRONMENTAL RISK

Some risks are well understood, such as driving and smoking. The vast quantity of information and statistics on automobile accidents allows us to draw conclusions about the risk of dying in a car crash. Similarly, voluminous data on smoking and disease demonstrate that smoking poses a health risk. However, some risks remain shrouded in uncertainty. Many environmental risks fall into this category. Environmental risk is a reality of today's world. The seemingly endless supply of synthetic chemicals, consumer goods, energy, and waste create new risks through chemical contamination, pollution, and environmental degradation. Environmental disasters such as chemical spills or explosions threaten millions of people living in the vicinity of manufacturing or storage facilities. The uncertain risks of global warming and ozone depletion loom ahead. The main characteristic of environmental risk is that it is usually involuntary. People do not choose to ingest chemical pollutants such as pesticides or industrial solvents in their food and water, suffer workplace exposures to dangerous chemicals, breathe polluted air, or expose themselves to radiation from nuclear fallout or faulty nuclear power plants. These environmental risks pose a unique problem to regulators charged with protecting the public's health. Limited information may be available on the health effects of these risks. Consequently, in an effort to protect the public's health, various government agencies study these potential hazards to determine the levels of risk they pose. This effort to understand these risks and to quantify their impact on human health is the field of risk analysis.

RISK CHARACTERISTICS

Risk has several defining characteristics. The two characteristics that most embody risk are uncertainty and the unknown. Risk can be defined as the likelihood of an unwanted occurrence coupled with an element of uncertainty about when the risk might occur.[4] Many other definitions of risk exist, all varying slightly, but focused on the concepts of uncertainty and the unknown. Other risk characteristics may include dread of a voluntary or involuntary nature that is immediate or latent, or a potential, threat to future generations. Risk often encompasses the ideas of morbidity and mortality, that is, the possibility of sickness or death.[5] Certain risk characteristics generate greater fear in people than risks that are known or voluntary. As many environmental risks have the characteristics of being unknown and inducing dread, they generate anxiety and concern about the consequences of exposure to the risk. Risk analysis is one method to allay this fear and anxiety, through developing knowledge about a risk, understanding its potential health effects, and devising methods to limit and manage the risk.

DEVELOPMENT OF RISK ANALYSIS

Risk analysis is the process of reviewing information on a hazard to characterize its impact on human health. This process involves a review of scientific studies, an understanding of the properties of a risk, an assessment of levels of human exposure and dose, and a conclusion about the likelihood, impact, and extent of a risk. By employing these methods, risk analysis allows researchers to arrive at conclusions regarding the severity and consequences of environmental risks. Risk analysis developed over the past 30 years from a need to understand modern, technology-based risks. It began with an increased interest in the environment on the part of the government, and scientific and public sectors. Rachel Carson's 1962 book *Silent Spring* spurred concern about environmental issues. Environmental crises such as Love Canal, Three Mile Island, and the chemical accidents in Seveso, Italy, and Bhopal, India, encouraged research into the effects of these environmental risks. The purpose was to assess the potential health impact of these hazards and to develop methods to cope with these risks.

The field of risk analysis developed as a response to several changes in the modern world. First, the number of risks society faces today has increased dramatically compared to times past. For example, over 70,000 synthetic chemicals are used today in industrial and manufacturing processes and in many consumer products and household items. Little detailed risk information is available for most of these substances.[6] Second, technology has improved our ability to measure risk. Through the continued development of risk analysis, we have begun to examine the hazards of all consumer goods and even products that have been taken for granted as being good for our health. Recent studies have looked at the effects of cow's milk on the human body. The main focus of the study examines intestinal irritation, intestinal bleeding, anemia, and allergic reactions in infants and children, as well as infections of salmonella. In a paper written by Robert M. Kradigan, MD, risk analysis of cow's milk examines the possibilities of viral infection and possible health hazards associated with milk.[7] Today, exposures are measured in the parts per million, billion, and even trillion, greatly increasing our capacity to accurately quantify exposure levels. Third, the number of government agencies directly or indirectly monitoring environmental risk have increased, such as the Food and Drug Administration, the Environmental Protection Agency, and state-run departments of environmental protection. Even towns and cities have agencies regulating local risks to public health. Fourth, the number of laws and regulations governing the environment and protecting the public health has increased dramatically over the past 30 years. Some of these laws have drastically changed the management of risks and dissemination of information. The Superfund Amendments and Reauthorization Act of 1986 (SARA), also known as the Emergency Planning and Community Right-to-Know Act, ensured that communities were prepared for chemical emergencies and had access to information on chemicals at sites in the area. Other examples of regulations include the Food, Drug and Cosmetic Act, the Toxic Substances Control Act, and the Clean Water Act. Fifth, an increase in public interest in environmental risks has spurred research in quantifying risks. Special-interest groups and grassroots movements have involved the public in environmental issues on both local and national levels.[8,9] The development of risk analysis has indeed led to the enlightenment of environmental hazards and risks associated with our lifestyles. Progress has been made in enlightening the masses on issues of risk analysis. Recent environmental risk analysis has led us to begin examining the effects of the population's neglect of their environmental responsibility. A recent article written by Radim J. ≥rám from the Institute of Experimental Medicine in the Czech Republic discusses how air pollution is affecting the reproductive health of the world population. The risk analysis conducted concurs with other current research that pollutants in the air cause reproductive health problems. For both men and women, polycyclic aromatic hydrocarbons are the most harmful and common pollutants, and are released into the air by both industrial and chemical processes. A growing awareness has developed over environmental issues because of the development of risk analysis.[10]

A detailed qualitative and quantitative description of a risk can be created by gathering the relevant study information, considering the important factors, and arriving at a conclusion about the risk. Risk analysis has many applications across a variety of fields. This risk description can be used in decision making and the development of environmental and public health policy. For instance, risk analysis can be used to estimate the likelihood of developing cancer from exposure to radiation, to determine the side effects of a medicine awaiting FDA approval, to ascertain whether a synthetic food product such as Olestra or saccharin is safe for human consumption, or if transgenically altered animals and plants pose a health risk. Risk analysis allows public groups to make informed decisions and weigh the risks and benefits in their community.

TOOLS OF RISK ANALYSIS

Risk analysis employs several scientific disciplines in its goal to characterize a risk (Figure 12.3). Significant amounts of information, both quantitative and qualitative,

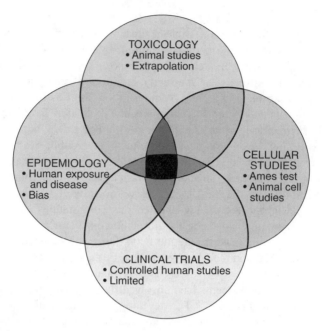

FIGURE 12.3 Tools of risk analysis.

are needed to develop an adequate picture of a risk. Toxicology, epidemiology, clinical trials, and cellular studies provide some of the information for a risk analysis. The types of information reviewed include characteristics of the chemical, type and length of exposure, dose, animal response, and human response. These disciplines are discussed in greater detail in the following text.

Toxicology

Toxicologists study chemicals to determine their physiological and health impacts on humans. Risk analysis relies heavily on toxicological studies and findings. A particular branch of toxicology, regulatory toxicology, aims at guarding the public from dangerous chemical exposures. For example, the health effects of nicotine, saccharin, and benzene have been explored and defined through toxicological studies. Toxicological studies usually involve controlled laboratory animal studies. The animals are evaluated for their responses to different doses of a substance. These controlled research conditions are crucial for procuring accurate information. Prior to a toxicology study, little may be known about a chemical, such as what amount causes minimal adverse health effects or death. In a study, scientists follow a series of steps to learn about a substance's activity in a living organism. A study can delineate both the lower and upper limits of a chemical's potency, that is, the amount that causes no effect in any animal and the amount at which all animals die. After these limits are established, scientists can work within this range to discover additional information about a chemical's health effects. As a chemical may have a myriad of health effects, the researchers will focus their interests on certain health responses. These specific physiological changes are called *endpoints*. Some common endpoints frequently studied include the effective dose of a chemical, named the no observable effect level (NOEL), the no observable adverse effect level (NOAEL), the lowest observable adverse effect level (LOAEL), and death. All these endpoints provide valuable information on the doses and effects of a chemical. This information can be incorporated into a risk assessment.[6,9] Do you want to learn more about risk assessment? Point your browser to http://www-unix.oit.umass.edu/~envhl565/weblinks.html and click on any of the Chapter 12 Web links associated with risk assessment, including:

- Welcome to Center for Applied Environmental Public Health
- Cumulative Index to Risk
- USEPA Superfund Risk Assessment
- International Society of Exposure Analysis
- The CRESP Homepage
- Risk Assessment Information System
- Society for Risk Analysis: Risk Assessment, Characterization, Management, Communication, and Policy

Dose

When studying a chemical agent, toxicologists use different doses to elicit different animal responses. They ascertain the threshold dose, which is the dose at which no effects are seen in the study animals. As the administered dose increases above the threshold amount, animals will begin to show adverse effects and some will die (Figure 12.4). The dose at which 50% of all the test animals die is called the *lethal dose*, or the LD_{50}. Similarly, the ED_{50}, or effective dose, is the dose at which 50% of the animals demonstrate a response to the chemical. A very toxic chemical will have a low LD_{50}, signifying a low dose sufficient to kill 50% of the test animals. A less toxic chemical will have a higher LD_{50}. The maximum tolerated dose (MTD) is a common measure of chemical exposure in animal studies. At this dose, researchers expect to see only up to a 10% loss in weight and no death, clinical toxicity, or pathological lesions in the animal population. Some critics of the MTD believe it is too high a dose. Chemicals that cause cancer at the MTD in test animals may not cause cancer at lower doses.[10]

Animal studies occur under controlled conditions. This control refers to exact doses, and specific lengths and times of exposure. By controlling the amount, timing, and duration of exposure, scientists can obtain increased quantity and quality of information on a chemical's health effects. For example, altering the amount, timing, or duration of a dose in a study can alter the chemical's capability to inflict damage. Large doses may cause one type of response; small doses, another. The timing of a dose may also be critical. For example, if exposure is timed to occur during a "window of vulnerability" in a fetus, minuscule amounts of an agent may cause severe damage. Outside this window, this amount would be harmless to the fetus. Clearly, control of the dose has a tremendous influence over the types of health effects seen. In addition, this control allows for variability

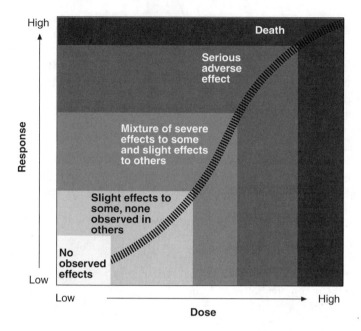

FIGURE 12.4 The dose–response curve. (Adapted from Covello, V.T. and Merkhofer, M.W., *Risk Assessment Methods,* Plenum Press, New York, 1993, chap. 1.)

in study design. Toxicological studies can occur over short, medium, or long periods of time. Short-term, or acute-exposure studies, occur over 2 weeks and generally involve high doses of the chemical substance. Medium-length or sub-chronic studies involve lower chemical doses over a longer time period, from 5 to 90 d. Long-term or chronic studies can last 2 years and use much smaller doses of chemicals. This variability improves understanding of a chemical's activity under different conditions.[11] For comparison purposes, animal studies establish a control group not exposed to the agent under consideration. This group provides the researchers a level of certainty that the physiological responses seen after exposure are in fact due to exposure and not to some unrelated, uncontrolled variable.[12]

Extrapolation

Extrapolation can be defined as using known information to infer something about the unknown.[13] In risk analysis, the term extrapolation refers to the use of animal data to predict human response to chemical exposure. Extrapolation is widely used in toxicology to deduce the risks humans face. The use of extrapolation is a vital tool of toxicology as human exposure studies are not widely accepted. Different types of animals can be used during laboratory studies, depending on the agent under consideration. Although dogs and monkeys have been used for animal studies, the most common laboratory animal is the rat. During laboratory studies, animals generally receive high doses of the chemical of interest for a measured span of time, often their lifetimes. Laboratory animals have shorter life spans than humans, allowing studies to document a full lifetime of exposure, from infancy to maturity to death. The results from these high-dose, short-duration studies are used to extrapolate human response to the longer-term, lower-level exposures we generally receive. The difference in dose level and duration of exposure generates some doubt about the validity of extrapolation. Furthermore, biological and metabolic differences between the test animal and humans may undermine the validity of extrapolations. For instance, the test animal may have one method for detoxifying a substance, whereas a human may use a different metabolic pathway. Alternatively, a test animal may lack an adequate detoxification pathway, whereas humans may be very effective at detoxifying a substance. Because of these differences, a substance that causes cancer in one animal may not cause cancer in another animal, or *vice versa*. This fact limits the accuracy in extrapolating from animal data to human response.[11] Despite these drawbacks, extrapolation is a crucial method in studying potential human outcomes based on the results of animal studies.

ACCEPTABLE DAILY INTAKES

The purpose of toxicological studies is often to establish an acceptable level of exposure or dose of a substance that is considered "safe." This level, which poses little risk, is termed the *acceptable daily intake* (ADI). In extrapolating from animal studies to humans, scientists incorporate safety factors when determining acceptable exposure (dose) levels. The NOEL established by the threshold dose in test animals is translated into the ADI or the reference dose for humans. These levels include safety factors from 10 to 10,000, depending on the protected group. The safety factor is added by dividing the threshold dose by some factor of ten, thereby reducing the acceptable level of exposure and risk from exposure and increasing safety. Again, which factor of ten is used depends on the level of protection sought, the uncertainty inherent in the study data, and who is being protected. Lower safety factors are used when high-quality human data are available. Higher safety factors, up to 10,000, are used as the uncertainty level goes up and the quality of the data diminishes. Although the ADIs are based on scientific studies and include a margin of safety, these numbers are not foolproof. As more data becomes available about a substance, the ADIs may change, either lower or higher. For example, as we have gained knowledge about the dangers of exposure to lead, the minimum level of exposure considered harmful to human health has decreased, as adverse health effects are seen at extremely low concentrations of lead.[4] We also have had to incorporate these ADIs into our diets. The nature of the environment has changed so that we must examine dietary ADI to set safety standards for dosages of daily allowances. Recent studies have examined ADI soy-based powders and liquids available as alternatives to milk. Although soy-based formulas have been found to offer most of the nutrients of breast milk, recent studies show that infants on soy formula are two to three times more likely to develop thyroid disease. With studies of this kind, we are better able to access risk factors associated with exposure to substances that may pose safety hazards.[13]

EPIDEMIOLOGY

What Is It?

Epidemiology is "the study of the distribution and determinants of disease frequency in the human population."[14] In an attempt to determine causality between an exposure and a disease, epidemiologists study human response to various environmental agents, such as tobacco smoke, pharmaceuticals, exercise, and even hair dye. These are not controlled laboratory studies, but rather, occur in the outside world, subject to fluctuating conditions. When an increase in the incidence of a disease is reported, epidemiologists try and account for the factors related to its occurrence. This was true when high incidences of morbidity were discovered among inner-city children with asthma, and it was hypothesized that this could be related to exposure to cockroach allergen. Children from eight major inner-city areas were studied. It was concluded that that this variable alone could not account for this situation. Many variables are introduced into these studies, such as personal choice, socioeconomic factors, age, demographics, and other factors. These factors, called *confounders*, can bias or skew the results of the study toward either more of less of an association than actually exists, depending on the confounding factor. Consequently, in designing a study, scientists will try to "control" for these confounding factors.[15]

Study Types

Several different types of epidemiological studies exist, including cross-sectional, cohort, and case-control studies. These study types differ in design by the methods categorization of study subjects and the timing of information gathering. Each type of study has its own strengths and weaknesses in finding an association between an exposure and a disease. A cross-sectional study is a type of descriptive study design in which both a subject's exposure and disease are determined at the same time. This type of study provides general information on the disease, such as who and where. Cohort studies and case-control studies are types of analytic studies, focused more on answering the question of why a particular disease develops. Cohort studies group people by exposure and then follow them over time for disease development. This type of prospective study is useful for studying rare exposures. Case-control studies group people by disease and then look retrospectively for a history of exposure. This type of study is useful for studying rare diseases. All these study types provide valuable information on disease occurrence in human populations.[12,15]

Bias

Epidemiology involves the study of humans. Consequently, it avoids the extrapolation problem seen in animal studies. However, it is not without drawbacks. A significant problem in epidemiology is bias. Bias is a type of error introduced into epidemiological studies through differential treatment of the cases and controls in a study, either during the selection process or information gathering. Selection bias, information bias, and nondifferential misclassification bias may lead to inaccurate study results. Researchers can address the problem of bias through careful study design. Despite this and other limitations, epidemiology remains an important tool for evaluating the link between exposure and disease.[12,15]

CLINICAL TRIALS

Clinical trials provide a direct means to evaluate human response to a substance under more controlled conditions than epidemiology studies. This type of risk assessment provides detailed information on dose–response, toxicity, efficacy, and side effects. In clinical drug trials, individuals volunteer to be exposed to a substance or to ingest a drug, and are assessed for their health response. The types of health responses measured are generally minor, short term, and reversible, because testing human subjects involves ethical consideration, and the trials are limited in duration. Volunteers must be informed of the risks, must give their consent, and may receive monetary compensation for their participation. Despite the volunteer status of subjects in human clinical trials, the ethics of human studies is a controversial area. Past studies on prisoners were halted because of ethical considerations. Recent controversy erupted when several pesticide manufacturers announced human testing of pesticides to determine safety. The companies resorted to human trials to demonstrate the safety of their products in the face of strict food quality legislation passed in 1996. Opponents felt the companies were taking advantage of volunteers, enticing them with significant monetary compensation, downplaying the inherent uncertainty of the exposure, and misleading them by calling pesticides "drugs." Clinical trials are used widely by pharmaceutical companies prior to releasing drugs to the public. Although they are limited in scope, depth, and duration, clinical human trials yield crucial information on human response.[10,16] Clinical trials provide us with information on the safety and efficacy of drugs. Human trials can provide great risk to participants, but they also offer a great service to humanity. The promise of drugs that may eradicate the flu, the AIDS virus, or the common cold cause a stir at the mere thought of the possibilities and lives that may be saved. The National Cancer Institute investigated the drug Tamoxifen in 1992 as a possible cure for previously treated patients and as a preventative agent for breast cancer. The drug was tested in a placebo-controlled study on 13,388 healthy women. Half the women were given Tamoxifen and the other half was given the placebo, with varying results. The clinical trial found that there were very satisfactory results from the study, with fewer cases of breast cancer than the placebo group of women; however, the women who took the drug had two to three times the incidence rate of uterine cancer. Only through clinical trials can we ascertain the health hazards associated with drugs that may be helpful or harmful. Their importance is crucial to the discovery of new drugs and their effect on the human body, both long and short term.[17]

CELLULAR TESTING

Cellular studies are used to test chemicals for mutagenicity and carcinogenicity. These tests can be done on bacteria or animal cells. A widely known bacterial test is the Ames test, which uses *Salmonella* bacteria lacking the ability to produce histidine, a vital growth chemical. These bacteria are exposed to the chemical in

FIGURE 12.5 Schematic of risk analysis process.

question, and assessed after 2 to 3 d for signs of growth. Indications of growth suggest a mutation occurred in the bacteria's genetic information, conferring the ability to produce histidine. Although relatively inexpensive and quick, cellular studies are not always accurate in identifying carcinogens. This problem relates to the difficulty of using single-celled organisms to predict human response to a substance. Other cellular studies include animal cellular studies, which are used to identify potential carcinogens. In these tests, the test chemical is applied to rodent or hamster cells. After a 1 to 3 week period, these cells are assessed for changes in growth. Unlike the Ames test, this type of study is fairly accurate in predicting carcinogenesis.[11]

THE PROCESS OF RISK ANALYSIS

With this arsenal of tools, the process of risk analysis can begin to identify and enumerate the risk of a particular hazard. This process has four steps: hazard identification, dose–response evaluation, exposure assessment, and risk characterization (Figure 12.5). These four steps describe the risk through qualitative and quantitative evaluations.

HAZARD IDENTIFICATION

The initial step in risk analysis, hazard identification, involves identifying chemicals that present a risk to human health. This step entails performing a qualitative assessment of a chemical's potential for negative health impacts on humans. A thorough evaluation of scientific research about the risk determines the likelihood of adverse effects on humans. A chemical's activity in living organisms is researched through a review of animal studies, cellular studies, epidemiological studies, and human trials. In addition, to ascertain the possibility of human exposure, a substance's boiling point, volatility, and solubility are considered. If a chemical has been identified as a potential threat to human health, the process of risk analysis continues.[3,18]

DOSE–RESPONSE ASSESSMENT

The next step, dose–response evaluation, provides a quantitative view of the risk. This step also involves a review of scientific studies and data. In this case, the magnitude of response is correlated with the dose (Figure 12.6). The health response is reviewed at each dose to estimate the differences in response. This correlation provides valuable information about levels of risk at different levels of exposure. In addition, epidemiological studies are performed to determine the agent's strength, or potency, and potential to produce negative health effects in humans. Although

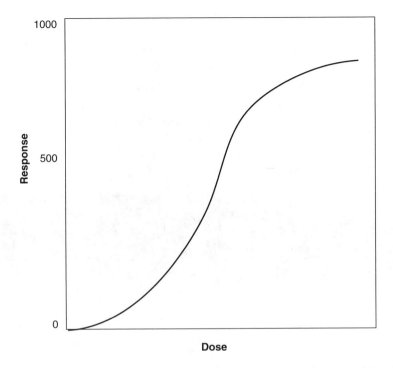

FIGURE 12.6 A typical simple dose–response curve. (Adapted from Cohrssen, J.J. and Covello, V.T., *Risk Analysis: A Guide to Principles and Method for Analyzing Health and Environmental Risks,* The Council on Environmental Quality, Washington, D.C., 1989, chap. 3.)

human studies are preferable for this evaluation, such data are not always available. In such cases animal data are used.[3,18]

EXPOSURE ASSESSMENT

After the dose–response assessment, an exposure assessment is performed. The purpose of the exposure assessment is to measure or estimate people's level of exposure. To accomplish this, the length of exposure, the amount of exposure, the route of exposure, the duration of exposure, and the number of people exposed are considered. Exposure is different from dose in that exposure refers to the amount of a substance in the environment, whereas dose is the level of a substance actually taken in by an organism. Dose can be influenced by many factors, such as how the substance enters the body, whether it is absorbed through the skin, ingested with food, or inhaled (Figure 12.7). This discrepancy between the amount of a substance measured and the actual dose is included in an exposure assessment. To perform an exposure assessment, specific information must be considered. The properties of the substance, such as its volatility, water solubility, ability to become airborne, or migrate through soil will influence its relationship with the environment and affect its routes to human exposure. Besides considering the substance's properties, the assessment should specify the group exposed, with special attention given to those

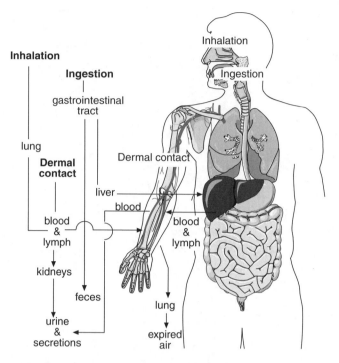

FIGURE 12.7 Routes of chemical exposure. (Adapted from Cohrssen, J.J. and Covello, V.T., *Risk Analysis: A Guide to Principles and Method for Analyzing Health and Environmental Risks,* The Council on Environmental Quality, Washington, D.C., 1989, chap. 3.)

particularly vulnerable, such as pregnant women, children, the elderly, and the health-compromised. Finally, the type of exposure must be noted, whether continuous, intermittent, short term, long-term, or chronic.[3,9] As is the case with studies looking at incinerators and the health risks associated with exposure to their emissions, cumulative effects should also be considered, taking into account the effect of all the sites in a given region. This is especially important with this particular hazard because of its ability to "travel long distances and persist in the environment."

RISK CHARACTERIZATION

The final step of risk assessment is risk characterization. Risk characterization provides a picture of the risk that addresses its severity, likelihood, and consequences. All the previous information from the hazard identification, the dose–response evaluation, and the exposure assessment are incorporated into the risk characterization. The risk characterization includes an estimate of the negative effects to exposed individuals, such as the number of cases of cancer or deaths per 100,000 people (Figure 12.8). Other types of risk estimates include an individual lifetime risk, such as a one in a million chance of cancer, a loss of life expectancy, as in days or years lost from exposure to a substance, and relative risk, that is, the risk of the exposed population vs. the unexposed. Probabilities of these effects are also provided. Any

Deaths per million people in a selected population (nation).

Deaths per million people within a certain radius of the released pollutant.

Deaths per weight (tons or pounds) of the toxic substance(s) released.

Loss of life expectancy associated with exposure to the toxic material(s).

FIGURE 12.8 Types of risk estimates. (Adapted from Stern, P.C. and Fineberg, HX, *Understanding Risk,* National Academy Press, Washington D.C., 1996, p. 50.)

drawbacks, limitations, and uncertainties inherent in the estimate are incorporated into the risk characterization.[3,11]

LIMITATIONS OF RISK ANALYSIS

The conclusions drawn from risk analysis are not foolproof. Limitations of risk analysis include uncertainty, variability, and effect of multiple exposures. Uncertainty exists in all estimates and recommendations about environmental risks. Generally, "zero risk" is not possible, because of high degrees of uncertainty involved in risk assessment. Although people want to be assured about a safe level of exposure to a chemical, sometimes too little is known to about the substance to provide any real assurance. A frightening example is represented by pharmaceuticals pulled from the market. Although a drug may undergo years of laboratory testing and even human trials, sometimes rare side effects are not seen until the drug is approved and released to the public. This scenario occurred recently with the popular diet drug fen-phen and the fast-acting analgesic Duract. In the case of both drugs, dangerous and even deadly side effects appeared in human users shortly after the drugs were available by prescription. Both drugs were subsequently taken off the market.[19] Even with rigorous laboratory testing and careful risk analysis, a degree of uncertainty remains. This uncertainty factor can erode the purpose of recommended doses, as people may doubt that any safe exposure level exists. In addition, interpersonal variability may strongly affect an individual's risks. Some people in a population are simply more sensitive to exposures than others and may be more likely to experience adverse health effects from a "safe" dose. Therefore, the built-in safety factors may not be enough to protect all individuals from harm. In addition, risk analysis rarely considers the fact that people are exposed to multiple risks at the same time. Rather, risk analysis focuses on a exposure to a single agent. Any synergistic or additive effects are ignored, providing an inaccurate picture of the true risks of real-world exposures. Despite these limitations, risk assessment is still a valuable tool for exploring and understanding the risks of the world.[10]

RISK MANAGEMENT AND COMMUNICATION

MANAGEMENT

Risk assessment provides us with tremendous amounts of information about risks. Risk management involves merging the results of risk analysis with various social

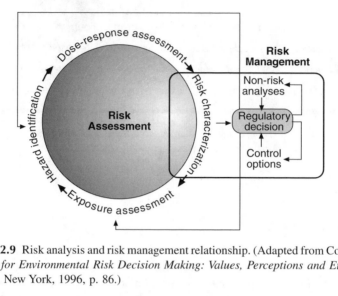

FIGURE 12.9 Risk analysis and risk management relationship. (Adapted from Cothern, C.R., *Handbook for Environmental Risk Decision Making: Values, Perceptions and Ethics,* Lewis Publishers, New York, 1996, p. 86.)

factors, such as socioeconomic conditions, political pressures, and economic concerns (Figure 12.9). Three avenues of risk management are educational, economic, and regulatory. These are not mutually exclusive, but can be used together to manage a risk. Educational risk management can include use of the media to inform the public about risk or the strategic placement of warnings to provide people with risk information, as on drugs and other chemical products. Economic risk management is accomplished through pollution taxes and permits that emphasize risk reduction through monetary incentives or disincentives. Regulatory risk management, also called command and control, is evident in the numerous laws governing potential environmental and health hazards in the U.S. The Clean Water Act, the Safe Drinking Water Act, and the Toxic Substances Control Act are examples of regulatory risk management, where laws restrict and limit pollution, chemical exposures, and releases.[21]

Risk management generally involves comparing the risk to some other factor such as the cost, or reducing the risk or the benefit gained from the risk (Figure 12.10). Several analytical methods include cost–benefit analysis, risk–benefit analysis, and contingent analysis. These comparisons allow regulators and agencies to determine the "best" course of action for managing or controlling a risk. Economics are a critical factor in risk management. Consequently, the "best" course of action is not always the one that reduces risk the most, but rather, is the most economically feasible option, reducing the greatest amount of risk per dollar spent. Risk management decisions do not solve all problems surrounding a risk or satisfy everyone's concerns: a decision that benefits one group may cause harm to another. An example is the process of selecting the location of an ash dump. A risk management decision is made to site the dump in a sparsely populated, rural area. This decision solves problem of where to dispose of incinerated waste. However, the residents of the rural community receive little or no benefit from the dump. Instead, they face the risk of

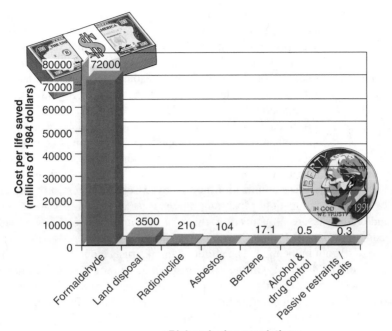

FIGURE 12.10 Comparison costs of risk management. (Adapted from Viscusi, W.K., *Rational Risk Policy*, Clarendon Press, Oxford, 1998, p. 98.)

land and water pollution from the ash, and a decrease in aesthetic value of their land from the existence of the dump and the noise of waste trucks through a previously low-traffic area.[22] Risk management also encompasses the idea of an acceptable level of risk and an acceptable number of cases. We cannot eliminate all risk, but we can hopefully reduce risk to some acceptable level. This idea of an "acceptable number" of cancer cases or an "acceptable number" of deaths is an uncomfortable but inherent aspect of risk management. Clearly, risk management is a complex undertaking that requires careful consideration of the risk characterization, political, economic, and social factors.[23,24] At times the multilevel process of risk management can result in indecision, delayed action, or in some incidences, no action, owing to the absence of funding. An example of this occurred in 1998 when a "panel convened by the National Institute of Environmental Health Sciences decided there was enough evidence to consider the invisible waves called electromagnetic fields, like those generated by power lines and electric appliances, a possible human carcinogen. Because no consensus could be reached regarding issues like safety thresholds, funding was withdrawn and nothing has been done with this information.

Risk Communication

Risk communication occurs when a dialogue occurs between any two or more parties involved in a risk. These parties may be government agencies or industry and citizen groups. Methods of risk communication include public hearings, emergency hotlines, and information pamphlets. The goal of risk communication is to effectively relay risk information developed through risk analysis to various interested groups. The World Resource Report can be used as an example of risk communication. It is a publication that is released jointly by the World Resources Institute, United Nations Environment Programme, the United Nations Development Programme, and the World Bank. Their collaborative message places an emphasis on prevention and further supports "coordination and communication among agencies that don't often interact."[25] Risk communication can be challenging, as it requires addressing people's different risk perceptions, biases, scientific knowledge, educational backgrounds, and even race and gender. Many of these areas require significant sensitivity and tact. Not all risk communication is effective. Certain factors are required for effective risk communication. Most importantly, trust and credibility are needed. As trust and credibility are influenced by the related factors of openness and honesty, knowledge and expertise, and concern and care, these factors also influence the success of risk communication. The public often mistrusts industry, so industry must act to earn the public trust. On the other hand, industry often perceives the public as lacking scientific knowledge, so the public must strive for a clear understanding of scientific principles. In addition to requiring trust, risk communication must often address differences in perception and meanings. Successfully addressing these differences allows for open, effective communication among parties subject to the risk.[25]

Other obstacles that hamper effective risk communication are the scientific terms and expressions used to discuss and describe risks. The public's vs. the expert's understanding and interpretation of risk information often differ significantly. For example, the phrase "one in a million" can be interpreted in two contradictory ways: the expert states the statistic to address society's risk as a whole. However, the individual, interested in his own chance of developing cancer, applies the probability to himself and attempts to determine his chance of being that "one in a million." In addition, scientific language with which the expert is comfortable may be foreign and meaningless to a layperson. A term such as *parts per billion* is difficult to interpret and may render any risk information useless to someone who cannot relate to the expression. Translating terms such as parts per billion into comprehensible and applicable measurements, such as 1 sec in 6.4 months, or 1.56 in. in 25,000 mi, or one half drop of liquid in 17,000 gal can facilitate comprehension of scientific language and increase risk communication (Figure 12.11). Risk communication involves building bridges among opposing groups, so that all parties can understand the relevant concepts and issues, although they may not agree on them. Risk communication is an essential aspect to managing environmental risks today.[27]

TIME: One second in 6.4 months.

LENGTH: 1.56 inches in 25,000 miles (the circumference of the earth at the equator).

AREA: One sq. yard in 324 square miles.

VOLUME: Approximately a half drop of water in a 17,000 gallon pool.

WEiGHT: A particle of dust (<0.01 grams) on a 2,000 pound automobile).

FIGURE 12.11 Parts per billion translated into comprehensible terminology. (Adapted from Peters, P.-G., Covello, V.T., and McCallum, D.B., The determinants of trust and credibility in environmental risk communication: an empirical study, *Risk Anal.,* 17, 43, 1997.)

REFERENCES

1. Simons, M., Big Sludge Spill Poisons Land in Southern Spain, *New York Times,* May 2, 1998, p. A3.
2. Jardine, C.G. and Hrudey, S.E., Mixed messages in risk communication, *Risk Anal.,* 1, 17, 1977.
3. Moeller, D.W., *Environmental Health,* revised ed., Harvard University Press, Cambridge, MA, 1997, chap. 16.
4. Covello, V.T. and Merkhofer, M.W., *Risk Assessment Methods,* Plenum Press, New York, 1993, chap. 1.
5. Derby, S.L. and Keeney, R.-L., Understanding "How safe is safe enough?" *Risk Anal.,* 1, 217, 1981.
6. Beck, B.X., Rudel, R., and Calabrese, E., The use of toxicology in the regulatory process, *Principles and Methods of Toxicology,* Hayes, A.W., Ed., Raven Press, Sacramento, CA, 1994, chap. 1.
7. Kradjian, Robert M., The Milk Letter: A Message to Parents, August 1998, http://www.afpafitness.com/milkdoc.htm.
8. U.S. Department of Health and Human Services, Task Force on Health Assessment, *Determining Risks to Health,* Auburn House Publishing, Dover, MA, 1986, pp. 3, 4.
9. Halbeib, W.T., *Emergency Planning and Community Right-to-Know Act,* Environmental Law Handbook, 14th ed., Sullivan, TER, Government Institutes, Inc., Rockville, MD, 1997, p. 481.
10. Malachowski, M.J., *Health Effects of Toxic Substances,* Government Institutes, Inc., Rockville, MD, 1995, chap. 4.
11. Cohrssen, J.J. and Covello, V.T., *Risk Analysis: A Guide to Principles and Method for Analyzing Health and Environmental Risks,* The Council on Environmental Quality, Washington, D.C.,1989, chap. 3.
12. Schettler, T., Solomon, G., Burns, P., and Valenti, M., *Generations at Risk — How Environmental Toxins May Affect Reproductive Health in Massachusetts,* MASSPIRG, Martin/VanderLoop Associates, 1996, Chap. 3.

13. *The Merriam Webster Dictionary*, Merriam-Webster, Inc., Springfield, MA, 1995, p. 183.
14. Hennekens, C.H. and Buring, J.E., *Epidemiology in Medicine*, Little Brown and Company, Boston, MA, 1987, chap. 1.
15. Hennekens, C.H. and Buring, J.E., *Epidemiology in Medicine*, Little Brown and Company, Boston, MA, 1987, chap. 2.
16. Stecidow, S., New Food Quality Act has Pesticide Makers Doing Human Testing, *Wall Street Journal*, September 28, 1998, p. Al.
17. U.S. Environmental Protection Agency, *Risk Assessment*, Washington, D.C., 1992, p. 3.
18. Stern, P.C. and Fineberg, HX, *Understanding Risk*, National Academy Press, Washington D.C., 1996, p. 50.
19. Sharpe, R., 20 Painkillers: How a Drug Approved by the FDA Turned into a Lethal Failure, *Wall Street Journal*, September 30, 1998, p. Al.
20. Cothern, C.R., *Handbook for Environmental Risk Decision Making: Values, Perceptions and Ethics*, Lewis Publishers, New York, 1996, p. 86.
21. Somers, E., Perspectives on risk management, *Risk Anal.*, 15, 677, 1995.
22. Viscusi, W.K., *Rational Risk Policy*, Clarendon Press, Oxford, 1998, p. 98.
23. Morgan, M.T. et al., *Environmental Health*, revised ed., Morton Publishing Co., Englewood, CO, 1997, p. 279.
24. Hallenbeck, W.H., *Quantitative Risk Assessment for Environmental and Occupational Health*, 2nd ed., Lewis Publishers, Ann Arbor, Ml, 1993, p. 2.
25. World Resources Institute (WRI), WRI Project — World Resources Report, World Resource Institute, Washing, D.C., Last updated Sept. 11, 2006, http://forests.wri.org/project_description2.cfm?project ID=86.
26. Peters, P.-G., Covello, V.T., and McCallum, D.B., The determinants of trust and credibility in environmental risk communication: an empirical study, *Risk Anal.*, 17, 43, 1997.
27. DiNardi, S.-P., *Calculation Methods for Industrial Hygiene*, Van Nostrand Reinhold, New York, 1995, p. 76.

13 Environmental Laws and Compliance

CHAPTER OBJECTIVES

A student, after reading this chapter, should be able to:

1. Discuss how a law is made and describe the system of environmental laws.
2. List and describe the major components of the major federal environmental laws including RCRA, CERCLA, EPCRA, SARA Title III, PPA, CAA, CWA, SDWA, stormwater regulations, pesticide regulations, and UST regulations.
3. Describe and discuss the major components of environmental compliance.

INTRODUCTION

Environmental laws are a maze of confusing and overlapping regulations that now cover several thousand pages of federal documents. It is not the purpose of this brief chapter to clarify or detail all of the environmental regulations, but to outline some of the major environmental legal initiatives and compliance issues. In pursuing this objective, the following areas will be covered:

- Environmental laws — some fundamentals.
- Resource Conservation and Recovery Act (RCRA).
- Comprehensive Environmental Responsibility, Compensation, and Liability Act (CERCLA).
- Emergency Planning and Community Right-to-Know Act (EPCRA SARA Title III), and transportation of hazardous materials and wastes.

- Pollution prevention and improved waste management programs and regulation of underground storage tanks (USTs).
- Pesticide regulation.
- Air-quality control.
- Water-quality control, stormwater management and Safe Drinking Water Act (SDWA).

ENVIRONMENTAL LAWS: SOME FUNDAMENTALS

THE MAKING OF A LAW

The average time for an environmental law to be passed at the federal level may take several years. Every federal law starts as a bill and passes through a complex system of checks and balances in the Congress (originally created by the framers of the constitution as a way of dispersing power and preventing tyrants from abusing that power). The bill is first introduced into the House. If it is passed, it goes to the Senate. The bill is then forwarded by the leadership to a standing committee (subcommittee) for review and support. Nearly 90% of bills never make it through this process. Recommended bills are brought forward through hearings for comments and opinions. The committee then meets to markup (discuss and amend) the bill and to vote on it. If the subcommittee and the parent committee in the Senate vote favorably for the bill, it is sent (reported) to the full chamber for debate and a vote. Before a bill is sent to the House, it goes first to the Rules committee, where a time limit is set for debate and indications are also given as to whether floor amendments are allowed. This is not so in the Senate, where riders are often attached to popular bills that may have no relationship to the original bill but often are passed along with it, forcing the president to accept it or veto the entire bill.[1]

If both the Senate and the House pass a bill with some differences (as is usually the case), then they go to a conference committee to resolve these differences. Should these differences be resolved, and the same version approved by both House and Senate, the bill moves forward to the president who may sign it into law or veto it. Although Congress may override a veto by a two-thirds majority vote in each chamber, it is very difficult to do so.

ENVIRONMENTAL LAWS ARE PART OF A SYSTEM

Environmental law is more than a collection of unrelated regulations; it is a system of statutes, regulations, executive orders, factual conclusions, case-specific interpretations, and accepted environmental principles that have evolved into environmental law.[2] Understanding the unifying principles of this system of law is both a challenge and the key to understanding its complexity. Environmental law is a system that uses all of the laws in our legal system to minimize, prevent, punish, or remedy the consequences of actions that damage or threaten the environment and public health and safety. Environmental law encompasses all the environmental protections that originate from (1) the U.S. Constitution and state constitutions; (2) federal, state, and local statutes; (3) regulations published by federal, state, and local agencies;

(4) presidential executive orders; (5) court decisions interpreting these laws and regulations; and (6) common law.[2]

Virtually all of the environmental laws require some form of compliance that falls within one or more of the following areas[2]:

1. *Notification requirements* are meant to deal with any intended or accidental releases of pollutants or hazardous wastes that need to be immediately reported to local and state and/or federal authorities. Failure to do so can significantly impact the environment, endanger public health, and cause significant liability to the persons associated with the release.
2. *Discharge or waste controls* are regulations designed to prevent or minimize the discharge, release, or disposal of wastes and pollutants into the environment.
3. *Process controls and pollution prevention* are regulations that promote minimizing of waste generation, reducing the quantities produced and released, and preventing the release or discharge of pollutants into the environment.
4. *Product controls* can reduce the generation and disposal of solid waste by altering the design or packaging of a product, along with the process by which it may be produced. Some regulations promote the use of less hazardous materials and reduced packaging to achieve these goals.
5. The *regulation of activities* that threaten natural habitat, fragile ecosystems, and vital resources may be necessary to provide protection for resources, endangered species, or habitat. Some of those activities include mining, harvesting of lumber, oil refining, and construction.
6. *Safe transportation requirements* recognize that accidents and spills along our highways occur at an enormous rate, and laws have been established to reduce the risks associated with the transport of hazardous materials. These laws provide for special packaging, labeling, placarding, operator training, and methods of response.
7. *Response and remediation laws* have been promulgated that regulate the cleanup of pollutants and hazardous wastes released to the environment. Laws such as CERCLA have specific provisions for the payment of cleanup costs.
8. These same laws provide for *compensation requirements* requiring responsible parties to pay for cleanup costs for damages done to the health or environment of the private sector. Compensation for damages done to public assets may be recovered by representatives of public interest.[1,2]

The system of environmental law includes federal statutes, executive orders, state laws, tax laws, local and municipal laws, business and regulatory law, environmental law and judicial decisions, common law, and torts.

Federal statutes include such regulations as the CAA or CWA, which establish federal and state regulatory programs that allow states to enact and enforce laws that meet federal minimum standards and achieve the regulatory objectives established by Congress. Most states have taken over regulatory programs, as allowed under federal regulations, and are often more stringent than their federal counterpart. These states are subject to federal intervention only when they fail to enforce the regulations. Executive orders are issued by the president and require federal facilities to provide leadership in protecting the environment and to comply with federal statutes and policies regarding the protection of the environment. They were first used in 1970 and include more than 18 executive orders, including such areas as federal compliance with right-to-know laws and pollution prevention requirements to federal actions that address environmental justice in minority populations and low-income populations.[3]

Tax laws are being used to create incentives for environmentally safe products and activities, and disincentives against environmentally detrimental products and activities. Among the approaches that have been adopted or seriously considered are recycling tax credits, taxes on use of virgin materials, taxes on hazardous waste generation, and excise taxes on various products. Many state attorney generals use their business regulatory authorities to control environmental claims made for products. Full disclosure of environmental liabilities in statements and reports has been required by the Securities and Exchange Commission for some time. Local and municipal laws include zoning and noise control ordinances, nuisance laws, air emission requirements, landfill restrictions or closures, local emergency planning, and product recycling incentives. These are not trivial ordinances and regulations, and must be incorporated into any compliance program. Such laws have been effectively used to control the operation and location of facilities.[2] Court interpretations of these laws and regulations are dynamic and changing and, therefore, judicial decisions have important impacts on the manner in which environmental laws are applied. The environmental system of laws also incorporates common law, which is a body of rules and principles that pertain to the government and the security of persons and property. These basic rules were originally developed in England and were then brought to the American colonies. After the American Revolution, these basic rules were formally adopted and enforced by the states.[2] Civil suits, in which the plaintiff seeks to remedy a violation of his rights, are common law actions.

Under common law, *tort* is a private wrong or wrongful act for which the injured party can bring forth a civil action. There is a general legal duty and responsibility to avoid causing harm to others by acts of omission or commission. When a person's rights have been violated by acts of carelessness, the injured party may seek compensation or restitution by means of a lawsuit. The 1990s have been called the era of "toxic torts" owing to the filing of tens of thousands of lawsuits involving asbestos and toxic chemical litigation cases. The three most common types of torts encountered in environmental law include:

1. *Nuisance*: Generally, a person may use his land or personal property as he sees fit with the limitation that the owner uses his property in a reasonable manner. Whenever a person uses his property to cause material injury or annoyance to his neighbor, his actions constitute a nuisance.
2. *Trespass*: An invasion of another's rights is the general definition of trespass. A more limiting definition of trespass is an injury to the person

or property or rights of another immediately resulting from an unlawful act. Unlawful intent is not necessary to constitute trespass.

3. *Negligence*: This is the portion of the tort law that deals with acts not intended to inflict injury. A case becomes one of criminal law if an intent to inflict injury exists. The degree of care and caution that would be taken or used by an ordinary person in similar circumstances is defined as the standard of care required by law in negligence cases. A defendant is liable if his actions cause injury, and without those actions the end result would not have happened. This involves a natural and continuous sequence of actions, unbroken by intervention. Persons harmed by the careless and improper handling or disposal of hazardous wastes can file suit to recover damages for their losses even if there has been full compliance with all government regulations and permit conditions.[2]

Do you want to learn more about environmental laws and compliance? Point your browser to http://www-unix.oit.umass.edu/~envhl565/weblinks.html and click on any of the Chapter 13 Web links associated with this topic:

- EPA Introduction to Laws and Regulation
- EPA Compliance and Enforcement
- CFR Title 40: Protection of Environment — Full Text
- FindLaw
- Environmental Law Net
- Environmental News Link
- The Environmental Law Institute
- The Practicing Attorney's Environmental Links
- Massachusetts DEP Enforcement Policies

FEDERAL ENVIRONMENTAL LAWS

Managing Hazardous Waste

Hazardous waste management has become a crucial issue. Human health has been threatened and water supplies have been polluted by the improper handling, release, and disposal of hazardous waste. Most hazardous waste is banned from community landfills. Just a few commercial hazardous waste disposal facilities exist, and their total capacity is limited. Liability and costs of waste disposal are increasing. Federal regulations require a cradle-to-grave approach for managing hazardous waste. These regulations establish a system to determine which wastes are hazardous, track wastes from generation to disposal, operate disposal and temporary storage facilities, and set up state hazardous waste programs.[4] An environmentally sound and economic approach to meeting legal requirements for hazardous waste management is reducing the amount of hazardous waste generated by the facility. Hazardous waste minimization is now the law. An integrated hazardous waste management program must be implemented with initiatives for reducing the use of hazardous waste in order to protect the facility and its employees.[5]

RCRA

The federal hazardous waste laws evolved after a series of amendments to the Solid Waste Disposal Act (SWDA) of 1965. The Resource Conservation and Recovery Act (RCRA) was passed as an amendment in 1976 to the SWDA (originally promulgated in 1965). General guidance in meeting hazardous waste requirements was provided to the EPA by RCRA. However, the EPA did not issue key hazardous waste regulations until May 1980. The discovery of hazardous waste problems at Love Canal, NY, in 1979/1980, and at many other locations thereafter, increased pressures for faster regulatory action.[6] The foundation for RCRA waste management regulations can be divided into three distinct programs or subchapters:

1. Subchapter C establishes a system for controlling hazardous waste from the point of generation to the ultimate disposal.
2. Subchapter D establishes systems for controlling solid or nonhazardous waste, such as household waste.
3. Subchapter I regulates underground tank storage of toxic substances and petroleum products.

Subchapter C of the RCRA is concerned with the everyday environmental management of hazardous wastes and involves a cradle-to-grave system that tracks hazardous waste from its origin through its ultimate disposal. Facilities must comply with regulations if they generate, transport, store, treat, and dispose of hazardous waste in amounts specified under the RCRA. Hazardous waste generators can be separated into three principal categories (state regulations may provide somewhat different designations and responsibilities to the three groups):

1. Large quantity generators (LQG) produce 1000 kg (2200 lb) or more of hazardous waste, or more than 1 kg (2.2 lb) of acutely hazardous waste (AHW) in a calendar month.
2. Small quantity generators (SQG) produce 100 to 1000 kg of hazardous waste and less than 1 kg of acutely hazardous waste in a calendar month.
3. SQGs that are conditionally exempt include those that generate no more than 100 kg per calendar month of hazardous waste or no more than 1 kg per month of AHW. These generators may be exempted from RCRA requirements provided their wastes are properly managed (e.g., sent to a solid waste disposal facility).

The administration of RCRA is done by the EPA, and state programs must be consistent with federal RCRA regulations. Comprehensive cradle-to-grave regulation of hazardous wastes is provided for in most state laws. The lead agency in implementing RCRA is usually the State Environmental Regulatory Agency. In general, most state hazardous waste regulations closely parallel federal provisions of the RCRA.[5] Within some states, special problems are addressed by the addition of more stringent requirements. Hazardous waste management has been changed in many ways. There is a ban on land disposal of certain wastes, SQGs are under tighter controls, and regulation of used oil as a hazardous waste has gone into effect in some states.

Identifying a Hazardous Waste

There are two types of hazardous wastes: (1) listed and (2) characteristic. Listed wastes include hundreds of hazardous wastes, including wastes from nonspecific sources (F wastes), manufacturing processes (K wastes), and discarded chemical products (P and U wastes), which are listed in the RCRA regulations. The Code of Federal Regulations (CFR), Title 40, Part 261, Subpart D, contains these lists of hazardous wastes. Products which are not listed may be characteristically hazardous. Under the direction of RCRA, the EPA developed generalized criteria for evaluating the potential hazard produced by a waste. The four criteria include:

1. *Ignitability*: easily catches fire, with a flash point of less than 140'F.
2. *Corrosivity*: easily corrodes materials or human tissue and very acidic or alkaline (pH less than 2 or greater than 12.5).
3. *Reactivity*: explosive, reacts with water or acid, and unstable.
4. *Toxicity*: causes local or systemic damages that result in adverse health effects in an organism, i.e., a poison (such as oil), an asphyxiant (such as carbon monoxide), a mutagen (a chemical that alters DNA), teratogen (causes birth defects), or a carcinogen (causes cancer). Toxicity is determined by a test known as the toxicity characteristics leachate procedure (TCLP).

The EPA has developed lists of hundreds of hazardous wastes using these criteria. Generators of nonlisted waste must analyze the waste to determine if it exceeds the EPA standards. If the waste exceeds any of the EPA's four criteria, it is considered a *characteristic hazardous waste*.

Tracking Hazardous Waste

All small and large quantity generators, transporters, and TSD (treatment, storage, and/or disposal) facilities must obtain an EPA identification number, as required by most state environmental agencies and RCRA. These ID numbers are used by state agencies and the EPA to track the flow or whereabouts of hazardous waste. This tracking is accomplished by using a paper trail created by shipping manifests. RCRA assures uniform record keeping for all shipments of hazardous wastes. A Uniform Hazardous Waste Manifest must be used for all shipments of RCRA-regulated hazardous wastes. State manifest forms, adapted from the EPA form, supplement this documentation in a number of states (Figure 13.1). Each shipment of hazardous waste requires an EPA manifest form.[6]

Other Requirements under RCRA

Accurate record keeping is essential to a successful hazardous waste management program. Complete and up-to-date records are necessary to ensure compliance with

FIGURE 13.1 The RCRA cradle-to-grave system tracks hazardous waste from its origin through its ultimate disposal by a manifest system.

environmental regulations. Shipments of hazardous waste need to be properly labeled, marked, and placarded according to the U.S. Department of Transportation (DOT) requirements. The facility managers remain responsible for the proper management of waste after it leaves the property and, thus, great care must be taken in selecting a waste hauler and TSD facility.[7]

CERCLA

The cleanup of inactive and abandoned hazardous waste sites is provided for by the Comprehensive Environmental Response, Compensation, and Liability Act (CERCLA), the Superfund Act of 1980. CERCLA and RCRA have the same basic objectives. However, the purpose of RCRA is to prevent hazardous waste problems at active sites, whereas CERCLA is designed to provide cleanup of existing problems at inactive sites. CERCLA also provides a mechanism of establishing financial liability to any or all parties responsible for the release of hazardous substances. These potentially responsible parties (PRPs) are liable for the costs of cleaning up those substances. These strict, joint, and liability provisions of CERCLA can result in costly investigative and cleanup activities at waste disposal sites. According to the EPA, the average cost per Superfund site is $35 million, and it is therefore important to manage wastes in a way that reduces the potential for exposure to these costs.[6]

The Superfund Amendments and Reauthorization Act (SARA) of 1986 amended and strengthened the CERCLA of 1980, the basic Superfund law. The regional EPA is the state's lead agency for implementation of the Superfund. Although crude oil, petroleum products, and natural gas products are excluded from this list, oil spills are addressed by the National Oil and Hazardous Substances Pollution Contingency Plan.[6]

Steps in Superfund: Find, Prioritize, and Clean

The first step in Superfund was to find the hazardous waste sites. The identification of hazardous waste sites is accomplished in three major ways. First, the EPA required all owners and operators (including former owners and operators) of hazardous substance treatment, stoage, and disposal (TSD) facilities to report the existence of these facilities by June 11, 1981. Second, the EPA regulated TSD facilities under the RCRA, and these reported facilities were added to the list. Third, state and local governments, as well as the public, have identified many additional hazardous waste dump sites and facilities. The CERCLA Information System (CERCLIS) database incorporates information for approximately 30,000 sites. Additionally, the EPA established specific reportable quantities (RQs) for the release of hazardous substances. Under the provisions of the Superfund, any releases of hazardous substances larger than these RQs must be reported to the National Response Center by the owners and operators of the facilities or vessels that release hazardous substances (except under federal permit). The center then notifies other appropriate agencies and begins any necessary emergency response or cleanup actions. Once sites are identified, the EPA is required to collect data and make a preliminary assessment report. The sites are then ranked using a risk assessment procedure for their likelihood of causing human health impacts or environmental damage. The sites with the highest ranking are placed on the national priority list (NPL) and are targeted for cleanup with monies coming from PRPs or from Superfund monies derived from taxes on chemical feedstocks and crude oil supplies. Therefore, the next step in the Superfund process is the identification of those parties responsible for site contamination. These PRPs and responsible parties identified by the EPA and state agencies can be required to finance cleanup activities, either directly or through reimbursement of federal Superfund expenditures. The final steps involve the cleanup of contaminated sites. The program includes extensive provisions for site investigations, selection of cleanup sites, and establishing the levels of cleanup to be attained.[6]

EPCRA SARA Title III

The possibility that mishaps at facilities handling hazardous chemicals could harm neighboring communities was brought to worldwide attention with the December 1984 toxic gas disaster in Bhopal, India. In response to this disaster, U.S. federal, state, and local governments created a variety of laws to improve accident prevention and emergency response planning activities by chemical-handling facilities and local

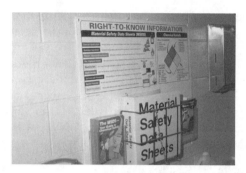

FIGURE 13.2 Material safety data sheets (MSDSs) on hazardous chemicals in the workplace must be made available to employees in a convenient place.

governments. Right-to-know laws increased public access to information about the storage and use of hazardous chemicals.

Title III of the SARA of 1986 was the first congressional response to the Bhopal accident. A federal program subtitled the EPCRA of 1986 was mandated by SARA Title III. Title III was one of those laws that was appended to the SARA legislation as a way of getting it passed, and it really has very little to do with the Superfund program. SARA Title III is focused on worker safety regulations. The Occupational Safety and Health Administration (OSHA) took the lead in developing the regulations because these new requirements were extensions of OSHA's traditional work safety roles. Identical requirements have been adopted by the EPA and are applied to state and local government employees (over whom OSHA has no jurisdiction). The purpose of Title III was to ensure the public and emergency response agencies that information regarding hazardous chemicals would be made available to them, thereby lessening the likelihood of Bhopal-like disasters in the U.S. The EPA and the individual states are responsible for these efforts. The EPA enforces SARA's requirements for compliance. Any facility that stores or handles extremely hazardous substances (EHS) at or in excess of EPA-prescribed threshold planning quantities (TPQs) must comply with SARA's release, planning, and release reporting requirements. This also covers unauthorized discharges (leaks or spills) of any EHS or of any substance already subject to emergency notification requirements under section 103(a) of the CERCLA. Any discharge of RQ must be immediately reported to appropriate local emergency response agencies (police, fire department, etc.) and national and state emergency response agencies, as well as to state and district Title III agencies.

Title III reporting requirements (SARA sections 311 and 312) and the federal Occupational Safety and Health Act of 1970 requires all facility owners or operators to prepare or have available material safety data sheets (MSDSs) (Figure 13.2). Businesses are required to provide either an MSDS for each on-site chemical in excess of its threshold quantity or a list of all such chemicals to the state and district Title III agencies and to the local fire department.

The reporting requirements in Title III (section 313) require selected businesses to file annual reports with the EPA and the state on the EPA Form R. These reports must include the following information for the current and previous calendar years:

- Estimated quantities of both routine and accidental releases of listed toxic chemicals
- The maximum amount of listed on-site chemicals during the calendar year
- The amount of listed chemicals contained in wastes transferred off-site

These reports estimating releases during the previous calendar year are due each July 1.

Beginning January 1, 2000, the EPA required industries to release fuller reports on 27 additional chemicals, including the endocrine disruptor, dioxin. These regulations required industries to report if they use 100 lb per year or more of these additional chemicals. In the case of dioxin, the industry must report if one-tenth gram is released.

TRANSPORTATION OF HAZARDOUS MATERIALS

DOT regulates the transportation of hazardous wastes and materials, and it is within this department that the Hazardous Materials Transportation Act (HMTA) is administered. States are allowed to adopt standards that are more stringent than federal regulations. State provisions are often similar to the federal standards. The regulations of the transportation of hazardous wastes can differ from state to state.[6]

What are hazardous materials under HMTA? As defined by HMTA, hazardous materials are those which are capable of creating an unreasonable risk to health, safety, or property during transportation. Any hazardous substance listed under DOT regulations (49 CFR sections 172.101, 172.102) that is transported in a package containing a quantity that equals or exceeds the listed RQs is considered a regulated material. Each hazardous material is designated by DOT and, depending on its risk to health, safety, or property, is assigned to classes and divisions. Each class has specific packaging, handling, labeling, and registration requirements.

Packaging, labeling, and construction requirements must now meet United Nations (UN) recommendations. There are three packaging groups under the UN-based regulations: Group I for the most hazardous to Group III for the least hazardous. Shipping papers must accompany every shipment. Basic information must be provided within the papers, including name of shipper, shipping names of materials, hazard class, standard ID numbers of materials under either the UN or NA (North American) system, packaging group (if there is one), quantity of each material being shipped, and the emergency response telephone number.

All containers being shipped must be appropriately marked and labeled. The labels must include the proper shipping name, ID number, and the consignee's name and address. The labels are distinctively diamond shaped and must clearly identify the hazard class. Even the size, color, and text of each label is specifically laid out by DOT (Figure 13.3). Transport vehicles must also display placards in a specified pattern. Emergency response information must also accompany each hazardous material being transported. The transporting company must also provide a monitored 24-h emergency response telephone number.[7]

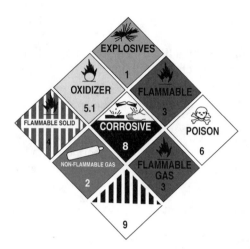

FIGURE 13.3 Examples of several distinctive diamond-shaped shipping labels as required under DOT.

POLLUTION PREVENTION AND IMPROVED WASTE MANAGEMENT PROGRAMS

The Pollution Prevention Act (PPA) of 1990 has become integrated into today's environmental regulations at both the state and federal levels. *Pollution prevention*, also referred to as *source reduction*, is defined as the deliberate decrease in the amounts of hazardous substances which enter the environment via recycling, treatment, or disposal.[7] The emphasis of the PPA focuses on remediation and treatment, rather than the prevention of pollution. Any pollution that cannot be prevented should be recycled, treated, and disposed of in a way that is not harmful to the environment. Mandatory requirements, as well as voluntary programs, to minimize pollution were established. The Act provides financial assistance to states for their role in prevention. Businesses are encouraged to share and transfer source reduction technology.[8]

For example, in response to the environmental problems that methyl tertiary butyl (MTBE) ether is causing in California, a new biorefinery is beginning to convert rice straw, orchard prunings, and other agricultural wastes into ethanol. This will be used as an additive in gasoline in the place of MTBE. The plan is to phase out MTBE and replace it with the cleaner alternative, ethanol.

Source reduction must be reported by facilities generating over a certain amount of toxic emissions during the previous calendar year. The form, referred to as Form R, must be filled out for each toxic chemical to include information such as the amount of toxic chemical recycled and treated. Also included is the amount of toxic chemical that entered a waste stream, percentage change from the previous year, and the amount anticipated for the following 2 years. Recycling efforts as well as source reductions must be reported for the previous calendar year.[4]

TOXIC SUBSTANCES CONTROL ACT (TSCA)[6]

There are over 70,000 chemicals listed in the TSCA inventory. TSCA mandates that manufacturers of chemicals develop safety and health data on chemicals and mixtures, and requires the EPA to regulate chemical substances and mixtures that present an unreasonable risk of injury to health and the environment. The EPA is also directed to act when chemical substances and mixtures pose imminent hazards, and is authorized to restrict the production and use of such hazardous materials. The EPA has done this by banning the use of polychlorinated biphenyls (PCBs) and asbestos-containing products. Congress intended that the EPA not abuse this authority and should not unnecessarily impede or create economic barriers to technological innovation.

There are a number of reporting and record-keeping activities and data collection procedures under TSCA. Additionally, there are mechanisms for public information and citizen suits to compel the EPA act when TSCA is violated. The basis of the reporting system is the TSCA inventory, originally compiled during 1977 and consisting of two lists: (1) a main inventory and (2) a confidential inventory for all regulated substances that are protected as trade secrets.

Since 1977, each new toxic substance falling under TSCA regulations needs to be added to the list. Before new chemical substances can be introduced into commerce, TSCA requires premanufacture notification, as well as notification of significant new uses of existing chemicals. TSCA regulations stipulate guidelines and standards for chemical testing and testing laboratories, rules for the reporting of chemical information, and the use of health and safety data.

TSCA gives the EPA the power to regulate, restrict, and even prohibit the manufacture, importation, and use of certain toxic chemicals. Under this mandate, the EPA has enacted strict regulations for:

- The phaseout, removal, and disposal of PCBs
- Asbestos and asbestos-containing materials
- Fully halogenated chlorofluorocarbons (CFCs; once used as aerosol propellants, responsible for ozone depletion in the upper atmosphere)
- Dibenzodioxins and dibenzofurans (contaminants present at Times Beach, MO[2])

REGULATION OF USTs

In the late 1970s and early 1980s, lawmakers became aware that thousands of tanks located underground and containing petroleum products such as gasoline were beginning to leak or were in danger of leaking and contaminating groundwater. The lawmakers developed standards for storage tank construction, installation, and removal, and defined a UST as any tank or underground piping connected to the tank that has a minimum of 10% of its volume located underground. The federal UST law is a component of the Hazardous and Solid Waste Amendments of 1984 (HSWA) under RCRA. This is known as *Subtitle* I, and this subtitle deals with USTs that store petroleum and other hazardous substances, but not hazardous wastes, which are regulated by other sections of RCRA. The EPA issued its final technical requirements for USTs on September 23, 1988. These regulations include detailed technical standards for new USTs installed after December 22, 1988, and retrofit requirements for tanks already in service on that date. These particular regulations only applied to the storage of petroleum or certain hazardous chemicals. The hazardous chemicals are the same as those listed under CERCLA, except for those chemicals listed as hazardous wastes. All of the UST regulations also applied fully to the underground piping, which is associated with the tanks in question. These standards also included tank registration and monitoring. Regulations were initially enforced secondary to tanks that were leaking, thereby causing contamination of groundwater. Aging storage tanks and steel associated with gas piping and oil become a hazard when these pipelines begin to leak and contaminate soil and water supplies.

There are several types of tanks and piping systems that are not included by the UST regulations, and they include septic tanks, tanks for heating oil for on-site use, and farm and residential fuel tanks under 1100 gal.

Leak detection devices had to be installed in all USTs as per federal mandates by December 1993. At the same time, federal and state regulations enforced the use of overfill protection or spill protection. Two categories of record keeping must be maintained including records generated at the point of old tank removal and new tank installation and records of proof of maintenance of monitoring devices, inventory reconciliation, and registration and permit renewal.

All tanks installed after December 22, 1988, are required to have the following construction standards: (1) new underground piping must be corrosion resistant, (2) new USTs must have both spill and overfill protection, and (3) leak detection provisions must be in place. Also included in the standards are regulations for new petroleum USTs. These USTs may continue to be single-lined, whereas all other USTs must now be double-lined for both primary and secondary containment (Figure 13.4).[4]

PESTICIDE REGULATION

The basic national framework for pesticide control is provided by the Federal Insecticide, Fungicide, and Rodenticide Act (FIFRA), which is administered by the EPA in cooperation with state and local agencies. Initially enacted in 1947, FIFRA has been amended several times. Pesticides are chemicals specifically formulated to be toxic to living things and are, therefore, subject to regulation. Pesticide regulations are intended to protect public health, livelihoods, and the environment by ensuring that workers are trained in proper application techniques, pesticides are properly handled and stored, and the location or content of chemicals is made known. Pesticides are known to emergency response units, medical personnel, and workers who may be exposed to the chemicals.[6]

Applicators of restricted-use pesticides must be certified based on demonstrated competency in the use of pesticides. Regulations also establish worker safety precautions, including requirements to warn field workers before pesticide applications, and to prevent exposure to workers not actually involved in pest control activities. The EPA uses the authority under FIFRA to collect information necessary to register and control the active ingredients in pesticides, whereas state and local agencies control the registration and actual use of the pesticides themselves. Congress has directed the EPA to develop regulations for the design of pesticide containers to promote safe storage and disposal of pesticides, and standards for the removal of

FIGURE 13.4 A typical underground storage tank (UST) installation meeting current regulatory requirements.

pesticides from containers prior to disposal. States that have been delegated primary FIFRA enforcement have the responsibility to incorporate and enforce these new requirements within 2 years.

AIR QUALITY CONTROL

A comprehensive national framework for protecting air quality was created by the FCAA. In 1990, amendments were enacted to measure compliance. There are eleven sections, referred to as *titles*. The first seven titles are most pertinent to most facilities and are briefly described below. The CAA was first enacted in 1963, but its scope was significantly expanded in 1970 and 1977. The 1990 CAA amendments include the following titles.[9]

Title I: Provisions for Attainment and Maintenance of the NAAQS

The EPA monitors six indicator pollutants for the primary protection of human health: (1) ozone, (2) carbon monoxide (CO), (3) particulate matter (PM_{10}), (4) sulfur oxides (SOx), (5) nitrogen oxides (NOx), and (6) lead (Pb). Areas where pollutant concentrations exceed the national ambient air quality standards (NAAQS) are referred to as *nonattainment areas*. These areas are categorized according to the severity of their air pollution problems. Deadlines by which NAAQS must be reached for these areas fall between 1993 and 2001. Later deadlines have been assigned to those areas with the most serious pollution problems. Nonattainment areas that move too slowly toward attainment, or those that fail to develop acceptable attainment plans, face the EPA sanctions. Currently, there are roughly 100 nonattainment areas, home to some 150 million people. Five classes of ozone violations (marginal, moderate, serious, severe, and extreme) and two categories (moderate and serious) for CO and PM_{10} are established by these amendments. Compliance deadlines and control requirements exist for each category. Severely polluted areas are given more time to comply, but they must adopt more stringent controls for ozone, CO, and PM_{10}. The CAA directs the EPA to set air quality and emissions standards at levels that protect public health and the environment.

Other Related Programs Established under Previous Air-Quality Regulations

The EPA gives states with approved *state implementation plans (SIPs)*, as required under the 1970 CAA, the authority to run federal clean air programs. The air quality in designated regions, basic strategies to achieve or maintain NAAQS, and enforcement strategies are described in a SIP and every state must submit a SIP, to the EPA. In various states, SIPs are the blueprints for achieving air-quality goals. Standards more stringent than those required by CAA may be imposed by the states. Once a state submits a SIP or amendment, the EPA either approves it or orders the state to revise it. If a state fails (or refuses) to submit an acceptable SEP, the EPA may either draft the SEP for that state or write (and implement) a federal implementation plan (FIP). SIPs are reviewed and revised continually. Each state must identify in its SIP, emission controls and other measures that prevent significant deterioration (PSD) of air quality with respect to two criteria of pollutants: SO_2 and PM_{10}. The PSD of air quality in pristine areas is the focus of this CAA program. Special attention is directed to national parks, national wilderness areas, national monuments, national seashores, and other areas of special natural, recreational, scenic, or historic value by the PSD program. A region's classification (Class I, II, or III) determines the allowable levels of SO_2 and PM_{10} for that region. The most severe restrictions are imposed on Class I regions; further construction of major stationary sources in these regions without a permit after PSD review is generally prohibited.

Emissions standards for new and modified sources of air contaminants are set by the EPA. New and modified sources are forced to achieve the greatest emissions reductions that are technologically and economically feasible by these new source performance standards (NSPSs). States are required by CAA to develop emissions standards for existing sources. If a state fails to submit a satisfactory SIP, the EPA may set these standards. New sources are typically subject to stricter emissions standards than existing sources, because new sources are considered to have greater flexibility to incorporate innovative emission control technologies. The 1990 CAA amendments add important new controls on stationary source.

The EPA announced new NAAQS for ground-level ozone and PM_{10} on July 17, 1997. Ground-level ozone is a major component of smog that is photochernically produced as a secondary pollutant of the stratosphere from the interaction of sunlight, NOx, and hydrocarbons. The previous ozone standard was last revised in 1979 and set at 0.12 ppm for 1 h. The previous standard did not adequately protect the public health, and the EPA replaced it with an 8-h standard set at 0.08 ppm. Similarly, the EPA announced a new standard for PM_{10} under the NAAQS based on the review of hundreds of scientific studies. The EPA has added a new annual $PM_{2.5}$ standard set at 15 $\mu g/m^3$ and a new 24-h $PM_{2.5}$ standard set at 65 $\mu g/m^3$. The EPA is retaining the current annual PM_{10} standard set at 50 $\mu g/m^3$.

Clear Skies Program

There has been a great deal of controversy since President Bush first announced the Clear Skies Act on February 14, 2002. Clear Skies legislation was first introduced in both Houses of Congress in July 2002, and reintroduced in February 2003. On January 24, 2005, Senators Inhofe and Voinovich introduced their version of the Clear Skies Act (S 131). The Senate Environment and Public Works Committee held hearings on this legislation on January 26 and February 2, 2005, respectively. The latest major action on this legislation at the writing of this text came on March 9, 2005, when the Committee on Environment and Public Works considered the bill and held a markup session.[10]

According to the EPA and Bush administration officials, the Clear Skies Act would:

- Establish federally enforceable emissions limits (or "caps") for three pollutants including sulfur dioxide (SO_2), NOx, and mercury. Advocates claim that the emissions reductions from Clear Skies would help to alleviate our nation's major air pollution–related health and environmental problems by targeting fossil-fueled electric generators and reducing fine particles, ozone, mercury, acid rain, nitrogen deposition, and visibility impairment. Mercury requirements would affect only those power plants that are coal fired.
- Use a regulatory approach known as *emission caps* and *trading* that provides power plants with some flexibility in reducing emissions in the least costly way.
- Maintain the authority of state and local government to set source-specific emissions limits for sources within their borders to ensure that NAAQS are met.

Administrators claim that Clear Skies would:

- Provide immediate benefits to human health and the environment when it is passed. EPA claims human health benefits would save $110 billion per year by 2020 in health-related costs, lost work, and other health considerations. This would be substantially greater than the annual costs of $6.3 billion associated with the costs of meeting the Clear Skies requirements.
- Help to avoid from 8,400 to 14,000 premature deaths by 2020, saving at least $21 billion in health benefits.
- Reduce hospital and emergency room visits by 30,000 and decrease the days with respiratory symptoms (including work loss days and school absences) by 12.5 million.

- Increase the number of people that breath air, meeting the national ozone and fine particle standards by 18 million.[11]

Environmental groups such as the Sierra Club and the National Resources Defense Council (NRDC) claim that the Clear Skies program would actually result in greater pollution than enforced current air pollution laws. The Sierra Club claims that by the 15th year of the Clear Skies program, there would be 450,000 tons more of NOx, 1 million tons more of SO_2, and 9.5 tons more of mercury if there were strong enforcement of existing CAA programs.[12,13]

Title II: Provisions Relating to Mobile Sources

Pollution reduction requirements for motor vehicles are defined by Title II. These requirements are of great significance to the facility and its large vehicle fleet. In areas with CO levels above 16 ppm or in areas designated as seriously, severely, or extremely above the limit for ozone, Title II requires centrally fueled fleets to bum only clean fuels.

Vehicle Emissions Requirements

The 1990 CAA pollution standards are stricter than the 1970 act for automobile and truck emissions. Tail pipe emissions of hydrocarbons, CO, and NO are expected to be reduced by these standards. New vehicles purchased by the facility (cars and light duty trucks up to 6000 lb) have stricter emission controls. The maintenance of appropriate emissions standards for nonmethane hydrocarbons, CO, and NO_x will affect fleet management.

Inspection or Maintenance Requirements

As mandated by the CAA, the Inspection and Maintenance (I&M) program requires periodic inspections of vehicles to ensure that emissions of specified pollutants are not exceeding established limitations. All ozone and CO nonattainment areas in the U.S. are required to have an I&M program. Annual inspection programs are required unless a state can demonstrate that a biennial program is just as effective.

Reformulated Gasoline Requirement

In 1995, reformulated gasoline with 2% minimum oxygen content was required during the winter months in nonattainment areas for CO. Presently being sold as gasoline, this oxygenated fuel contains 10% ethanol or 15% MTBE as additives.

Title III: Hazardous Air Pollutants

Hazardous air pollutants are defined as those substances that are hazardous to human health or the environment, but are not specifically covered under other portions of the CAA. A list of 189 toxic air pollutants for which emissions must be reduced is included in this law. A more comprehensive program for regulating HAP emissions was introduced by the 1990 amendments. All new and existing stationary sources that emit HAPs must comply with air toxic emissions standards to be adopted by the EPA.

Maximum achievable control technology (MACT) will have to be applied by stationary sources to control their HAP emissions. MACT is a standard that the EPA uses to define various categories of sources. The EPA was to produce a list of all categories of stationary sources that emit HAPs by November 1992. The EPA must adopt MACT standards for at least 40 categories, and must then adopt MACT standards for 25% of the remaining categories by November 1994, with another 25% due by November 1997, and the balance by November 2000.[6] Area-wide control strategies must be formulated by the EPA by the year 2000. The strategy must identify at least 30 HAPs emitted by area sources presenting the greatest threat to public health in most areas, and identify source categories emitting 90% or more of each of the 30 HAPs. The strategy must also show how the EPA intends to reduce by at least 75%, the cancer risk posed by the release of HAPs from stationary sources. Sources will have 9 years to comply with the strategy. Reports to Congress that identify high-risk metropolitan areas and describe actions being taken to reduce health risks posed by HAP emissions are due in 1998 and 2002.[6]

Title IV: Acid Deposition Control

Title IV regulates the sources of acid deposition. Most of the acid deposition in the U.S. comes from the burning of fossil fuels. The operation of facility boilers is affected by these Title IV regulations. Emissions of SO_2 and NO from fossil-fuel-fired electric utility plants are the leading cause of acid rain. An ambitious program to reduce and control these emissions is included in the 1990 amendments. However, the program does not affect existing units with a generating capacity of 25 MW or less, qualifying cogeneration units, or qualifying small power producers.

The reduction of SO_2 and NO_X emissions is to be done in two phases, the first beginning in 1995 and the second in 2000. Each SO_2 source will be given tradable allowances equal to the number of tons of SO_2 it is permitted to emit. If emissions are reduced below this amount, a source can sell its extra allowances to another source. New SO_2 sources will have to purchase allowances from existing sources for their emissions. Certain boiler units in Phase I and others in Phase II must meet the EPA emissions standards to achieve NO_X reductions.

Plans for achieving the required emission reductions must be prepared and submitted to the EPA by units covered by the SO_2 and NO_X regulations. These same units must obtain air emissions permits (see following section), install and operate continuous emission monitoring systems, and submit detailed records of their SO_2 and NO_X emissions. Clean coal technology demonstration projects are exempt from NSPSs and nonattainment requirements, therefore, SO_2 and NO_X sources are encouraged to switch to clean coal technologies.

Title V: Permits

Title V requires all major sources of air pollutants to obtain permits, and extends emissions controls to thousands of sources in many unregulated areas. For the first time, a nationwide program of air emissions permits had been established by the 1990 amendments. Following the EPA guidelines, the states were encouraged to set up their own permit programs, which are subject to the EPA oversight. Each state proposed a permit program, and faced sanctions if they failed to submit acceptable permit programs covering nonattainment areas.

Permit holders must certify that the facility complies with the terms of its permit on an annual basis. Under certain conditions, compliance with the permit terms will be considered compliance with CAA. Monitoring progress reports must be submitted at least once every 6 months by permit holders.[6]

Title VI: Stratospheric Ozone Protection

A complete production phaseout of ozone-depleting chemicals (especially CFCs and halons) is required by Title VI. A new national program designed to minimize human impact on the stratospheric ozone layer by phasing out the use of CFCs was introduced by Title VI of the 1990 CAA amendments. Congress originally directed that parts of the phaseout be completed by the year 2000. The Bush administration accelerated the phaseout schedule for certain listed CFCs in 1995, based on new evidence concerning the deterioration of the ozone layer. The U.S. signed an international treaty on preventing ozone depletion in 1987 known as the *Montreal Protocol*. The accelerated phaseout schedule keeps the U.S. in terms with this ozone treaty. Ozone-depleting chemicals are divided into Class I and Class II substances, and phaseout schedules for each class are prescribed by Title VI. The greatest threat to the ozone layer is Class I chemicals, which were not to be produced after December 31, 1995. Methyl chloroform is an exception, and its production may continue until January 1, 2002. Meanwhile, specific annual percentage reductions are required. Beginning January 1, 2030, all production of Class II substances is banned by Title VI.

Title VII: Provisions Relating to Enforcement

The amendments contain a broad array of provisions that brings the law up to date with the other major environmental statutes, and thus makes the law more readily enforceable.

WATER-QUALITY CONTROL

The basic framework of the Federal Water Pollution Control Act (FWPCA), passed by the Congress over President Nixon's veto in 1972, remains today as a part of the modern CWA. Effluent limitation guidelines, water quality requirements, and the permit program are continually enforced. The goals of the CWA are to eliminate the discharge of pollutants into surface waters and to achieve water quality that "provides for the protection and propagation of fish, shellfish and wildlife," and "for recreation in and on the water."[10] The act also incorporates the prohibition of the "discharge of toxic pollutants in toxic amounts."

The CWA only applies to surface waters and those groundwater, which is hydrostatically connected to surface water. The SDWA provides federal protection of groundwater sources. One priority of the CWA is the discharge of mainly industrial pollutants into our nation's waters, referred to as *point sources*. These controls are also known as *technology-based effluent limitations*. The CWA ensures that the water-quality standards for the waters receiving the discharge are met, and that the discharger applies the required technology-based standards to the facility through application of a comprehensive permitting scheme known as the *National Pollutant Discharge Elimination System* (NPDES).[6]

The second major focus is that of water quality–based controls. The CWA describes three categories of pollutants including (1) conventional pollutants (biochemical oxygen demand, total suspended solids, pH, fecal coliform, oil, and grease); (2) toxic pollutants, including the 65 pollutants that are identified by the EPA; and (3) nonconventional pollutants (all pollutants not categorized as either toxic or conventional).[4]

The CWA allows states to apply to the EPA for authorization to administer various aspects of the federal NPDES program. Generally, for a state to obtain the EPA authorization to administer all or part of the permitting program, it must match the requirements mandated by CWA or apply more stringent ones. The owner or operator of a facility must obtain a permit under the NPDES program requirements prior to discharging any pollutant into the waters of the state from any point source. Both CWA and most state acts broadly define these terms to regulate almost all activities that result in the release of pollutants from a discrete point (e.g., a pipe).[6]

Stormwater[14]

Water quality may be degraded and pollutant concentrations may exceed water quality standards as a result of polluted stormwater. Pollutants may dissolve in the stormwater, become suspended, or float on the surface. Pollutants that could potentially occur in stormwater draining from sites include sediments, solids, nutrients, oxygen-demanding materials, bacteria, hydrocarbons, heavy metals, herbicides, pesticides, and other organic substances. The runoff and pollutants enter storm drains (Figure 13.5), and are then discharged into receiving waters, such as streams and lakes. During dry weather periods, pollutants can also enter the storm drain system as a result of improper connections or illegal dumping.

In 1987, Congress amended the CWA, requiring the EPA to establish phased NPDES requirements for stormwater discharges. NPDES permits are required for the discharge of pollutants from any point source into waters of the U.S., which includes almost any surface water or wetland. These permits have provisions for monitoring the discharge of pollutants into U.S. waters and for establishing source controls where necessary. On November 16, 1990, the EPA published a final rule in the Federal Register (55 CFR 47990) that contains permit requirements for stormwater discharges. Facilities regulated under the new federal stormwater program must comply with the state and/or federal stormwater regulations. There are three distinct types of stormwater permits established by the federal regulations: individual, general, and group.

Oil and Hazardous Substance Spill and Reporting Requirements[6]

The CWA contains specific provisions regulating the handling of oil and hazardous substances. Specific penalties and rules of liability for unauthorized releases of these materials are established by CWA. Similar to the requirements outlined under Super-

FIGURE 13.5 Stormwater runoff is a significant contributor to water pollution and is regulated under amendments to the CWA.

fund, these provisions focus on reporting unauthorized leaks, spills, and discharges into water (discharges without a permit). Under the CWA, any person in charge of a vessel or an onshore or offshore facility who has knowledge of an unauthorized release of oil or any hazardous substance must immediately report the release in a manner consistent with the National Oil and Hazardous Substances Pollution Contingency Plan (commonly called the *National Contingency Plan*).

The Oil Pollution Act of 1990 was created in response to the Exxon Valdez oil spill in 1989, which dumped 11 million gal of oil into the waters off South Central Alaska. Media coverage of black oily beaches and the resulting deaths of thousands of birds, sea otters, whales among other wildlife created public outrage. This, in addition to the size of spills and location, led to OPA 1990. This act provides tougher penalties and liabilities for oil spills, allocates more funds and other resources for dealing with spill cleanup, and places more responsibility on the executive branch to respond to these incidents.

COMPLIANCE STRATEGIES[15]

Since the 1970s, employers have been given increased responsibilities. Managers are becoming more responsible for environmental health and employee safety. Various agencies have implemented regulations that are enforced to further protect the employee and the environment. Employers have developed worker safety programs, and the method in which these programs are enforced can directly impact worker health and safety, liability and insurance premiums, worker's compensation, remediation costs and fines, and costs for meeting regulatory requirements. However, it is not uncommon that such programs result in a negative company image owing to the inadequacy of the implemented programs. Unfortunately, these programs are often not given enough attention and support. Additionally, supervisor and worker training programs are frequently insufficient.

There are criminal penalties for noncompliance with environmental health and safety programs. There has been a tremendous increase in liability for both the federal agencies and for individuals from the civil and criminal provisions of recently reauthorized environmental legislation. Even though the thousands of pages of regulations are difficult to interpret, they still apply. Some of the regulatory initiatives include pollution prevention and waste minimization in facilities. The regulations are enforceable throughout all facets of a facility's operation including air, water, and solid waste disposal. The Community Right-to-Know Act plays a large role in this process, requiring facilities to publicly report their waste production for hazardous wastes, whereas the PPA ensures that waste is reduced through raw material substitution, process modification, and recycling. The RCRA targets recycling programs.

After gaining a basic understanding of the systems of environmental laws, employers need to apply their knowledge in order to remain in compliance. The development of such strategies for compliance is based on the premise that the most effective protection against aggressive enforcement and other efforts to assess liability is aggressive compliance.

The government institutes have published a variety of books that discuss and detail the elements of appropriate compliance organization. There are several basic principles with which one should become familiar: (1) The first principle is that environmental law compliance is the responsibility of everyone; (2) the second principle describes the need to demonstrate a concern for compliance by providing appropriate education and training; (3) the third principle focuses on the prevention of violations and minimization of liability through aggressive implementations of environmental objectives; and (4) the fourth principle recommends periodic audits to verify compliance and identify areas that can be improved upon. These audits will be conducted to address environmental issues and controls, such as asbestos, indoor air quality, and UST regulations.[2]

TRENDS IN REGULATORY COMPLIANCE

The states are responsible for nearly 90% of regulatory inspections involving air, water, and hazardous wastes, with the EPA conducting the remainder of the inspections. Routine inspections usually consist of sampling analysis and record keeping. Inspections may also include investigations of operation and maintenance programs. The EPA has developed target strategies, based on the quantity and severity of toxic materials discharged, emitted, or disposed off by a facility, as reported under the Community Right-to-Know Act. This data is compared to other self-reported data required by other acts and with information stored in the EPA files such as manifest reports and permit applications. The EPA will likely focus on repeated violations and evidence of suspicious self-reporting.

Traditionally, regulators have focused largely on the use of standard enforcement tools, such as sanctions, to persuade industries found violating environmental laws to comply. However, another approach is being explored by some environmental experts. It is being suggested that this approach be combined with an effort to educate industry officials regarding environmental laws in hopes of inducing increases in the standard of current practice, and thus promoting a healthier environment for the future.

A situation that has recently arisen is the question of whether U.S. companies can be sued for environmentally damaging acts committed in other countries. In San Carlos, Ecuador, carcinogenic by-products of extraction methods were poured into open pits and left to seep into the ground. They also contributed significantly to air pollution, when the pits were set fire to burn off their contents. As a result of this issue 30,000 Ecuadorians are filing a class action suit against Texaco. It is being said that Texaco executives violated human rights and should be held accountable for their actions.

On the other side of the world, China has begun to deal with the issue of environmental destruction that is resulting from its own development. Over the past two decades, China has implemented several pollution control programs to curb the amount of pollution produced by its growing industries. These include enacting emissions taxes and setting a general goal of stop environmental deterioration before the year 2015.

REFERENCES

1. Burns, I.A., Peltason, J., Cronin, T.E., and Magleby, D.B., *Government by the People,* 15th ed., Prentice Hall, Englewood Cliffs, NJ, 1993, chap. 14.
2. Sullivan, T.E.P., *Fundamentals of Environmental Law,* Government Institutes, Rockville, MD, 1995, chap. 1.
3. U.S. Postal Service, A Proposal to Establish Environmental Coordinators in District Offices of the Northeast Area, Northeast Area Environmental Compliance Staff, Northeast Area Processing and Distribution, 6 Griffin Road North, Windsor, CT 06006–7030, February, 25, 1993.
4. Vance, B., Weinsoff, D.J. et al., *Toxics Program Commentary: Massachusetts,* prepared by Touchstone Environmental, Specialty Technical Publishers, North Vancouver, B.C., Canada, V7M IA5, 1992.
5. U.S. Postal Service, *Waste Reduction Guide,* Handbook AS-552, Northeast Area Environmental Compliance Staff, Northeast Area Processing and Distribution, 6 Griffin Road North, Windsor, CT, 06006–7030, February 1992.
6. Vance, B., Weinsoff, D.J., Henderson, M.A., and Elliot, J.E., *Toxics Program Commentary: Massachusetts,* prepared by Touchstone Environmental, and published by Specialty Technical Publishers, North Vancouver, B.C., Canada V7M I A5, 1992.
7. U.S. Postal Service, *Hazardous Waste Management Guide,* Handbook AS-553, Northeast Area Environmental Compliance Staff, Northeast Area Processing and Distribution, 6 Griffin Road North, Windsor, CT, 06006-7030, May 1992.
8. Moeller, D.W., *Environmental Health,* Harvard University Press, Cambridge, MA, 1997, chap. 13.
9. U.S. Postal Service, *Clean Air Act Compliance,* Handbook AS-553, Northeast Area Environmental Compliance Staff, Northeast Area Processing and Distribution, 6 Griffin Road North, Windsor, CT, 06006–7030, April 1992.

10. The Library of Congress, Thomas, S.313, http://thomas.loc.gov/cgi-bin/bdquery/z?d109:s.00131, June 2006.

11. U.S. Environmental Protection Agency, Clear Skies, last updated on Thursday, March 2, 2006, http://www.epa.gov/air/clearskies/legis.html.

12. Mund, N., Clean Air: Facts About the Bush Administration's Plan to Weaken the Clean Air Act, Sierra Club, retrieved June 2006, http://www.sierraclub.org/cleanair/clear_skies.asp.

13. National Resources Defense Council (NRDC), Dirty Skies, The Bush Administration's Air Pollution Plan, last revised October 5, 2003, http://www.nrdc.org/air/pollution/qbushplan.asp.

14. Sullivan, T.F.P., *Environmental Law Handbook,* Government Institutes, Rockville, MD, 1997, chap. 4.

15. U.S. Postal Service, *NPDES: Stormwater Guide,* Handbook AS-554, Northeast Area Environmental Compliance Staff, Northeast Area Processing and Distribution, 6 Griffin Road North, Windsor, CT, 06006-7030, May, 1992.

14 Emergency Preparedness

CHAPTER OBJECTIVES

A student, after reading this chapter, should be able to:

1. Discuss the history and origins of Incident Command System (ICS), the National Incident Management System (NIMS), and the National Response Plan (NRP).
2. List and describe the major components of ICS, NIMS, and NRP.
3. Describe and discuss the role of environmental health (EH) professionals in emergency and disaster response.

INTRODUCTION

PRE 9/11

It is not terrorist bombs or the release of anthrax spores that spawned the birth of modern-day emergency preparedness. It was not the 9/11 attack on the World Trade Center twin towers in New York that began this effort to prepare the U.S. for disaster response. It began with a wildfire in Southern California in 1970 that swept through several counties during a 13-d period, burning more than half a million acres, destroying over 700 structures, and taking 16 lives. The awesome fury of this fire swept across county lines and created chaos among responding agencies, which could not coordinate efforts because there was no clear person in charge, no common easily understood language among police, fire, and emergency response agencies, and no previously established coordination plan among these multiple agencies.[1]

Because of this tragedy, the 92nd Congress approved funding for the U.S. Forest Service Research group to design a system or process to significantly improve the capacity of the agencies in Southern California to provide protection against wildfires. Multiple state and federal agencies in Southern California joined to form a workable interagency system that could deliver resources and persons to the site of a major fire in a coordinated fashion. This agency became known as Firefighting Resources of California Organized for Potential Emergencies (FIRESCOPE). The participating agencies determined that there were five major program components that required planning and attention. These included:

1. The coordination of the resources among many agencies during a large incident
2. Creating a standard and easily understood terminology by all participants to avoid the use of professional jargon that would differ among police, fire, and emergency medical responders
3. Develop systems and technical capacity that would permit communication among many participating agencies
4. Provide training to all members of the participating agencies in communication, terminology, and command structure
5. Develop better methods for predicting the behavior of fires

These objectives were ultimately condensed into two components of the FIRESOPE system designated as:

1. Incident Command System (ICS), which provided a specific command structure and coordinated incident management tool
2. Multi-Agency Coordination System (MACS), which improved coordination among multiple agencies for large incidents that demanded massive resources

In 1976, the agencies involved in FIRESCOPE agreed on common terminology and procedures for the ICS structure and successfully employed ICS on a number of wildland and urban fires in 1978. As the success of the ICS management of fires responses grew, the use of the system expanded to nonfire incidents. It became clear that the management structure of ICS was well adapted to numerous types of incidents to which fire agencies must respond, including hazardous materials accidents, earthquakes, floods, plane crashes, and more. The term the agencies began to apply to the management of these incidents was *all-hazard* as they employed the well-documented and proven command structure of ICS to a wide variety of incidents. The single-incident management system of ICS was being successfully applied on an all-hazards basis.[1,2]

According to a report by the FIRESCOPE agencies, efforts were made to "extend these proven systems to other public safety services as the foundation of a uniform and comprehensive emergency management system." Some of the initial steps to extend this ICS management structure began in 1983 with the creation of the National Interagency Incident Management System (NIIMS). The agency made a few minor organizational and procedural modifications and small changes in terminology that allowed improved adaptation to a multiagency, all-hazards environment.[2] In practice, these applications were generally limited to regional or multistate applications involving responses to natural disasters. Natural disasters of national significance such as the flooding of New Orleans or the terrorist attacks on the World Trade Center towers in New York had not yet occurred to test the system at the national level.

Post 9/11

The importance of an organized response to terrorist threats to our country was made very clear on September 11, 2001, when terrorists took control of several airplanes filled with passengers and used those planes as missiles. Two of the planes crashed into the World Trade Center towers in New York, causing their collapse. Another plane crashed into the Pentagon in Washington, D.C., and destroyed one side, whereas a fourth plane dived into a Pennsylvania field after brave passengers tried to retake control of the aircraft. More than 3000 people were killed by these brutal attacks. The Bush administration declared these as "acts of war," and asked Congress for $20 billion to help rebuild and recover from these attacks.[3]

In swift reaction to these events, the Homeland Security Act of 2002 (H.R. 5005–8) was approved with the purpose of establishing a Department of Homeland Security (DHS) as an executive department of the U.S. The mission of DHS is to prevent terrorist attacks within the U.S., reduce the vulnerability of the U.S. to terrorism, and to minimize damage and assist in recovery from those terrorist attacks that do occur in the U.S. The DHS was officially established on January 23, 2003, causing a massive reorganization of the government with the intent of providing a unifying core for a vast national network of institutions and organizations that are important to national security and emergency response.

More than 20 federal agencies were directed to become part of the DHS and to be compartmented into one of four directorates, including (1) the Border and Transportation Directorate; (2) the Emergency Preparedness and Response Directorate; (3) the Science and Technology Directorate; and (4) the Information Analysis and Infrastructure Protection Directorate.[4,5]

President Bush issued Homeland Security Presidential Directive 5 (HSPD-5) on February 28, 2003. HSPD-5 directed Secretary of Homeland Security (Tom Ridge) to develop and then administer a National Incident Management System (NIMS) to provide a consistent management tool for government, private, and nongovernmental organizations to function together during a major incident. NIMS represents a national approach to incident management that scales up and modifies ICS to be applicable to a broad range of possible natural and man-made disasters of varying sizes and complexities.

ICS and NIMS

Among the most likely reasons for response failure in an emergency includes confusion about who is in charge, unclear lines of authority, and lack of clear communication. The management model used successfully for more than 30 years to overcome these problems is ICS. This is a standard, on-scene, all-hazard man-

TABLE 14.1
Titles For Supervisory Levels Under ICS

Organizational Level	Title
Incident Command	Incident Commander
Command Staff	Officer
General Staff	Chief
Branch	Director
Division/Group	Supervisor
Unit	Leader

agement system that is highly flexible, interdisciplinary, and able to adjust to one or more incidents that vary in size and complexity. The following is a brief description of ICS. (Those wishing to learn more and/or take qualified courses may go to http://training.fema.gov/EMIWeb/IS/is100.asp.[6]) ICS position titles are distinct and recognizable anywhere in the country and at any agency or organization using the ICS system. There is only one incident commander at each incident and only that person is given the title of commander. Members of the command staff are called officers, and members of the general staff are called chief, as described in Table 14.1 and Figure 14.1.

ICS provides a very clear reporting structure so that each response person knows to whom to report and what they are expected to do. All response personnel answer to one immediate supervisor and can be assured that communication will be understandable and in plain English. Typical ICS management structure as shown in Figure 14.1 may be reduced to only an incident commander if the incident is very small, whereas ICS structure may be greatly expanded if the incident grows or is already complex and large. As the incident grows, there may be a need to add supervisory layers to the organizational structure. As an example, the incident commander can add branches, divisions, groups, units and single resources to the management structure as needed (Figure 14.2).

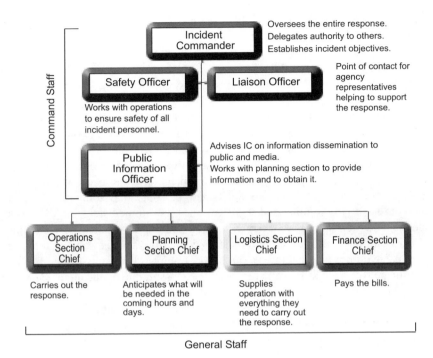

FIGURE 14.1 Typical ICS management structure and brief description of responsibilities.

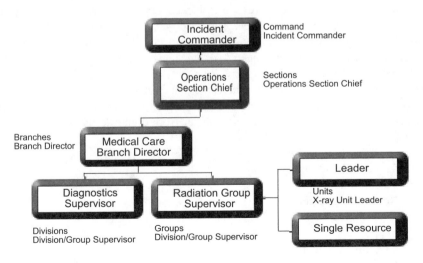

FIGURE 14.2 Example of options for expanding incidents that require added supervisory layers to the ICS organizational structure.

Another important element of ICS is the use of a manageable span of control. Effective span of control refers to the number of people or elements answering to an individual such as a commander or officer. This number is usually maintained between three to seven reporting elements with a desired ratio of one supervisor to five reporting elements. When the numbers exceed this range, expansion of the organization should occur.

There are more elements to ICS, and the reader is referred to http://training.fema.gov for more complete information on ICS. It is understood that the uniformed services of communities, such as fire and police, function under ICS on a daily basis and so they are intimately familiar with it as second nature. Public health does not normally function within an ICS command structure because most public health departments and agencies are quite small, personnel have defined roles, communication is usually simple, and there are very few immediately life-threatening incidents. However, the threat of larger, natural, and human-made disasters requires local and state public health to coordinate with multiple agencies and so they must communicate within ICS and the overall NIMS framework.[7] Public health may play a subordinate role such as surveillance of water to assure the public an incident such as a flood or hurricane has not contaminated the water, or provide information to the public about preventative measures to protect one's health. Public health may also play a major role in a multiagency response such as a pandemic flu outbreak. Whether large or small incidents, the response of public health needs to be within the NIMS/ICS framework. (To better understand how ICS might work within a public health structure, a detailed description of the public health incident command system is available online at http://www.Ualbanycphp.org/pinata/phics/guide/default.cfm.)

How does the NIMS fit into this picture? NIMS recognizes that ICS will be used at the local level. However, until the appearance of NIMS, there have been no standards for domestic incident response that encompassed the governments and agencies at all state and federal levels. The purpose of NIMS is to provide a unified national response to incident management for all potential incidents regardless of size and complexity. A major objective of NIMS is also to improve coordination and cooperation among public and private agencies and organizations throughout a whole range of domestic incident management activities.

NIMS is composed of six related components that function together as a system to prepare for, prevent, respond to, and recover from domestic incidents.[8] These include:

- Command and management
- Preparedness
- Resource management
- Communications and information management
- Supporting technologies
- Ongoing management and maintenance

The *command and management* elements of NIMS during an incident is based on the use of (1) ICS for establishing the operation and structure of incident management organizations; (2) MACS which defines the operation and structure of

FIGURE 14.3 Example of unified command structure.

supporting entities; and (3) public information systems that establishes the processes, communication systems, and procedures for delivering timely and accurate information to the public.

NIMS compliancy also requires participating entities to *be prepared* and to provide continuing preincident training, planning, exercises, personnel qualification checks, certification standards, and equipment acquisition and certification standards.

NIMS requires effective *resource management* so that there are standard methods for managing equipment, supplies and personnel throughout the incident. Similarly, there needs to be a standardized system for communications information management, including systems and technology procedures that work across all agencies and a mechanism of ensuring that information flows smoothly through a commonly accepted architecture.[8]

This description of NIMS is incomplete in this text, and the interested student is directed to www.fema.org for more complete information. There are differences between NIMS and ICS, based on the level of response required. When there is more than one responding agency or there are incidents that cross-jurisdictional boundaries, a unified command may be employed, whereby command decisions and planning are completed by senior representatives of the participating organizations who comprise the unified command (Figure 14.3).

An *area command* (1) may be employed if there are multiple incidents that are being managed by an ICS organization, (2) a large incident crosses jurisdictional boundaries, (3) no specific site can be identified as are many public health emergencies, (4) there is no identifiable cause, and (5) an incident is geographically dispersed and evolving over time, such as a widespread foodborne outbreak. The area command is responsible for setting overall strategy and priorities, allocating critical resources, ensuring proper management of the incidents, and that objectives are met and strategies are followed. Incidents that fall within these categories require coordination at higher jurisdictional levels (Figure 14.4).

An area command is based on an ICS structure, but without an operations component, because those functions are performed at each site (Figure 14.5).

When large- or wide-scale emergencies require significant resources and complex responses, a MACS is usually employed. An example would be a statewide

FIGURE 14.4 Area command is used for coordinating multiple incidents with different incident command posts (ICPs).

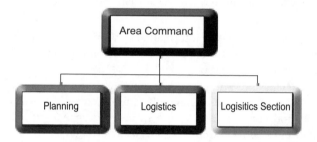

FIGURE 14.5 An area command structure does not require an operations section, as those functions are normally performed at each site.

response to a natural disaster such as a major hurricane. Typically, emergency operations centers (EOCs) established by the emergency management agency at the state level performs this function. See the following link for information on your state emergency management association: http://www.emergencymanagement.org/states/. Such EOCs coordinate multiple agencies that provide a broad array of resources in response to an emergency or disaster. EOC staffing and organization may include personnel from multiple jurisdictions and functional disciplines, including police, fire, public health, public works, environmental management, medical response units, and possibly many others. The numbers and disciplines of EOC staff will depend on the expected incident needs, size of the jurisdiction, and the resources available (Figure 14.6).

FIGURE 14.6 An example of an emergency operations center (EOC) management structure.

NATIONAL RESPONSE PLAN (NRP)

The HSPD-5 led to the creation of the NRP in establishing a "single, comprehensive approach to domestic incident management to prevent, prepare for, respond to, and recover from terrorist attacks, major disasters, and other emergencies."[9] The NRP is constructed around the previously described NIMS and is intended to be an all-hazards plan that provides the structure for incident management at all levels of federal, state, and local government and for any major type of incident including a major public health emergency. The NRP is designed to coordinate activities among all levels of government, tribal, private-sector, and nongovernmental agencies. The NRP is designed to be scalable and flexible to meet the needs of incident management ranging from multijurisdictional to incidents of national significance (INS) such as massive hurricane. The Secretary of Homeland Security will consult with other departments, agencies, and the White House before declaring INS. The NRP serves as the core operational plan for national incident management and establishes the process, protocols, and structures for such responses. The NRP and NIMS are companion documents that function together to integrate the capabilities and resources at all government levels, nongovernment entities, tribal, and private-sector organizations into a coordinated national response for domestic incident management.

How NRP Should Work

When an incident occurs, it is managed at the lowest jurisdictional level possible. If it is a train derailment with the release of a toxic gas as an example, it may be handled locally with the assistance of state agencies and response teams. The NRP does alter the typical ICS response of first responders to carry out their functions or responsibilities. Should the derailment incident expand to multiple jurisdictions or states with greater potential consequences, then more than one federal, state, or local agency will have jurisdiction. The incident commanders from each agency will form a unified command to coordinate resources, develop a common incident action plan with a unified set of objectives. An incident command post (ICP) will likely be designated and the unified command develops the NIMS/ICS command structure from the top–down according to the size and complexity of the incident. The ICP serves as the reporting center for responders to receive their assignments. EOCs will be notified and receive direction from the unified command. The EOCs will coordinate support functions and provide resources. Should the Secretary of Homeland Security or his or her designee determine there is a need to implement one or more components of the NRP, then one or more components of the NRP may be implemented. If the secretary of DHS determines the incident to be of national significance, then the secretary will manage the federal response. Otherwise, the incident will be managed by other authorities in coordination with the NRP framework.

(Independent study courses on NIMS [NIMS: An Introduction IS-700] and NRP [The NRP: An Introduction IS-800] are available through the Federal Emergency Management Agency's [FEMA] Emergency Management Institute of DHS at http://www.training.fema.gov/emiweb/IS/crslist.asp. You may also obtain copies of these plans at http://www.dhs.gov/dhspublic/interapp/editorial/editorial_0566.xml, or by calling (800) 368-6498.)

THE ROLE OF THE ENVIRONMENTAL HEALTH (EH) PROFESSIONAL IN EMERGENCY/DISASTER RESPONSE

BACKGROUND

EH professionals must have a large arsenal of skills and knowledge to competently perform their day-to-day activities. Much of this material is available through a Web-based course developed by the Twin Cities Metro Advanced Practice Center and produced by the University of Minnesota, Center for Public Health Education and Outreach and available at the following site: http://www.umncphp.umn.edu/preparedness/site/index.htm.

This online course provides on EH concepts at the awareness level, likely roles for EH staff during an emergency, some essential skills in responding to these emergencies, a guide to reducing health impacts, and some instruction on enhanced communication and collaboration with other responders.

The American Public Health Association (APHA) collaborated with EH practitioners to create three major EH competency areas now in use by the Centers for Disease Control and Prevention (CDC) and the Food and Drug Administration (FDA).[10] These three major EH competencies include:

1. Assessment — gathering information, data analysis, evaluation
2. Management — record keeping, problem solving, project management
3. Communication — risk communication, conflict resolution, education

The EH professional will use these competencies in carrying out their daily duties as well as responding to emergencies. The role they ultimately pursue during an emergency will depend on the incident type and gravity and the emergency response plans developed by the health department and the community. EH personnel should participate in the development of the emergency response plans because the plans provide direction for EH professionals during an incident response. The EH professional will have responsibilities for identifying, monitoring, and reducing environmental hazards that may impact on the health of the community during a disaster. Conducting these activities requires EH personnel to work in partnership with other responding organizations in the emergency management structure including police, fire, public works, Red Cross, medical response teams, and so forth. This coordinated effort will focus on restoring secure sources of water, food, and shelter, providing sanitation and waste management.[10]

These are the roles that EH personnel normally pursue on a daily basis, but some activities may be more pronounced during an emergency response, and must be done within the structure of ICS. Therefore, the EH person will report to a single person within the incident command structure and carry out assignments directed by that person according to the incident action plan (IAP). The EH personnel will be

Reviewed: 6/06

AnyPlace Department of Health
Incident Command System (ICS)
Emergency Response
Job Action Sheet

Position Title

Operations

Reports to	**Epidemiology Branch Director**

Reports to: <u>Operations Section Chief</u>
Operations Command Center Location: _____ Telephone: _____
Name of Event:_____Date and Time ICS role instituted:____

◄── **Mission**

Mission: *Interpret pattern of disease, coordinate investigation, develop appropriate standardized tool to use in case investigations.*

Immediate: Tasks must be completed first when reporting for duty

- ☐ Read entire Job Action Sheet
- ☐ Obtain briefing from Operations Section Chief
- ☐ Assign patient Tracking, Investigation, and/or Syndromic Surveillance Group

Intermediate: Tasks to be completed after immediate ones

- ☐ Update and/or tailor standardized questionnaire or tool(s) as needed
- ☐ Relate information back to Medical Consultant and Operations Section Chief
- ☐ Establish and maintain ongoing contact with Planning Section Chief to coordinate data and analysis of information

Extended: Tasks to be completed later on or on a continuig basis during work shift

- ☐ Document all action decisions and intervention
- ☐ Observe all staff for signs of stress. Report issues to Operations Section Chief.
- ☐ Provide rest periods and relief for staff, as needed.

EPI Branch Director   1 of 1

FIGURE 14.7 An example of a typical job action sheet (JAS) in public health.

expected to perform their assigned duties often based a predefined set of activities defined in a job action sheet (JAS) created as part of the emergency response plan for their agency. A JAS describes the functional role of a position/person in an emergency response. Tasks on the JASs may be amended by the addition or deletion of tasks according to the needs of the incident. The modification of tasks would be performed by the section chief or unit leader who is issuing the JAS (Figure 14.7).

The EH professional is expected to function within the incident command structure during an emergency response to (1) assess hazards and assist in risk analysis, (2) act through providing services and applying regulations to reduce hazards, and (3) ensure public awareness by working with others to communicate hazards and methods to minimize those hazards. As an example, a natural disaster such as a hurricane with significant flooding would require EH professionals to evaluate the safety of food and water. The EH professional could act by imposing regulations that close down unsafe food service establishments, closing water supplies, and ordering the importing of safe water until adequate disinfection of water supplies can be reestablished. EH personnel would also be expected to help create accurate and timely information for the public on what the environmental issues are and what steps to take for personal protection. EH professionals will be expected to function within their traditional role of inspection and regulatory enforcement to make certain that licensed facilities such as food service establishments, lodging, camps, public swimming pools, and mobile home parks meet the sanitation standards of local and state ordinances and regulations. They will also be expected to work as partners with multiple agencies to provide environmental information on general sanitation, food safety, drinking water distribution, waste disposal, inspection of temporary housing, and mass feeding centers.[10]

IMMEDIATE ACTIONS

A set of immediate actions suggested by the Minnesota Department of Health to rapidly restore environmental health services during and after a disaster would include[10]:

1. Obtaining information on the movements of populations at or near the incident site; identifying the location of facilities used to house displaced persons, hospitals, and medical facilities; and determining where relief workers will be housed to determine if those living quarters meet sanitary and housing codes.

2. Determining what capacities remain for delivering basic environmental health services including inspection and regulation enforcement of food, water, and housing.

3. Quickly assessing what damage has occurred to those food, water, housing, solid waste, and other environmental entities that fall under the public health regulations and ordinances. The EH professional should also attempt to discover what the immediate needs are for safe water, housing, food and sanitation. These conditions should apply to any temporary settlements where displaced persons and relief volunteers may be housed. The major response and recovery elements likely to be present themselves during or after a disaster requiring prompt attention by environmental professionals include[10,11]:
   • Water safety
   • Food safety
   • Liquid waste disposal (sewage)
   • Solid waste disposal
   • Housing/mass care shelters
   • Vector control
   • Hazardous materials and household hazardous waste
   • Medical and infectious waste
   • Radiological materials

The California Conference of Directors of Environmental Health (CCDEH) Disaster Preparedness Technical Advisory Committee prepared a manual as a field guide to help EH professionals respond to disasters. The manual provides an overview of the key responses and recovery actions for each category that an EH professional take when protecting the public's health. It is designed to complement and not replace existing disaster preparedness plans and is not intended to serve as a disaster response plan.[11] The pocket-size manual may be ordered at http://www.ccdeh.com/products/default.htm.

The CDC National Center for Environmental Health has also produced a Web site that serves as a clearinghouse of information resources related to emergency and terrorism preparedness for EH practitioners found at http://www.cdc.gov/nceh/ehs/ETP/default.htm.

The CDC Web site provides resources for emergency and terrorism preparedness specifically for EH practitioners on the subjects of general preparedness, food protection, water security, hazardous materials, building environment, radiation, vector control, and biological and infectious waste.[12] As an example, the CDC section related to food protection provides specific information on how practitioners can effectively prevent and respond to foodborne disease outbreaks and the site emphasizes food-security related issues. Examples of some of the information available at this site include:

- CDC's public health emergency preparedness and response — offers information on food safety after disasters such as floods, hurricanes, or tornadoes.
  - Keep food and water safe after a natural disaster or power outage
  - Hand hygiene after a disaster
- CDC's foodborne outbreak response and surveillance — includes tools for investigating and responding to food borne disease.
  - Outbreak investigation toolkit
  - Foodborne diseases A–Z
  - Epidemiologic case studies
  - Epi info software
- Emergency handbook for food managers (PDF, 397 KB) — provides food service establishments, food safety trainers, and government EH staff with a quick-reference handbook that is easy to use for emergency preparedness and response.
  - Handbook companion discussion guide (PDF, 177 KB)
  - Photo lessons (PDF, 964 KB)
  - Photo lessons in Spanish (PDF, 992 KB)
- Food safety self-inspection checklist (PDF, 776 KB) — provides EH professionals and food managers with the tools for routine food security and safety evaluation by food service establishments. The two checklists address food security concerns, such as inspecting products for signs of tampering, restricting access to records and food areas, and evaluating general food safety.

- In an emergency — equipment and supplies list for food service fact sheet (PDF, 978 KB)
- In an emergency — discard or salvage fact sheet (PDF, 40 KB)

Do you want to learn more about emergency preparedness? Point your browser to http://www-unix.oit.umass.edu/~envhl565/weblinks.html and click on any of the Chapter 14 Web links associated with this topic:

- FEMA training for IS 100
- Main FEMA site
- Public Health Incident Command System
- State Emergency Management Associations
- Independent Study Courses on NIMS and NRP
- Source for Disaster Field Manual for EH Professionals
- Training on Emergency Preparedness and Response Fundamentals for Environmental Health Professionals
- CDC Emergency and Terrorism Preparedness for Environmental Health Practitioners

REFERENCES

1. FIRESCOPE (Firefighting Resources of California Organized for Potential Emergencies), Past, Current and Future Directions: A Progress Report, October 1988, http://www.firescope.org.
2. Irwin, B., FIRESCOPE — A Record of Significant Decisions, June 1981, http://www.firescope.org.
3. September 11 News, Today's 9/11 News Headlines and Archives of the September 11, 2001 Attack on America, http://www.September11News.com.
4. Department of Homeland Security (DHS), DHS Organization, http://www.dhs.gov/dhspublic/interapp/editorial/editorial_0413.xml.
5. U.S. Environmental Protection Agency (USEPA), Homeland Security: Our History, http://www.epa.gov/homelandsecurity/htm/ohs-history.htm, last updated February 27, 2006.
6. Department of Homeland Security, Federal Emergency Management Institute (FEMA), FEMA IS Course Material Download: IS-100 Introduction to Incident Command System, IS-100, http://training.fema.gov/EMIWeb/IS/is100.asp.
7. Qureshi, K., Gebbie, K., and Gebbie, E.N., *A Guide for the Management of Emergencies or other Unusual Incidents within Public Health Agencies, PHICS: Public Health Incident Command System (Volume 1)*, October 2005, http://www.Ualbany-cphp.org/pinata/phics/guide/default.cfm.
8. Department of Homeland Security, Federal Emergency Management Agency, FEMA Independent Study Program: IS-700 National Incident Management System (NIMS), An Introduction, May 12, 2006, http://www.training.fema.gov/EMIWeb/IS/IS700.asp.
9. The Department of Homeland Security (DHS), Quick Reference Guide for the National Response Plan, Version 4.0, May 22, 2006, http://www.dhs.gov/dhspublic/interapp/editorial/editorial_0566.xml.

10. Olson, D., Carlstrom, C., Bezdliek, S., Tichich, T., and Hargreaves, A., Emergency Preparedness and Response Fundamentals for Environmental Health Professionals, Online Course, prepared by Twin Cities Metro Advanced Practice Center, Hennepin County, Minnesota, http://www.umncphp.umn.edu/preparedness/site/index.htm.

11. California Conference of Directors of Environmental Health (CCDEH) Disaster Preparedness Technical Advisory Committee, Disaster Field Manual for Environmental Health Specialists, California Association of Environmental Health Professionals, 1998.

12. CDC, National Center for Environmental Health, Environmental Health Services, Emergency and Terrorism Preparedness Specifically for Environmental Health Practitioners, July 24, 2006, http://www.cdc.gov/nceh/ehs/ETP/default.htm#gen.

Glossary

Abiotic Physical and chemical components of the ecosystem, such as air, sunlight, minerals, etc., and their interaction to produce climate, salinity, turbulence, and other conditions that influence the physical conditions under which organisms survive.

Abortion Refers to the medical means of terminating a pregnancy.

Absorption Passing through the skin.

Acceptable daily intake (ADI) An acceptable level of exposure or dose of a substance that is considered "safe," or poses little risk.

Acid A material that has a pH of less than 7.0. Acids are corrosive to human tissue and some metals.

Acidic deposition Characterized by emissions of nitrogen and sulfur oxides that are partially converted in the atmosphere to nitric and sulfuric acids that return to the earth in rain, snow, fog, and on dry particles.

Active transport A process that involves ATP in conjunction with special carrier proteins to move molecules through a membrane against a concentration gradient (i.e., high concentration to low).

Acute effect Immediate effect from a high-intensity chemical exposure that results in a change in health usually within 24 h. Many diseases such as measles are acute and have a rapid onset, are usually self-limiting, and are of relatively short duration.

Acutely hazardous waste Any hazardous waste with an EPA Waste Code beginning with the letter "P" or any of the following F codes: F020, F021, F022, F023, F026, and F027. These wastes are subject to stringent quantity standards for accumulation and generation.

Additive acute toxicity The ability of a substance to cause poisonous effects resulting in severe biological harm or death soon after a single exposure or dose. Also, any severe poisonous effect resulting from a single short-term exposure to a toxic substance. Occurs when the effects of two substances taken together are equal in the sum of the two substances taken separately.

Adduct Carcinogenic residues bound to DNA.

Adsorb To collect the molecules of a chemical on the surface of another material.

Age distribution Describes the percentage of the population in various age categories from birth to death.

Age–sex composition Sex ratio and age distribution of a population. The age–sex composition of the population has a profound effect on the birth and death rates of a country.

Air exchange rate The rate at which air is replaced in the structure by external air.

Air pollutant Refers to any substance that enters the atmosphere from human activities and have the capacity to cause human disease and welfare effects resulting in disturbances to plant life and physical structures.

Albedo The combined reflective ability of cloud cover and ground surfaces.

Alkali A material that has a pH of greater than 7.0. Generally corrosive to human tissue.

Allowance Defined under 1990 CAAA as the right to emit 1 ton of sulfur dioxide.

Anaerobically In the absence of oxygen.

Anaphylactic A type I hypersensitivity that is associated with hay fever, asthma, and anaphylactic shock from insect bites, certain foods such as shellfish, and a variety of chemicals. The reactions are usually immediate, occurring within moments in persons previously sensitized by exposure to the offending antigen.

Angiogenesis A process whereby malignant cells grow into a small mass that develops nutrient-bearing blood vessels.

Antagonistic A combined effect of two or more substances that is lesser than the sum of those substances acting independently.

Anthropogenic Generated or created by humans.

Anticodon The recognition site for nucleotides on the sRNA molecule. Apoenzyme. Protein component of an enzyme.

Apoptosis A process of cellular self-destruction.

Area Command An organization established to oversee the management of (1) multiple incidents each of which has its own operating Incident Command System organization; or (2) a very large incident that has multiple incident management teams assigned to it. Area Command sets the unifying strategy and priorities, assigns critical resources based on priorities, makes certain the incidents are properly managed, and ensures that objectives are being met and that strategies are followed.

Ash The residue that remains after burning.

Asphyxiant A substance that causes suffocation.

Atoms The basic units of elements, consisting of a small dense center called a nucleus surrounded by a cloud of negatively charged electrons.

Autotrophic Refers to organisms that synthesize their own food from inorganic substances, and obtain energy from the sun. Most plants are autotrophs.

Base The physical location where primary logistics functions for an incident are coordinated and administered. There is only one Base per incident. The Incident Command Post may be at the same place as the Base.

Best management practices Schedules of activities, prohibitions of practices, maintenance procedures, and other management practices to prevent or reduce the pollution of U.S. waters, including treatment requirements, recycling, reduction, reuse, operating procedures, and practices to control facility site runoff, spillage or leaks, sludge or waste disposal, or drainage from raw material storage.

Bias A type of error introduced into epidemiologic studies through differential treatment of the cases and controls in a study, either during the selection process or information gathering.

Bioaccumulation Refers to the storage of substances in the body at levels above what would be normally expected.

Biodegradable material Waste material that is capable of being broken down by microorganisms into simple stable compounds such as carbon dioxide and water.

Biogeochemical cycling Refers to the recycling of nutrients.

Biological oxygen demand (BOD) A standard test to determine the amount of dissolved oxygen in water used by microorganisms that feed on organic matter in the water. The test is performed over a 5-d period under controlled conditions.

Biological transmission A process in which the disease agent goes through some reproductive stage in the vector and is transmitted by a bite, fecal matter, or vomit of the vector in an active process.

Biological vector Refers to vectors (usually arthropods) in which a microbial disease agent goes through development or multiplication prior to becoming infective.

Biomarkers These are mutations occurring along a gene appearing at characteristic sites and with unique base sequences.

Biomass The accumulation of organic material in an ecosystem.

Biomes Based on the dominant types of vegetation that are strongly correlated with regional climate patterns.

Biosphere That part of the earth and its atmosphere that can support actively metabolizing life.

Biotic Components of the ecosystem include living organisms and the products of these organisms, such as waste products. Examples are urine, feces, decaying leaves and twigs, bones, and flesh.

Biotic potential The unrestricted growth of populations resulting in the maximum growth rate for a particular population.

Biotransformation A process in which metabolic processes alter the structure and characteristics of a chemical.

Birthrate Refers to the number of individuals added to a population through reproduction (live births) and is normally expressed as the number of live births per 1000 population (counting the population at the midpoint of the year).

Bog Usually a stagnant, acidic, nutrient-poor ecosystem filled with peat and sphagnum moss. Its abundance of decaying material consumes enormous amounts of carbon dioxide.

Boiling point The temperature at which a liquid turns to a vapor at ambient pressure.

Bottom ash Larger residue pieces that fall through the bottom of the boiler grate after burning is completed.

Bottomland hardwood forests These are deciduous forested wetlands consisting of gum and oak trees adapted to areas of seasonal flooding or covered with water most of the year and found along streams and rivers in the Southeast and Southcentral flood plains of the U.S.

Building-related illness (BRI) A phenomenon characterized by well-defined illnesses that occur in a building, which can be traced to specific building problems.

Buttress A shallow root system formed by trees with bases that spread at the bottom of the tree. This growth provides support in thin soil and may enhance the ability of the tree to secure nutrients over a broader area.

Bypass waste Waste that cannot be recycled, reused, or burned but must be buried at the landfill.

Canopy An almost unbroken sea of green produced by leaves of trees at 70–100 ft above the forest floor.

Carcinogen Cancer-producing substance.

Carnivore Animal that eats primarily animal flesh. These animals include lions, tigers, hyenas, frogs, or even ladybugs (they feed on aphids).

Carriers Persons who exhibit no symptoms but harbor and transmit the disease organisms.

Carrying capacity The sum of these limiting factors that serve to control the numbers of a species that can survive in defined area over time.

Ceiling limit The concentration of chemicals in air that should never be exceeded during any part of a working exposure.

Chain of command A succession of control, control, executive, command, or management positions in hierarchical order of authority.

Chaparral Area of moderately dry climate and limited summer precipitation further characterized by small (3-15 ft) shrubs with leathery leaves that contain aromatic and flammable substances.

Characteristic waste A waste classified as hazardous because it is ignitable, corrosive, reactive, or toxic as determined by the toxicity characteristic leaching procedure. It has an EPA waste code in the range D0001 to D043. Each of these four characteristics is defined in 40 CFR 261.20 Subpart C.

Chemical formula A listing of the atoms in a molecule.

Chlorinated solvent An organic solvent containing chlorine atoms, for example, methylene chloride and 1,1,1-trichloromethane, that is used in aerosol spray containers and in certain paints.

Chlorofluorocarbons (CFCs) A family of inert, nontoxic, and easily liquefied chemicals used in refrigeration, air conditioning, packaging, or insulation or as solvents or aerosol propellants. Because CFCs are not destroyed in the lower atmosphere, they drift into the upper atmosphere, where the chlorine is released and destroys ozone.

Chronic effects Effects of repeated, low-dosage chemical exposure that may not show up for many years or may last for many years. Diseases having a slow onset and lasting for extended periods are said to be chronic and may include cancer, emphysema, and some forms of heart disease.

Clean Water Act (CWA) Redesignated name for the Federal Water Pollution Control Act following the 1977 amendments.

Climate Can be viewed as average weather within a geographical area over years, or even centuries.

Climax community Forms in an undisturbed environment and continues to grow and perpetuate itself in the absence of further disturbance.

Clonal evolution A process whereby a single cell develops a genetic mutation that enables it to divide even when normal cells do not replicate.

Cloning vector A fragment of self-replicating DNA molecule.

Code of Federal Regulations (CFR) The detailed regulations, written by federal agencies, to implement the provisions of laws passed by Congress. Regulations in the CFR have the force of federal law.

Co-dominant Refers to genes that are partly expressed if present as a single allele (an allele is a pair of genes situated at the same site on paired chromosomes) and fully expressed if present on both.

Codon A group of three consecutive (triplet) mononucleotide units in the mRNA that specifies the position of one amino acid in a protein molecule.

Command Staff The Command Staff in an incident management organization or incident command system (ICS) that includes the Incident Command and the special staff positions of Public Information Officer, Safety Officer, Liaison Officer, and other positions as required, who report directly to the Incident Commander. They may have an assistant or assistants, as needed.

Commensals A relationship in which two kinds of organisms live in or near each other with one obtaining benefits, and the other is not damaged or benefited.

Complement A system composed of several (at least 20) serum proteins that follow a sequence or cascade of actions beginning with the Ag-Ab binding. This sequence of action enhances phagocytosis, inflammation, and cell lysis.

Composting A controlled process of degrading organic matter by microorganisms into a humuslike material.

Conditionally exempt generator A hazardous waste generator that meets the following criteria: in every single month of the year, the site generates no more than (1) 100 kg (220 lb) of hazardous waste, (2) 1 kg (2.2 lb) of acute hazardous waste, and (3) 100 kg (220 lb) of material from the cleanup of spillage of acute hazardous waste.

Congenital defects The defects occurring during the embryonic stages of development.

Contaminant Any physical, chemical, biological, or radiological substance or matter that has an adverse affect on air, water, or soil.

Corrosives Chemicals that are acidic or basic in pH and are destructive to human tissue and many metals.

Crossbreeding This involves the selection of genetic and morphologic variants that possess one or more of the desired characteristics and then crossing the plants/animals in the hope that the most desired characteristics of the different species will appear in the new crossbred organisms.

Cytogenetic defects Abnormalities in the number or structure of chromosomes.

Death rate Refers to the number of individuals removed from a population through death and is normally expressed as the number of deaths per 1000 population (counting the population at the midpoint of the year).

Decomposers Insects, bacteria, fungi, and protozoans that sequentially break down complex organic materials into low-energy mineral nutrients that once again may be reabsorbed and used by plants.

Definitive host The host in which the parasite reaches sexual maturity.

Deforestation The permanent decline in crown cover of trees to a level that is less than 10% of the original cover.

Demographic transition The simultaneous decline of death and birth rates in a country, resulting in a diminishing difference between death and birth rates and a very low rate of natural increase, permitting a stable population with very long doubling times.

Denitrifying bacteria Bacteria that live in the mud and sediment of lakes, ponds, streams, and estuaries and recycle organic nitrogen back to atmospheric nitrogen.

Deserts Areas defined by their arid climates which average less than 10 in. of precipitation a year and where evaporation tends to exceed this precipitation, causing a negative annual water budget.

Developmental disease A disease that occurs when faults or mistakes occur within the genes (or chromosomes), or stages in development of the fetus are disturbed.

Diffusion A passive process that occurs when molecules move from areas of high concentration to low concentration.

Direct contact transmission Occurs when the disease is passed directly from a source to the host by physical contact without an intervening object.

Director The ICS title for individuals responsible for supervision of a Branch.

Discharge The discharge of a pollutant when used without qualification.

Disposal Final placement or destruction of toxic, radioactive, or other wastes; surplus or banned pesticides or other chemicals; polluted soils; and drums containing hazardous materials from removal actions or accidental releases. Disposal may be accomplished through use of approved secure landfills, surface impoundments, land farming, deep well injection, ocean dumping, or incineration.

Division Divisions are used to separate an incident into geographical areas of operation and organizationally is located within the ICS organization between the Branch and the Task Force/Strike Team.

DOT The U.S. Department of Transportation.

Droplet infections A process whereby the organisms are contained within mucous droplets expelled to the air by sneezing, coughing, or even talking and travel short distances to be inhaled or ingested by persons very near by (usually less than a meter).

Doubling time The length of time required for a population to double its size.

Dry adiabatic lapse rate A theoretical condition in which a parcel of warm, dry air will cool at the rate of 10°C/km of altitude.

Dynamic equilibrium Processes that tend to balance each other so that there is little net loss over long periods of time.

Ecosystems Refers to identifiable areas within nature where the organisms interact among themselves and their physical environment and exchange nutrients largely within that system.

Effective demand The "demand level" corresponding to purchasing power without regard to food requirements.

Effective dose The amount of a chemical that reaches the general circulation and, ultimately, the location where a particular effect is seen or felt.

Effluent The water and the quantities, rates, and concentrations of chemical, physical, biological, and other constituents that are discharged from a point source.

Emergency Operations Center (EOC) The physical location where coordination of information and resources to support domestic incident management activities usually occurs. An EOC may be a temporary facility or may be located in a more central or permanently established facility.

Emergent trees Taller trees that penetrate the lower canopy to form an umbrella-like canopy over the trees below. These trees reach heights up to 130 ft.

Emigration This is a process by which some species disperse or migrate out of the area.

Emissions Gases and particulates that reach the air from combustion or industrial processes.

Endemic Refers to diseases that already exist in a community, where it is maintained in a low but constant incidence.

Endocrine disruptors Synthetic chemicals that have been referred to as xenoestrogens, estrogenic, hormone-mimicking, and endocrine-disrupting chemicals that have the ability to mimic or block hormones; cause unwelcome changes in hormone receptor sites; cause changes in the baseline production of hormones; or replace hormones on the carrier proteins. This makes some proportion of the population of native functional hormone unavailable and produces adverse health consequences that may be barely noticeable or very severe.

Endpoints The final or targeted result of a health event such as death or recovery. Researchers will focus their interests on certain health responses or endpoints from exposure to biological, physical, or chemical agents.

Energy recovery The generation of power in the form of steam or electricity from the burning of garbage.

Enforcement Efforts made by official agencies to ensure agreement with environmental laws and regulations.

Environment In its broadest definition, the term refers to personal and cultural behavior including smoking, diet, alcohol consumption, sexual and reproductive patterns, workplace, infections, food additives, and pollution along with the strictly physical environment.

Environmental disease Refers to any pathological process having a characteristic set of signs and symptoms that are detrimental to the well-being of the individual and are the consequence of external factors including exposure to physical or chemical agents, poor nutrition, and social or cultural behaviors.

Environmental lapse rate The actual change of temperature with altitude.

Environmental resistance Refers to those pressures that limit population and may include such factors as disease, wars, predatory behavior, toxic waste accumulation, or species competition.

Environmental tobacco smoke (ETS) The tobacco smoke consisting of side-stream smoke that originates from the burning end of the cigarette, and exhaled smoke.

EPA (also USEPA) The U.S. Environmental Protection Agency. Responsible for making and enforcing the laws and regulations regarding protection of the environment.

EPA identification number A 12-character number assigned by either EPA or the authorized state to each hazardous waste generator, transporter, and treatment, disposal, or storage facility.

Epidemic A situation in which there is a marked increase in the incidence of disease within a limited area affecting ever-increasing numbers of people.

Epiphytes Air plants that grow on branches high in the trees. These plants are supported by the tree limbs and extract moisture from the air, trapping leaf fall and wind-blown dust for nutrients.

Estuaries Inland marine waterways found worldwide. They are dominated by grass or grasslike plants and lie mostly between barrier islands and beaches.

Exoskeleton A hard outer skin that protects internal organs of arthropods, and serves as an attachment site for muscles to provide movement.

Exponential growth phase That part of the growth curve where the population increases at a rate proportional to its size, or exponentially.

Extrapolation Refers to the use of animal data to predict human response to chemical exposure.

Extremely hazardous substances (EHS) Any of 406 chemicals identified by EPA on the basis of toxicity and listed under SARA Title III. The list is subject to revision.

Facilitated diffusion A process in which molecules such as amino acids and sugars require specialized carrier proteins to be transported across a membrane.

Fallowing An agricultural process whereby fields remain unplanted every few years, and this resting permits moisture and nutrients to return.

Family planning Programs that are directed at assisting couples in having the number of children they desire.

Flammable limits The percentage concentrations of gas or vapor in air that will support combustion.

Flashpoint The temperature at which a flammable liquid will flash when provided with an ignition source.

Fly ash Very small solid particles of ash and soot generated by burning and carried in gases that fly up into the environmental control equipment.

Foodborne pathogens Agents that include parasites, viruses, fungi, and bacteria that have the potential to contaminate a food supply.

Food security A situation characterized by all people having physical and economic access to the basic food they need to work and function normally.

Food web The complex and interrelated patterns of consumption among many trophic levels.

Form R Toxic Chemical Release Inventory (TRI) Report Form. This form is required by Section 313 of the Emergency Planning and Community Right-to-Know Act (Title III of the Superfund Amendments and Reauthorization Act of 1986), Public Law 99-99. The information on this form was last collected for calendar year 1990 with the report due by July 1, 1991.

General Staff The command personnel reporting to the Incident Commander. The General Staff consists of: Operations Section Chief, Planning Section Chief, Logistics Section Chief, and Finance/Administration Section Chief.

Generation Refers to the weight of materials and products as they enter the waste management system from residential, commercial, institutional, and industrial sources and before materials are recovered or combusted.

Generator A person, company, site, or mobile source that produces solid or liquid waste.

Gene transfer The transfer of genetic information from one organism to another by artificial means in order to create new organisms with desired characteristics.

Greenhouse effect A process in which heat energy (infrared) reflected from the earth is absorbed by infrared-absorbing gases such as carbon dioxide, chlorofluorocarbons (CFCs), methane, and water vapor, thereby trapping the warmth.

Halon Bromine-containing compounds with long atmospheric lifetimes whose breakdown in the stratosphere cause depletion of ozone. Halons are used in firefighting.

Hazardous material A substance or compound that is capable of producing adverse health and safety effects.

Hazardous waste A by-product of society that, because of its quantity, concentration, or physical, chemical, or infectious characteristics, may cause, or significantly contribute to, an increase in serious, irreversible illness or pose a substantial present or potential hazard to human health, safety, or welfare or to the environment when improperly treated, stored, transported, used, or disposed of, or otherwise managed. Possesses at least one of four characteristics (ignitability, corrosivity, reactivity, or toxicity) or appears on special EPA lists.

Herbivore Animal that eats only plants.

Holoenzyme Refers to a complete and active enzyme consisting of an apoenzyme and its cofactor.

Humus Organic litter that decays further to a rich organic mat.

Hypersensitivity An exaggerated immune response to the presence of an antigen.

Immigrate The process in which species settle in a new area.

Incident Action Plan An oral or written plan containing general objectives clarifying the overall strategy for managing an incident. It may include the identification of operational resources and assignments. When written, the plan may have a number of forms as attachments (e.g., traffic plan, safety plan, communications plan, map, etc.).

Incident Commander The individual responsible for the management of all incident activities, including the development of strategies and tactics and the ordering and the release of resources.

Incident Command Post (ICP) The physical place where the primary command functions are carried out. The ICP may be colocated with the incident base or other incident facilities.

Incident Command System (ICS) A standardized on-scene emergency management construct specifically designed to provide for the adoption of an integrated organizational structure that reflects the complexity and demands of single or multiple incidents, without being hindered by jurisdictional boundaries. ICS is the combination of facilities, equipment, personnel, procedures, and communications operating within a common organizational structure, designed to aid in the management of resources during incidents. It is used for all kinds of emergencies and is applicable to small as well as large and complex incidents. ICS is used by various jurisdictions and functional agencies, both public and private, to organize field-level incident management operations.

Incident Management Team The Incident Commander and appropriate Command and General Staff personnel assigned to an incident.

Incineration Refers to the burning of certain types of solid, liquid, or gaseous materials; or a treatment technology involving destruction of waste by controlled burning at high temperatures.

Incomplete demographic transition A decrease in deaths, but without the time for adjustment in birthrates. The number of births remains high, and the combination of high birthrates and lowered death rates results in very high rates of natural increase.

Incomplete metamorphosis Refers to insects such as roaches, body lice, and grasshoppers that go through three developmental stages including egg, nymph, and adult stages.

Indirect contact transmission A process in which the disease agent is transmitted from a reservoir to a host by a nonliving object.

Infectious diseases Result from the pathological process occurring when a microbial agent invades the body.

Information Officer A Command Staff member who is responsible for interfacing with the public and media or with other agencies requiring information directly from the incident. The Information Officer may have assistants, but there is only one Information Officer per incident.

Ingestion Swallowing.

Integrated solid waste management The practice of using several alternative waste management techniques to manage and dispose of specific components of the postal waste stream. Waste management alternatives include source reduction, recycling, composting, energy recovery, and landfilling.

Intermediate host Host in which a parasite passes through its larval or asexual stages of development.

Inversion A meteorological phenomenon in which a stable, slow-moving air mass results in the formation of a dense, cool layer of air near the earth with warmer air above. The warming of air with increasing elevation is a reversal of the normal tropospheric cooling process.

Ionizing radiation The energy resulting from radiant energy that interacts with matter to form charged particles, and may include electromagnetic radiation (gamma and x-radiation) or particle radiation (alpha, beta, and neutron).

Isotopes Atoms of the same element that have different mass numbers.

K-strategy A reproductive strategy in which large organisms with relatively long life spans have only a few offspring, but devote their energies to protecting and nurturing the offspring to enhance their individual survival until they can reproduce.

Kwashiorkor A disease of dietary insufficiency usually seen in children 1 to 3 years old.

Lag phase The initial portion of the growth curve in which the organisms show no increase in growth rate but are undergoing changes that ready them for rapid proliferation.

Landfill A huge basin ranging in size from 20 to 70 acres wide scooped out of the earth. The sides and bottom are lined with layers of clay and geotextiles. Inside are also 12 to 14 in. layers of sand in which is buried a piping system that collects the leachate that runs through the garbage.

Large quantity generator (LQG) A generator of hazardous waste that meets any of the following criteria: (1) the site generates in one or more months during a year 1000 kg (2200 lb) or more of RCRA hazardous waste; (2) the site generates in one or more months during the year, or accumulates at any time 1 kg (2.2 lb) of RCRA acute hazardous waste; or (3) the site generates or accumulates at any time more than 100 kg (220 lb) of spill cleanup material contaminated with RCRA acute hazardous waste.

LD$_{50}$ Defined as the dose of a chemical which will cause death in 50% of a defined animal population.

Leachate The liquid formed as moisture and rainfall pass through refuse in landfills.

Liaison Officer A Command Staff member who is responsible for coordinating with representatives from cooperating agencies.

Lianas Woody vines that grow quickly up the tree trunks when a temporary gap opens in the canopy. Here, they flower and fruit in the treetops.

Lichens Mixtures of fungi and algae that grow together.

Listed wastes Wastes specifically named in 40 CFR 261.3. These wastes are listed as hazardous under RCRA but have not been subjected to the toxic characteristics listing process because the dangers they present are considered self-evident. They bear EPA waste codes beginning with the letters F, P, U, or K.

Loam Soils best suited to agriculture that consist of sand, silt, and some clay in a homogeneous mixture.

Logistics Section The ICS section responsible for providing services, facilities, and materials for an incident.

Macronutrients Nearly all life consists of large percentages of sulfur, phosphorous, carbon, oxygen, hydrogen, and nitrogen. These are chemical elements necessary for life in amounts usually larger than 1.0 ppm.

Marasmus Nutritional marasmus may appear in children 1 year of age who lack sufficient food and show signs of wasting, failure to grow, and complications of infections with multiple vitamin deficiencies.

Material recovery facility (MRF) A type of solid waste recycling facility that separates out all usable products and usable parts from individual products.

Material Safety Data Sheet (MSDS) An information sheet, required under the OSHA Communication Standard, that provides information on a specific chemical health and physical hazards, exposure limits, and precautions. Section 311 of SARA requires facilities to submit MSDSs under certain circumstances.

Mass number The total number of protons and neutrons in an atomic nucleus.

Materials recycling facilities (MRFs) Facilities that perform the function of preparing recyclables for marketing.

Maximum mixing depth (MMD) Refers to the height a column of air can rise. Generally, the higher the column of rising air, the better the mixing or dispersion of pollutants.

Mechanical injury Injury that results from a transfer of a damaging excess of kinetic injury to tissues resulting in abrasions, lacerations, punctures, contusions, broken bones, and projectile wounds.

Mechanical transmission A process of passive transmission by which organisms such as arthropods transmit a disease by simply carrying the disease organisms on the feet or other body part to a person's food or an open sore on the person.

Metastasis A process in which malignancies travel, producing tumors at distant sites.

Monitoring The measurement, sometimes continuous, of air or water quality.

Monoculture The planting of single crops.

Mucociliary streaming A process whereby particles such as dust and pollen of 10 g or larger are removed by the constant streaming of mucous propelled from the bronchial and tracheal passages by the cilia beating at over 1300 times per minute.

Multi-Agency Coordination System (MACS) The system that integrates personnel, equipment, procedures, facilities, and communications. When activated, MACS coordinates the resources of assisting agencies in a multiagency or multijurisdictional environment.

Mutagens Agents in the environment including chemicals or radiation that promote mutations.

Mutation A change in the nucleotide base sequence of DNA.

National Ambient Air Quality Standards (NAAQS) Refers to standards that have been set under the Federal Clean Air and its amendments for criteria pollutants, thereby establishing maximum concentrations allowed in the ambient or outdoor air environment.

National Incident Management System A system mandated by HSPD-5 designed to provide a unified national response to incident management for all potential incidents regardless of size and complexity. A major objective of NIMS is to improve coordination and cooperation among public and private agencies and organizations throughout a whole range of domestic incident management activities.

National Interagency Management System (NIIMS) An program developed by NWCG that includes five major subsystems providing a unified systems approach to all-hazard incident management. The subsystems include The Incident Command, Qualifications and Certification, System, Training Supporting Technologies, and Publications Management.

National Pollutant Discharge Elimination System (NPDES) A provision of the Clean Water Act that prohibits discharge of pollutants into waters of the U.S. unless a special permit is issued by EPA, a state, or (where delegated) a tribal government on an Indian Reservation.

Necrosis A process in which lesions are produced that are local and often result in killing areas of tissue.

Neoplasia New and uncontrolled growth of abnormal tissue from the transformation of normal body cells (tumors).

NIOSH National Institute for Occupational Safety and Health.

Nonattainment areas Those Air Quality Control Regions (AQCRs) as designated by the EPA that exceed the levels of one or more of the criteria pollutants during the year.

Nonpoint source Refers to pollutants entering the environment from a broad area and may include scattered sources.

Normal lapse rate A characteristic loss of temperature as air rises. Air will lose heat from such mechanisms as radiative cooling at an average of $-6.50°C/km$.

Off-site facility A hazardous waste treatment, storage, or disposal area that is located at a place away from the generating site.

Omnivore Animal that eats plants and animals, and includes rats, bears, humans, hogs, foxes, and many others.

Oncogenes Cancer-causing genes.

On-site facility A hazardous waste treatment storage, or disposal area that is located on the generating site.

OSHA Occupational Safety and Health Administration. Designed to monitor the working conditions of individuals and ensure that they are safe.

Oxidant A substance that readily gives up an oxygen atom, or removes hydrogen from a compound.

Packaging A product container or product cushion or wrapping around the product or container. Packaging may also refer to preconsumer shipping and storage containers as well as packaging on the shelf

Pandemic An epidemic that spreads throughout the world, as in the case of various outbreaks of influenza.

PCBs Polychlorinated biphenyls. A known carcinogen.

Permafrost Lies as a permanent layer of ice or frozen soil beneath a thin layer of topsoil.

Permanent threshold shift (PTS) A permanent hearing loss resulting from damage and loss of hair cells in the hearing apparatus.

Permissible exposure limit (PEL) An exposure limit to a chemical, as determined by OSHA, below which most individuals can work day after day with no harmful effects. This limit is legally enforceable.

Pesticide Substance or mixture of substances intended for preventing, destroying, repelling, or mitigating any pest. Also, any substance or mixture of substances intended for use as a plant regulator, defoliant, or desiccant. A pesticide can accumulate in the food chain or contaminate the environment.

pH A unit for measuring hydrogen ion concentrations. A pH of 7 indicates a neutral water or solution. At a pH lower than 7, a solution is acidic; at a pH higher than 7, a solution is alkaline.

Phagocytosis This process occurs when specialized cells such as phagocytes extend out streams of cytoplasm much like an amoeba and engulf solid particles to form a sac around the particle(s).

Photochemical Refers to the initiation of chemical reactions by sunlight.

Picocurie The equivalent of one trillionth of a curie representing slightly more than 2.0 radioactive disintegrations per minute.

Pinocytosis A process in which cells absorb nutrients and liquids. A small drop adheres to the cell membrane surface where an indentation forms to surround the drop and form a vesicle which then detaches from the membrane.

Pioneer plants These are early plants which establish the beginning stages of secondary succession, and may include wildflowers, followed by tall grasses and compact woody bushes.

Point mutation Single-gene defect.

Point source Refers to pollutants entering the environment from a specific point such as a pipe or a specific source such as a factory or treatment plant.

Pollution Refers to the presence of a foreign substance — organic, inorganic, radiological, or biological — that tends to degrade the quality of the environment so as to create a health hazard.

Pollution prevention The reduction of multimedia pollutants at the source and by the use of environmentally sound recycling. This prevention includes all regulated toxic and nontoxic substances.

Polymorphisms These are genetic differences reflecting the existence of several variations of the same gene.

Population Considered to be the breeding group for an organism. Different species not only have differing physical attributes, but also differ in population characteristics.

Population control Government-directed programs that set a policy for establishing an optimum population size.

Population momentum A situation in which a population has grown so much that even smaller growth rates lead to additions of larger numbers of people to the population.

Prairie potholes Glacially created freshwater marshes found primarily in the upper Midwestern U.S.

Pulmonary Having to do with lungs and breathing.

Transporter A person engaged in the off-site transportation of hazardous waste by air, rail, road, or water.

Radioisotope A radioactive isotope.

Rate of natural increase Determined by subtracting the death rate from the birthrate.

Recovered materials Waste material and by-products that have been recovered or diverted from solid waste, excluding materials and by-products generated from, and commonly used within, an original manufacturing process.

Recruitment failure Refers to the failure of eggs, larvae, and juvenile fish to thrive at pH levels of water below 5.5. Fish populations decline with the most sensitive, juvenile fish first disappearing.

Recyclable materials Materials that still have useful physical or chemical properties after serving their original purpose and that can be reused or remanufactured into new products.

Recycled materials Materials, which otherwise would have been destined for disposal, that have been collected, reprocessed or remanufactured, and made available for reuse.

Refuse-derived fuel (RDF) A pelletized fuel derived from MSW that can be mixed with coal and burned in a standard boiler.

Refuse reclamation Conversion of solid waste into useful products, for example, composting organic wastes to make soil conditioners or separating aluminum and other metals for melting and recycling.

Replacement level Defined as a total fertility rate (TFR) that corresponds to a population exactly replacing itself.

Reservoir Living organisms or inanimate objects that provide the conditions in which organisms may survive, multiply, and also provide the conditions necessary for transmission.

Resource Conservation and Recovery Act (RCRA) The federal statute that regulates the generation, treatment, storage, disposal, or recycling of solid and hazardous waste.

Resource recovery The act of removing materials from MSW for a productive use.

Resource recovery facility Any place, equipment, or facility designed and/or operated to separate or process solid waste into usable materials or to incinerate solid waste to create heat and energy.

Respiration A process in which the respiratory system supplies oxygen to the body's cells and expels carbon dioxide from the body.

Restriction enzymes These are microbial enzymes that are used to cut DNA into several predictable and reportable pieces.

Retrogression A process resulting in a new ecosystem state with fewer overall species.

Rotation Agricultural practices that traditionally have included the planting of nitrogen-using (corn) and nitrogen-fixing (beans) crops in alternative years in the same space or planting them in alternating bands in the same year.

R-strategy A reproductive strategy characterized by populations comprising typically small, short-lived organisms, which produce large numbers of offspring and receive little or no parental care.

Safety Officer A Command Staff member who is responsible for assessing and monitoring safety hazards or unsafe situations, and for developing measures that protect the ICS personnel.

Saltwater intrusion A condition in which saltwater can seep into an aquifer as freshwater water is pumped out of it, thereby polluting the freshwater.

Sanitary sewer A channel or conduit that carries household, commercial, and industrial wastewater from the source to a treatment plant or receiving stream.

Saprophytic Organisms that feed primarily on dead organic matter. As these organisms process components of the dead tissues, they leave behind undigested materials that are in turn processed by the next group of decomposers until finally there is nothing left but energy-poor mineral nutrients.

Scavengers Organisms that prefer to feed on the dead remains of animals.

Scurvy Develops from a lack of vitamin C (ascorbic acid) in the diet and, consequently, the synthesis of collagen and mucopolysaccharides is disturbed.

Sedimentary Nutrients originating from soil or rocks.

Sedimentary cycle A process by which substances are gradually leached from sedimentary rock by the actions of rain or erosion from water.

Sick building syndrome (SBS) A phenomenon in which occupants of a building display acute symptoms without a particular pattern and the varied symptoms cannot be associated with a particular source.

Span of control The supervisory ratio under ICS management that ideally should be three-to-seven individuals, with five-to-one being established as optimum.

Replication A process whereby dividing cells provide each daughter cell with one complementary strand of the parent and then synthesize a complementary strand to produce double-stranded DNA that is a duplicate of the parent DNA.

Sewer A channel or conduit that carries wastewater and stormwater runoff from the source to a treatment plant or receiving stream. Sanitary sewers carry household, industrial, and commercial wastes. Storm sewers carry runoff from rain or snow. Combined sewers are used for both purposes.

Sex ratio Refers to the number of males relative to the number of females in the population.

Sludge A semisolid residue from any number of air or water treatment processes. Sludge can be a hazardous waste.

Small quantity generator (SQG) A hazardous waste generator that meets all the following criteria: in one or more months of the year the site generates more than 100 kg (220 lb) of hazardous waste, but in no month (1) generates 1000 kg (2200 lb) or more of acute hazardous waste, (2) generates 1 kg (2.2 lb) or more of acute hazardous waste, or (3) generates 100 kg (220 lb) or more of material from the cleanup of a spillage of acute hazardous waste.

Solid waste Nonliquid, nonsoluble materials ranging from municipal garbage to industrial wastes that contain complex, and sometimes hazardous, substances. Solid wastes also include sewage sludge, agricultural refuse, construction and demolition wastes, and mining residues. Technically, solid waste also refers to liquids and gases in containers.

Solid waste management Supervised handling of waste materials from their source through recovery processes to disposal.

Solvent Substance (usually liquid) capable of dissolving or dispersing one or more other substances. Solvents include, but are not limited to, the nonspent materials listed by the EPA.

Source reduction Any action taken before waste is generated that reduces its volume and toxicity.

Species Normally considered to be a group of organisms that can breed together with the production of a viable and fertile offspring.

Spermicides Chemicals that are designed to kill sperm.

Stable ecosystems Ones in which materials are constantly recycled within the system through growth, consumption, and decomposition.

Storage Temporary holding of waste pending treatment or disposal. Storage methods include containers, tanks, waste piles, and surface impoundments.

Stratosphere A region of the atmosphere extending from the tropopause to about 50 km above the earth. This region contains the ozone layer extending from 24 to 40 km.

Succession Refers to the predictable and gradual progressive changes of biotic communities toward the establishment of a climax community.

Succulents Plants that store water in their tissues to be used during dry spells. The water may be stored in different parts of the plant including stem, leaf, fruit, or roots. The familiar American cactus is a stem succulent. The aloe plant is an African leaf succulent.

Superfund The program operated under the legislative authority of the Comprehensive Environmental Response, Compensation, and Liability Act (CERCLA) and Superfund Amendment Reauthorization Act (SARA) that funds and carries out the EPA solid waste emergency and long-term removal remedial activities. These activities include establishing the National Priorities List, investigating sites for inclusion on the list, determining their priority level on the list, and conducting and supervising the ultimately determined cleanup and other remedial actions.

Survivorship curve A mathematical plot in which the fate of young individuals is followed to the point of death to describe mortality at different ages.

Swamps Often-transitional environments between large open water and drier land. They are home to species from both extremes and are marked by flood-tolerant trees and shrubs.

Synergistic Term used when the combined effect of two or more substances is much greater than the sum of the those substances acting alone.

Taiga Boreal forest that is an almost continuous band of coniferous (cone-bearing) trees extending in a giant arc from Alaska, through Canada, Europe, and Siberia.

Temperate broadleaf deciduous forest (TBDF) Forests that are located in western and central Europe, eastern Asia and eastern North America. These are climax deciduous forests including oak, maple, hickory, chestnut, elm, linden, and walnut trees.

Temperate evergreen forest Area where soil is poor and droughts and fires are frequent, the predominant species tend to be evergreens including coniferous, needle-leafed trees with some broad-leafed evergreens. In such areas, spruce, fir, and ponderosa pine proliferate.

Temperate rain forest In cool coastal climates where there is considerable rainfall or frequent heavy fogs, one is likely to see forests of giant trees such as redwoods which characteristically reach heights of 198 ft (60 m) to 297 ft (90 m). Such areas exist along the Pacific coast of North America.

Temporary threshold shift (TTS) A reversible hearing loss normally lasting from moments to hours, although longer periods of temporary hearing loss are possible.

Teratologic defect Defect that arises during the embryonic period of development.

Terracing An agricultural process in which farmers must create level fields along the contour of the slope when they are driven to plant on hillsides.

Thermal injury Injury that results when tissues are exposed to extremely low or extremely high temperatures.

Threshold limit value (TLV) The maximum concentration of chemical which a worker can be exposed to day after day as determined by ACGIH. This value is not legally enforceable.

Tipping fee The cost for a truck to unload its waste at a landfill.

Total fertility rates (TFR) Represents the number of children a woman in a given population is likely to bear during her reproductive lifetime providing that birthrates remain constant for at least a generation.

Toxic Harmful to living organisms.

Toxic Characteristic Leachate Procedure (TCLP) This is a process described by the EPA that requires acidified liquid to be percolated through waste in a laboratory-created landfill condition for 24 h. Samples of the leachate are tested for the amounts of the target substance released.

Toxicity test The means to determine the toxicity of a chemical or an effluent using living organisms. A toxicity test measures the degree of response of an exposed test organism to a specific chemical or effluent.

Toxic substances Those that (1) can produce reversible or irreversible bodily injury; (2) have the capacity to cause tumors, neoplastic effects, or cancer; (3) can cause reproductive errors including mutations and teratogenic effects; (4) produce irritation or sensitization of mucous membranes; (5) cause a reduction in motivation, mental alertness, or capability; (6) alter behavior; or (7) cause death of the organism.

Trace element Element required in tiny amounts for life such as zinc, manganese, chlorine, iron, and copper.

Tragedy of the commons A behavior in which society will likely pass on the consequences of their destructive actions if they will benefit in the short term and receive little or no negative consequences from that action.

Transstadial A form of transmission in which disease organisms are passed from infected adult ticks through the egg, larval, nymphal, and adult stages.

Treatment Any method, technique, or process, including neutralization, designed to change the physical, chemical, or biological character of composition of any hazardous waste so as to neutralize such wastes, to recover energy or material resources from the waste, or to render such waste nonhazardous or less hazardous; safer to transport, store, or dispose; or amenable to recovery, storage, or reduction in volume.

Treatment, storage, and disposal (TSD) facility Site where a hazardous substance is treated, stored, or disposed. TSD facilities are regulated by the EPA and states under RCRA.

Trophic level Refers to each stage of a food web through which energy travels, consisting of producers, primary consumers (herbivores), and secondary consumers (carnivores). It is a system of categorizing organisms by the way they obtain food.

Tropical rain forest The tropical rain forest is complex in terms of species diversity and structure. The combination of constant warmth, with average monthly temperatures above 17.8°C and precipitation greater than 100 in. per year, encourages rapid plant growth throughout the year.

Troposphere A region of the atmosphere extending from the earth's surface to approximately 10 km. The troposphere is greater over the equator (12 km), and lesser over the poles (8 km).

TSDR Treatment, storage, disposal, or recycling.

Tubal ligation A surgical procedure requiring small incisions near the navel and just above the pubic area in which the fallopian tubes are tied off or cauterize the tubes. This procedure blocks the entry of eggs into the uterus.

Tundra A biome limited to the upper latitudes of the Northern Hemisphere and forms a belt around the Arctic Ocean.

Understory Area on the floor of the rain forest where there is sparse vegetation in the inner sections of the undisturbed rain forest, and these sections have to compete intensely in an area with little sunlight and poor soil.

Unified Area Command A Unified Area Command is established when incidents under an Area Command are multijurisdictional.

Unified Command Unified Command is a management system under ICS that promotes a common response, set of objectives, and incident action plan among all the participating agencies with responsibility for the incident without giving up agency authority, responsibility, or accountability.

Uniform Hazardous Waste Manifest The shipping document (EPA Form 8700-22 or 8700-22a) that pertains to hazardous waste and is duly signed by the generator.

Unity of Command The practice within an organization whereby each person reports to one and only one designated person.

Urbanization The mass migration of people to the cities.

Vapor density The weight of a gas or vapor compared to an equal volume of air. Gases with a vapor density of greater than one settle to the ground; gases that have a vapor density less than one rise in air.

Vapor pressure The pressure of vapor above a liquid.

Vasectomy Male sterilization accomplished by making an incision on either side of the scrotum and snipping out a piece of the vas deferens.

Vectors Refers to animals that carry pathogens from one host to another, either from another human that is infected or from an infected animal.

Vehicle transmission The transmission of disease agents by various media such as air, water, food, intravenous fluids or blood, and drugs.

Very small quantity generator (VSQG) A conditionally exempt generator of hazardous waste.

Volume reduction The processing of waste materials so as to decrease the amount of space the materials occupy, usually by compacting or shredding (mechanical), incineration (thermal), or composting.

Waste Any material discarded as worthless, defective, or of no further use such that when disposed of, may pose a threat to human health or the environment.

Waste management costs The costs to procure, handle, treat, and dispose of wastes; a key factor in deciding to introduce source reduction techniques.

Waste minimization The reduction, to the extent feasible, of hazardous waste that is generated or subsequently treated, stored, or disposed. It includes any source reduction or recycling activity undertaken by a generator that results in (1) the reduction of total volume or quantity of hazardous waste; (2) the reduction of toxicity of hazardous waste; or (3) both, as long as the reduction is consistent with the goal of minimizing present and future threats to human health and the environment.

Waste reduction Any change in a process, operation, or activity that results in the economically efficient reduction in waste material per unit of production without reducing the value output of the process, operation, or activity, taking into account the health and environmental consequences of such change.

Waste stream A specific type of waste leaving a facility or operation.

Waste-to-energy (WTE) Combustion of refuse with the production of energy.

Water stress A term defined as the ratio of water withdrawal to water availability.

Wetlands Those areas of land where water saturation is the major factor influencing the nature of soil development and the communities of plants and animals that live in the soil and on the surface.

Windrow A composting process in which rows of material are placed next to each other outdoors and periodically mixed to incorporate air into the mix.

Zoonoses Diseases that are spread mostly from animals to humans.

Acronyms and Abbreviations

ABERA	(Federal) Asbestos Hazard Emergency Response Act
AHW	Acutely hazardous waste
ASHAA	(Federal) Asbestos School Hazard Abatement Act
AST	Above ground storage tank
ATSDR	Agency for Toxic Substances and Disease Registry
BAT	Best available technology economically achievable
BOD	Biochemical oxygen demand
BPT	Best practicable control technology currently available
CAA	(Federal) Clean Air Act
CERCLA	(Federal) Comprehensive Environmental Response, Compensation, and Liability Act of 1980 (Superfund)
CERCLIS	CERCLA Information System
CFC	Chlorofluorocarbon
CFR	Code of Federal Regulations
CO	Carbon monoxide
CSA	Clear Skies Act
CWA	(Federal) Clean Water Act
DEP	Department of Environmental Protection
DOT	(U.S.) Department of Transportation
DPH	Department of Public Health
DWPL	Drinking Water Priority List
EHS	Extremely hazardous substances/Extraordinarily hazardous substances
EIS	Environmental impact statement
EOC	Emergency operations center
EPA (USEPA)	(U.S.) Environmental Protection Agency
EPCRA	Emergency Planning and Community Response Act
FIFRA	Federal Insecticide, Fungicide, and Rodenticide Act
FWPCA	Federal Water Pollution Control Act
HAP	Hazardous air pollutant
HCFC	Hydrochlorofluorocarbon
HIV	Human immunodeficiency virus
HMTA	(Federal) Hazardous Materials Transportation Act
HRS	Hazard ranking system
HSA	(Federal) Hazardous Substances Act
HSWA	(Federal) Hazardous and Solid Waste Amendments 1984
HW	Hazardous waste
IAP	Incident action plan
ICP	Incident command post
ICS	Incident Command System
IPM	Integrated Pest Management
LAER	Lowest achievable emission rate
LEPC	Local emergency planning committee
LEPD	Local emergency planning district
LOAEL	Lowest observable adverse effect level

LQG	Large quantity generator
LUFF	Leaking underground fuel tank
LUST	Leaking underground storage tank
MACT	Maximum achievable control technology
MCL	Maximum contaminant level
MSDS	Material safety data sheet
NA	North America
NAAQS	National ambient air quality standards
NESHAPs	National Emission Standards for Hazardous Air Pollutants
NIIMS	National Interagency Incident Management System
NIMBY	Not in my backyard
NIMS	National Incident Management System
NO_2	Nitrogen dioxide
NOx	Nitrogen oxides
NOAEL	No observable adverse effect level
NOEL	No observable effect level
NPDES	National Pollution Discharge Elimination System
NPL	National Priorities List
NRP	National Response Plan
NSPS	New source performance standard
NSR	New source review
OEM	Original equipment market
OPP	Office of Pollution Prevention
OSHA	(U.S.) Occupational Safety and Health Administration
PCB	Polychlorinated biphenyl
PEL	Permissible exposure limit
PIO	Public information officer
$PM_{2.5}$	Particulates <2.5 microns in size
PM_{10}	Particulates <10 microns in size
POTW	Publicly owned treatment works
PPA	(Federal) Pollution Prevention Act
ppb	Parts per billion
ppm	Parts per million
PRP	Potentially responsible parties
PSD	Prevention of significant deterioration
RCRA	(Federal) Resource Conservation and Recovery Act
PWS	Public water system
R&D	Research and development
Reg	Regulations
ROD	Record of decision
RP	Responsible party
RQ	Reportable quantity
SDWA	(Federal) Safe Drinking Water Act
SIC	Standard Industrial Classification
SIP	State implementation plan
SITE	Superfund Innovative Technology Evaluation
SOP	Standard operating procedure
SO_2	Sulfur dioxide
SPCC	Spill Prevention Control and Countermeasure (plan)
SQG	Small quantity generator

Stats	Statutes
SWDA	(Federal) Solid Waste Disposal Act
TCLP	Toxicity characteristic leaching (leachate) procedure
TPQ	Threshold planning quantity
TPY	Tons per year
TSCA	(Federal) Toxic Substances Control Act
TSD	Treatment, storage, and disposal
UIC	Underground injection control
UN	United Nations
USA	United States of America
UST	Underground storage tank
VOC	Volatile organic compound
VSQG	Very small quantity generator

Index

F